Microfluidic Cell Culture Systems

Microfluidic Cell Culture Systems

Editors

Christopher Bettinger

Jeffrey T. Borenstein

Sarah L. Tao

ELSEVIER

AMSTERDAM • BOSTON • HEIDELBERG • LONDON
NEW YORK • OXFORD • PARIS • SAN DIEGO
SAN FRANCISCO • SINGAPORE • SYDNEY • TOKYO
William Andrew is an imprint of Elsevier

William Andrew is an imprint of Elsevier
The Boulevard, Langford Lane, Kidlington, Oxford OX5 1GB, UK
225 Wyman Street, Waltham, MA 02451, USA

First edition 2013

British Library Cataloguing-in-Publication Data
A catalogue record for this book is available from the British Library

Library of Congress Cataloging-in-Publication Data
A catalog record for this book is available from the Library of Congress

ISBN: 978-1-4377-3459-1

For information on all Elsevier publications
visit our web site at books.elsevier.com

Typeset by MPS Limited, Chennai, India
www.adi-mps.com

Printed and bound by CPI Group (UK) Ltd, Croydon, CR0 4YY

Transferred to digital print 2012

Contents

PART 2 TISSUE ENGINEERING STRATEGIES

Preface

The concurrent emergence of microfabrication technologies and the biotechnology revolution have produced a vibrant intellectual landscape for the invention of new technologies to culture and manipulate cells using engineered *in vitro* environments. Advances in microfabrication, materials, and processing have provided a basis to design and control the presentation of multimodal cues to cells. Parallel advances in cell biology, gene sequencing, imaging, and microscopy empower scientists and engineers to extract rich data sets that aim to understand the inherent complexity of biology. There are several motivating factors for productive interdisciplinary research activities at the intersection of microfabrication technologies with cell culture strategies. First, the ability to manipulate fluids with a high degree of precision is highly advantageous. This is accomplished through novel materials and new micron-scale fluid handling components such as pumps, mixers, and valves. In addition, systems with a characteristic length scale of microns ensure ordered laminar flow, which can expedite mixing and separation of fluids. Another important reason for the productive interaction between microfabrication and cell culture technologies lies in the congruence of length scales. Microfabrication techniques are able to create unique structures that can achieve cellular and subcellular dimensions. These structures can be used to manipulate cells and their organelles with a high degree of precision. Nanometer-scale structures can be used in parallel to influence related processes such as protein adsorption and cellular processes such as adhesion and migration. When combined with additional advances in biotechnology, cell biology, and tissue engineering, microfabrication technologies have the potential to broadly impact many aspects of biomedical engineering. New fundamental knowledge may be gained through hypothesis-driven research that is enabled by precise control of the cellular microenvironment. Technological advances and discovery-driven research may include engineering new tissue constructs for potential applications in regenerative medicine and drug discovery. These and related topics will be the primary focus of this text.

The development of new small molecules and biologics for therapeutic applications has slowed significantly in recent years, due in large part to limitations in existing approaches for assessing the safety and efficacy of these compounds in preclinical models. Conventional cell culture assays and animal models have historically provided the foundation for preclinical models, but each suffers from shortcomings that may be overcome by the emergence of new technologies. Species differences present an enormous challenge for toxicity evaluation in preclinical studies, but conventional cell culture assays using human primary cells or stem cells in two-dimensional formats that do not replicate organ microenvironment and do not incorporate interactions between tissues and organs are likewise not sufficiently predictive of clinical responses [1]. Therefore, the opportunity exists to incorporate human-derived cells in three-dimensional microenvironments representative of tissue and organ physiology, and further to integrate these functional units to reflect organ crosstalk, in a manner that enables more rapid, efficient, and accurate preclinical assessment of emerging molecular compounds, biologics, and other medical countermeasures [2]. The emergence of microscale fabrication tools enables the construction of three-dimensional microenvironments that are more representative of tissue and organ physiology in appropriate dimensional scales [3], while microfluidic techniques provide the capability to control critical functions such as dosing and sampling in these systems. These new capabilities are clearly of great interest to industrial researchers in the pharmaceuticals and biologics industry, but also to the US Department of Defense and the Department of Health and Human Services as part of a broader national initiative aimed at bolstering capabilities for dealing with emerging infectious diseases and bioengineered threats [4]. Ultimately, it is envisioned that these microphysiological systems will contribute to dramatic reductions in the cost and time required to develop and assess the safety and efficacy of new treatments for a host of diseases ranging from cancer to pandemic influenza outbreaks.

The focus of this book is to demonstrate the potential for rational material design in solving a diverse range of biomedical problems where the functional utility of cells and tissues are cornerstones. As such, the book is divided into three sections, comprising 17 chapters written by recognized experts from leading groups around the world. It covers a spectrum of topics and highlights more recent work that represents the current state of the art in the field. The first section of the book is intended to blend elements of theory and key concepts with novel materials and fabrication techniques to impart physiological relevance of microfluidic systems. The second and third sections are designed to demonstrate further how these microfluidic principles can be utilized for two distinct applications, tissue engineering strategies and *in vitro* systems. Each of these applications requires a unique microfluidic toolset. Through the examples presented, it is demonstrated that microfluidics can provide highly controlled microenvironments which can be utilized to regulate cell behavior in a variety of applications from the recapitulation of native tissues for clinical therapies to high-throughput screening systems which are scalable and easy to operate. The enabling nature of microfluidic technology may result in numerous future advances in basic biological understanding, clinical therapeutics, and diagnostics.

The sections of the book are organized as follows:

- Materials and Fabrication Methods

Materials and associated microfabrication processes form the cornerstone for advanced microfluidic devices that are dedicated for cell culture applications. These two essential topics are discussed concurrently in this section. One prevalent theme is the evolution of materials that are used in microfluidic devices. The composition of microfluidic devices has gradually migrated from traditional engineering polymers toward biopolymers. Novel materials require new materials processing and fabrication strategies to produce biologically relevant microfluidic devices. Microfluidic devices with biologically active materials and biomimetic structures can be designed to influence cell behavior through the coordinated presentation of multimodal cues such as mechanical properties, integrin-binding domains, topographic structures, and soluble factors. These considerations and related topics will be the subject of this section.

- Tissue Engineering Strategies

Tissue engineering is the process in which engineering principles are utilized to assemble cells for the purpose of forming new tissues. Over the past few decades, the field has focused on design of both synthetic and native materials to direct cells toward formation of physiologically relevant tissues. In this section, the use of microfluidic technologies in tissue engineering applications is explored. Two major concepts are examined through the examples outlined in these chapters. First is the use of microfluidics to dynamically manipulate the chemical and mechanical cellular environment in order to direct and regulate cellular behavior. Second is the use of microfluidics to serve as perfusion networks to sustain long-term culture of engineered tissues through continuous flow of oxygen, nutrients, or withdrawal of waste products. The utility of microfluidics in tissue engineering is delineated in this section through specific tissues including vascular, pulmonary, and kidney.

- *In Vitro* Models

Emerging *in vitro* models for specific tissue and organ systems such as the liver and cardiovascular system, and tools aimed at assessing dynamic processes within these systems, are the focus of this section. Microfluidic approaches capable of mimicking the microstructure of the liver and the specific arrangement and interactions involving liver parenchyma are described, and microscale techniques for manipulating and sorting cells and applying shear dynamics to cell populations are also covered. Another area undergoing rapid development is the use of microfluidic approaches to generate gradients and flows within physiologically relevant matrix materials to simulate processes such as angiogenesis and metastasis. These and related topics are also covered in this section.

We hope that our readers find this book to be a useful and stimulating reference that provides key information regarding the application of microfluidic technologies toward cell culture and tissue engineering applications, and we would like to thank all of our authors for sharing their recent results and advances in this exciting and potentially transformative field.

Jeffrey T. Borenstein

References

[1] L.G. Griffith, M.A. Swartz, Capturing complex 3D tissue physiology *in vitro*, Nat. Rev. Mol. Cell. Biol. 7 (2006) 211–224.

[2] M.B. Esch, T.L. King, M.L. Shuler, The role of body-on-a-chip devices in drug and toxicity studies, Ann. Rev. Biomed. Eng. 13 (2011) 55–72.
[3] D. Huh, Y.-S. Torisawa, G.A. Hamilton, H.J. Kim, D.E. Ingber, Microengineered physiological mimicry: organs-on-chips, Lab Chip 12 (2012) 2156–2164.
[4] NIH News. <http://www.nih.gov/news/health/sep2011/od-16.htm>, (accessed 16.09.11).

List of Contributors

Molly Allen
CellASIC Corp., CA, USA

Hojae Bae
Center for Biomedical Engineering, Department of Medicine, Brigham & Women's Hospital, Harvard Medical School, Boston MA, USA

Leon M. Bellan
Koch Institute for Integrative Cancer Research, Massachusetts Institute of Technology, MA, USA

Jeffrey T. Borenstein
Biomedical Engineering Center, Draper Laboratory, Cambridge MA, USA

Holly Chamberlain
Koch Institute for Integrative Cancer Research, Massachusetts Institute of Technology, MA, USA

Joseph L. Charest
Department of Biomedical Microsystems, Draper Laboratory, MA, USA

Seok Chung
School of Mechanical Engineering, Korea University, Korea

Carolyn G. Conant
Fluxion Biosciences, CA, USA

Nicholas J. Douville
Department of Biomedical Engineering, University of Michigan, Ann Arbor MI, USA

Else M. Frohlich
Department of Biomedical Microsystems, Draper Laboratory, Cambridge MA, USA

Brian M. Gillette
Department of Biomedical Engineering, Columbia University, NY, USA

Stefan Giselbrecht
Institute for Biomedical Technology and Technical Medicine, University of Twente, The Netherlands

Eric Gottwald
Institute for Biomedical Technology and Technical Medicine, University of Twente, The Netherlands

Linda G. Griffith
Department of Biological Engineering, Massachusetts Institute of Technology, MA, USA

James B. Grotberg
Department of Biomedical Engineering, University of Michigan, Ann Arbor MI, USA

Jen-Huang Huang
Artie McFerrin Department of Chemical Engineering, Texas A&M University, TX, USA

Paul Hung
CellASIC Corp., CA, USA

Anne Hsieh
Department of Chemical Engineering and Applied Chemistry, University of Toronto, Ontario, Canada

Niraj K. Inamdar
Department of Mechanical Engineering, Massachusetts Institute of Technology, Cambridge MA, USA

Cristian Ionescu-Zanetti
Fluxion Biosciences, CA, USA

Arul Jayaraman
Artie McFerrin Department of Chemical Engineering, Texas A&M University, TX, USA

Jessie S. Jeon
Department of Mechanical Engineering, Massachusetts Institute of Technology, Cambridge, MA, USA

Mohammad R.K. Mofrad
Department of Bioengineering, University of California, Berkeley, CA, USA

Hirokazu Kaji
Department of Bioengineering and Robotics, Tohoku University, Sandai, Japan

Roger D. Kamm
Cecil and Ida Green Distinguished Professor of Biological and Mechanical Engineering, Massachusetts Institute of Technology, Cambridge, MA, USA

Dmitry Kashanin
Cellix Ltd., Ireland

Ali Khademhosseini
Center for Biomedical Engineering, Department of Medicine, Brigham & Women's Hospital, Harvard Medical School, Boston MA, USA

Young Kyun Kim
Department of Bioengineering, University of California, Berkeley, CA, USA

Robert Langer
Koch Institute for Integrative Cancer Research, Massachusetts Institute of Technology, MA, USA

Philip Lee
CellASIC Corp., CA, USA

Kam W. Leong
James B. Duke Professor, Department of Biomedical Engineering, Duke University, NC, USA

Cristina Martin
Department of Chemical Engineering, University of Salamanca, Spain

Christopher Moraes
Department of Biomedical Engineering, University of Michigan, Ann Arbor MI, USA

Shashi K. Murthy
Department of Chemical Engineering, Northeastern University, MA, USA

J. Tanner Nevill
Fluxion Biosciences, CA, USA

Hesam Parsa
Department of Biomedical Engineering, Columbia University, NY, USA

Toby Paul
Cellix Ltd., Ireland

Sara Pizzato
Cellix Ltd., Ireland

Brian D. Plouffe
Department of Chemical Engineering, Northeastern University, Boston, MA, USA

Milica Radisic
Department of Chemical Engineering and Applied Chemistry, Institute of Biomaterials and Biomedical Engineering, Heart & Stroke/Richard Lewar Centre of Excellence, University of Toronto, Ontario, Canada

Arne Schleunitz
Institute for Biomedical Technology and Technical Medicine, University of Twente, The Netherlands

Šeila Selimović
Center for Biomedical Engineering, Department of Medicine, Brigham & Women's Hospital, Harvard Medical School, Boston MA, USA

Samuel K. Sia
Department of Biomedical Engineering, Columbia University, NY, USA

Shuichi Takayama
Department of Biomedical Engineering, University of Michigan, MI, USA

Nimalan Thavandiran
Institute of Biomaterials and Biomedical Engineering, University of Toronto, Ontario, Canada

Joe Tien
Department of Biomedical Engineering, Boston University, MA, USA

Roman Truckenmüller
Institute for Biomedical Technology and Technical Medicine, University of Twente, The Netherlands

James G. Truslow
Department of Biomedical Engineering, Boston University, Boston, MA, USA

Victor M. Ugaz
Artie McFerrin Department of Chemical Engineering, Texas A&M University, TX, USA

Joshua B. White
Department of Biomedical Engineering, University of Michigan, Ann Arbor MI, USA

Vivienne Williams
Cellix Ltd., Ireland

Keith H.K. Wong
Department of Biomedical Engineering, Boston University, Boston, MA, USA

Diana Wu
Koch Institute for Integrative Cancer Research, Massachusetts Institute of Technology, MA, USA

Yun Xiao
Department of Chemical Engineering and Applied Chemistry, Institute of Biomaterials and Biomedical Engineering, University of Toronto, Ontario, Canada

Yong Yang
Department of Chemical Engineering, West Virigina University, Morgantown, WV, USA

Sang-Hee Yoon
Department of Bioengineering, University of California, Berkeley, CA, USA; Current address: Wyss Institute for Biologically Inspired Engineering, Harvard University, Cambridge MA, USA

Boyang Zhang
Department of Chemical Engineering and Applied Chemistry, Institute of Biomaterials and Biomedical Engineering, University of Toronto, Ontario, Canada

Beili Zhu
Department of Chemical Engineering, Northeastern University, Boston, MA, USA

PART

1

Materials and Fabrication Methods

CHAPTER

Microfluidic Cell Culture Platforms with Embedded Nanoscale Features

1

Yong Yang[1] and Kam W. Leong[2]

[1]*Department of Chemical Engineering, West Virigina University, Morgantown, WV, USA*

[2]*Department of Biomedical Engineering, Duke University, NC, USA*

1.1 Introduction

In most living tissues, cells typically reside in a microenvironment where cells interact with the extracellular matrix (ECM) as well as the neighboring cells. The ECM is composed of diverse biomacromolecules including glycosaminoglycans, fibrous proteins such as collagen, elastin, and fibronectin, and nonfibrous proteins such as growth factors and cytokines, with size ranging from several to hundreds of nanometers. For example, collagens show a hierarchical structure of collagen fibrils of 10–300 nm in diameter to collagen fibers up to several microns in diameter [1]. The ECM constructed from these biomacromolecules often includes significant topography at the nanoscale. Basement membrane, for example, is a ubiquitous component of ECM that plays an important role in tissue development and organization. Basement membranes manifest a complex mixture of pores, ridges, and fibers with sizes in the nanometer range [2–6]. In an ECM of connective tissues, the fibrous proteins penetrate through a hydrophilic matrix composed of proteoglycans and interstitial fluids to form a 3-D network with nonfibrous proteins dispersed inside [7]. Although the interstitial fluid flows, driven by dynamic stresses or/and pressure gradients, can be extremely slow (between 0.1 and 4.0 μm/s [8]), they nevertheless play an important role in nutrient transport, tissue maintenance, and remodeling, as well as establishment of the microenvironment [9,10]. Together, the ECM serves as a structural support for cells and provides, in concert with the spatiotemporally arranged signaling molecules and external stimuli, topographical and mechanical cues that serve to regulate the phenotypes and function of mammalian cells.

A number of recent findings underscore the phenomenon that mammalian cells have the capacity to respond to environmental features at the nanoscale on synthetic surfaces [11–15]. In addition to inducing pronounced changes to cell

Microfluidic Cell Culture Systems. ISBN: 978-1-4377-3459-1

morphology, and consequently gene expression, nanotopographical cues could potentially help induce the differentiation of stem cells into certain lineages [11,13,15]. For instance, by manipulating levels of nanoscale order of nanopits (120 nm in diameter, 100 in. depth), Dalby et al. [11] have shown that the near-square topography stimulates human mesenchymal stem cells (hMSCs) to produce bone mineral *in vitro*, in the absence of osteogenic supplements at levels similar to the cells cultured with osteogenic media; highly ordered nanotopography permits prolonged retention of multipotency of hMSCs [12]. Another study shows that hMSCs cultured on small (~30 nm diameter) TiO_2 nanotubes exhibit promoted adhesions without noticeable levels of differentiation, whereas larger (~70–100 nm) nanotubes elicit a dramatic cell elongation (~10-fold increase compared with the flat control), which induces cytoskeletal stress and selective differentiation into osteoblast-like cells [13]. In addition, fluid shear stress can remodel cytoskeletal organization, alter gene and protein expression, and influence cell proliferation [16]. For instance, hMSCs seeded on silicate-substituted tricalcium phosphate scaffolds demonstrate improved proliferation and osteogenesis in the flow perfusion culture compared with the static cell culture. Flow perfusion culture also facilitates homogenous distribution of cells and ECM proteins throughout the entire scaffold, whereas only a peripheral layer is obtained after static culture [17]. Intriguingly, it has been reported that shear stress promoted the differentiation of smooth muscle cells and embryonic stem cells into endothelial cells [18,19].

It is thus vital to integrate these external cues and engineer a dynamic cell culture platform with embedded nanoscale features for cellular studies [20]. Microfluidic systems, in particular the ones based on poly(dimethylsiloxane) (PDMS), have been extensively applied to the studies of cell stimulation and selection [21,22], cell lysis and biochemical analysis [23], cell manipulation [24,25], and cell culture [26–30]. In cell culture, the microfluidic setup can work as a circulatory system, enhance mass transfer of nutrients, gases, and metabolites, provide a spatiotemporal control of delivery of signaling molecules, and create mechanical strain in the physiological range. However, it is challenging to integrate nanoscale features into a microfluidic platform [31].

The two main challenges of fabricating a microfluidic platform with embedded nanoscale features are creating a large area of nanopatterned surface for cell culture and preserving the fidelity of the nanotopography after assembly. In this chapter, we first review techniques for engineering nanoscale features. Particular emphasis is placed on generating a large area of nanopatterned surface. The assembly techniques suitable for PDMS-based microfluidic platforms are then discussed. Several studies that engineer microfluidic platforms with embedded nanoscale features for cancer detection and stem cell research are discussed.

1.2 Engineering of nanoscale features

Nanotechnology is referred to a length scale of 1–100 nm in the physical realm [32]. However, considering that cellular structures are built from biomolecules, it is appropriate to extend its length scale beyond 100 nm and upward to the submicrometer range. Here, we review the conventional techniques for fabricating irregular and regular nanoscale features, and generating large area of nanopatterned surfaces.

1.2.1 Fabrication of irregular nanoscale features

Chemical etching, in addition to modifying the surface chemistry of cell culture substrates [33,34], may generate irregular nanoscale features [34]. By combining silicon etching and silver deposition, Peng et al. [35] synthesized large silicon nanowire arrays. The diameters of the nanowires range from 30 to 150 nm and their lengths are up to 50 μm. These nanowire arrays provide numerous surface area and have been utilized to capture circulating tumor cells (CTCs) [36].

Self-assembly offers a simple and low-cost process to make large-area periodic nanostructures. It requires only the mixing of components such as colloidal nanoparticles and polymers, which can be spontaneously assembled into a nanotextured structure. Colloidal lithography forms nanopatterns based on the colloid–colloid and colloid–substrate interactions. The resulting nanopatterns can either be used in the current form or as masks. Lipski et al. [37] aminosilated silica nanoparticles of 50 and 100 nm and self-assembled them onto metal substrates, thus demonstrating a single-step fabrication of nanoscale surface with controllable surface characteristics of charge, roughness, and chemistry. Some polymer blends spontaneously undergo phase separation during spin coating. This approach, often called polymer demixing, can produce different topographies such as pits, islands, or ribbons of varying height. The ratio of the polymers used varies with the topography, and the concentration of polymer solution changes the feature sizes [38,39]. Demixing the polystyrene (PS) and poly(4-bromostyrene) (PS/PBrS) system can form nanoislands with a variety of sizes, and PS segregates to the surface upon annealing [40–42]. Another example is PS–poly(*n*-butylmethacrylate) (PS/PnBMA), in which the PnBMA segregates spontaneously to cover the surface at ambient temperature without annealing [43,44]. Both systems can provide the nanotopography with a single chemistry at the top surface.

Self-assembly as a stand-alone method for nanofabrication is presently unable to produce structures with precise spatial positioning and arbitrary shapes [45]. Higher density of defects compared to conventional nanofabrication techniques is also a drawback. Templated self-assembly, which uses top-down lithographic approaches to provide the topographical and/or chemical templates to guide the bottom-up assembly of colloidal particles and macromolecules, may get the best out of the two approaches. For instance, Yin et al. [46] demonstrated the templated self-assembling of PS beads and silica colloids (≥150 nm in diameter) into complex aggregates that include polygonal or polyhedral clusters, linear or zigzag chains, and circular rings by geometric confinement. Block copolymers can self-assemble into periodic microdomains on the order of ~10–100 nm. Templates can be used to control the ordering and orientation of the microdomains [47–49].

Cheng et al. [50] used a sparse chemical pattern with alternating pinning/neutral regions to guide the self-assembly of block copolymers down to ~10 nm.

1.2.2 Fabrication of regular nanoscale features

1.2.2.1 Focused beam lithography

Highly defined nanoscale features are usually generated by using focused beam (electron, ion, photon) lithography (FBL).

In electron beam lithography (EBL), predefined nanopatterns are generated by scanning a focused electron beam across an electron-sensitive resist (tone) coated on a substrate and the resist subsequently undergoes solvent development. The nanopattern can be further transferred into underneath substrate via deep reactive ion etching. A high resolution can be achieved by using high-energy exposure electrons (greater than 75 kV), thinnest usable resist (~50–100 nm), and low atomic number substrate feasible. For example, isolated poly(methyl-methacrylate) (PMMA) lines of 3–5 nm have been developed [51], and an even higher resolution of 1–2 nm has been demonstrated on 50 nm thick metal halide resists by using a 5 Å diameter focused (100 keV) electron beam [52].

In focused ion beam lithography (FIBL), liquid-metal ion sources are focused onto fine spots on the order of 10 nm and the electrostatic beam deflector controls the landing location of the ions on the substrate. The feature size is largely dictated by the beam size and the interaction of the beam with the target material. A resolution of 3 nm has been demonstrated [53] and complex 2-D or 3-D nanostructures can be created [54].

In two-photon lithography (TPL), the focal point of a laser is scanned within the bulk of a photosensitive material to induce a chemical change [55], allowing the formation of arbitrarily complex 3-D patterns [56]. As a result of the combination of optical and chemical nonlinearity, TPL can achieve a sub- to single-micrometer resolution [57,58]. Using a confocal laser scanning microscope equipped with a femtosecond mode-locked near-infrared (IR) laser, one can microfabricate structures of any shape with a resolution less than 100 nm [59]. For example, suspended lines with widths as small as 30 nm have been fabricated in SU-8 [60].

FBL has different applications depending on the beam used. The electron or photon beam is mainly applied for writing on soft materials, while FIBL uses much heavier ions and can strike with greater energy at relatively short wavelengths to directly transfer patterns onto hard materials (such as semiconductors, metals, or ceramics). Although the FBL techniques enable generation of very fine topography, the capital investment, the accessibility, and the complexity of processes restrict their widespread use in academic research.

1.2.2.2 Replication techniques

Soft lithography and nanoimprinting lithography (NIL) are commonly used techniques to transfer the nanostructures to final substrates, usually polymeric substrates. Through these inexpensive processes, the nanostructures can be replicated in great quantity, resulting in the initial expenditure of the mold being an insignificant part of the overall cost of the device.

Soft lithography involves casting liquid polymer precursor such as PDMS against a topographically patterned mold. After solidification, the polymeric substrate is transferred with the reversed pattern [61]. Feature sizes as small as 2 nm using single-walled carbon nanotubes as templates have been demonstrated [62]. In principle, the ultimate limit can be pushed to less than 0.5 nm [45].

NIL involves pressing a nanopatterned mold into a polymeric substrate at elevated temperature and pressure [63,64]. Features as small as ~5 nm [65,66] and aspect ratios up to ~20 (channels of 100 nm width and 2 μm depth) [67] have been demonstrated. Among the operation parameters, temperature and pressure are the most important. Becker et al. suggested that the thermal cycle, which is the temperature range between imprinting and demolding, should be 25–40°C to minimize thermally induced stresses in the materials and replication errors due to a difference in thermal expansion coefficients between the mold and the substrate [68]. However, using a lower temperature requires a larger compression force to imprint the substrate, which leads to higher flow-induced stresses [69] and incomplete pattern transfer into the substrate [70]. The alternatives are to use liquid UV-curable precursors [71] or supercritical fluids as processing aids [72,73].

1.2.2.3 Stitching technique

For cellular studies, a large area of nanopatterned surface is desirable so that enough cells can be seeded for subsequent biochemical and molecular biology analyses. However, because the commonly used lithography techniques such as EBL are time consuming and expensive, the cost of a large master with highly defined nanoscale features is prohibitively expensive for a typical research laboratory. As discussed earlier, self-assembly is presently unable to produce defect-free, nanostructures with precise spatial positioning and arbitrary shapes as a stand-alone method [45]. Alternative nanofabrication techniques such as Langmuir–Blodgett lithography [74], step-and-flash lithography [75], and a roll-to-roll NIL [76] have been developed but have not been widely adopted because of their sophisticated process or requirement of specialized equipments.

A simple and cost-effective stitching technique has been developed to generate a large area of nanopatterned surface [77]. As illustrated in Figure 1.1A, multiple PDMS molds, replicated from an EBL-written mold via soft lithography could be stitched into a single large nanopatterned surface. These PDMS molds with the nanopattern face-down are aligned and placed on the mirror side of a Si wafer, thus ensuring that all the nanopatterned surfaces are at the same level. A PDMS prepolymer (uncured mixture of PDMS resin and curing agent) layer spin-coated on a glass plate is pressed onto the unpatterned side of PDMS molds. After curing, the multiple PDMS molds are thereby stitched into a single mold. The thicknesses of the multiple PDMS molds may vary from batch to batch. Gluing the backside of these molds with a PDMS prepolymer layer can compensate the

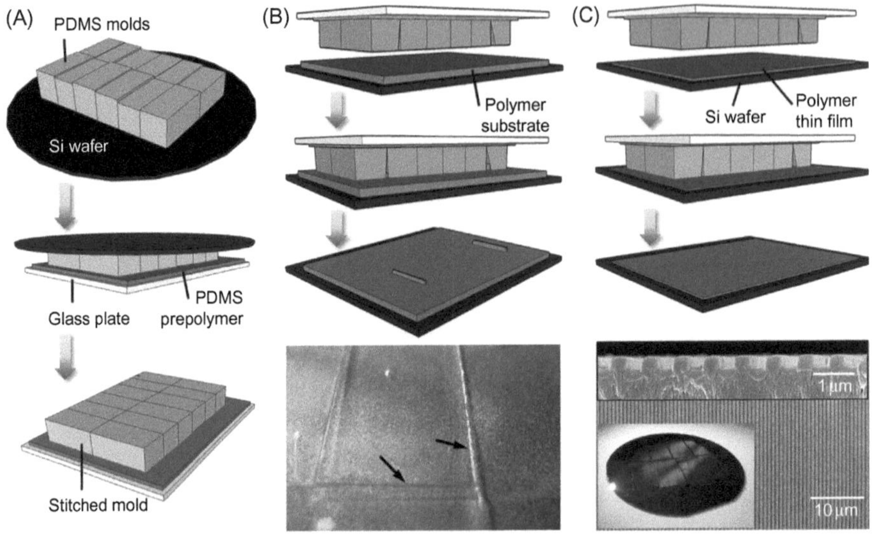

FIGURE 1.1

Stitching technique. (A) Schematic illustration of stitching technique. (B) When the stitched mold was imprinted into a thick polymer substrate, the polymer was squeezed into interstices as indicated by the arrows in the optical image. (C) A large area of nanopattern on PS thin film spin-coated on a 75 mm wafer (the inset optical image). SEM images show both the top and the cross-sectional views of final transferred PS nanogratings.

Source: *Reproduced from Ref. [77] with permission from the Royal Society of Chemistry.*

variation in the thickness of these PDMS molds. It is also understandable that multiple PDMS molds cannot be stitched seamlessly. The PDMS prepolymer layer can be partially cured to eliminate the possibility of the prepolymer being pulled to the patterned surface via the interstice by the capillary force.

Because of the interstice between the PDMS molds, the stitched mold is usually used as a master to replicate more nanopatterned substrates. However, when the stitched mold is used for pattern transferring such as hot embossing and casting, the polymer substrate might flow into the interstice and result in an uneven surface (Figure 1.1B). These rises will cause an incomplete sealing when microchannels are sealed against this patterned substrate, thus failing the microfluidic chip. With polymer thin film technology, this problem can be readily solved. A polymer thin film is spin-coated on a Si wafer. The stitched mold is imprinted into the thin film (Figure 1.1C). The polymer rises in interstices can be minimized by tuning the film thickness, which is also determined by the height of the nanoscale features. Figure 1.1C shows a large area (4.8 cm × 4.8 cm) of PS nanogratings. It is imprinted using a mold stitched from eight pieces of PDMS molds with gratings of 350 nm width, 700 nm pitch, and 280 nm depth. This technique exhibits a high fidelity, as indicated in the SEM images (Figure 1.1C). The width, spacing, and depth of the nanogratings changed from the original 355.1 ± 0.6, 327.1 ± 8.2, and 274.5 ± 1.7 to 366.9 ± 8.5, 335.9 ± 24.6, and 266.7 ± 7.4, respectively, on the final PS substrate. The dimensional change is less than 5%.

This nanopatterned PS film can be used as cell culture substrates directly or as a mold to replicate more nanopatterned PDMS substrates. To accommodate a short working distance and facilitate microscopic observations at high magnification, PDMS prepolymer can be spin-coated on the nanopatterned PS mold and a microscope slide is placed on the top. After curing, a thin layer (~1 mm thick) of PDMS with the nanopattern replica bonded to the glass slide can be generated.

This stitching technique is versatile. Patterns of various dimensions, shape, and spacing can be stitched. Moreover, the patterns can be precisely arranged as needed.

1.3 Assembly of PDMS-based microfluidic platforms

PDMS-based microfluidic systems have seen widespread use because they afford advantages such as easy fabrication, low cost, optical transparency, and high gas permeability. Multiple PDMS components are usually assembled into a complete microfluidic system reversibly or irreversibly.

1.3.1 Reversible assembly

A PDMS component can be brought into contact with another PDMS component or a glass substrate to form conformal sealing. This sealing, results from van der Waals force between two flat surfaces, is reversible and cannot withstand a high pressure (>5 psi).

Mechanical clamping has also been used to secure the conformal sealing, thus allowing injection of fluids at a higher pressure. A unbalanced clamping force on the PDMS components may pose a risk of leakage on the microfluidic system.

1.3.2 Irreversible assembly

To overcome the high pressure caused by fast flows in the microscale channels, irreversible assembly of PDMS microfluidic platforms can be achieved by treating the surface of PDMS components chemically or using adhesives.

A commonly used surface treatment of PDMS is oxygen (O_2) plasma treatment. O_2 plasma treatment converts the silane ($Si{-}CH_3$) groups at PDMS surface to silanol ($Si{-}OH$) groups, which can condense with appropriate groups (OH, COOH, ketone) on another surface to form an irreversible seal and can withstand pressure of 30–50 psi [78]. The resultant silanol groups at the same surface might also condense to form a brittle layer where silicon might be bonded to three or four oxygen atoms (SiO_x) [79], which leads to a decrease in bond strength.

In addition, a systematic investigation of O_2 plasma treatment of PDMS substrates reveals that O_2 plasma treatment changes the surface chemistry, consequently, the topography and mechanical properties; the latter effect, in particular, has pronounced effects on cell behavior (Yang, unpublished data).

For PDMS-based microfluidic systems, one (the curing agent) [80] or two components [81] of PDMS, or partial cured PDMS [82] have been used as adhesive for assembly of microfluidics. For instance, the stamp-and-stick technique directly spin-coats viscous PDMS prepolymer as adhesive layer, which is 1–1.5 μm thick even at 8000 rpm for 8–9 min [81]. This assembly technique demonstrates a substantial bond strength up to 100 psi [81] because of their same material chemistry. However, the uncured PDMS from surrounding areas tends to flow into the features [83].

1.3.3 Microtransfer assembly

A simple and flexible assembly technique, called microtransfer assembly (μTA) technique, has been developed based on polymer thin film and microcontact printing technologies. As illustrated in Figure 1.2A, PDMS prepolymer is diluted in hexane to a 5 wt% solution and spin-coated onto a Si wafer to form a thin film (1). A PDMS mold with microchannels (2) is brought into contact with the film. After 1 min, the PDMS mold is taken off from the wafer, aligned, and brought into contact with another PDMS substrate (3) with or without nanogratings, and kept at 80°C for 2 h to complete the assembly. During the μTA process, good solvent such as hexane [84] is used to dilute the PDMS prepolymer, and the film thickness can thus be tuned from several microns to less than 100 nm [85]. Moreover, the thin film can be partially cured through prebaking at a lower temperature, for example, at 50°C for 5 min, to increase its viscosity. The viscous thin PDMS film is restricted to contacted areas without flowing in and filling the surrounding nanostructures when there were nanostructures on the final PDMS substrate (3).

Figure 1.2B shows microchannels embedded with nanogratings which are perpendicular to the flow direction. Figure 1.2B and C compares the assembly performance of μTA and O_2 plasma assembly. The microchannels sealed by μTA have defined boundaries, while in O_2 plasma-assembled microchannels the nanogratings are visible even outside the microchannels (indicated by the light reflection). When dyed water flows through the microchannels, no leakage is observed in microchannels sealed using μTA (the enlargement of Figure 1.2B). The PDMS prepolymer fills the nanogratings in the contact area and prevents leakage. After peeling off the microchannel mold from the nanopatterned substrate, it was observed that the nanogratings within the microchannel remained intact while those in the contacted area were smoothed out. In contrast, the fluid leaks through the nanogratings in the O_2 plasma-assembled microchannels (the enlargement of Figure 1.2C). In addition, O_2 plasma assembly requires a plasma generator or reactive ion etcher in a clean room, while the μTA technique does

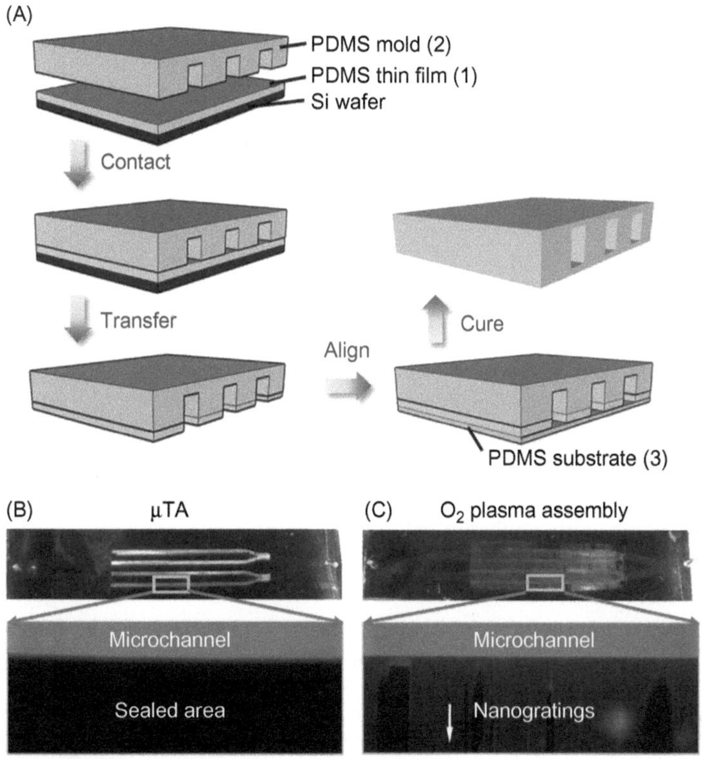

FIGURE 1.2

Microtransfer assembly technique. (A) Schematic illustration of μTA process. Optical images of rhodamine B-dyed water flowing through nanograting-embedded microchannels sealed using (B) μTA and (C) O_2 plasma-assisted assembly. Nanogratings are perpendicular to the flow direction. The enlargements show the assembly performance of these two assembly techniques.

Source: *Reproduced from Ref. [77] with permission from the Royal Society of Chemistry.*

not have any specific equipment requirement but a spin coater. Contaminants such as dust will not affect the performance of μTA, but may be fatal to O_2 plasma assembly.

Apparently, μTA can improve the sealing performance as well as keep the nanotopography intact. By controlling the prebaking temperature and time, the assembly can be reversible or irreversible, depending on the extent of partial curing of PDMS prepolymer. This flexibility is important, catering to whether the assembly should be strong enough to withstand a high pressure or the microfluidic platform needs to be opened up for cellular analysis. For example, it would be advantageous to open up the microchannel to observe the immunostained cells under confocal microscope at a high magnification.

1.4 Microfluidic platforms with embedded nanoscale features for cell studies

With advances in nanofabrication and assembly techniques, nanoscale features can be integrated into a microfluidic platform. Recently, microfluidic platforms with embedded nanoscale features have been reported for cancer cell detection and stem cell research.

1.4.1 Cancer cell detection

Inspired by enhanced interactions between silicon nanowires and nanoscale components of the cellular surface (e.g., microvilli, microridges, and cilia) [86,87], Wang et al. [88] integrated nanostructured silicon substrates with chaotic micromixers to achieve high-efficient capture of CTCs. As shown in Figure 1.3A.1, silicon nanopillars (SiNP) of 100–200 nm in diameter and 12–15 μm in height are fabricated on patterned silicon wafers using a wet chemical etching method [35], and then coated with epithelial-cell adhesion-molecule antibody (anti-EpCAM) for recognizing/capturing EpCAM-expressing tumor cells [36]. A PDMS chip with embedded chevron-shaped micromixers on the channel roof (Figure 1.3A.2)

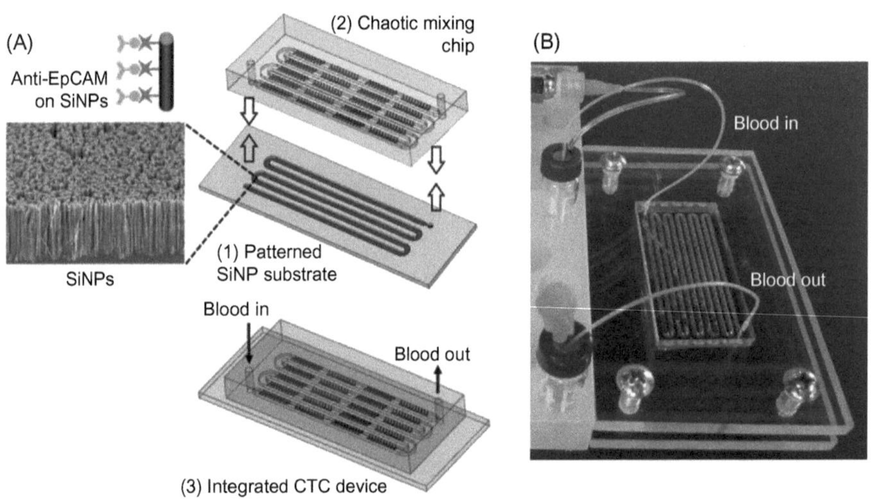

FIGURE 1.3

A nanostructured microfluidic device integrated with chaotic micromixers for capturing CTCs. (A) Schematic representation of the CTC-capture device. The device is composed of two functional components: (i) patterned SiNP substrate with anti-EpCAM-coating exhibiting vastly enhanced CTC-capture affinity and (ii) an overlaid microfluidic chaotic mixing chip (the channel is 1000 μm wide, 100 μm deep, and 88 cm long) capable of promoting cell–substrate contact frequency. (B) A photograph of the CTC-capture device.

Source: *Reproduced from Ref. [88] with permission from Wiley-VCH Verlag GmbH & Co. KGaA.*

is overlaid onto the patterned SiNP substrates (Figure 1.3A.3). The two PDMS components are assembled by mechanically clamping them between two polyacylate chip holders (Figure 1.3B). This chaotic mixing channel induces vertical flow in the channel and increases cell–SiNP contact frequency, thus significantly improving sensitivity in detecting rare CTCs from whole blood. In 17 out of 26 patient blood samples, this platform captures significantly greater CTC numbers compared to the commercial CellSearch assay. This new CTC-capture technology is envisioned to enable early detection of cancer metastasis and isolation of rare populations of cells that the existing technologies cannot do.

Different from the irregular SiNPs, Kwon et al. [21] fabricated defined 400 nm pillars and 400 nm lines on ultraviolet (UV)-curable polyurethane acrylate (PUA) substrate and embedded them into a PDMS microchannel by applying O_2 plasma assembly technique. The microfluidic channel consists of four branch channels (600 μm in width and 50 μm in height). Flat or nanoscale PUA features (400 nm pillars, 400 nm perpendicular, or 400 nm parallel lines) are embedded into each microchannel (Figure 1.4); each patterned area is 0.6 mm × 4 mm. The adherent cells including human breast epithelial cells (MCF10A) and cancer cells (MCF7) on these flat control and nanotopography are counted with increasing

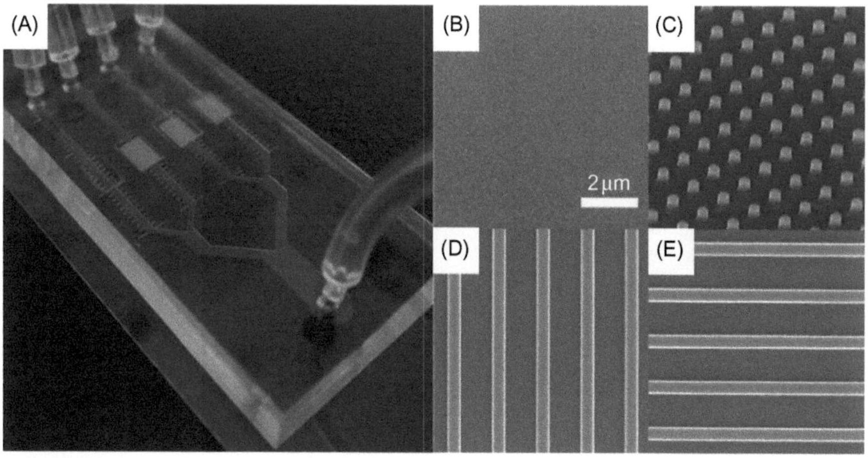

FIGURE 1.4

A microfluidic device for separation and enrichment of human breast cancer cells. (A) A photograph showing the fabricated microfluidic channel having four branch channels with patterned substrate (each branch channel is 600 μm in width, 50 μm in height, and 15 mm in length). Different diffraction colors represent different patterns. (B)–(E) SEM images of flat or three nanostructures embedded inside a microfluidic channel: (C) flat surface, (D) 400 nm pillars, (E) 400 nm perpendicular lines, and (F) 400 nm parallel lines. (For interpretation of the references to color in this figure legend, the reader is referred to the web version of this book.)

Source: *Reproduced from Ref. [21] with permission from the Royal Society of Chemistry.*

flow rate after a preculture for a period of time (e.g., 1, 2, and 4 h). On all the nanotopography, MCF10A show higher adhesion than cancer cells (MCF7) regardingless of culture time and flow rates tested, thus resulting in enrichment of cancer cells. The cell-adhesion behavior is opposite to what is observed by Wang et al. [88], probably due to the surface coating of anti-EpCAM on SiNPs. They also applied the nanostructured microfluidic platform for adhesion assays of endothelial cells [89].

1.4.2 Stem cell regulation

With the enabling stitching and μTA techniques, microfluidic platforms embedded with nanoscale features are engineered to regulate stem cell behavior with topographical and fluidic cues [77]. Figure 1.5A shows the design and fabrication of such a microfluidic platform. Microchannels of 500 μm in width, 52 μm in height, and 2 cm in length are generated via conventional photolithography, followed by soft lithography. Near the inlet and outlet of the microchannels, perfusion channels of 12 μm in width and 52 μm in height are fabricated to help retain the cells within the main microchannels (Figure 1.5C). Nanogratings of 350 nm linewidth, 700 nm pitch, and 280 nm depth are written in a PMMA thin film spin-coated onto a Si wafer using EBL, and then transferred onto PDMS layer supported on a glass slide. The nanogratings are aligned in directions parallel or perpendicular to the flow direction, respectively. Figure 1.5B shows a photograph of the microfluidic platform with embedded nanogratings perpendicular

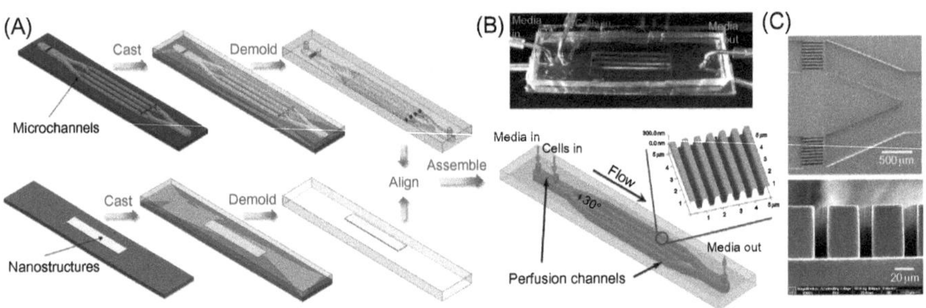

FIGURE 1.5

A typical microfluidic platform with embedded nanoscale features. (A) Design and fabrication of the microfluidic platform. Nanogratings of 350 nm in width, 700 nm in pitch, and 280 nm in depth (atomic force microscopic image) were perpendicular to the microchannels of 500 μm in width and 50 μm in height. The divergent angles are 30°. (B) Optical image of a microfluidic platform. The nanogratings within the microchannels reflect light and are visible. (C) SEM images of top (upper panel) and side view (lower panel) of perfusion channels.

Source: *Reproduced from Ref. [77] with permission from the Royal Society of Chemistry.*

to the flow direction. The nanogratings within the microchannels reflect light and are visible.

With the very low height (h) to width ratio (b), $53/500 < 0.1$, the microchannels are treated as a parallel plate system for the steady laminar flow. Shear stress is calculated using equation $\tau_{wall} = 6\mu Q/bh^2$, where Q is the flow rate for each microchannel and μ is the viscosity of the culture media (the value of water, 0.001 Pa s, is used). The effects of nanogratings on the shear stress are assessed using the computational fluid dynamics (CFD) simulation. The results show that the shear stresses at the front and rear edges of the perpendicularly arranged nanogratings were higher than that on the top of the nanogratings. However, this increase only existed in the region very near to the grating surface (5% increase at 500 nm away from the grating surface) and diminished to 3% when it is 5 μm away. Considering that cells adhered to the nanogratings, the wall shear stress calculated based on the flat surface was used in this study. The flow rates were varied from 0.005 to 0.025 ml/h; the corresponding shear stress were from 0.006 to 0.03 N/m^2 (0.3 dyn/cm^2, 1 N/m^2 = 10 dyn/cm^2). In this study, when the shear stress is lower than 0.015 N/m^2 (0.15 dyn/cm^2) no obvious influence of the flow on hMSCs was observed compared with the static condition; while most cells detached when the shear stress was increased beyond 0.03 N/m^2(0.3 dyn/cm^2). Therefore, the shear stresses studied were between 0.015 and 0.3 N/m^2.

hMSCs are cultured on both flat surface and nanogratings under static and dynamic conditions using the microfluidic platform. hMSCs exhibit different morphology, actin cytoskeleton, and focal adhesion organization under various conditions shown in Figure 1.6. Under static condition, the cells adhere well to the flat surface and have a well-spread, multipolar morphology and well-defined stress fibers (by labeling F-actin with phalloidin) oriented randomly (Figure 1.6A). On nanopatterned surface, the cells align along the nanogratings and appear to have a smaller size with diffuse stress fibers (Figure 1.6E). In addition, on the flat surface, clear and distinct focal adhesions (by using an anti-paxillin antibody followed by an Alexa Fluor 594-conjugated secondary antibody) are observed at the periphery of the cell and distributed throughout the cell without preferential orientation (Figure 1.6B–D). On the nanogratings, the focal adhesions are mainly found in the lamellipodia region, aligned along the nanogratings (Figure 1.6F–H). When a shear stress of 0.03 N/m^2 (0.3 dyn/cm^2) is applied, the focal adhesions as well as the stress fibers are steered toward the flow direction on the flat surface (Figure 1.6I–L). However, hMSCs cultured on the perpendicular nanogratings are still aligned along the nanogratings, but spread in the flow direction accompanied with a significant increase of cell spreading area (Figure 1.6M–P). Interestingly, the focal adhesions keep oriented along the nanogratings (Figure 1.6O).

Cell–substrate interactions begin with formation of focal adhesions via binding and clustering of integrins onto the adhesive ligands present in the substrate. Focal adhesion formation can be altered and directed using nanotopography, depending on dimension, shape, spacing, and symmetry of the nanoscale features [90].

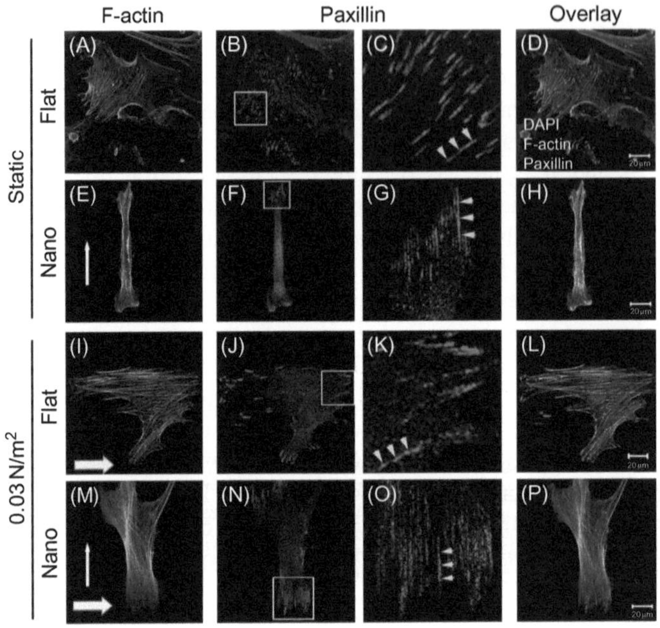

FIGURE 1.6

Confocal images of hMSCs grown on (A–D, I–L) flat and (E–H, M–P) nanopatterned PDMS under static (A–H) and dynamic (I–P) conditions. The nuclei were stained with DAPI in blue, the actin filaments were stained with phalloidin in green, and focal adhesions were stained with paxillin in red. The narrow white arrows indicate the nanograting orientation and the wide yellow arrows denote direction of the flow with a shear stress of 0.03 N/m^2. These images are at the same scale with the scale bar of 20 μm except images (C), (G), (K), and (O), which enlarge the paxillin arrangement in the boxed regions of the images (B), (F), (J), and (N), respectively. The yellow arrowheads point to the focal adhesion. (For interpretation of the references to color in this figure legend, the reader is referred to the web version of this book.)

Source: *Reproduced from Ref. [77] with permission from the Royal Society of Chemistry.*

Nanogratings constrain the formation of focal adhesions and consequently the stress fibers preferentially align along the nanograting direction, thus guiding the spreading and migration of cells. In this study, the flow-induced shear stress, ranging from 0.006 to 0.03 N/m^2 is not high enough to alter the directional formation of focal adhesions. As a result, the cells migrate preferentially along the nanograting direction although the flow deviated the migration slightly. In contrast, on the flat surface the cells migrate randomly under the low flow rates applied [77].

Of particular interest is the alignment and deformation of nuclei, which would result in changes of gene and protein expression [91,92] and is strongly associated with the lineage commitment of stem/progenitor cells [93]. Shown in Figure 1.7 are the effects of nanotopography and shear stress on nuclear alignment and

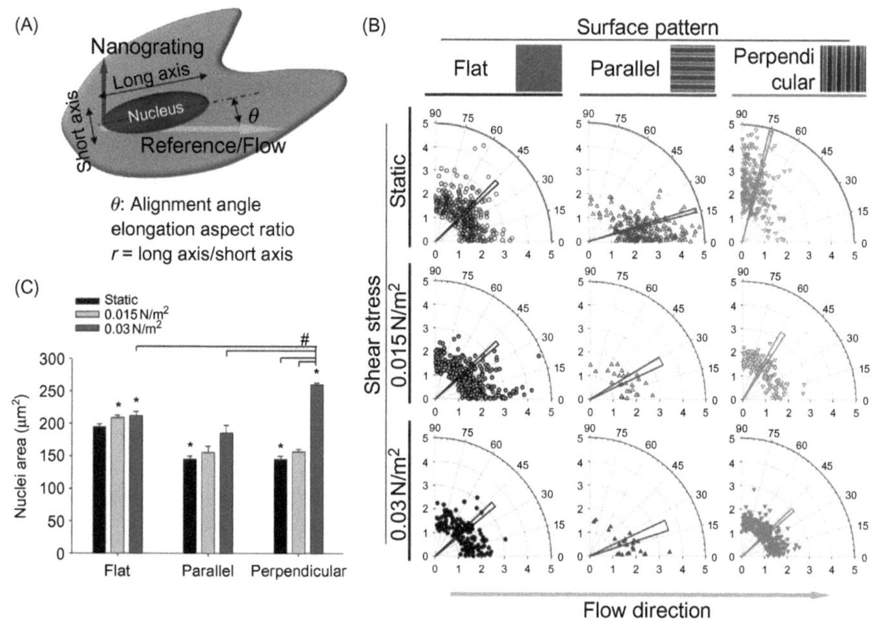

FIGURE 1.7

The effects of nanotopography and shear stress on nuclear alignment and deformation. (A) Definition of alignment angle θ and elongation aspect ratio r. (B) Polar plots of nuclear alignment (angular coordinate) and elongation (radial coordinate) of hMSCs grown on flat and nanopatterned PDMS surfaces under static and dynamic conditions: 0.015 and 0.03 N/m^2. The yellow horizontal arrow gives the reference direction, e.g., the flow direction. Each data point represents a nucleus. In each panel, the narrow triangle on the top of the data points the mean θ, the length of the short side represents the standard deviation, and the length is defined as two times of the elongation aspect ratio of the nucleus r for clarity. (C) Effects of nanogratings and shear stress on the nuclear area of hMSCs. *: significant difference ($p<0.01$) from Flat-stat; #: significant difference ($p<0.01$) between two groups. (For interpretation of the references to color in this figure legend, the reader is referred to the web version of this book.)

Source: *Reproduced from Ref. [77] with permission from the Royal Society of Chemistry.*

deformation. To describe the nuclear alignment and deformation, the alignment angle (θ) and elongation aspect ratio (r) are measured (Figure 1.7A). In the polar plots (Figure 1.7B), the alignment angle and elongation aspect ratio are presented by the angular and radial coordinates, respectively. Each polar plot is named by a combination of the x-axis label (flat or orientation of nanogratings) and the y-axis label (static or shear stress level) for clarity. For instance, "Flat-Stat" stands for the flat surface under static condition, and "Perp-0.015" for the nanogratings perpendicular to the flow with a shear stress of 0.015 N/m^2. Under static condition, the alignment angles distribute equally in the range of 0–90° on the flat surface, indicating a random orientation of the nuclei. On nanogratings, 62.0% of the nuclei orient in a 30° wide sector (or 15° alignment angle) centered on the reference line. The nanogratings elongate the nuclei to an r range of 1–5 compared with the range of 1–3 on the flat surface. On the flat surface, when the shear stress is applied the nuclei are forced to orient toward the flow direction slightly. The influence is not significant. On the nanogratings in both parallel and perpendicular directions to the flow, an increase in shear stress favors the nuclear alignment along the flow direction. An increase in shear stress from 0.015 to 0.03 N/m^2 increases the percentages of nuclei in the 30° wide sector from 35.7% and 9.2% to 45.8% and 25.7%, respectively. However, on the parallel nanogratings, the cells are easily detached. On the perpendicular nanogratings, the alignment angles are 74.6 ± 1.1 ($n = 250$), 57.3 ± 2.1 ($n = 153$), and 41.0 ± 1.5 ($n = 315$) for static, 0.015, and 0.03 N/m^2 conditions, respectively. The shear stress forces the nuclei to deviate from the nanograting direction toward the flow, thus resulting in a more uniform distribution.

The flow effects depend on cell adhesion to the substrate. It was reported that a flow parallel to 15 μm wide fibronectin strips elongated the endothelial cells constrained on the strips with enhanced stress fibers and focal adhesions, and a reduction in apoptosis, but perpendicular flow did not [94]. In the present study, the apparent surface that hMSCs can "sense" is reduced by the nanogratings; the cell membrane elasticity might prevent a radical bending and hence the plasma membrane bridges over the top of nanogratings rather than reaching the bottom [95]. Different from the previous report where the endothelial cells are in intimate contact with the substrate [94], in the current study, the cells might not fill the nanogratings, the fluid could penetrate into the nanogratings underneath the cells on the parallel nanogratings. In a synergetic way, the nanotopography and shear stress weaken the cell adhesion and detach cells. However, on the perpendicular nanogratings, the fluid might not penetrate underneath the cells; the competitive effects of nanotopography and shear stress reorient the nuclei.

These two physical cues alter the projected nuclear area as well (Figure 1.7C). On the flat surface, the fluid shear stress increases the nuclear area significantly, from 194.8 ± 4.8 (Flat-Stat, $n = 303$) to $208.7 \pm 4.7\ \mu m^2$ (Flat-0.015, $n = 391$) and $211.8 \pm 6.7\ \mu m^2$ (Flat-0.03, $n = 155$). Under static condition, the average nuclear area decreases significantly: from $194.8 \pm 4.8\ \mu m^2$ on the flat surface to $145.1 \pm 4.1\ \mu m^2$ ($n = 250$) on nanogratings. When the low shear stress (0.015 N/m^2) is applied to parallel and perpendicular nanogratings, there is no significant difference in the nuclear area compared with those under static condition. Interestingly, when the shear stress is doubled, compared with the flat surface, parallel nanogratings reduces the nuclear area while the perpendicular nanogratings significantly extend the nuclear area to $260.3 \pm 2.7\ \mu m^2$ ($n = 315$, $p < 0.0001$). It is clear that nanotopography constrains the nuclear area; the shear stress amplifies the reduction when it works with nanotopography in a synergetic manner. In other words, nanotopography and shear stress can be used independently or in concert to manipulate orientation and deformation of cell body and cell nucleus.

This control of morphology may be useful to regulate cell phenotypes and functions. Cell shape is a potent regulator of cell growth and differentiation [96,97]. It is generally thought that an increase in cell spreading leads to an increased proliferation. Nanotopography, very often, reduces the effective surface area cells can sense, hence decreasing the proliferation although the cell alignment, elongation, and migration may be enhanced [15, 98–100]. For example, the proliferation rate of bovine pulmonary artery smooth muscle cells cultured on nanogratings similar to those in this study is significantly reduced to $35.5 \pm 8.6\%$ of the bromodeoxyuridine (5-bromo-2′-deoxyuridine, BrdU) incorporation compared to $56.6 \pm 10.7\%$ on the flat controls [98]. Moreover, nanotopography may affect cell differentiation, depending on the dimension, geometry (e.g., ridges, grooves, pillars, and pits), and symmetry of nanotopography [11,13,15]. For instance, hMSCs elongated on 350 nm PDMS nanogratings exhibit significant upregulation of neuronal markers such as microtubule-associated protein 2 and β-tubulin III (Tuj1) compared to flat controls [15]. Another example is that hMSCs cultured on TiO_2 nanotubes of 70–100 nm diameter perpendicular to the surface show a dramatic cell elongation (~10-fold increased compared with the flat control), which induces cytoskeletal stress and selective differentiation into osteoblast-like cells [13].

Cell morphology and cytoskeleton organization influence cells via actin stress fibers [101–103]. Alteration in the actin-mediated tension affects cell spreading, contraction, and migration, thus modulating nuclear shape and plasticity. For instance, the differentiation of stem cells into a chondrocytic phenotype requires a rounded cell shape. Direct comparison of cell and nuclear shape of hMSCs indicates that a more rounded nuclear shape is associated with the greatest expression of molecular markers associated with chondrogenesis [104]. Nuclear deformation strongly depends on the spatial organization and orientation of the actin cytoskeleton. The transmission of mechanical stresses from the cytoskeleton to the nucleus may not only depend on cytoskeletal tension but also on the orientation of the cytoskeletal filaments and the way the filaments are connected to the nucleus [105].

It is evident that the stress fiber-mediated tension plays an important role in directing the cell fate. Variation in the feature size, shape, and configuration of nanotopography may influence integrins and other cell-adhesion molecules, thus altering the size and distribution of focal adhesions [106]. Shear stress, on the other hand, can modify organization of stress fibers and focal adhesions [94]. Together, both nanotopography and fluid shear stress can exert effects on cell body and nucleus. It is expected that the orchestrated interplay of these topographical and mechanical cues can guide intracellular signaling events, formation, and organization of focal adhesions and stress fibers to regulate the fate of stem cells.

1.5 Summary

In this chapter, we discuss the nanofabrication techniques for both regular and irregular features, in particular the patterning of a large area. We also review the assembly techniques for PDMS-based microfluidic platforms. Convergence of these techniques enables the integration of nanoscale features into a microfluidic cell culture platform. Although at an early stage of development, nanostructured microfluidic platforms have shown promise in advancing cancer research and stem cell studies. In cancer cell isolation, nanoscale features with or without the aid of specific cell-adhesion protein coated on the channels may improve the capture efficiency of cancer cells under fluidic forces. In stem cell studies, leveraging the interplay of well-defined nanotopographical and fluid-induced mechanical cues may modulate the stem cell behavior and improve our mechanicastic understanding of stem cell–substrate interactions. We therefore envision that nanostructured microfluidic platforms, by more faithfully mimicking the cellular microenvironment *in vivo*, will help advance diverse fields such as cancer diagnostics, drug discovery, and stem cell expansion.

Acknowledgment

This work is partially supported by the National Institute of Health. The authors would also like to thank Dr. Weixiong Wang for CFD simulation on shear stress distribution.

References

[1] S. Mwenifumbo, M.M. Stevens, ECM interactions with cells from the macro- to nanoscale, in: K.E. Gonsalves, C.R. Halberstadt, C.T. Laurencin, L. Nair (Eds.), Biomedical Nanostructures, John Wiley & Sons, 2007, 225–260.

[2] R.G. Flemming, C.J. Murphy, G.A. Abrams, S.L. Goodman, P.F. Nealey, Effects of synthetic micro- and nano-structured surfaces on cell behavior, Biomaterials 20 (6) (1999) 573–588.

[3] I. Shirato, Y. Tomino, H. Koide, T. Sakai, Fine structure of the glomerular basement membrane of the rat kidney visualized by high-resolution scanning electron microscopy, Cell Tissue Res. 266 (1) (1991) 1–10.

[4] K. Hironaka, H. Makino, Y. Yamasaki, Z. Ota, Renal basement membranes by ultrahigh resolution scanning electron microscopy, Kidney Int. 43 (2) (1993) 334–345.

[5] G.A. Abrams, S.L. Goodman, P.F. Nealey, M. Franco, C.J. Murphy, Nanoscale topography of the basement membrane underlying the corneal epithelium of the rhesus macaque, Cell Tissue Res. 299 (1) (2000) 39–46.

[6] H. Gong, T.F. Freddo, M. Johnson, Age-related changes of sulfated proteoglycans in the normal human trabecular meshwork, Exp. Eye Res. 55 (5) (1992) 691–709.

[7] Z. Ma, S. Ramakrishna, Nanostructured extracellular matrix, Encycl. Nanosci. Nanotechnol. 7 (2004) 641–655.

[8] H. Dafni, T. Israely, Z.M. Bhujwalla, L.E. Benjamin, M. Neeman, Overexpression of vascular endothelial growth factor 165 drives peritumor interstitial convection and induces lymphatic drain: magnetic resonance imaging, confocal microscopy, and histological tracking of triple-labeled albumin, Cancer Res. 62 (22) (2002) 6731–6739.

[9] J.M. Rutkowski, M.A. Swartz, A driving force for change: interstitial flow as a morphoregulator, Trends Cell Biol. 17 (1) (2007) 44–50.
[10] M.V. Hillsley, J.A. Frangos, Bone tissue engineering: the role of interstitial fluid flow, Biotechnol. Bioeng. 43 (7) (1994) 573–581.
[11] M.J. Dalby, N. Gadegaard, R. Tare, A. Andar, M.O. Riehle, P. Herzyk, et al., The control of human mesenchymal cell differentiation using nanoscale symmetry and disorder, Nat. Mater. 6 (12) (2007) 997–1003.
[12] R. McMurray, N. Gadegaard, P. Tsimbouri, K. Burgess, L. McNamara, R. Tare, et al., Nanoscale surfaces for the long-term maintenance of mesenchymal stem cell phenotype and multipotency, Nat. Mater. 10 (8) (2011) 637–644.
[13] S. Oh, K.S. Brammer, Y.S.J. Li, D. Teng, A.J. Engler, S. Chien, et al., Stem cell fate dictated solely by altered nanotube dimension, Proc. Natl. Acad. Sci. U.S.A. 106 (7) (2009) 2130–2135.
[14] G.A. Silva, C. Czeisler, K.L. Niece, E. Beniash, D.A. Harrington, J.A. Kessler, et al., Selective differentiation of neural progenitor cells by high-epitope density nanofibers, Science 303 (5662) (2004) 1352–1355.
[15] E.K.F. Yim, S.W. Pang, K.W. Leong, Synthetic nanostructures inducing differentiation of human mesenchymal stem cells into neuronal lineage, Exp. Cell Res. 313 (9) (2007) 1820–1829.
[16] S. Chien, Molecular basis of modulation of vascular functions by mechanical forces, Adv. Ser. Biomech. 4 (2008) 79–97.
[17] L. Bjerre, C.E. Buenger, M. Kassem, T. Mygind, Flow perfusion culture of human mesenchymal stem cells on silicate-substituted tricalcium phosphate scaffolds, Biomaterials 29 (17) (2008) 2616–2627.
[18] N. Wang, H. Miao, Y.-S. Li, P. Zhang, J.H. Haga, Y. Hu, et al., Shear stress regulation of Krueppel-like factor 2 expression is flow pattern-specific, Biochem. Biophys. Res. Commun. 341 (4) (2006) 1244–1251.
[19] K. Yamamoto, T. Sokabe, T. Watabe, K. Miyazono, J.K. Yamashita, S. Obi, et al., Fluid shear stress induces differentiation of Flk-1-positive embryonic stem cells into vascular endothelial cells in vitro, Am. J. Physiol. 288 (4, Pt. 2) (2005) H1915–H1924.
[20] E. Martines, K. McGhee, C. Wilkinson, A. Curtis, A parallel-plate flow chamber to study initial cell adhesion on a nanofeatured surface, IEEE Trans. Nanobioscience 3 (2) (2004) 90–95.
[21] K.W. Kwon, S.S. Choi, S.H. Lee, B. Kim, S.N. Lee, M.C. Park, et al., Label-free, microfluidic separation and enrichment of human breast cancer cells by adhesion difference, Lab Chip 7 (11) (2007) 1461–1468.
[22] J. El-Ali, S. Gaudet, A. Guenther, P.K. Sorger, K.F. Jensen, Cell stimulus and lysis in a microfluidic device with segmented gas–liquid flow, Anal. Chem. 77 (11) (2005) 3629–3636.
[23] J.T. Nevill, R. Cooper, M. Dueck, D.N. Breslauer, L.P. Lee, Integrated microfluidic cell culture and lysis on a chip, Lab Chip 7 (12) (2007) 1689–1695.
[24] Y.-S. Torisawa, B.-H. Chueh, D. Huh, P. Ramamurthy, T.M. Roth, K.F. Barald, et al., Efficient formation of uniform-sized embryoid bodies using a compartmentalized microchannel device, Lab Chip 7 (6) (2007) 770–776.
[25] A. Khademhosseini, J. Yeh, G. Eng, J. Karp, H. Kaji, J. Borenstein, et al., Cell docking inside microwells within reversibly sealed microfluidic channels for fabricating multiphenotype cell arrays, Lab Chip 5 (12) (2005) 1380–1386.

[26] R. Gomez-Sjoeberg, A.A. Leyrat, D.M. Pirone, C.S. Chen, S.R. Quake, Versatile, fully automated, microfluidic cell culture system, Anal. Chem. 79 (22) (2007) 8557–8563.

[27] K.-J. Jang, K.-Y. Suh, A multi-layer microfluidic device for efficient culture and analysis of renal tubular cells, Lab Chip 10 (1) (2010) 36–42.

[28] K.-I. Kamei, S. Guo, Z.T.F. Yu, H. Takahashi, E. Gschweng, C. Suh, et al., An integrated microfluidic culture device for quantitative analysis of human embryonic stem cells, Lab Chip 9 (4) (2009) 555–563.

[29] N. Korin, A. Bransky, U. Dinnar, S. Levenberg, A parametric study of human fibroblasts culture in a microchannel bioreactor, Lab Chip 7 (5) (2007) 611–617.

[30] Y.-C. Toh, C. Zhang, J. Zhang, Y.M. Khong, S. Chang, V.D. Samper, et al., A novel 3D mammalian cell perfusion-culture system in microfluidic channels, Lab Chip 7 (3) (2007) 302–309.

[31] B.P. Helmke, A.R. Minerick, Designing a nano-interface in a microfluidic chip to probe living cells: challenges and perspectives, Proc. Natl. Acad. Sci. U.S.A. 103 (17) (2006) 6419–6424.

[32] M.C. Roco, Nanotechnology: convergence with modern biology and medicine, Curr. Opin. Biotechnol. 14 (3) (2003) 337–346.

[33] L.E. Niklason, J. Gao, W.M. Abbott, K.K. Hirschi, S. Houser, R. Marini, et al., Functional arteries grown in vitro, Science 284 (5413) (1999) 489–493.

[34] D.C. Miller, A. Thapa, K.M. Haberstroh, T.J. Webster, Endothelial and vascular smooth muscle cell function on poly(lactic-co-glycolic acid) with nano-structured surface features, Biomaterials 25 (1) (2004) 53–61.

[35] K. Peng, Y. Yan, S. Gao, J. Zhu, Synthesis of large-area silicon nanowire arrays via self-assembling nanoelectrochemistry, Adv. Mater. 14 (16) (2002) 1164–1167.

[36] S. Wang, H. Wang, J. Jiao, K.J. Chen, G.E. Owens, K. Kamei, et al., Three-dimensional nanostructured substrates toward efficient capture of circulating tumor cells, Angew. Chem. Int. Ed. Engl. 48 (47) (2009) 8970–8973.

[37] A.M. Lipski, C. Jaquiery, H. Choi, D. Eberli, M. Stevens, I. Martin, et al., Nanoscale engineering of biomaterial surfaces, Adv. Mater. 19 (4) (2007) 553–557.

[38] M.J. Dalby, G.E. Marshall, H. Johnstone, S. Affrossman, M.O. Riehle, Interactions of human blood and tissue cell types with 95-nm-high nanotopography, IEEE Trans. Nanobiosci. 1 (1) (2002) 18–23.

[39] M.J. Dalby, M.O. Riehle, H. Johnstone, S. Affrossman, A.S.G. Curtis, In vitro reaction of endothelial cells to polymer demixed nanotopography, Biomaterials 23 (14) (2002) 2945–2954.

[40] S. Affrossman, G. Henn, S.A. O'Neill, R.A. Pethrick, M. Stamm, Surface topography and composition of deuterated polystyrene-poly(bromostyrene) blends, Macromolecules 29 (1996) 5010–5016.

[41] M.J. Dalby, G.E. Marshall, H.J.H. Johnstone, S. Affrossman, M.O. Riehle, Interactions of human blood and tissue cell types with 95-nm-high nanotopography, IEEE Trans. Nanobiosci. 1 (2002) 18–23.

[42] S. Affrossman, S.A. O'Neill, M. Stamm, Topography and surface composition of thin films of blends of polystyrene with brominated polystyrenes: effects of varying the degree of bromination and annealing, Macromolecules 31 (18) (1998) 6280–6288.

[43] S. Affrossman, R. Jerome, S.A. O'Neill, T. Schmitt, M. Stamm, Surface structure of thin film blends of polystyrene and poly(n-butyl methacrylate), Colloid Polym. Sci. 278 (10) (2000) 993–999.

[44] C.C. Berry, M.J. Dalby, R.O.C. Oreffo, D. McCloy, S. Affrosman, The interaction of human bone marrow cells with nanotopographical features in three dimensional constructs, J. Biomed. Mater. Res. A. 79A (2) (2006) 431–439.

[45] B.D. Gates, Q. Xu, M. Stewart, D. Ryan, C.G. Willson, G.M. Whitesides, New approaches to nanofabrication: molding, printing, and other techniques, Chem. Rev. 105 (4) (2005) 1171–1196.

[46] Y. Yin, Y. Lu, B. Gates, Y. Xia, Template-assisted self-assembly: a practical route to complex aggregates of monodispersed colloids with well-defined sizes, shapes, and structures, J. Am. Chem. Soc. 123 (36) (2001) 8718–8729.

[47] K. Shin, H. Xiang, S.I. Moon, T. Kim, T.J. McCarthy, T.P. Russell, Curving and frustrating flatland, Science 306 (5693) (2004) 76.

[48] Y. Tada, S. Akasaka, H. Yoshida, H. Hasegawa, E. Dobisz, D. Kercher, et al., Directed self-assembly of diblock copolymer thin films on chemically-patterned substrates for defect-free nano-patterning, Macromolecules 41 (23) (2008) 9267–9276.

[49] L. Rockford, Y. Liu, P. Mansky, T.P. Russell, M. Yoon, S.G.J. Mochrie, Polymers on nanoperiodic, heterogeneous surfaces, Phys. Rev. Lett. 82 (12) (1999) 2602–2605.

[50] J.Y. Cheng, C.T. Rettner, D.P. Sanders, H.-C. Kim, W.D. Hinsberg, Dense self-assembly on sparse chemical patterns: rectifying and multiplying lithographic patterns using block copolymers, Adv. Mater. 20 (16) (2008) 3155–3158.

[51] C. Vieu, F. Carcenac, A. Pepin, Y. Chen, M. Mejias, A. Lebib, et al., Electron beam lithography: resolution limits and applications, Appl. Surf. Sci. 164 (2000) 111–117.

[52] A. Muray, M. Scheinfein, M. Isaacson, I. Adesida., Radiolysis and resolution limits of inorganic halide resists, J. Vac. Sci. Technol, B. 3 (1) (1985) 367–372.

[53] T. Nagase, K. Gamo, T. Kubota, S. Mashiko, Direct fabrication of nano-gap electrodes by focused ion beam etching, Thin Solid Films 499 (1-2) (2006) 279–284.

[54] A.A. Tseng, Recent developments in nanofabrication using focused ion beams, Small 1 (10) (2005) 924–939.

[55] W. Denk, J. Strickler, W. Webb, Two-photon laser scanning fluorescence microscopy, Science 248 (4951) (1990) 73–76.

[56] C. LaFratta, J. Fourkas, T. Baldacchini, R. Farrer, Multiphoton fabrication, Angew. Chem. Int. Ed. Engl. 46 (33) (2007) 6238–6258.

[57] S. Maruo, O. Nakamura, S. Kawata, Three-dimensional microfabrication with two-photon-absorbed photopolymerization, Opt. Lett. 22 (2) (1997) 132–134.

[58] B.H. Cumpston, S.P. Ananthavel, S. Barlow, D.L. Dyer, J.E. Ehrlich, L.L. Erskine, et al., Two-photon polymerization initiators for three-dimensional optical data storage and microfabrication, Nature 398 (6722) (1999) 51–54.

[59] S.J. Jhaveri, J.D. McMullen, R. Sijbesma, L.-S. Tan, W. Zipfel, C.K. Ober, Direct three-dimensional microfabrication of hydrogels via two-photon lithography in aqueous solution, Chem. Mater. 21 (10) (2009) 2003–2006.

[60] S. Juodkazis, V. Mizeikis, K. Seet, M. Miwa, H. Misawa, Two-photon lithography of nanorods in SU-8 photoresist, Nanotechnology 16 (6) (2005) 846–849.

[61] Y. Xia, G.M. Whitesides, Soft lithography, Annu. Rev. Mater. Sci. 28 (1998) 153–184.

[62] F. Hua, Y. Sun, A. Gaur, M.A. Meitl, L. Bilhaut, L. Rotkina, et al., Polymer imprint lithography with molecular-scale resolution, Nano Lett. 4 (12) (2004) 2467–2471.

[63] S.Y. Chou, P.R. Krauss, P.J. Renstrom, Imprint of sub-25 nm vias and trenches in polymers, Appl. Phys. Lett. 67 (21) (1995) 3114–3116.

[64] S.Y. Chou, P.R. Krauss, P.J. Renstrom, Imprint lithography with 25-nanometer resolution, Science 272 (5258) (1996) 85–87.

[65] S.Y. Chou, P.R. Krauss, W. Zhang, L. Guo, L. Zhuang, Sub-10 nm imprint lithography and applications, J. Vac. Sci. Technol, B. 15 (6) (1997) 2897–2904.

[66] M.D. Austin, H. Ge, W. Wu, M. Li, Z. Yu, D. Wasserman, et al., Fabrication of 5 nm linewidth and 14 nm pitch features by nanoimprint lithography, Appl. Phys. Lett. 84 (26) (2004) 5299–5301.

[67] K. Ansari, J.A. van Kan, A.A. Bettiol, F. Watt, Fabrication of high aspect ratio 100 nm metallic stamps for nanoimprint lithography using proton beam writing, Appl. Phys. Lett. 85 (3) (2004) 476–478.

[68] H. Becker, U. Heim, Polymer hot embossing with silicon master structures, Sens. Mater. 11 (5) (1999) 297–304.

[69] N. Roos, T. Luxbacher, T. Glinsner, K. Pfeiffer, H. Schulz, H.-C. Scheer, Nanoimprint lithography with a commercial 4-in. bond system for hot embossing, Proc. SPIE-Int. Soc. Opt. Eng. 4343 (2001) 427–435.

[70] F. Gottschalch, T. Hoffmann, C.M.S Torres, H. Schulz, H.-C. Scheer, Polymer issues in nanoimprinting technique, Solid State Electron. 43 (6) (1999) 1079–1083.

[71] L.J. Guo, Nanoimprint lithography: methods and material requirements, Adv. Mater. 19 (4) (2007) 495–513.

[72] Y. Wang, Z. Liu, B. Han, Y. Huang, J. Zhang, D. Sun, et al., Compressed-CO_2-assisted patterning of polymers, J. Phys. Chem. B. 109 (25) (2005) 12376–12379.

[73] S. Fujita, D. Ono, M. Ohshima, H. Iwata, Supercritical CO_2-assisted embossing for studying cell behaviour on microtextured surfaces, Biomaterials 29 (34) (2008) 4494–4500.

[74] S. Lenhert, M.-B. Meier, U. Meyer, L. Chi, H.P. Wiesmann, Osteoblast alignment, elongation and migration on grooved polystyrene surfaces patterned by Langmuir-Blodgett lithography, Biomaterials 26 (5) (2005) 563–570.

[75] M. Colburn, S. Johnson, M. Stewart, S. Damle, T.C. Bailey, B. Choi, et al., Step and flash imprint lithography: a new approach to high-resolution patterning, Proc. SPIE-Int. Soc. Opt. Eng. 3676 (Pt. 1) (1999) 379–389.

[76] S.H. Ahn, L.J. Guo, High-speed roll-to-roll nanoimprint lithography on flexible plastic substrates, Adv. Mater. 20 (11) (2008) 2044–2049.

[77] Y. Yang, K. Kulangara, J. Sia, L. Wang, K. Leong, Engineering of a microfluidic cell culture platform embedded with nanoscale features, Lab Chip 11 (9) (2011) 1638–1646.

[78] D.C Duffy, J.C. McDonald, O.J.A. Schueller, G.M. Whitesides, Rapid prototyping of microfluidic systems in poly(dimethylsiloxane), Anal. Chem. 70 (23) (1998) 4974–4984.

[79] H. Hillborg, J.F. Ankner, U.W. Gedde, G.D. Smith, H.K. Yasuda, K. Wikstrom, Crosslinked polydimethylsiloxane exposed to oxygen plasma studied by neutron reflectometry and other surface specific techniques, Polymer 41 (18) (2000) 6851–6863.

[80] B. Samel, M.K. Chowdhury, G Stemme, The fabrication of microfluidic structures by means of full-wafer adhesive bonding using a poly(dimethylsiloxane) catalyst, J. Micromech. Microeng. 17 (8) (2007) 1710–1714.

[81] S. Satyanarayana, R.N. Karnik, A. Majumdar, Stamp-and-stick room-temperature bonding technique for microdevices, J. Microelectromech. Syst. 14 (2) (2005) 237–245.

[82] M.A. Unger, H.-P. Chou, T. Thorsen, A. Scherer, S.R. Quake, Monolithic microfabricated valves and pumps by multilayer soft lithography, Science (Washington, DC) 288 (5463) (2000) 113–116.

[83] M.A. Eddings, M.A. Johnson, B.K. Gale, Determining the optimal PDMS-PDMS bonding technique for microfluidic devices, J. Micromech. Microeng. 18 (6) (2008) 067001/1-067001/4.

[84] J.N. Lee, C. Park, G.M. Whitesides, Solvent compatibility of poly(dimethylsiloxane)-based microfluidic devices, Anal. Chem. 75 (23) (2003) 6544–6554.

[85] A.L. Thangawng, R.S. Ruoff, M.A. Swartz, M.R. Glucksberg, An ultra-thin PDMS membrane as a bio/micro-nano interface: fabrication and characterization, Biomed. Microdevices 9 (4) (2007) 587–595.

[86] S. Iyer, R.M. Gaikwad, V. Subba-Rao, C.D. Woodworth, I. Sokolov, Atomic force microscopy detects differences in the surface brush of normal and cancerous cells, Nat. Nanotechnol. 4 (6) (2009) 389–393.

[87] K.E. Fischer, B.J. Alemán, S.L. Tao, R.H. Daniels, E.M. Li, M.D. Bünger, et al., Biomimetic nanowire coatings for next generation adhesive drug delivery systems, Nano. Lett. 9 (2) (2009) 716–720.

[88] S. Wang, K. Liu, J. Liu, Z. Yu, X. Xu, L. Zhao, et al., Highly efficient capture of circulating tumor cells by using nanostructured silicon substrates within tegrated chaotic micromixers, Angew. Chem. Int. Ed. 50 (13) (2011) 3084–3088.

[89] S.Y. Hwang, K.W. Kwon, K.J. Jang, M.C. Park, J.S. Lee, K.Y. Suh, Adhesion assays of endothelial cells on nanopatterned surfaces within a microfluidic channel, Anal. Chem. 82 (7) (2010) 3016–3022.

[90] A.S. Curtis, B. Casey, J.O. Gallagher, D. Pasqui, M.A. Wood, C.D. Wilkinson, Substratum nanotopography and the adhesion of biological cells. Are symmetry or regularity of nanotopography important? Biophys. Chem. 94 (3) (2001) 275–283.

[91] M.J. Dalby, M.O. Riehle, S.J. Yarwood, C.D.W. Wilkinson, A.S.G. Curtis, Nucleus alignment and cell signaling in fibroblasts: response to a micro-grooved topography, Exp. Cell Res. 284 (2) (2003) 274–282.

[92] C.H. Thomas, J.H. Collier, C.S. Sfeir, K.E. Healy, Engineering gene expression and protein synthesis by modulation of nuclear shape, Proc. Natl. Acad. Sci. U.S.A. 99 (4) (2002) 1972–1977.

[93] J.D. Pajerowski, K.N. Dahl, F.L. Zhong, P.J. Sammak, D.E. Discher, Physical plasticity of the nucleus in stem cell differentiation, Proc. Natl. Acad. Sci. U.S.A. 104 (40) (2007) 15619–15624.

[94] C.-C. Wu, Y.-S. Li, J.H. Haga, R. Kaunas, J.-J. Chiu, F.-C. Su, et al., Directional shear flow and Rho activation prevent the endothelial cell apoptosis induced by micropatterned anisotropic geometry, Proc. Natl. Acad. Sci. U.S.A. 104 (4) (2007) 1254–1259.

[95] P.T. Ohara, R.C. Buck, Contact guidance in vitro: a light, transmission, and scanning electron microscopic study, Exp. Cell Res. 121 (2) (1979) 235–249.

[96] C.S. Chen, M. Mrksich, S. Huang, G.M. Whitesides, D.E. Ingber, Geometric control of cell life and death, Science 276 (5317) (1997) 1425–1428.

[97] R. McBeath, D.M. Pirone, C.M. Nelson, K. Bhadriraju, C.S. Chen, Cell shape, cyto-skeletal tension, and RhoA regulate stem cell lineage commitment, Dev. Cell 6 (4) (2004) 483–495.

[98] E.K.F. Yim, R.M. Reano, S.W. Pang, A.F. Yee, C.S. Chen, K.W. Leong, Nanopattern-induced changes in morphology and motility of smooth muscle cells, Biomaterials 26 (26) (2005) 5405–5413.

[99] S Gerecht, C.J. Bettinger, Z. Zhang, J.T. Borenstein, G. Vunjak-Novakovic, R. Langer, The effect of actin disrupting agents on contact guidance of human embryonic stem cells, Biomaterials 28 (28) (2007) 4068–4077.

[100] C.J. Bettinger, Z. Zhang, S. Gerecht, J.T. Borenstein, R Langer, Enhancement of in vitro capillary tube formation by substrate nanotopography, Adv. Mater. 20 (1) (2008) 99–103.

[101] C.B. Khatiwala, S.R. Peyton, A.J. Putnam, Intrinsic mechanical properties of the extracellular matrix affect the behavior of pre-osteoblastic MC3T3-E1 cells, Am. J. Physiol. Cell Physiol. 290 (6) (2006) C1640–C1650.

[102] H.J. Kong, T.R. Polte, E. Alsberg, D.J. Mooney, FRET measurements of cell-traction forces and nano-scale clustering of adhesion ligands varied by substrate stiffness, Proc. Natl. Acad. Sci. U.S.A. 102 (12) (2005) 4300–4305.

[103] N. Wang, I.M. Tolic-Norrelykke, J. Chen, S.M. Mijailovich, J.P. Butler, J.J. Fredberg, et al., Cell prestress. I. Stiffness and prestress are closely associated in adherent contractile cells, Am. J. Physiol. 282 (3, Pt.1) (2002) C606–C616.

[104] S.H. McBride, M.L.K. Tate, Modulation of stem cell shape and fate A: the role of density and seeding protocol on nucleus shape and gene expression, Tissue Eng. A 14 (9) (2008) 1561–1572.

[105] P. Roca-Cusachs, J. Alcaraz, R. Sunyer, J. Samitier, R. Farre, D. Navajas, Micropatterning of single endothelial cell shape reveals a tight coupling between nuclear volume in G1 and proliferation, Biophys. J. 94 (12) (2008) 4984–4995.

[106] V. Vogel, M. Sheetz, Local force and geometry sensing regulate cell functions, Nat. Rev. Mol. Cell Biol. 7 (4) (2006) 265–275.

CHAPTER

Microvascular Networks for Tissue Engineering 2

Jen-Huang Huang, Arul Jayaraman and Victor M. Ugaz

Artie McFerrin Department of Chemical Engineering, Texas A&M University, TX, USA

2.1 Introduction

Interest in developing new capabilities to engineer artificial tissue and organ structures suitable for implantation has intensified during the past decade owing to their enormous untapped life saving potential. Health care costs related to organ failure are staggering, easily extending into the hundreds of billions of dollars in the United States alone—a situation compounded by the fact that transplantation is often the only viable therapeutic option in many cases (e.g., acute physical trauma and severe burns) [1]. A well-known example is cardiovascular disease, a condition impacting over 80 million adults in the United States, contributing to 1 out of every 3 deaths in 2007—many of which occurred in patients awaiting a transplant [2]. These problems are expected to grow over the coming years as overall life expectancies increase, as evident by the fact that the waiting list for donor organs in the United States has doubled in size during the past 20 years. In 2010, out of a total of 110,942 patients on the transplantation waiting list in the United States, only 22,104 received organs and 6521 died awaiting a suitable donor. Engineered organ replacements offer a compelling alternative to address this growing need.

A major barrier to production of organ-sized tissue constructs is that it has proven challenging to design scaffold microenvironments that are capable of providing sufficient nutrient delivery and waste removal throughout extended 3-D volumes [3–7]. Incredible progress has been made toward development of porous biodegradable scaffolds for tissue culture [8–10], but transport within these scaffolds primarily occurs by molecular diffusion—a passive and inherently slow process. This fundamental limitation effectively restricts the maximum thickness of engineered tissues to 100–200 μm [11,12], beyond which cells in the interior remain nutrient limited and die [13,14]. In nature, vascular networks have evolved to elegantly address this problem by incorporating highly branched fractal-like architectures that are efficiently space filling while minimizing the energy required to sustain transport [15,16]. Metabolically active cells in the body, for example, are often located no more than 50–100 μm away from a capillary bed.

Microfluidic Cell Culture Systems. ISBN: 978-1-4377-3459-1

In addition to functioning as local nutrient and waste sinks, the specific microstructure and positioning of these capillary networks provides cues that help orchestrate the organization and differentiation of cells within the tissue structure. Consequently, the ability to embed functional capillary networks within tissue engineering scaffolds would represent a significant step forward toward realizing the goal of manufacturing organ-sized constructs *in vitro* [17,18].

2.2 Characteristics of branched vascular networks

Naturally occurring vascular networks are notoriously challenging to replicate artificially owing to their unique complexity, characterized by a well-defined global architecture that persists despite the fact that no two networks are completely identical at all structural levels. The extent of this self-similarity is generally expressed in terms of parameters related to a network's branching characteristics (average branching angle, diameter ratio and symmetry between parent and daughter branches, fractal dimension, etc.). A key quantity more directly related to transport efficiency can be obtained by considering a balance between the energy needed to sustain flow through a conduit versus the energy required to produce enough fluid to fill its volume. This analysis yields a relationship known as Murray's law, predicting that an optimal system-wide state of uniform shear stress corresponds to a global scaling of flow rate with the cube of microchannel diameter [16,19–21]. Specifically, in the case of a parent channel of diameter d_0 bifurcating into two daughter channels with diameters d_1 and d_2, Murray's law implies that $d_0^k = d_1^k + d_2^k$, where a value of $k = 3$ is predicted under ideal conditions (e.g., laminar and nonpulsatile flow). In physiological systems, values of the branching exponent k generally range between 2 and 3 depending on location within the network (e.g., data from porcine vasculature display scaling exponents of $k = 2.82$ for $d_0 \leq 40\ \mu m$, $k = 2.50$ for $40\ \mu m < d_0 \leq 200\ \mu m$, and $k = 2.35$ for $d_0 > 200\ \mu m$) [22–24]. Other notable examples include human pulmonary vasculature ($k \sim 2.5$ [25]) and the pig coronary arterial tree ($k \sim 2.2$ [26]). These deviations from the ideal $k = 3$ scaling arise in part because pulsatile flow generally dominates in the larger diameter vessels (e.g., aorta and major arteries), where area preserving branching ($k = 2$) helps to minimize wave reflections at the branching points [27]. Smaller vessels, on the other hand, maintain exponents closer to $k = 3$ because the need to maintain the blood flow with minimal energy consumption becomes a dominant consideration [16].

2.3 Fabrication of 2-D microvascular networks

Although methods currently exist to artificially construct analogs of individual large blood vessels, the formation of the much smaller diameter capillaries is more challenging. Photolithography-based micromachining methods have recently emerged as valuable tools to enable construction of microfluidic and microelectromechanical systems. These same processes offer a natural avenue to construct vascular-mimicking networks by allowing complex microscale geometries to be fashioned in planar substrates (Table 2.1). In the early 2000s, Vacanti and coworkers [36,39,55] pioneered some of the first applications of these methods to produce microvascular topologies in silicon wafers, which subsequently served as rigid molds to imprint the patterned networks into biodegradable polymers like poly(lactic-co-glycolic acid) (PLGA). This process, typically referred to as *soft lithography*, involves the following basic sequence of steps: (i) spin coating a photoresist onto the surface of clean silicon wafers, (ii) patterning the photoresist by exposure to ultraviolet (UV) light through a photomask, (iii) developing the pattern by exposure to a chemical that selectively removes the photoresist in areas exposed to the UV light, (iv) etching the network pattern into the silicon substrate, and (v) stripping away the remaining photoresist. The resulting silicon wafer now contains a negative imprint of the desired microchannel pattern that then serves as a rigid mold or *master* against which a positive replica can be cast in the desired substrate material (in this case by pouring molten PLGA onto the surface and peeling it away after solidification). The patterned polymer is then bonded to a flat surface to create enclosed microchannels [56].

Variations of this basic soft-lithographic method have enabled microvascular-mimicking flow networks to be constructed in a variety of substrate materials suitable for cell culture. Poly(dimethyl siloxane) (PDMS) is one of the most widely used substrates owing to a number of favorable properties including optical transparency and gas permeability [37,38], but more recent studies have explored a wider range of materials. For example, Fidkowski et al. [35] synthesized a rubber-like biodegradable polymer, poly(glycerol sebacate) (PGS), and applied it to construct an endothelialized capillary network using human umbilical vein endothelial cells (HUVECs) (Figure 2.1A). A number of hydrogel substrates have also been investigated, including agarose [40], alginate (Figure 2.1B) [41], and collagen/alginate (Figure 2.1C) [53]. Ling et al. [40] demonstrated that flow networks embedded in agarose can efficiently provide nutrient and waste exchange for surrounding cells. Choi et al. [41] investigated metabolite (e.g., calcein-acetoxymethyl ester (AM)) delivery in cell-seeded alginate hydrogels, while Gillette et al. [53] demonstrated *in situ* collagen fiber assembly in a cell-seeded hydrogel. Additional complexity can be introduced by machining channels with variable depths, accomplished by processes such as laser machining into silicon followed by acid etching (Figure 2.1D) [48–51]. Although not fully 3-D, the ability to vary the depth enables construction of networks with architectures that more closely follow the scalings predicted by Murray's law [57,58]. Another novel approach was demonstrated by Cui and Boland [52], who constructed micron-sized fibrin channels using drop-on-demand polymerization implemented via an inkjet printing process, and subsequently demonstrated successful cell culture for up to 21 days in the resulting microchannel network (Figure 2.1E).

Table 2.1 Summary of Literature Reporting Construction of 2-D Microvascular Networks

Channel Cross Section	Channel Size Range	Micromachining Method	Construction Time	Substrate Material(s)	Substrate Size	References
Rectangular	20–60 μm	Electroforming	–	Polycarbonate, polylactide-coglycolides	1×2 cm^2	[28]
Rectangular	20–60 μm	Photoresist-based molding	–	PMMA, PDMS	1×2 cm^2	[29]
Circular	60–120 μm	Photoresist-based molding and bonding	–	PDMS	1×2 cm^2	[30]
Rectangular	5–300 μm	Micropatterning	–	Collagen–glycosaminoglycan membrane	1.7 mm thickness	[31]
Rectangular	100 μm–5 mm	Soft lithography	–	PDMS	1×2 cm^2	[32]
Semicircular	40–120 μm	Wet etching	5 h	Silicon	7.5×2.5 cm^2	[33]
Rectangular	6–50 μm	Soft lithography with sacrificial element	5 h	Gelatin	85 cm^2	[34]
Rectangular	45 μm–3 mm	Soft lithography	~5 days	PGS	$3.8 \times 7.5 \times 0.1$ cm^3	[35]
Rectangular	10–500 μm	Wet/plasma etching	–	Silicon and Pyrex	10 cm^2	[36]
Rectangular	35 μm–5 mm	Lithographic and etching followed by soft lithography	4 h	PDMS	2×4 cm^2	[37,38]
Rectangular	–	Wet/plasma etching	–	Silicon mold/PLGA	5×5 cm^2	[39]
Rectangular	50–1000 μm	Soft lithography, molding agarose to wafer then sealing two layers of agarose together	–	Agarose with or without cell encapsulation	1 cm thickness	[40]
Rectangular	100 μm	Soft lithography, injection of alginate and cells, sealing of hydrogel layers to form channels	–	Calcium alginate hydrogel seeded with chondrocytes	1×1 cm^2	[41]
Rectangular	6–63 μm	Soft lithography	–	PDMS	–	[42]

Rectangular	267–727 μm	Electric discharge machining of stainless steel mold then soft lithography to cast in PDMS	–	PDMS	Circle, 130 mm diameter	[43]
Rectangular	20 μm–2.73 mm	Micromachining of stainless steel mold then soft lithography to cast in PDMS	–	PDMS	5 × 5 cm^2	[44]
Rectangular	1.197 mm	Soft lithography	–	PDMS	2 × 2 cm^2	[45]
Round	80–190 μm	Electroplating of silicon master mold then micropatterning	–	Polystyrene	1 × 3 cm^2	[46]
Rectangular	50–400 μm	UV exposure to polymer through mask pattern	–	pCLLA acrylate	–	[47]
Rectangular	125–660 μm	Laser nanosecond ablation followed by chemical wet etching	–	Silicon	1.5 cm	[48,49]
Semicircular	50 μm–1 mm	Laser ablation followed by chemical wet etching	8 h	Silicon mold/PDMS channel	1 cm	[50]
Rectangular	90–600 μm hydraulic diameter	Femtosecond laser ablation followed by chemical wet etching	5 h	Silicon	3 × 6 cm^2	[51]
Round	100 μm	Inkjet printing of cells	–	Fibrin	2 × 8 mm^2	[52]
Rectangular	40–60 μm	Hydrogel constructs cast from PDMS mold	–	Alginate, collagen I, Matrigel I, fibrin	1 × 2 mm^2	[53]
Round	<10 μm or larger if cocultured	Laser-guided direct writing	Weeks, if coculturing with hepatocytes	Matrigel, collagen	0.5 mm thickness	[54]

Publications from the same laboratory are grouped together.

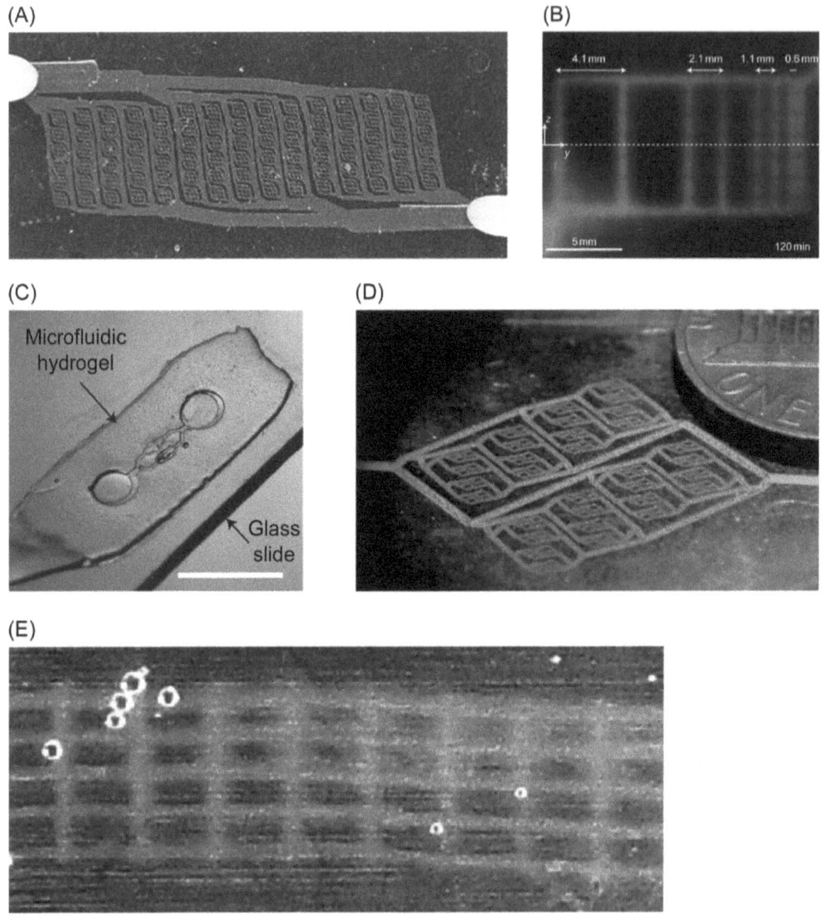

FIGURE 2.1

Construction of 2-D microvascular networks. Soft-lithographic methods enable networks to be constructed in a variety of substrates including (A) PGS [35], (B) alginate [41], and (C) collagen/alginate [53]. Other construction methods include (D) laser machining [48] and (E) inkjet printing [52].

2.4 Fabrication of 3-D microvascular networks

Although vascularized microfluidic networks are readily produced in 2-D by photolithographic or soft-lithographic techniques, comparable progress toward constructing 3-D networks has occurred only recently (Table 2.2). This capability is of interest because it is generally recognized that 3-D culture environments more closely replicate physiological conditions *in vitro* than 2-D cultures. The simplest approach to generate 3-D architectures involves stacking multiple 2-D vascularized

Table 2.2 Summary of Literature Reporting Construction of 3-D Microvascular Networks

Channel Cross Section	Channel Size Range	Micromachining Method	Construction Time	Substrate Material(s)	Substrate Size	References
Varied	100 μm	Soft lithography	–	PDMS	1 mm thick	[59]
Rectangular	10 μm–1 cm	Wet/plasma etching	–	Silicon mold/PLGA	5×5 cm^2	[55]
Rectangular	35 μm–5 mm	Soft lithography	–	PDMS	$3 \times 3 \times 5$ cm^3	[60]
Circular	0.9–2.5 mm	Deposition of bioink particles into tubular structures using bioprinter	2–4 days	Agarose		[61]
Circular	Varied	Cell sheet engineering	3–7 days	NHDF and HUVEC sheets	60 μm	[62]
Circular	30 μm	Co-seeding HUVEC and 10T1/2 precursor cells *in vivo*	Up to 2 weeks	Collagen gel	165 μm	[63]
Circular	30–45 nm	Culturing HUVEC on ionic self-assembling peptide gels	2 days	Peptide gel	500 μm	[64]
Varied	Varied	Modular assembly of HUVEC coated cylinders	3 days	Collagen–HepG2 cylinders	0.57 mm diameter, 1.8 mm long	[65]
Circular	230 μm	Direct-write assembly	–	Epoxy resin		[66]
Circular	200 μm	Direct-write assembly	–	Epoxy resin	$20 \times 20 \times 20$ mm^3	[67]
Circular	100–500 μm	Direct-write assembly	–	Epoxy resin	$40 \times 40 \times 5$ mm^3	[68]
Circular	200–600 μm	Direct-write assembly	–	Pluronic F127-diacrylate (F127-DA)	$40 \times 20 \times 1$ mm^3	[69]

(Continued)

Table 2.2 (Continued)

Channel Cross Section	Channel Size Range	Micromachining Method	Construction Time	Substrate Material(s)	Substrate Size	References
Rectangular	200 μm	Direct-write assembly	–	Epoxy resin	$45 \times 7 \times 10\ mm^3$	[70]
Circular	Vertical: 100–410 μm	Dual ink deposition and vertical ink writing	–	Epoxy resin	$50 \times 12 \times 6\ mm^3$	[71]
Trapezoidal	200 μm	Photolithographic and plasma etching to produce silicon master	–	PGS	$10 \times 5 \times 1\ mm^3$	[72]
Circular	1–100 μm	Sacrificial sugar structures	–	PCL	$1 \times 2 \times 0.5\ cm^3$	[73]
Rectangular	400 μm	Layer-by-layer direct printing	–	Collagen	$5 \times 5 \times 1\ mm^3$	[74]
Circular	10 μm–1 mm	Electrostatic discharge	Seconds	PMMA, PLA	$25 \times 10 \times 10\ mm^3$	[75]
Varied	15–20 μm	Angiogenesis in three layers of fibroblast sheets *in vitro*	15–31 days	Fibroblasts, keratinocytes, HUVEC or HMVEC	$3 \times 3\ cm^2$	[76,77]

Publications from the same laboratory are grouped together.

layers in a fluidically interconnected arrangement [55,60,72] (Figure 2.2A–C). For example, Bettinger et al. [72] assembled five microfluidic layers constructed in a biodegradable substrate to create a 3-D tissue culture scaffold, which was then seeded with a high density of HepG2 cells that formed aggregates throughout the microchannels. The capacity for long-term viability was assessed by measuring albumin secretion rates. Kniazeva et al. [60] employed a similar approach to construct individual blood-gas subunits that could be stacked together to form a microfluidic respiratory assist device to study gas permeance in artificial lung applications.

Lewis and coworkers [69] have pioneered a novel approach involving direct writing of organic inks to create 3-D microvascular networks in a Pluronic F127-diacrylate (DA) hydrogel (Figure 2.2D). In this approach, the ink is precisely dispensed using a fine nozzle positioned by a robotically controlled multiaxis stage

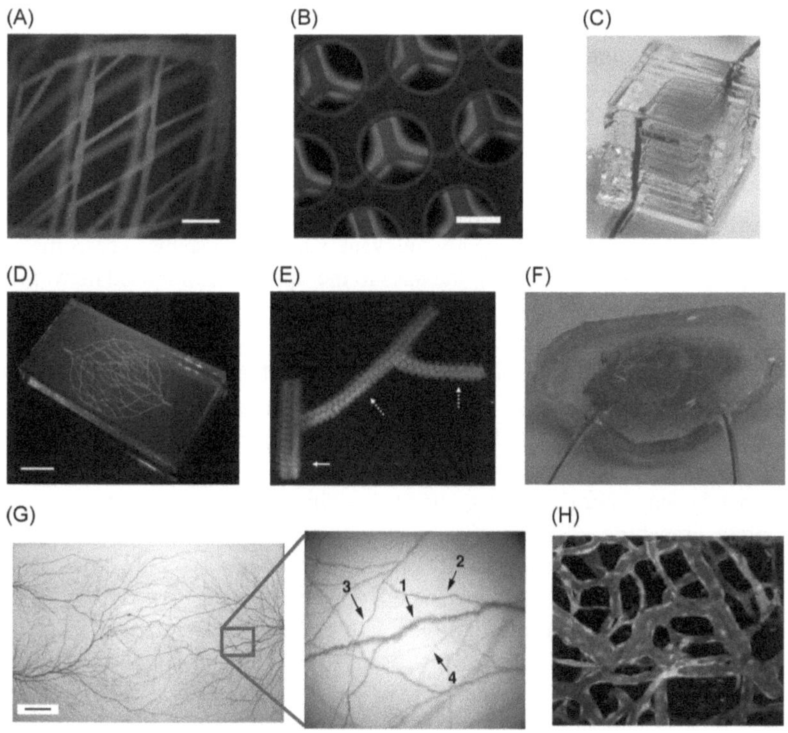

FIGURE 2.2

Construction of 3-D microvascular networks. (A) Stacking of 2-D PLGA-based micropatterns [55]. (B) Stacking of 2-D PGS-based micropatterns [72]. (C) Stacking of 2-D PDMS-based micropatterns [60]. (D) Embedding of vasculature within a Pluronic F127-based construct by direct-write assembly [69]. (E) Deposition of bioink particles into tubular structures using a bioprinter [61]. (F) Network constructed using a sacrificial sugar process [73]. (G) Tree-like branched network fabricated in PLA using electrostatic discharge [75]. (H) Vasculogenesis by co-seeding HUVEC and 10T1/2 precursor cells *in vivo* [63].

so that 3-D structures can be produced in a layer-by-layer fashion. After encapsulation in a desired substrate, the fugitive ink is liquefied and removed by application of a modest vacuum, leaving a microchannel network embedded in the surrounding matrix [66–68,70,71]. Other researchers have applied similar direct writing methods based on fusing multicellular spheroid aggregates containing various cell types (e.g., smooth muscle and fibroblasts) into tubular structures which serve as building blocks within a molding template (Figure 2.2E) [61,78]. Layer-by-layer deposition of these components permits the resulting tube diameter, wall thickness, and branching pattern to be accurately controlled. These scaffold-free constructs offer an alternative approach to enable tissue culture in expanded 3-D volumes.

Other sacrificial techniques have also been developed to construct microvascular networks. Golden and Tien [34] demonstrated the ability to embed complex 2-D networks within hydrogels by lithographic patterning of gelatin as a sacrificial element. Lee et al. [74] extended this concept into 3-D by printing a heated gelatin solution between collagen layers. A novel albeit unconventional approach was employed by Bellan et al. [73] who embedded microvascular networks within PDMS, epoxy, and polycaprolactone (PCL) matrices using melt-spun sugar fibers as the sacrificial element (Figure 2.2F).

A key drawback associated with many of these methods is that the microchannels produced generally possess a relatively narrow range of diameters, rendering them less representative of physiologically relevant vascular systems where diameters become progressively smaller from the base to the tips of the branched network. One novel approach to mimic this architecture involves using electrostatic discharge (Figure 2.2G) [75]. This process harnesses electron beam irradiation to implant a high level of electric charge inside an insulating polymeric substrate so that the energy released upon discharge will be sufficiently intense to locally vaporize and fracture the surrounding material. In this way, networks of highly branched tree-like microchannels with diameters ranging from approximately 10 μm to 1 mm are produced that become permanently embedded within the substrate. These networks incorporate branching characteristics quantitatively similar to naturally occurring vasculature and have been successfully produced in both poly(methyl methacrylate) (PMMA) and biodegradable poly(lactic acid) (PLA) substrates.

In addition to manufacturing scaffolds with embedded vasculature, progress has also been made toward inducing formation of microvascular networks by directed differentiation of vascular endothelial cells and vascular progenitors (e.g., angioblasts), a process called *vasculogenesis* [79]. Although the resulting blood vessels sometimes become immature and unstable [80], Jain and coworkers [63] have demonstrated the ability to produce a long-lasting blood vessel network (functional for 1 year) by coculturing vascular endothelial cells and mesenchymal precursor cells (Figure 2.2H). Finally, a promising approach based on decellularization of allogeneic or xenogeneic donor organs has been explored to reconstruct a variety of tissues including heart [81], liver [82], lung [83–85], and blood vessels [86]. The decellularization process yields an acellular 3-D scaffold directly replicating its natural counterpart that can subsequently be re-seeded with either functional parenchymal cells or selected progenitor cell populations. Initial results have shown promise, however challenges associated with fibrosis of the *in vivo* cultured constructs after implantation as well as the need to rely on donor organs to provide a source for each decellularized construct still remain to be overcome.

2.5 Microchannel topologies

Ideally, microfluidic channels comprising vascular networks should incorporate circular cross-sectional profiles mimicking those in natural vasculature. But conventional lithographic and molding techniques are generally only capable of creating square, rectangular, or trapezoidal cross sections. Despite these limitations, the ease with which 2-D microchannel networks can be constructed using soft lithography has led it to become one of the most widely used micromachining methods [35–39,55]. Other approaches including electroforming [28], laser ablation [51], and sacrificial elements [34] have also been used to produce 2-D branched networks with rectangular cross-sectional profiles. These studies have been proven useful to reveal key features associated with metabolite and oxygen exchange, waste dispersal, flow behavior, and extracellular matrix (ECM) formation that are critical to understand in order to undertake long-term cell culture.

Despite these successes, however, the use of noncircular cross sections for these studies poses challenges because the sharp corners are difficult to seed uniformly [87], making it difficult to form the confluent layers necessary to produce fully endothelialized networks. Moreover, wall shear stresses (WSS) can vary widely across the width of microchannels with rectangular or trapezoidal cross sections [88], as well as in the vicinity of flow obstructions [89]. Since these local shear stresses play an important role in governing functional phenotype and gene expression in endothelial cells (ECs) [90–92], the extent to which experiments using these geometries faithfully replicate the physiological microenvironment must be considered. One area where microfluidic networks have provided useful insights is in studies investigating blood clotting phenomena (both in initial [93] and in propagation [94] phases) that contribute to venous and arterial thrombosis according to Virchow's triad describing venous thrombosis in terms of an interplay among effects related to (i) vessel wall damage, (ii) blood flow changes, and (iii) alterations in the blood's fluidic characteristics [95].

Recent advances in micromachining technology have made it possible to construct semicircular topologies by applying methods such as photoresist-based molding [30], wet etching [33], electroplated molding [46], laser ablation [50], and micromilling [96]. Microchannels with circular cross sections are then constructed by aligning two substrates imprinted with complementary semicircular patterns and bonding them together. Direct-ink writing methods have also been employed to construct 3-D branched networks with circular cross-sectional profiles in hydrogel matrices [69]. The electrostatic discharge method also shows promise to enable 3-D networks with circular cross-section topologies to be embedded in larger substrates [75].

In addition to the cross-sectional profile of individual microchannels, the branching architecture of the global network should match that of the physiological system of interest (Table 2.3). This can be accomplished by applying Murray's law to determine hierarchy of changes in diameter that should be experienced by microchannels entering and exiting each branch point. These concepts have been applied to design and construct branched microvascular networks in both 2-D [33,43,44,48–51,97] and 3-D [60,68,75]. But while providing an important guide, Murray's law is not the only consideration because in reality the WSS values display variability in physiological systems depending on the particular location within the flow network. For example, mean WSS values were found to range on average between 1.1 and 1.4 Pa in the common carotid artery, whereas values in the 0.5–5.0 Pa range are found in arterioles larger than 15 μm in diameter. In smaller arterioles (capillaries) in the 6 μm range, WSS values above 10 Pa can occur owing to nonregulatory properties of mesenteric arterioles [99]. These kinds of local variations, coupled with realistic flow conditions (Table 2.4) must be incorporated into the design of artificial vascular networks in order to ensure that physiological conditions are reproduced.

2.6 Engineering meets biology: toward tissue engineering applications

Construction of microvascular networks, while important, is only one piece of a much broader puzzle that must be solved in order to artificially manufacture complex tissues. These insights have been developed through extensive studies dating back to Weinberg and Bell's [101] pioneering demonstration of the first *in vivo* tissue-engineered blood vessels in 1986, cultured from a combination of bovine endothelial cells, smooth muscle cells (SMCs), and fibroblasts. Through these efforts, it is now generally appreciated that organ-sized artificial tissues require (i) an embedded vascular network to sustain transport of nutrients and waste, (ii) a 3-D matrix to support cell seeding and culture, and (iii) incorporation of appropriate cell types into the construct.

The range of cell types that need to be incorporated into functional tissues is illustrated by considering the case of capillaries, the most abundant blood vessels, consisting of ECs surrounded by basement membrane (composed of sheet-like depositions of ECM material) and a sparse layer of pericytes (a type of perivascular cell found in small vessels) embedded within the EC basement membrane (Figure 2.3A). This wall structure and large surface area-to-volume ratio enables capillaries to function as the primary pathway for exchange of nutrients between blood and tissue. Arterioles and venules are characterized by an increased coverage of mural cells compared with capillaries. Precapillary arterioles are completely invested with vascular SMCs that form their own basement membrane and are circumferentially arranged, closely packed, and tightly associated with the endothelium. More complex structures can also include fibroblasts, precursor cells able to synthesize ECM material [102].

Table 2.3 Summary of Literature Reporting Construction of Branched Microvascular Networks

Channel Cross Section	Channel Size Range	Micromachining Method	Based on Murray's Law	Spatial Geometry	References
Square	35 μm–5 mm	Lithography and etching followed by soft lithography	No	2-D	[35–39,55]
Rectangular	1.197 mm	Soft lithography	No	2-D	[45]
Square, rectangular, or trapezoid	44.2–250 μm	Simulation	Yes	2-D	[49,97]
Rectangular	125–660 μm	Laser ablation followed by chemical wet etching	Yes	2-D	[48]
Rectangular	>50 μm	Simulation	No	2-D	[98]
Rectangular	100 μm–5 mm	Soft lithography	No	2-D	[32]
Rectangular	20 μm–2.73 mm	Micromachining of stainless steel mold then soft lithography to cast in PDMS	Yes	2-D	[43,44]
Circular	80–190 μm	Electroplating of silicon master mold then micropatterning	No	2-D	[46]
Rectangular	40–60 μm	Hydrogel constructs cast from PDMS mold	No	2-D	[53]
Semicircular	40–120 μm	Wet etching	Yes	2-D	[33]
Rectangular	20–60 μm	Electroforming	No	2-D	[28]
Circular	60–120 μm	Photoresist-based molding and bonding	No	2-D	[30]
Semicircular	50 μm–1 mm	Laser ablation followed by chemical wet etching	Yes	2-D	[50]
Rectangular	90–600 μm	Laser ablation followed by chemical wet etching	Yes	2-D	[51]

(*Continued*)

Table 2.3 (Continued)

Channel Cross Section	Channel Size Range	Micromachining Method	Based on Murray's Law	Spatial Geometry	References
Rectangular	6–50 μm	Soft lithography with sacrificial element	No	2-D	[34]
Rectangular	35 μm–5 mm	Soft lithography	Yes	3-D	[60]
Circular	100–500 μm	Direct-write assembly	Yes	3-D	[68]
Circular	200–600 μm	Direct-write assembly	No	3-D	[69]
Circular	10 μm–1 mm	Electrostatic discharge	Yes	3-D	[75]

Publications from the same laboratory are grouped together.

Table 2.4 Characteristic Flow Properties in Blood Vessels Under Human Physiological Conditions [100]

	Artery	Arteriole	Capillary	Venule	Vein
Diameter (μm)	4000	30	6	20	5000
Average flow rate (ml/s)	1.3	4×10^{-6}	8×10^{-9}	9×10^{-7}	1.0
Mean pressure (mmHg)	90	60	25	15	12

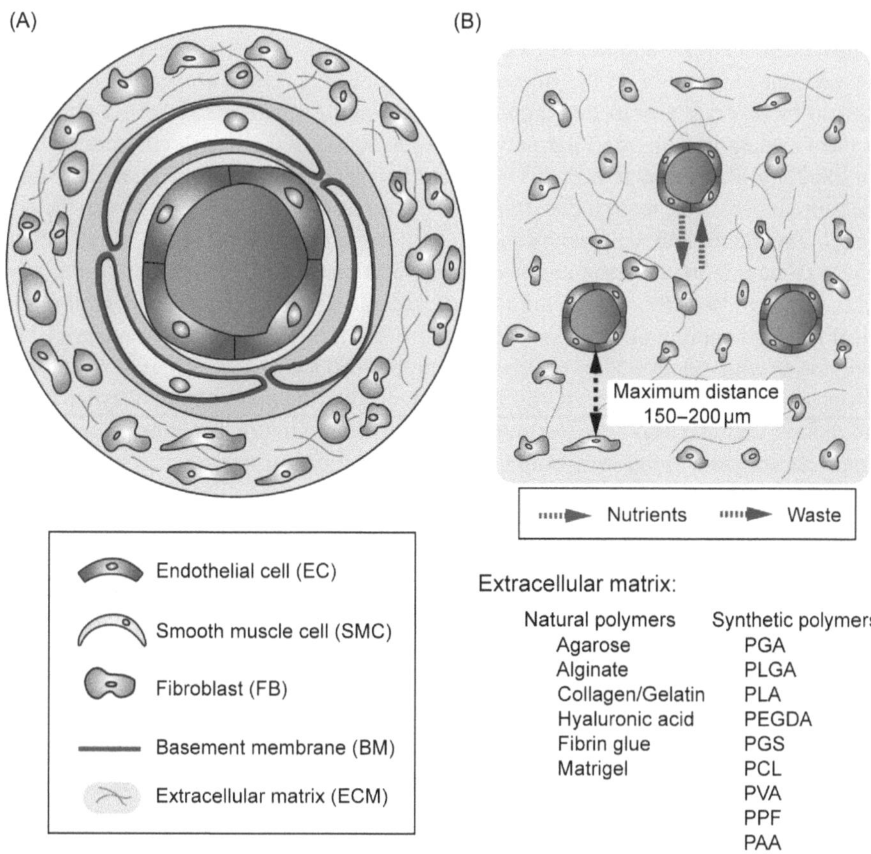

FIGURE 2.3

Cellular composition of blood vessels. (A) The walls of larger vessels consist of three major cellular layers: ECs, SMCs surrounded by basement membrane, and fibroblasts incorporated with ECM. (B) Artificial ECM materials can be broadly classified into natural and synthetic polymers. Mass transport between the cells and the nutrient blood flow source is conducted within the ECM, but cells can survive by diffusion alone only if they are located within ~150–200 μm of a vascular element.

The role of vascularization can now be appreciated in the context of a general strategy to construct organ-sized artificial tissues *in vitro*. Ordinarily, the process of vasculogenesis begins to unfold upon cell seeding when the ECM environment coupled with hypoxia (low oxygen) experienced by the cells stimulates endogenous release of angiogenic growth factors responsible for growth of new blood vessels. Unfortunately, this spontaneous vascularization process occurs too slowly to supply the seeded cells with adequate oxygen and nutrients during the time needed for a mature network to form [103]. Furthermore, it is not yet fully understood how to direct the process so that well-defined interconnected branched networks are produced. These obstacles can be overcome by incorporating embedded 3-D vasculature within the scaffold to both provide transport of nutrients and waste throughout the cell-seeded scaffold, and act as a template for construction of artificial vasculature by seeding the interior of the microchannels with ECs. These embedded networks should incorporate a branched architecture containing small arteries (1–2 mm diameter) leading into an arteriolar network (100–1000 μm diameter), and terminating in capillary-like vascular bed (10–15 μm diameter). This network should in turn be surrounded by a matrix that functions as a reservoir to deliver growth factors (e.g., vascular endothelial growth factor), and a scaffold that serves as a temporary ECM (Figure 2.3B; in tissue engineering applications, ECM materials are generally classified into two groups: (i) natural biodegradable polymers that are inherently biodegradable, nonimmunogenic, and biocompatible and (ii) synthetic biodegradable polymers with controllable mechanical properties and cross-linkable functional groups). The resulting scaffold can then be seeded with appropriate cell types (e.g., adult stem cells derived from the patient) whose differentiation could be orchestrated to produce mature blood vessels and ultimately functional tissue.

2.7 Outlook and future challenges

Incredible progress has been made to develop technologies capable of producing 3-D microchannel networks mimicking the features of physiological vasculature, and these methods are likely to play a key role in enabling culture of functional organ-sized engineered tissues. A key future challenge will be to ensure convergence between the science and engineering associated with these vascular network construction approaches and the biology associated with incorporating this technology into the broader scope of *in vitro* tissue culture applications. This shift toward the biological aspects of the problem introduces new considerations. For example, as the complexity of tissue engineering constructs continues to evolve, it becomes increasingly important to maintain some degree of standardization with respect to experimental conditions. This is evident when considering the array of conditions already employed in microchannel-based cell culture studies (Table 2.5). Meaningful comparisons between results obtained by different research groups are possible only if common cell types and culture conditions are

Table 2.5 Summary of Literature Reporting Cell Culture Studies in Microchannel Networks

Cell Type	Culture Conditions	Perfusion Rate	Results	References
Human umbilical vein endothelial cells (HUVECs)	Cells were cultured on the surface of devices and incubated at 37°C in a 5% CO_2	N/A	Cell viability was maintained after 21 h	[33]
Human microvascular endothelial cells (HMEC-1)	Cells were seeded in the microchannels and maintained in a humidified atmosphere of 5% CO_2 at 37°C	0.5 ml/h	Culture was continued for up to 2 weeks, and all channels remained covered with a HMEC-1 monolayer. The HMEC-1 cells lined the channels as a monolayer and expressed markers for CD31 and von Willebrand factor	[37]
Human umbilical vein endothelial cells (HUVECs)	Cells were seeded in the microchannels and maintained in a humidified atmosphere of 5% CO_2 at 37°C	100 μl/h	Cell expansion slowed after 10 days and parts of the capillary networks were covered by nearly confluent ECs within 14 days. The cells remained stable under these culture conditions for at least 4 weeks	[35]
Hepatoma cell line HepG2/C3A	Cells were seeded in microchannels and maintained in humidified atmosphere of 5% CO_2 at 37°C	0.5 ml/h	The device can sustain both human hepatoma cells and primary rat hepatocytes by continuous *in vitro* perfusion of medium, allowing proliferation and maintaining hepatic functions such as serum protein synthesis and metabolism	[38]

(*Continued*)

Table 2.5 (Continued)

Cell Type	Culture Conditions	Perfusion Rate	Results	References
HepG2, human hepatocarcinoma cells, HUVEC, 3T3-L1 mouse fibroblasts	Humidified atmosphere of 5% CO_2 at 37°C	5 μl/min	HepG2 cells could be kept alive for 2 weeks with appropriate sterilization and collagen coating steps	[47]
AML-12 hepatocytes	Cells were embedded in agarose hydrogel and cultured in humidified atmosphere of 5% CO_2 at 37°C	10 μl/min	Delivery of media through the channel allowed cells in close proximity to the channels to remain viable for at least 3 days	[40]
Chondrocyte	Cells were embedded in alginate hydrogel with microfluidic network	10 μl/min	Zone of metabolic activity decays over a shorter distance at higher densities of cells	[41]
Human umbilical vein endothelial cells (HUVECs)	Cell was cultured in cylindrical microchannel networks	N/A	Cells formed a confluent monolayer and remained viable for at least 24 h of culture within the microchannels	[46]
	Cells were embedded in alginate hydrogel with microfluidic network	N/A	Microfabricated 3-D matrices maintain stable interfaces over at least 3 weeks	[53]
Bovine endothelial cells (BEC)	Cells were cultured in the microchannels and incubated at 37°C in a 5% CO_2	8 μl/min	Cells can survive in the scaffold for up to 4 weeks	[30]
	Cells were cultured in the microchannels and incubated at 37°C in a 5% CO_2	8 μl/min	After 96 h of seeding, cells completely adhere to the microchannels	[28]

Human dermal fibroblasts and human dermal microvascular endothelial cells (HDMECs)	Human fibroblasts embedded within microfluidic collagen gels and HDMECs seeded on the walls of collagen channels	292 μl/h	After 5 days of culture, seeded cells remained localized to the internal surfaces of collagen networks	[34]
Fibroblasts, keratinocytes, HUVEC or HMVEC	Two inferior sheets were plated with or without HUVECs, whereas a superior third sheet was seeded with keratinocytes to provide the epidermal layer	N/A	Development of functional blood vessels. Also promoted vascularization of the implanted tissues	[76,77]
Human umbilical vein endothelial cells (HUVECs) and neonatal normal human dermal fibroblasts (NHDFs)	Stacked multilayered cell sheets were incubated at 37°C for 3 or 7 days	N/A	Prevascular networks composing of HUVECs were formed in all the triple layer constructs	[62]
Human dermal fibroblasts	Cells were embedded in the collagen hydrogel	4 μl/min	Viability of cells was maintained with channel diffusion after a week of culture, while the viability dropped without channel perfusion	[74]
Human hepatocarcinoma cells (HepG2)	Cells were cultured in the microchannels and incubated at 37°C in a 5% CO_2	5–20 μl/min	Without culture change cells rapidly died. Cells in good condition for 10 days because O_2 permeability of PDMS and optimum flow rate	[59]
Hepatocyte carcinoma cells (HepG2)	Cells were cultured in the microchannels and incubated at 37°C in a 5% CO_2	280 μm/h (volumetric flow rate)	Reactors were perfused for a period of up to 1 week to demonstrate long-term viability within the devices	[72]
Human umbilical vein endothelial cells (HUVECs) and 10T1/2 mesenchymal precursor cells	Cells were seeded in 3-D constructs and then implanted in mice	N/A	Coimplantation of HUVEC and 10T1/2 allowed the 10T1/2 cells to differentiate into mural cells and form stable vessels after 4 months while HUVEC alone construct rarely survived	[63]

(*Continued*)

Table 2.5 (Continued)

Cell Type	Culture Conditions	Perfusion Rate	Results	References
Human hepatoma cell (HepG2) and human umbilical vein endothelial cells (HUVECs)	Collagen–HepG2 modules were seeded and incubated with HUVEC and maintained in the flow circuit at 37°C in a 5% CO_2	0.334 ml/min	HepG2 cells remained viable within an assembled construct over 7 days. HUVEC retained their nonthrombogenic phenotype and delayed clotting times and inhibited the loss of platelets associated with perfusion of whole blood through the construct	[65]
Human umbilical vein smooth muscle cells (HUVSMCs) and human skin fibroblasts (HSFs)	Multilayered tubular vascular grafts were fabricated by bioprinter and incubated at 37°C with 5% CO_2	N/A	Viability upon implantation remains problematic as infiltration of surrounding vessels	[61]

consistently employed. Improved methods to seed and culture cells in extended 3-D spaces also need to be developed and refined. Current microscopy-based approaches to observe cell culture in thin planar multiwell plate formats are inadequate to probe conditions deep within the interior of centimeter-sized constructs. Other key challenges involve embedding vasculature into porous scaffold substrates and developing methods amenable to mass production. Overcoming these barriers promises to set the stage for a host of exciting developments in the field of tissue engineering during the coming years.

Acknowledgments

We thank Molly Morin for assistance in compiling the literature for this chapter. This work is supported in part by the US National Science Foundation under grant DMR-1106005.

References

[1] B. Palsson, S. Bhatia, Tissue Engineering, Pearson Prentice Hall, Upper Saddle River, NJ, 2004.

[2] V.L. Roger, Heart disease and stroke statistics-2011 update: a report from the American Heart Association (123 (2011) 18), Circulation 123 (2011) E240.

[3] L.G. Griffith, M.A. Swartz, Capturing complex 3D tissue physiology *in vitro*, Nat. Rev. Mol. Cell Biol. 7 (2006) 211–224.

[4] M.C. Peters, P.J. Polverini, D.J. Mooney, Engineering vascular networks in porous polymer matrices, J. Biomed. Mater. Res. 60 (2002) 668–678.

[5] G. Vozzi, C. Flaim, A. Ahluwalia, S. Bhatia, Fabrication of PLGA scaffolds using soft lithography and microsyringe deposition, Biomaterials 24 (2003) 2533–2540.

[6] X. Wu, E. Rabkin-Aikawa, K.J. Guleserian, T.E. Perry, Y. Masuda, F.W.H. Sutherland, et al., Tissue-engineered microvessels on three-dimensional biodegradable scaffolds using human endothelial progenitor cells, Am. J. Physiol. Heart Circ. Physiol. 287 (2004) H480–H487.

[7] E. Lavik, R. Langer, Tissue engineering: current state and perspectives, Appl. Microbiol. Biotechnol. 65 (2004) 1–8.

[8] F. Van Eijk, D.B.F. Saris, J. Riesle, W.J. Willems, C.A. Van Blitterswijk, A.J. Verbout, et al., Tissue engineering of ligaments: a comparison of bone marrow stromal cells, anterior cruciate ligament, and skin fibroblasts as cell source, Tissue Eng. 10 (2004) 893–903.

[9] G. Vunjak-Novakovic, G. Altman, R. Horan, D.L. Kaplan, Tissue engineering of ligaments, Annu. Rev. Biomed. Eng. 6 (2004) 131–156.

[10] S.D. Waldman, C.G. Spiteri, M.D. Grynpas, R.M. Pilliar, R.A. Kandel, Long-term intermittent compressive stimulation improves the composition and mechanical properties of tissue-engineered cartilage, Tissue Eng. 10 (2004) 1323–1331.

[11] L.E. Freed, F. Guilak, X.E. Guo, M.L. Gray, R. Tranquillo, J.W. Holmes, et al., Advanced tools for tissue engineering: scaffolds, bioreactors, and signaling, Tissue Eng. 12 (2006) 3285–3305.

[12] L.G. Griffith, G. Naughton, Tissue engineering—current challenges and expanding opportunities, Science 295 (2002) 1009–1014.

[13] K. Kellner, G. Liebsch, I. Klimant, O.S. Wolfbeis, T. Blunk, M.B. Schulz, et al., Determination of oxygen gradients in engineered tissue using a fluorescent sensor, Biotechnol. Bioeng. 80 (2002) 73–83.

[14] C.K. Colton, Implantable biohybrid artificial organs, Cell Transplant. 4 (1995) 415–436.

[15] M. Labarbera, Principles of design of fluid transport-systems in zoology, Science 249 (1990) 992–1000.

[16] G.B. West, J.H. Brown, B.J. Enquist, A general model for the origin of allometric scaling laws in biology, Science 276 (1997) 122–126.

[17] R.M. Nerem, D. Seliktar, Vascular tissue engineering, Annu. Rev. Biomed. Eng. 3 (2001) 225–243.

[18] J.G. Rheinwald, H. Green, Formation of a keratinizing epithelium in culture by a cloned cell line derived from a teratoma, Cell 6 (1975) 317–330.

[19] C.D. Murray, The physiological principle of minimum work. I. The vascular system and the cost of blood volume, Proc. Natl. Acad. Sci. U.S.A. 12 (1926) 207–214.

[20] M. LaBarbera, Principles of design of fluid transport systems in zoology, Science 249 (1990) 992–1000.

[21] D.R. Emerson, K. Cieslicki, X. Gu, R.W. Barber, Biomemetic design of microfluidic manifolds based on a generalized Murray's law, Lab Chip 6 (2006) 447–454.

[22] J.G. Restrepo, E. Ott, B.R. Hunt, Scale dependence of branching in arterial and bronchial trees, Phys. Rev. Lett. 96 (2006) 128101–128104.

[23] G.S. Kassab, Scaling laws of vascular trees: of form and function, Am. J. Physiol. Heart Circ. Physiol. 290 (2006) H894–H903.

[24] E. Vanbavel, J.A.E. Spaan, Branching patterns in the porcine coronary arterial tree—estimation of flow heterogeneity, Circ. Res. 71 (1992) 1200–1212.

[25] W. Huang, R.T. Yen, M. McLaurine, G. Bledsoe, Morphometry of the human pulmonary vasculature, J. Appl. Physiol. 81 (1996) 2123–2133.

[26] N. Mittal, Y. Zhou, C. Linares, S. Ung, B. Kaimovitz, S. Molloi, et al., Analysis of blood flow in the entire coronary arterial tree, Am. J. Physiol. Heart Circ. Physiol. 289 (2005) H439–H446.

[27] V.V. Gafiychuk, I.A. Lubashevsky, On the principles of the vascular network branching, J. Theor. Biol. 212 (2001) 1–9.

[28] G.J. Wang, C.L. Chen, S.H. Hsu, Y.L. Chiang, Bio-MEMS fabricated artificial capillaries for tissue engineering, Microsyst. Technol. Micronanosyst. Inf. Storage Process. Syst. 12 (2005) 120–127.

[29] G.J. Wang, Y.F. Hsu, S.H. Hsu, R.H. Horng, JSR photolithography based microvessel scaffold fabrication and cell seeding, Biomed. Microdevices 8 (2006) 17–23.

[30] G.J. Wang, K.H. Ho, S.H. Hsu, K.P. Wang, Microvessel scaffold with circular microchannels by photoresist melting, Biomed. Microdevices 9 (2007) 657–663.

[31] V. Janakiraman, B.L. Kienitz, H. Baskaran, Lithography technique for topographical micropatterning of collagen-glycosaminoglycan membranes for tissue engineering applications, J. Med. Device 1 (2007) 233–237.

[32] V. Janakiraman, S. Sastry, J.R. Kadambi, H. Baskaran, Experimental investigation and computational modeling of hydrodynamics in bifurcating microchannels, Biomed. Microdevices 10 (2008) 355–365.

[33] J.P. Camp, T. Stokol, M.L. Shuler, Fabrication of a multiple-diameter branched network of microvascular channels with semi-circular cross-sections using xenon difluoride etching, Biomed. Microdevices 10 (2008) 179−186.

[34] A.P. Golden, J. Tien, Fabrication of microfluidic hydrogels using molded gelatin as a sacrificial element, Lab Chip 7 (2007) 720−725.

[35] C. Fidkowski, M.R. Kaazempur-Mofrad, J. Borenstein, J.P. Vacanti, R. Langer, Y.D. Wang, Endothelialized microvasculature based on a biodegradable elastomer, Tissue Eng. 11 (2005) 302−309.

[36] S. Kaihara, J. Borenstein, R. Koka, S. Lalan, E.R. Ochoa, M. Ravens, et al., Silicon micromachining to tissue engineer branched vascular channels for liver fabrication, Tissue Eng. 6 (2000) 105−117.

[37] M. Shin, K. Matsuda, O. Ishii, H. Terai, M. Kaazempur-Mofrad, J. Borenstein, et al., Endothelialized networks with a vascular geometry in microfabricated poly(dimethyl siloxane), Biomed. Microdevices 6 (2004) 269−278.

[38] A. Carraro, W.M. Hsu, K.M. Kulig, W.S. Cheung, M.L. Miller, E.J. Weinberg, et al., *In vitro* analysis of a hepatic device with intrinsic microvascular-based channels, Biomed. Microdevices 10 (2008) 795−805.

[39] J.T. Borenstein, H. Terai, K.R. King, E.J. Weinberg, M.R. Kaazempur-Mofrad, J.P. Vacanti, Microfabrication technology for vascularized tissue engineering, Biomed. Microdevices 4 (2002) 167−175.

[40] Y. Ling, J. Rubin, Y. Deng, C. Huang, U. Demirci, J.M. Karp, et al., A cell-laden microfluidic hydrogel, Lab Chip 7 (2007) 756−762.

[41] N.W. Choi, M. Cabodi, B. Held, J.P. Gleghorn, L.J. Bonassar, A.D. Stroock, Microfluidic scaffolds for tissue engineering, Nat. Mater. 6 (2007) 908−915.

[42] S.S. Shevkoplyas, S.C. Gifford, T. Yoshida, M.W. Bitensky, Prototype of an *in vitro* model of the microcirculation, Microvasc. Res. 65 (2003) 132−136.

[43] D.M. Hoganson, H.I. Pryor, I.D. Spool, O.H. Burns, J.R. Gilmore, J.P. Vacanti, Principles of biomimetic vascular network design applied to a tissue-engineered liver scaffold, Tissue Eng. A 16 (2010) 1469−1477.

[44] D.M. Hoganson, H.I. Pryor, E.K. Bassett, I.D. Spool, J.P. Vacanti, Lung assist device technology with physiologic blood flow developed on a tissue engineered scaffold platform, Lab Chip 11 (2011) 700−707.

[45] R.J. Gilbert, H. Park, M. Rasponi, A. Redaelli, B. Gellman, K.A. Dasse, et al., Computational and functional evaluation of a microfluidic blood flow device, ASAIO J. 53 (2007) 447−455.

[46] J.T. Borenstein, M.M. Tupper, P.J. Mack, E.J. Weinberg, A.S. Khalil, J. Hsiao, et al., Functional endothelialized microvascular networks with circular cross-sections in a tissue culture substrate, Biomed. Microdevices 12 (2010) 71−79.

[47] E. Leclerc, K.S. Furukawa, F. Miyata, Y. Sakai, T. Ushida, T. Fujii, Fabrication of microstructures in photosensitive biodegradable polymers for tissue engineering applications, Biomaterials 25 (2004) 4683−4690.

[48] D. Lim, Y. Kamotani, B. Cho, J. Mazumder, S. Takayama, Fabrication of microfluidic mixers and artificial vasculatures using a high-brightness diode-pumped Nd: YAG laser direct write method, Lab Chip 3 (2003) 318−323.

[49] D.R. Emerson, K. Cieslicki, X.J. Gu, R.W. Barber, Biomimetic design of microfluidic manifolds based on a generalised Murray's law, Lab Chip 6 (2006) 447−454.

[50] D.H. Kam, J. Mazumder, Three-dimensional biomimetic microchannel network by laser direct writing, J. Laser Appl. 20 (2008) 185–191.
[51] D.H. Kam, L. Shah, J. Mazumder, Femtosecond laser machining of multi-depth microchannel networks onto silicon, J. Micromech. Microeng. 21 (2011) 045027.
[52] X.F. Cui, T. Boland, Human microvasculature fabrication using thermal inkjet printing technology, Biomaterials 30 (2009) 6221–6227.
[53] B.M. Gillette, J.A. Jensen, B.X. Tang, G.J. Yang, A. Bazargan-Lari, M. Zhong, et al., *In situ* collagen assembly for integrating microfabricated three-dimensional cell-seeded matrices, Nat. Mater. 7 (2008) 636–640.
[54] Y. Nahmias, R.E. Schwartz, C.M. Verfaillie, D.J. Odde, Laser-guided direct writing for three-dimensional tissue engineering, Biotechnol. Bioeng. 92 (2005) 129–136.
[55] K.R. King, C.C.J. Wang, M.R. Kaazempur-Mofrad, J.P. Vacanti, J.T. Borenstein, Biodegradable microfluidics, Adv. Mater. 16 (2004) 2007–2012.
[56] D.C. Duffy, J.C. McDonald, O.J.A. Schueller, G.M. Whitesides, Rapid prototyping of microfluidic systems in poly(dimethylsiloxane), Anal. Chem. 70 (1998) 4974–4984.
[57] C.D. Murray, The physiological principle of minimum work. II. Oxygen exchange in capillaries, Proc. Natl. Acad. Sci. U.S.A. 12 (1926) 299–304.
[58] C.D. Murray, The physiological principle of minimum work applied to the angle of branching of arteries, J. Gen. Physiol. 9 (1926) 835–841.
[59] E. Leclerc, Y. Sakai, T. Fujii, Cell culture in 3-dimensional microfluidic structure of PDMS (polydimethylsiloxane), Biomed. Microdevices 5 (2003) 109–114.
[60] T. Kniazeva, J.C. Hsiao, J.L. Charest, J.T. Borenstein, A microfluidic respiratory assist device with high gas permeance for artificial lung applications, Biomed. Microdevices 13 (2011) 315–323.
[61] C. Norotte, F.S. Marga, L.E. Niklason, G. Forgacs, Scaffold-free vascular tissue engineering using bioprinting, Biomaterials 30 (2009) 5910–5917.
[62] N. Asakawa, T. Shimizu, Y. Tsuda, S. Sekiya, T. Sasagawa, M. Yamato, et al., Pre-vascularization of *in vitro* three-dimensional tissues created by cell sheet engineering, Biomaterials 31 (2010) 3903–3909.
[63] N. Koike, D. Fukumura, O. Gralla, P. Au, J.S. Schechner, R.K. Jain, Creation of long-lasting blood vessels, Nature 428 (2004) 138–139.
[64] A.L. Sieminski, A.S. Was, G. Kim, H. Gong, R.D. Kamm, The stiffness of three-dimensional ionic self-assembling peptide gels affects the extent of capillary-like network formation, Cell. Biochem. Biophys. 49 (2007) 73–83.
[65] A.P. McGuigan, M.V. Sefton, Vascularized organoid engineered by modular assembly enables blood perfusion, Proc. Natl. Acad. Sci. U.S.A. 103 (2006) 11461–11466.
[66] D. Therriault, S.R. White, J.A. Lewis, Chaotic mixing in three-dimensional microvascular networks fabricated by direct-write assembly, Nat. Mater. 2 (2003) 265–271.
[67] D. Therriault, R.F. Shepherd, S.R. White, J.A. Lewis, Fugitive inks for direct-write assembly of three-dimensional microvascular networks, Adv. Mater. 17 (2005) 395–399.
[68] W. Wu, C.J. Hansen, A.M. Aragon, P.H. Geubelle, S.R. White, J.A. Lewis, Direct-write assembly of biomimetic microvascular networks for efficient fluid transport, Soft Matter 6 (2010) 739–742.

[69] W. Wu, A. DeConinck, J.A. Lewis, Omnidirectional printing of 3D microvascular networks, Adv. Mater. 23 (2011) H178–H183.

[70] K.S. Toohey, N.R. Sottos, J.A. Lewis, J.S. Moore, S.R. White, Self-healing materials with microvascular networks, Nat. Mater. 6 (2007) 581–585.

[71] C.J. Hansen, W. Wu, K.S. Toohey, N.R. Sottos, S.R. White, J.A. Lewis, Self-healing materials with interpenetrating microvascular networks, Adv. Mater. 21 (2009) 4143–4147.

[72] C.J. Bettinger, E.J. Weinberg, K.M. Kulig, J.P. Vacanti, Y.D. Wang, J.T. Borenstein, et al., Three-dimensional microfluidic tissue-engineering scaffolds using a flexible biodegradable polymer, Adv. Mater. 18 (2006) 165–169.

[73] L.M. Bellan, S.P. Singh, P.W. Henderson, T.J. Porri, H.G. Craighead, J.A. Spector, Fabrication of an artificial 3-dimensional vascular network using sacrificial sugar structures, Soft Matter 5 (2009) 1354–1357.

[74] W. Lee, V. Lee, S. Polio, P. Keegan, J.H. Lee, K. Fischer, et al., On-demand three-dimensional freeform fabrication of multi-layered hydrogel scaffold with fluidic channels, Biotechnol. Bioeng. 105 (2010) 1178–1186.

[75] J.H. Huang, J. Kim, N. Agrawal, A.P. Sudarson, J.E. Maxim, A. Jayaraman, et al., Rapid fabrication of bio-inspired 3D microfluidic vascular networks, Adv. Mater. 21 (2009) 3567–3571.

[76] A.F. Black, F. Berthod, N. L'Heureux, L. Germain, F.A. Auger, *In vitro* reconstruction of a human capillary-like network in a tissue-engineered skin equivalent, FASEB J. 12 (1998) 1331–1340.

[77] L. Gibot, T. Galbraith, J. Huot, F.A. Auger, A preexisting microvascular network benefits *in vivo* revascularization of a microvascularized tissue-engineered skin substitute, Tissue Eng. A 16 (2010) 3199–3206.

[78] K. Jakab, A. Neagu, V. Mironov, R.R. Markwald, G. Forgacs, Engineering biological structures of prescribed shape using self-assembling multicellular systems, Proc. Natl. Acad. Sci. U.S.A. 101 (2004) 2864–2869.

[79] W. Risau, I. Flamme, Vasculogenesis, Ann. Rev. Cell. Dev. Biol. 11 (1995) 73–91.

[80] R.K. Jain, Molecular regulation of vessel maturation, Nat. Med. 9 (2003) 685–693.

[81] H.C. Ott, T.S. Matthiesen, S.K. Goh, L.D. Black, S.M. Kren, T.I. Netoff, et al., Perfusion-decellularized matrix: using nature's platform to engineer a bioartificial heart, Nat. Med. 14 (2008) 213–221.

[82] B.E. Uygun, A. Soto-Gutierrez, H. Yagi, M.L. Izamis, M.A. Guzzardi, C. Shulman, et al., Organ reengineering through development of a transplantable recellularized liver graft using decellularized liver matrix, Nat. Med. 16 (2010) 814–820.

[83] T.H. Petersen, E.A. Calle, L.P. Zhao, E.J. Lee, L.Q. Gui, M.B. Raredon, et al., Tissue-engineered lungs for *in vivo* implantation, Science 329 (2010) 538–541.

[84] A.P. Price, K.A. England, A.M. Matson, B.R. Blazar, A. Panoskaltsis-Mortari, Development of a decellularized lung bioreactor system for bioengineering the lung: the matrix reloaded, Tissue Eng. A 16 (2010) 2581–2591.

[85] H.C. Ott, B. Clippinger, C. Conrad, C. Schuetz, I. Pomerantseva, L. Ikonomou, et al., Regeneration and orthotopic transplantation of a bioartificial lung, Nat. Med. 16 (2010) 927–933.

[86] C. Quint, Y. Kondo, R.J. Manson, J.H. Lawson, A. Dardik, L.E. Niklason, Decellularized tissue-engineered blood vessel as an arterial conduit, Proc. Natl. Acad. Sci. U.S.A. 108 (2011) 9214–9219.

[87] J.V. Green, T. Kniazeva, M. Abedi, D.S. Sokhey, M.E. Taslim, S.K. Murthy, Effect of channel geometry on cell adhesion in microfluidic devices, Lab Chip 9 (2009) 677–685.

[88] H. Lu, L.Y. Koo, W.C.M. Wang, D.A. Lauffenburger, L.G. Griffith, K.F. Jensen, Microfluidic shear devices for quantitative analysis of cell adhesion, Anal. Chem. 76 (2004) 5257–5264.

[89] O.F. Khan, M.V. Sefton, Endothelial cell behaviour within a microfluidic mimic of the flow channels of a modular tissue engineered construct, Biomed. Microdevices 13 (2011) 69–87.

[90] C.L. Ives, S.G. Eskin, L.V. Mcintire, Mechanical effects on endothelial-cell morphology—*in vitro* assessment, In Vitro Cell. Dev. Biol. 22 (1986) 500–507.

[91] G. Helmlinger, R.V. Geiger, S. Schreck, R.M. Nerem, Effects of pulsatile flow on cultured vascular endothelial-cell morphology, Trans. ASME J. Biomech. Eng. 113 (1991) 123–131.

[92] A.M. Malek, S. Izumo, Mechanism of endothelial cell shape change and cytoskeletal remodeling in response to fluid shear stress, J. Cell Sci. 109 (1996) 713–726.

[93] F. Shen, C.J. Kastrup, Y. Liu, R.F. Ismagilov, Threshold response of initiation of blood coagulation by tissue factor in patterned microfluidic capillaries is controlled by shear rate, Arterioscler. Thromb. Vasc. Biol. 28 (2008) 2035–2041.

[94] M.K. Runyon, C.J. Kastrup, B.L. Johnson-Kerner, T.G. Van Ha, R.F. Ismagilov, Effects of shear rate on propagation of blood clotting determined using microfluidics and numerical simulations, J. Am. Chem. Soc. 130 (2008) 3458–3464.

[95] E.F. Mammen, Pathogenesis of venous thrombosis, Chest 102 (1992) S640–S644.

[96] M.E. Wilson, N. Kota, Y. Kim, Y.D. Wang, D.B. Stolz, P.R. LeDuc, et al., Fabrication of circular microfluidic channels by combining mechanical micromilling and soft lithography, Lab Chip 11 (2011) 1550–1555.

[97] R.W. Barber, D.R. Emerson, Optimal design of microfluidic networks using biologically inspired principles, Microfluid. Nanofluid. 4 (2008) 179–191.

[98] V. Janakiraman, K. Mathur, H. Baskaran, Optimal planar flow network designs for tissue engineered constructs with built-in vasculature, Ann. Biomed. Eng. 35 (2007) 337–347.

[99] R.S. Reneman, A.P. Hoeks, Wall shear stress as measured *in vivo*: consequences for the design of the arterial system, Med. Biol. Eng. Comput. 46 (2008) 499–507.

[100] W.F. Boron, E.L. Boulpaep, Medical Physiology: A Cellular and Molecular Approach, Saunders/Elsevier, Philadelphia, PA, 2009.

[101] C.B. Weinberg, E. Bell, A blood-vessel model constructed from collagen and cultured vascular cells, Science 231 (1986) 397–400.

[102] C.J. Drake, J.E. Hungerford, C.D. Little, Morphogenesis of the first blood vessels, Morphogen. Cell. Interact. 857 (1998) 155–179.

[103] M.W. Laschke, Y. Harder, M. Amon, I. Martin, J. Farhadi, A. Ring, et al., Angiogenesis in tissue engineering: breathing life into constructed tissue substitutes, Tissue Eng. 12 (2006) 2093–2104.

CHAPTER

3 Microfluidics for Engineering 3D Tissues and Cellular Microenvironments

Brian M. Gillette, Hesam Parsa and Samuel K. Sia
Department of Biomedical Engineering, Columbia University, NY, USA

3.1 Introduction

Cellular microenvironments in native tissues are 3D, inhomogeneous, anisotropic, and dynamic in terms of their composition of cells, extracellular matrix (ECM) components, soluble factors, and physical forces (e.g., fluid flow and mechanical stress). Therefore, methods to recapitulate and control various components of cellular microenvironments *in vitro* and *in vivo* are highly useful for both studying and engineering biological systems. Microfluidics-based technologies, which involve fluid flow in microscale (typically 1–500 μm) channels, enable precision spatiotemporal control over mass transport and are thus well suited to control the assembly and dynamic culture of engineered cellular microenvironments.

The relative ease with which cells can be seeded on the surfaces of microfluidic devices has enabled broad applications of microfluidic devices for biological studies. For example, microfluidic devices have been used to exert controlled shear stresses and gradients of soluble factors over cell monolayers [1,2]. Though several cell types have monolayer-like or floating native phenotypes (e.g., epithelial and blood cells, respectively), most cells reside in a 3D environment *in vivo*. Traditionally, most microfluidic culture systems were designed to culture all cell types in 2D monolayers regardless of the native phenotype; however, recently there has been tremendous effort to mimic the natural phenotype of the cells in 3D environments in microfluidic devices.

3D ECM and hydrogel fabrication technologies have been merged with microfluidics in numerous ways. For example, microfluidic devices can be used for casting of 3D native ECM hydrogels (such as collagen) with precisely defined geometries [3] and for fabricating composite 3D hydrogel microstructures by *in situ* photopolymerization [4]. Incorporation of 3D ECMs and hydrogels within polydimethlysiloxane (PDMS)-based microfluidic devices can enable control over soluble factor gradients in 3D gels [5] or dynamic communication between

Microfluidic Cell Culture Systems. ISBN: 978-1-4377-3459-1

discrete tissue constructs [6]. 3D hydrogels can also be fabricated with embedded microfluidic channels for perfusion with solutions (e.g., to mimic vascular networks) [7–11] or for patterning of multiple 3D ECM phases [12,13]. Thus, microfluidics methods have thus been developed to precisely control most facets of 3D microenvironments, including the structure and composition of 3D ECM, delivery and removal of soluble factors, and application of forces (fluid shear stress); such capabilities are highly useful for both engineering tissue scaffolds for regenerative medicine applications or engineering 3D cellular microenvironments for basic biological studies.

In this chapter, we highlight several different ways that microfluidics can be utilized in engineering 3D tissues and cellular microenvironments, with particular emphasis on methods developed in our laboratory. In Section 3.2, we discuss methods to microfabricate 3D tissue scaffolds using microfluidics. In Section 3.3, we discuss methods to assemble and dynamically culture 3D microenvironments within PDMS-based microfluidic devices. In Section 3.4, we discuss methods which incorporate microfluidic channels directly within engineered 3D tissue scaffolds, for both perfusion of the tissue construct with media and assembly of multiphase 3D tissues.

3.2 Fabricating 3D tissue scaffolds using microfluidics

The inherent spatial and temporal controls over delivery of solutions in microfluidics provide an excellent means for precisely controlling the composition and structure of 3D tissue scaffolds. By flowing gel precursor solutions into microfluidic devices and gelling *in situ*, the geometry of the resulting hydrogel or ECM scaffold can be defined precisely by the structure of microfluidic chambers. Further, by combining microfluidic casting with spatial control over polymerization (e.g., using light exposure through photomasks or laser patterning for photopolymerizable materials), fabrication of complex composite scaffold consisting of multiple materials or compositions can be achieved. Our laboratory has developed methods for microfluidics-based fabrication of both natural ECM gels (such as collagen) and photopolymerizable hydrogels (such as PEG–DA).

3.2.1 Microfluidic casting for engineering the microstructure of natural ECM scaffolds

Microfluidic casting offers a simple method for microfabricating tissue scaffolds consisting of natural ECM components such as collagen. ECM precursor solutions can be flowed into casting chambers which define the geometry of the scaffold with high precision, and then gelled *in situ* (e.g., by increasing temperature to induce ECM self-assembly). In contrast to other techniques to control scaffold microstructure such as lyophilization, the mild conditions for casting in microfluidic chambers permit the encapsulation of cells within the scaffold during the casting step and the collagen fibers also retain their native fibrillar structure. Scaffolds can be removed from the casting chamber and seeded with cells or integrated with additional tissue phases. Alternatively, the contraction of natural ECM by encapsulated cells can create space to permit patterning of subsequent cell–ECM phases; this approach has been used to mimic the spatial organization of cells and ECM of arterial walls [14].

Our group has used microfluidics to fabricate engineered skin constructs with precise control over the outer geometry, pore size, and void fraction of collagen–glycosaminoglycan (GAG) scaffolds [3] (Figure 3.1). Control over cell invasion into engineered skin grafts is a critical parameter for inducing proper

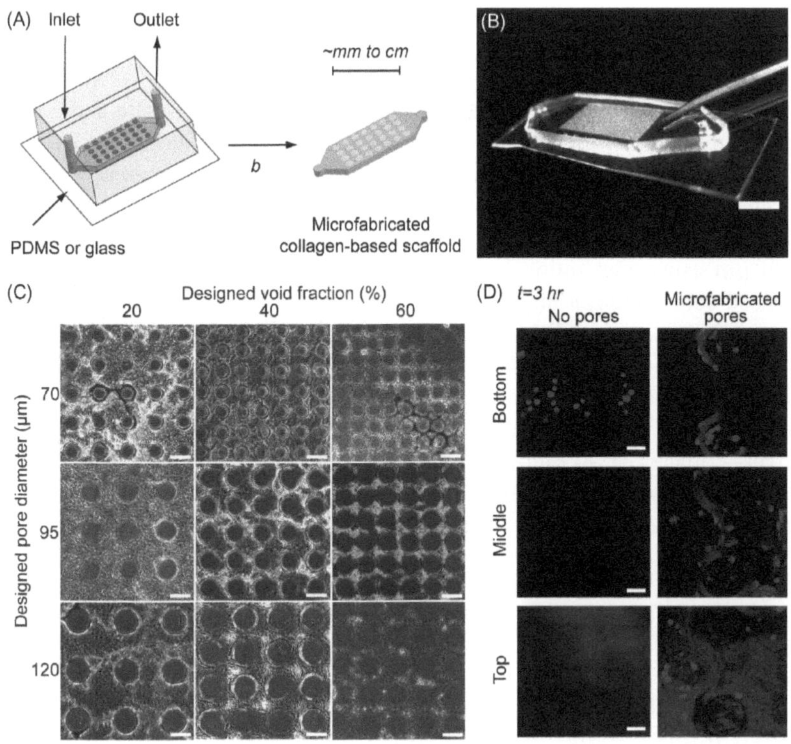

FIGURE 3.1

(A) Schematic diagram of microfluidic device used to fabricate porous collagen–GAG dermal scaffolds. The chamber dimensions define the outer scaffold geometry and an array of pillars defines the pore size and pore density. (B) Image of the PDMS device. (C) Phase contrast images of dehydrated collagen–GAG scaffolds after removal from the casting chamber. (D) Migration of fibroblasts in porous and nonporous collagen–GAG scaffolds. A fibroblast-seeded layer of Matrigel was integrated on the bottom of the collagen–GAG scaffold, and migration of the fluorescent fibroblasts was observed for 5 days. Confocal images show slices of the construct at the Matrigel–collagen–GAG interface (bottom), in the collagen–GAG layer (middle), and at the interface of the collagen–GAG layer and PDMS chamber. Images were recorded 3 h following construct assembly. Nonporous control scaffolds are shown on the left, with the porous scaffold shown on the right. Scale bar is 50 μm.

healing and integration of the graft. Commonly used techniques such as lyophilization can control average pore size, but often the scaffolds have a large pore distribution that is suboptimal for precisely controlling cellular infiltration. In this work, we designed a microfluidic chamber to define the overall scaffold geometry (thickness and lateral dimensions) and the pore size and void fraction using an array of posts (Figure 3.2A, B). To form the scaffolds, collagen and chondroitin-6-sulfate (a GAG commonly used in engineered skin constructs) solutions were mixed and held on ice, then flowed into the chamber at cold temperature and gelled *in situ* by increasing the temperature to 37°C for 1 h. The constructs could then be removed from the chamber, dehydrated, and attached to a thin PDMS membrane to aid handling.

We fabricated a range of pore sizes (70–125 μm) and void fractions (20–60%) hypothesized to be optimal for cellular ingrowth (Figure 3.2C). We found that constructs with void fractions 70% or larger were fragile and could not be easily removed from the chamber, and the pores smaller than 30 μm were difficult to produce. We used image analysis of segmented microscopy images to precisely measure pore sizes across conditions and found that this technique produced highly uniform pores with standard deviations of ±3–4 μm, compared to ±30 μm using conventional lyophilization. To test how the presence of pores influenced cell migration through the scaffold in an *in vitro* model, we integrated a layer of fluorescently labeled fibroblasts seeded in Matrigel on the open surface of the collagen–GAG scaffold (Figure 3.2D). Confocal image slices across the constructs indicated that fibroblasts could rapidly traverse the porous scaffolds within 3 h, while in control scaffolds without pores the fibroblasts remained at the Matrigel–collagen–GAG interface.

Overall, this technique demonstrated a simple way in which microfluidics can be used to precisely control the structure of ECM-based tissue scaffolds. Precision control of scaffold structures which guide cellular migration behavior could eventually lead to improved functioning and integration of tissue scaffolds and biomaterials for wound-healing applications.

3.2.2 Fabrication of composite photopolymerizable hydrogel microstructures using microfluidics

By coupling photopolymerizable hydrogels with microfluidic casting, enhanced spatial control of scaffold structure and composition can be achieved. In contrast to the previous examples in which the gel precursor polymerizes throughout the entire casting chamber, spatial patterning of gel polymerization using light (via photomasks or laser patterning) can be used to fabricate composite microstructures through sequential delivery and polymerization within the chamber.

Our group has developed a technique to efficiently assemble surface-bound composite microstructures consisting of photopolymerizable hydrogels using microfluidics (Figure 3.2) [4]. The experimental setup consists of a PDMS chamber with microfluidic inlets and outlets that can be mounted on a microscope

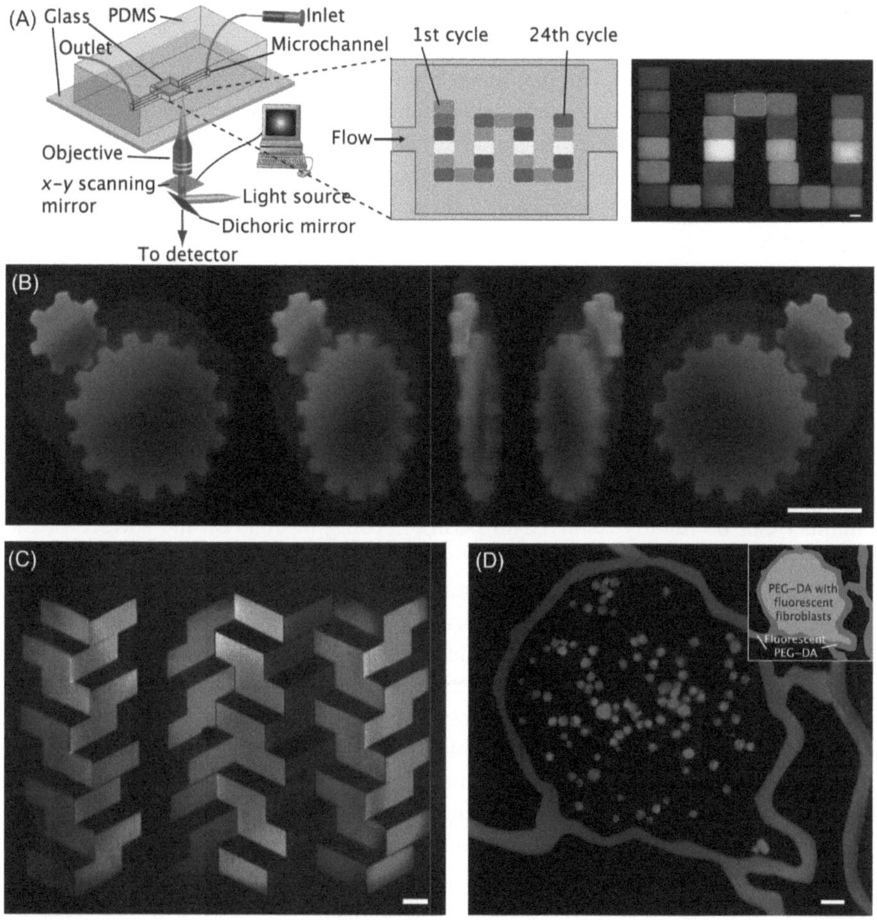

FIGURE 3.2

(A) Schematic diagram of the microfluidic device used in fabrication of 3D composite microstructures and an example of 3D microstructures made from 24-cycle fabrication. A microfluidic device is placed onto the stage of an inverted fluorescence microscope (shown is the setup for a confocal microscope; for an epifluorescence microscope, an automated *x–y* stage is used instead of an *x–y* scanning mirror). Photopolymerizable reagents or washes (with PBS or ethanol) are delivered into the chamber by injection into the inlet. Each cycle consists of exposure (1–2 s), washing of unpolymerized prepolymer (1 min), injection of prepolymer for subsequent cycle (12 s), and alignment (less than 1 min, including the time needed for manual fine adjustment for a conventional fluorescence microscope). Here, the microstructures consist of 40% w/v of PEG–DA, with 2% w/v of Irgacure 2959 as photoinitiator and a combination of fluorescent dyes (scale bar = 200 μm). (B, C) Fabrication of 3D composite microstructures using an argon laser on a confocal laser scanning microscope. (B) 3D images from confocal microscopy

(*Continued*)

stage (Figure 3.2A). The chamber is enclosed with a glass coverslip that is treated with 3-(trimethoxysilyl) propyl methacrylate to enhance bonding of photopolymerizable acrylate modified poylmers such as poly(ethylene glycol)-diacrylate (PEG–DA). The microscope serves as a light source for photopolymerization and is also used to visualize the fabricated structures. We used both inverted fluorescence microscopes and confocal microscopes to pattern different types of features. For example, we used a rectangular field diaphragm on an inverted fluorescence microscope to expose a selected region of the objective field of view, and used a motorized stage to adjust the exposure position over 24 cycles of photopolymerization (Figure 3.2A right panel). Each colored rectangle shown consists of PEG–DA doped with different ratios of colored dyes. Alternatively, using region-of-interest scanning on a confocal microscope, more arbitrary shapes could be fabricated, such as the gear and staircase shapes shown in Figure 3.2B, C. The resolution of features using confocal microscopy was approximately 3 μm. The mild fabrication conditions can support encapsulation of cells within the 3D hydrogel microstructures with over 95% viability (Figure 3.2D). Further, the chamber can be readily removed to permit seeding of cells on the surface of the microstructures [15].

We identified three technical parameters that were critical for accomplishing consistent and rapid fabrication: (i) washing flow rates must be low enough and use at least 1 ml of solution for sufficient washing of unpolymerized prepolymer without shearing off the polymerized structures, (ii) concentration of photoinitiator should be high in order to minimize polymerization time, decrease diffusion of radicals, and increase the spatial fidelity and resolution, and (iii) focusing of the objective lens within the chamber needs to be accurate, which can be assisted by flowing a fluorescent dye into the chamber. Potential limitations of the method include: difficulty of fabricating thick structures (greater than hundreds of microns) with high spatial fidelity due to light scattering; potentially long fabrication time for large structures (several millimeters) of many different types of materials (hundreds); and limitations in choice of materials, which ideally share compatible sets of solvents (to avoid swelling and collapsing of polymerized structures), and exhibit sufficiently low viscosity (less than tens of centipoise) for reliable microfluidic flow.

◀ of a gear design fabricated from PEG–DA prepolymer doped with fluorescein (green) and rhodamine B (red). Total fabrication time was 9 min. (C) Merged fluorescence image (red, blue, and green channels) of a 12 by 8 array of parallelograms, fabricated via three lithographic cycles, over a large surface area (1 mm^2). Total fabrication time was 24 min. (D) Fluorescence image of mammalian cells inside a 3D composite microstructure. The patterned microstructure consists of PEG–DA that encapsulated 3T3 fibroblasts (labeled with CellTracker orange) interlinked with a vessel-like acellular hydrogel network (PEG–DA labeled with rhodamine B). (For interpretation of the references to color in this figure legend, the reader is referred to the web version of this book.)

One significant application of this technique is for studying cell behavior in inhomogeneous and anisotropic 3D environments that mimic the complexity of biological microenvironments. For example, we have used this method to construct composite 3D PEG–fibrinogen hydrogels with distinct stiffness patterns for studying how cells migrate in response to discrete stiffness changes [15]. Since the stiffness of the surrounding environment dictates numerous cell behaviors including migration and differentiation, the ability to pattern materials with varying mechanical properties can be useful for studying such cellular processes or potentially to pattern engineered tissue formation.

In the examples in this section, microfluidics is primarily used to aid in fabrication of the 3D environments and scaffolds, which are then to be removed from the microfluidic device and used for their intended application. Microfluidic devices can also be highly useful for dynamically culturing arrays of 3D microenvironments for biological studies. In the next section, we describe a method which incorporates 3D ECMs in a PDMS-based microfluidic device that uses pneumatic valves to control fluid flow to individually addressable culture chambers.

3.3 Dynamic 3D cell cultures within PDMS microfluidic devices

One of the primary uses for microfluidic cell culture devices is to control fluid flow and the soluble factor environment surrounding cells in the device. The ability to create large arrays for testing many different conditions in parallel using minimal reagent volumes are significant advantages of microfluidic platforms. Such setups could involve an array of cell-seeded chambers with inlets that can be interfaced with gradient generators [16] or a system of valves which permits chambers to be individually addressed [17].

Our laboratory has developed a microfluidic device that dynamically controls the soluble factor microenvironments of an array of individually addressable culture chambers filled with 3D cell-seeded ECMs (Figure 3.3) [6]. We incorporated pneumatically actuated valves between the culture chambers to control fluid flow and permit or restrict soluble factor exchange across adjacent cell chambers (Figure 3.3A). We cast the 3D hydrogels directly within the culture chambers by flowing a thermally gelling precursor solution (collagen or Matrigel in this study) into the device under cold temperature, sequentially filling the chambers with different cell–ECM combinations by actuating different valves. One critical issue in culturing 3D gels within microfluidic devices is to permit fabrication of the gels *in situ* while keeping fluidic access ports open for delivery of culture media. To accomplish this, we used pneumatic actuation of a thin ring-shaped PDMS membrane under high pressure to form an open microchannel in the 3D hydrogel during the casting step (Figure 3.3B–D). After gelling, we reduce the pressure to open the fluid flow path. Another critical issue is stably attaching the 3D matrix to

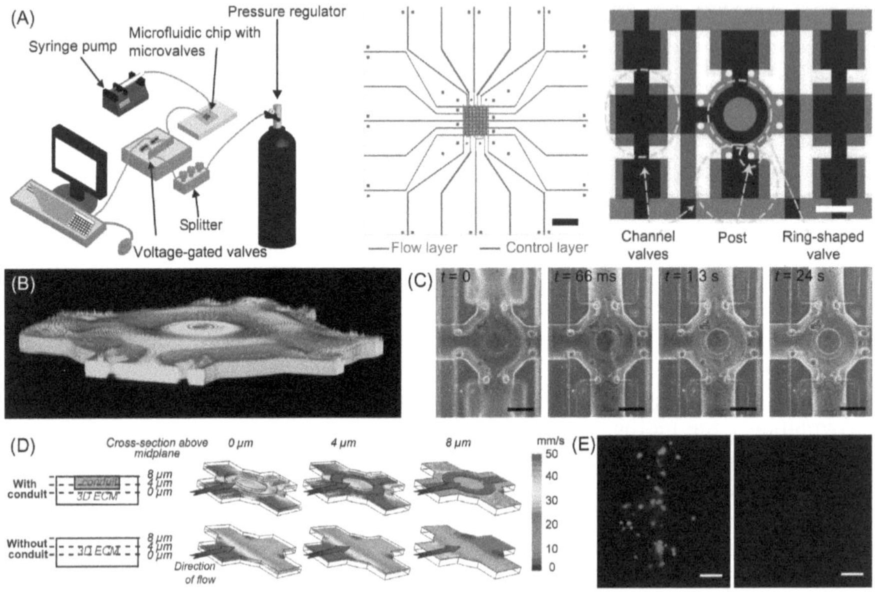

FIGURE 3.3

(A) Schematic diagram of the experimental setup and chip design for a dynamic 3D cell culture device. A two-layered PDMS chip is connected to a regulated gas tank via voltage-gated valves. For media perfusion through all microchambers, a syringe pump is connected to the chip. For injection of reagents or cell–matrix mixtures into specific microchambers, the path of the fluid flow is controlled through the switching of voltage-gated valves. The design of the microfluidic chip is shown by an overlay of both the control and the flow layers for the whole chip (middle) and a single chamber (right). (B) 3D reconstruction of a confocal image of a chamber loaded with 3D Matrigel, with visible conduits above the 3D gel created by partial deflection of a ring-shaped valve positioned above the chamber. (C) Time-lapse brightfield images showing perfusion of water through a single chamber loaded with a 3D Matrigel and red food coloring dye (some debris remained lodged in the gel). When water flows into the chamber, the red dye is washed away in the conduit (as shown at $t = 66$ ms). The remaining dye in other regions slowly washes away and is completely cleared after 24 s. (D) Flow velocities across the microchambers with and without a 6 μm tall conduit. Vertical cross sections of the channel indicate the position of the three horizontal cross-sectional planes relative to the 3D Matrigel and open conduit, in chambers with and without the conduit. The images on the right show the local velocity distribution for the cross-sectional planes at 8, 4, and 0 μm above the midplane with a pressure drop of 20 kPa between the inlets and the outlets. Chambers with conduits exhibit flow rates of about fivefold higher and more uniform velocity distribution than chambers without conduits. (E) Fluorescence images of stem cells expressing GFP under OCT4 promoter indicating that the cells express GFP (remaining undifferentiated) and have uptaken CellTracker blue that diffused from an adjacent chamber. Scale bar is 50 μm. (For interpretation of the references to color in this figure legend, the reader is referred to the web version of this book.)

the device to prevent distortion or detachment from the device under forces of fluid flow. Here, we incorporated a set of microposts at the edges of the chamber to help secure the gel in place (shown in Figure 3.3A). A 3D reconstruction of confocal images confirmed the gel within the chamber maintained the fluid conduit after casting and fully filled the chamber around the microposts (Figure 3.3B).

The open channel gel design enabled rapid exchange of soluble factors (within seconds) in the individual 3D gels (Figure 3.3C, D). The rapid perfusion enabled encapsulated mouse embryonic stem cells (mESC) expressing green fluorescent protein (GFP) under the OCT4 promoter to maintain an undifferentiated state (as indicated by GFP expression) for up to 5 days culture when leukemia inhibitory factor was included in the media (Figure 3.3E left). Further, by selectively opening or closing valves between adjacent chambers, communication between cells in adjacent chambers via soluble factors can be permitted or restricted. For example, we demonstrated that keeping the valves closed between chambers prevents diffusion of a small molecule fluorescent cell tracking dye (Celltracker blue), as mESC only fluoresced blue after opening the valve, indicating diffusion of the dye between chambers (Figure 3.3E right).

Overall, this microfluidic platform enables dynamic manipulation of the soluble factor microenvironment and cell–cell paracrine communication across an array of 3D microenvironments with specific cell and ECM types in each chamber. Cells in each environment can be quantitatively and noninvasively monitored through microscopy or analysis of media extracted from individual chambers. The ability to dynamically control and observe 3D microenvironments could reveal new insights into the temporal aspects of processes such as differentiation, proliferation, migration, and tumorigenesis in response to dynamic environmental changes. In the future, the development of different 3D organoids within the chambers could be used to mimic interconnected tissue and organ systems.

While the method described in this section incorporates microfluidic channels in hydrogels to enhance perfusion of the 3D microenvironment, PDMS-based channels and valves are still needed to regulate flow and diffusive transport throughout the array. In the next section, we discuss methods which construct microfluidic devices from all hydrogel and ECM materials, with a focus on the use of microfluidic hydrogels for patterning 3D multiphase tissues.

3.4 Hydrogel-based microfluidic culture devices and tissue scaffolds

The recent development of microfabrication processes for hydrogel and ECM biomaterials has allowed unprecedented control of engineered tissue structure and composition at the microscale [7,9,10,18,19]. Such techniques can allow fabrication of multiphase tissue constructs with precision control over the geometry and composition of integrated 3D ECMs. Control over ECM microstructure and composition can strongly influence cellular responses such as differentiation and tissue morphogenesis. For example, control over ECM surface topology has been used to mimic skin rete ridges and enhance engineered skin epithelialization [20]. Open microstructures such as pores and microfluidic channels can be used to enhance media perfusion [7,8,10,11] and cellular migration through engineered tissue constructs [3,21].

The incorporation of microfluidic channels directly in hydrogel and ECM scaffolds has proven highly useful for engineering *in vitro* microenvironments and complex tissue constructs. Microfluidic channels have been successfully fabricated in a broad range of natural and synthetic hydrogel/ECM materials including agarose [10], alginate [7,8], collagen [9,22], gelatin [11], fibrin [9], silk [23], and PEG [21]. Hydrogel-based microfluidic devices, unlike PDMS, permit transport of water and soluble factors and are thus well suited for 3D cell culture. Applications of microfluidic hydrogels include controlling gradients of soluble factors across 2D/3D cell cultures [24], perfusing engineered tissue scaffolds to enhance viability and tissue function [8,25], and patterning 3D ECM to fabricate integrated multiphase tissues [12,13].

Microfluidic patterning of 3D ECMs within microfluidic tissue scaffolds is one way to precisely control local cell and ECM properties in a multiphase tissue construct. Instead of perfusing microchannels with media, ECM precursor solutions are flowed into the channels and gelled *in situ* to form distinct tissue phases. Importantly, such patterned 3D ECMs must be stably integrated to prevent mechanical separation due to cell contractile forces or handling. Our group has developed a method for tissue patterning that uses collagen fibers to integrate distinct tissue phases, which we review in the following sections.

3.4.1 Assembly of natural ECM multiphase tissues using microfluidics

Native tissues often consist of multiple types of cells and ECMs that are integrated together in microscale spatial patterns (e.g., the vascular and parenchymal compartments of a tissue). Our group has developed microfluidic hydrogels for patterning multiphase tissues consisting of fibrous natural ECMs [12,13]. In this method, a microfluidic hydrogel is perfused with ECM precursor solutions which are then gelled *in situ* to form pattered 3D ECM phases integrated within the microfluidic hydrogel bulk phase (Figure 3.4).

Importantly, because many cell types exert contractile forces on fibrous ECMs such as type I collagen, the interfaces between tissue phases must be integrated together to prevent separation by cell contraction. To integrate the interfaces of microfluidically patterned ECMs with the bulk phase, we form a continuous network of type I collagen fibers across the interfaces by the nucleation and

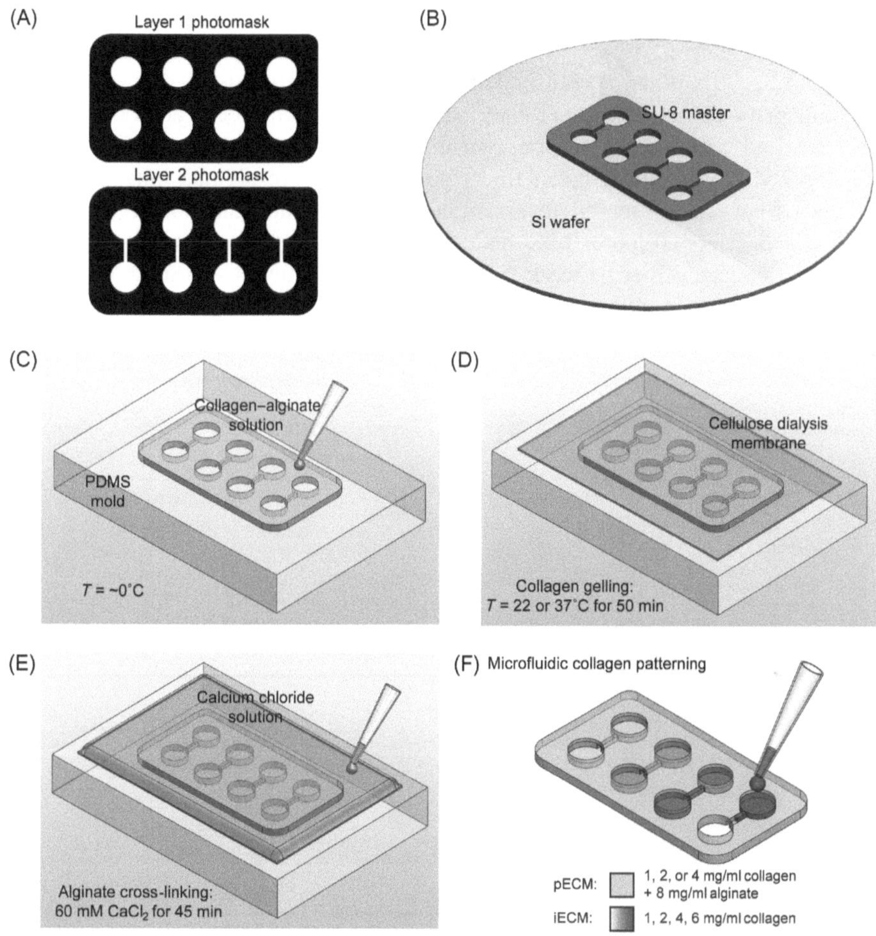

FIGURE 3.4

Schematic of microfluidic hydrogel fabrication. (A) Two photomasks are used to define the microfluidic hydrogel features: the first layer defines the hydrogel outer boundaries and inlets and the second mask defined the geometry of the microchannels joining the inlets. (B) A master mold for the microfluidic hydrogel is created in SU-8 on a silicon wafer. (C) Cold collagen–alginate solutions are pipetted into the molds while maintaining cold temperature. (D) A cellulose dialysis membrane is flattened over the top of the mold, and the temperature is increased to gel the collagen for 50 min. (E) A 60 mM $CaCl_2$ supplemented HEPES buffer solution is pipetted on top of the membrane to cross-link the alginate solution for 45 min. (F) The hydrogel is then transferred onto a glass coverslip, and then collagen solutions are flowed into the channels and gelled *in situ*.

assembly of new collagen fibers in the integrated ECM (iECM) from preformed collagen fibers in the preformed ECM (pECM) bulk phase (Figure 3.5A, B). When a cold collagen precursor solution is placed adjacent to the collagen–alginate hydrogel interface, new collagen fibers are observed to nucleate and assemble from the preformed fibers upon temperature increase (Figure 3.5C). Without collagen fibers present at a pure alginate interface, new collagen fibers are observed to assemble in the channel and not preferentially from the interface (Figure 3.5C bottom). We similarly observed formation of integrated collagen-fiber networks when other ECMs such as fibrin and Matrigel were patterned along with collagen.

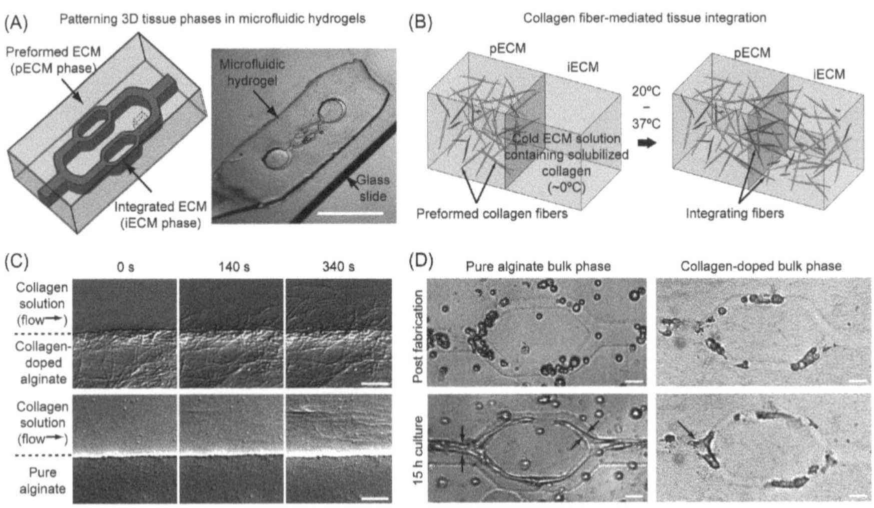

FIGURE 3.5

Microfluidic patterning and collagen-fiber integration method for fabricating integrated multiphase tissue constructs. (A) Left: schematic image showing the preformed microfluidic bulk hydrogel phase (pECM) and the patterned iECM phase. The boxed region indicates the magnified interface shown in panel (B). Right: a stereoscope image of a microfluidic hydrogel prior to ECM patterning. (B) Schematic image of collagen-fiber mediated tissue integration. (C) Time series of collagen-fiber assembly at the phase interface. Collagen fibers in the iECM assemble from collagen fibers at the pECM interface into the polymerizing iECM solution (upper panel), but do not nucleate from a pure alginate pECM phase interface (lower panel). Scale bars are 10 μm. (D) Stabilization of iECM against cellular contractile forces. Left: images showing separation of patterned EC-seeded collagen from a pure alginate pECM interface due to cell contraction after a few hours in culture (arrows in bottom panel indicate boundaries of contracted collagen). Right: images showing a stable interface between patterned EC-seeded collagen and a collagen-doped alginate pECM after the same culture period. ECs migrated within the patterned collagen and formed multicellular structures (arrow) within the channel area. Scale bars are 50 μm.

The creation of integrated collagen-fiber networks across the interface permitted stable long-term 3D culture of microfluidically patterned cells (Figure 3.5D). When endothelial cell (EC)-seeded collagen was patterned in the absence of preformed collagen fibers in a pure alginate microfluidic gel, rapid contraction of the collagen by cells resulted in separation of the two phases (Figure 3.5D left). By contrast, when integrated collagen-fiber networks were created across the interface, the patterned ECM remained stable and supported cell migration and networking behavior of ECs undergoing *in vitro* capillary formation (Figure 3.5D right). Importantly, the presence of cross-linked alginate at the boundary restricted migration of endothelial cells (due to the small pore size of cross-linked alginate) to the microchannel-defined patterns. However, since alginate ionic cross-links can be removed using a chelator, the ability for the alginate to permit or restrict cell migration can be reversibly switched; this property can be utilized to investigate cellular responses to dynamic spatial constraints, for example [26].

Fibrillar ECM proteins (such as type I collagen) impart structural, mechanical, and signaling functions to tissues, and the microstructure (size, density, and orientation) of collagen and other ECM fibers varies among different tissues and at tissue interfaces. Thus, engineering ECM fiber microstructures can be useful for controlling local tissue development in integrated multiphase tissues and for studying how transitions in ECM fiber structure influence cell behaviors. The microstructure of collagen fibers can be varied over a wide range by altering the conditions for self-assembly. For example, by varying the concentration and gelling temperature of collagen solutions patterned within collagen–alginate microfluidic channels, we fabricated tissue interfaces with a wide range of transitions in fiber structures (Figure 3.6). We also found that the collagen assembly conditions could influence the bonding strength of collagen–alginate slabs glued together using the collagen-fiber integration method [13].

Cell morphogenesis processes, such as endothelial capillary formation, can be highly dependent on the ECM fiber microstructure and mechanical properties. Thus, the structure of ECM fibers throughout an integrated multiphase engineered tissue must be optimized to properly support tissue morphogenesis across the individual phases. By applying this method as an example in the next section, we describe the optimization of collagen-fiber structure to support endothelial capillary morphogenesis within stably integrated patterned 3D collagen ECMs.

3.4.2 Optimization of collagen-fiber interfacing for directing capillary morphogenesis within microfabricated tissue constructs

As the size scale of microfluidic channels corresponds to that of native microvascular networks, a potentially significant use of microfluidics in tissue engineering is to control the structure of microvascular networks in engineered tissues. In this section, we present an application of the microfluidic patterning and collagen-based interfacing techniques toward the development of prevascularized tissue constructs. Numerous groups have investigated using microfluidic scaffolds for

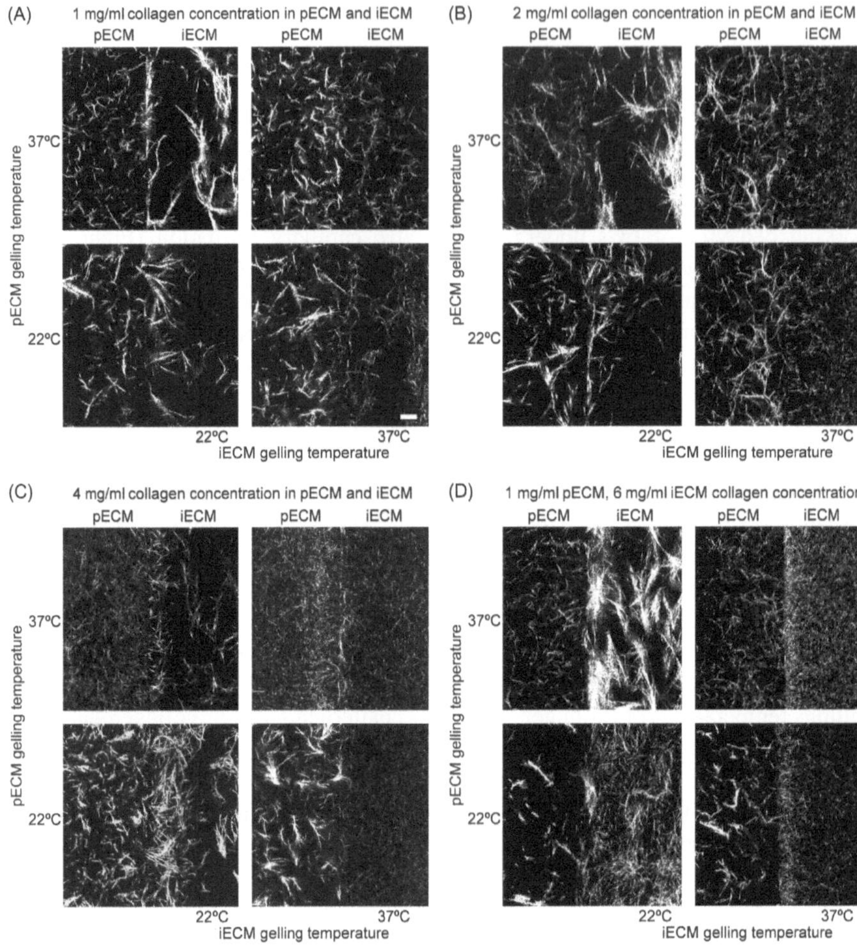

FIGURE 3.6

Representative images of interface structure for: (A) 1 mg/ml collagen in both pECM and iECM, (B) 2 mg/ml collagen in both pECM and iECM, (C) 4 mg/ml collagen in both pECM and iECM, and (D) 1 mg/ml collagen in pECM and 6 mg/ml collagen in iECM for all combinations of gelling temperature for each phase. Scale bar is 10 μm. Note that each image is one slice (0.25 μm thick) out of 41 images in the *z*-stack. A single fiber typically crosses multiple *z*-slices.

engineering vascular network geometry *in vitro* [9,27–29]. Previous methods to microfabricate microvascular networks have focused on coating the walls of microchannels with ECs [9,29]. Such configurations are able to investigate the behavior of a monolayer of ECs in response to fluid flow within the artificial lumen, and can be used to assess many physiologic functions such as inflammatory responses and vessel permeability [30,31]. However, endothelial lumenogenesis in 3D environments (*in vitro* and *in vivo*) proceeds by the formation and intercellular fusion of intracellular vacuoles [32]. Numerous studies have employed this native mechanism to form "prevascularized" tissues that have the capacity to integrate with host microvascular networks and become rapidly perfused postimplantation [33–37]. Thus, by combining microfluidic spatial patterning with 3D capillary morphogenesis in microfabricated tissue constructs, we hypothesize that the development of prevascular networks with the capacity to form anastomoses with host vasculature could be directed in computer-designed patterns. Enhanced control over the structure of embedded microvascular networks may lead to improvements in the survival, integration, and function of engineered tissue grafts, as the varying structure of microvascular networks across different tissue types are intimately related with tissue-specific functions (e.g., muscle microvascular networks contain long parallel segments running along muscle fibers, while adipose and skin microvessels are more reticular).

In an effort to create computer-designed vessel networks formed using native 3D capillary morphogenesis, we utilized the microfluidic tissue patterning and collagen interfacing method to spatially control the *in vitro* development of microvessel structures in engineered tissue constructs [38]. Endothelial capillary formation *in vitro* is highly dependent on ECM properties, including collagen concentration, mechanical properties, and gel boundary conditions (i.e., free floating or constrained) [39–41]. Thus, in order to support physiologic formation of endothelial tube structures within micropatterned 3D collagen matrices, a stable interface and structurally/mechanically appropriate microenvironment must be created for encapsulated cells. To optimize collagen assembly parameters for interface stability and EC functions, we used a two-step, two-temperature gelling scheme [42] to engineer the fiber structure of patterned collagen matrices. We then further developed the microfluidic patterning method to create stably integrated multilayer scaffolds with fully enclosed channels.

We first aimed to optimize the structure and composition of collagen in the microchannels for interface stability and EC capillary formation. Using multichannel microfluidic hydrogels (e.g., as shown in Figure 3.4 [13]), we patterned different concentrations of collagen (4 and 6 mg/ml) seeded with endothelial cells (1 or 2 million per ml) within the channels, and applied a two-step, two-temperature gelling scheme to modulate collagen-fiber structure of the patterned matrix [42]. Specifically, we initiated collagen assembly at room temperature for varying time periods (0, 2, 5, 8, or 10 min) at $\sim 22°C$, then completed assembly at 37°C for 40 min to vary the fiber structure of the patterned matrix. To control collagen assembly as consistently as possible, we ensured temperatures were maintained cold throughout the microfluidic patterning process by first cooling the gels in a refrigerator and maintaining them on a cold plate while flowing collagen. Since the viscosity of collagen increases rapidly upon nucleation, we wanted to pattern the collagen as quickly as possible to allow easy flow of the collagen through the microchannels. Because some collagen nucleation occurs even at very low temperatures [43], collagen solutions were mixed immediately prior to patterning, always maintained cold, and freshly prepared for each construct.

We found that when we rapidly transferred the samples to the incubator (0 min at room temperature), the collagen appeared grainy with few discernible fibers (Figure 3.7A). Previous studies of microfluidically patterned collagen within PDMS-based microchannels have observed similar structural results under 37°C gelling conditions [43]. This work suggested that when collagen is gelled in microscale channels, the large surface-to-volume ratio results in much faster heating of the collagen than in typical bulk preparations. Consequently, it was posed that such a rapid increase in temperature may not allow adequate nucleation time for collagen assembly, resulting in a nonfibrous matrix. In our experiments, although the collagen and cells initially filled the channels, it appeared that the matrix was rapidly torn apart by the cells. This resulted regions devoid of collagen throughout the channels, with clusters of cells and collagen scattered throughout (Figure 3.7B). Under these conditions, the EC could not form capillaries, since there was no stable matrix through which they could migrate and network.

At the other extreme, at 10 min room temperature incubation, collagen fibers became quite large (similar to fibers assembled entirely ~22°C) (Figure 3.7C). Under these conditions, endothelial cells were unable to contract the matrix and generate open spaces as was the case for immediate 37°C gelling, but the ECs did not form tube networks, and remained mostly rounded. It is possible that the larger fibers may have been too rigid for the ECs to manipulate, as seen in previous studies using glioma cells [42].

We qualitatively observed a range of intermediate fiber structures using intermediate ~22°C incubation times (Figure 3.8). We found that 2 and 5 min room temperature nucleation resulted in fairly similar matrix structures (Figure 3.8A, B). Under these conditions, we observed stable interfaces of the surrounding collagen, compaction of the collagen fibers by ECs, migration, networking, and vacuole formation. At 8 min (Figure 3.8C), the fibers began to become large and sparse. Under these conditions, we observed more rapid cell spreading in some cells (likely due to the more open mesh), while others remained rounded. ECs did not effectively form capillary structures, similar to 10 min ~22°C incubation. The nature of cell–cell and cell–ECM interactions and cell behavior were thus modulated by fine tuning the structure of patterned collagen matrices.

In order to observe the cells and collagen at high resolution (100× 1.43 NA oil immersion objective), we had sealed the bottom of the channels were sealed by a cover glass. However, the cover glass did not support stable bonding of the patterned collagen, and we observed considerable contraction of the matrix away from the surface (Figure 3.9). After 24 h, the EC had typically contracted the collagen up to ~45 μm (about 40% of the channel height) away from the cover glass, and the cells were localized near the top of the channel (Figure 3.9 bottom row). Over several days in culture, this collagen continued to collapse, and EC tube structures regressed. Although significant contraction from the free channel

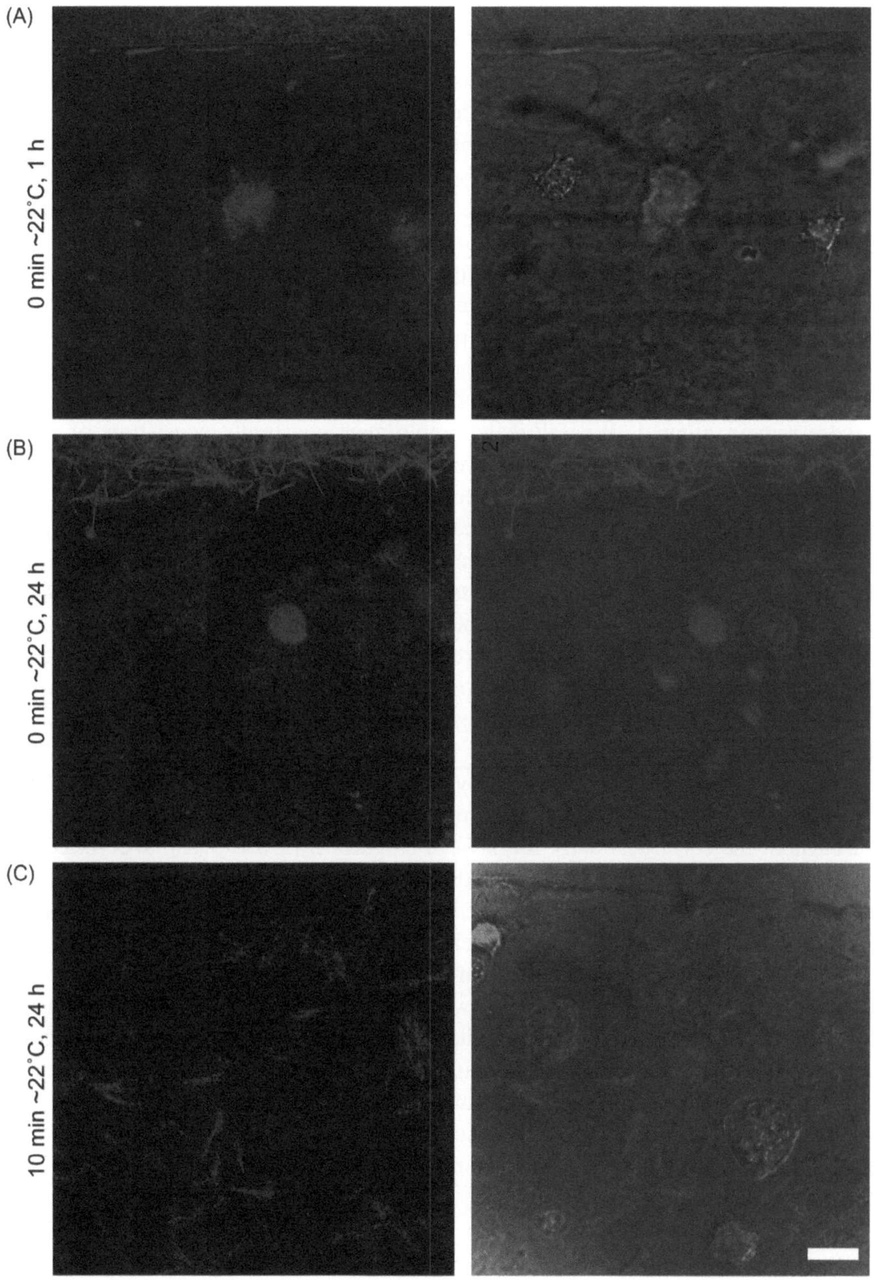

FIGURE 3.7

Patterned collagen structure under immediate 37°C or 10 min ~22°C assembly conditions. (A, B) Collagen structure for immediate 37°C assembly at 1 and 24 h after

(Continued)

surface did not prevent capillary structure formation in our previous work, in those experiments the channels had smaller dimensions and were sealed against tissue culture polystyrene; these factors could potentially limit the extent of contraction. To promote enhanced long-term vascular morphogenesis within the channels, we aimed to securely interface the collagen on all surfaces of the channel by stacking multiple collagen–alginate layers together.

We fabricated multilayer microfluidic hydrogel constructs by first aligning and stacking gels on a PDMS support structure prior to patterning collagen in the channels (Figure 3.10). The PDMS supports were molded from a micromachined acrylic master (Figure 3.10E, F). The supports used a "button" mechanism to snap the constructs in place through two slots on each end, which prevents the constructs from floating away from each other before sealing the layers together (Figure 3.10A–D). Guiderails along the sides of the supports ensured that the layers remain aligned. We fabricated multilayer constructs consisting of three layers: a flat base layer and two layers containing a set of six microchannels (2.5 mm in length, and either 100 μm × 100 μm or 200 μm × 200 μm W × H). The individual layers were approximately 250 μm thick. In this configuration, the channel walls are completely surrounded by collagen–alginate, and thus should support collagen-fiber interfacing on all channel surfaces. Two autoclaved PDMS supports were adhered to a 60 mm petri dish and were plasma treated to make them hydrophilic. Gels were then manually stacked and fastened to the supports under an inspection microscope within the culture hood. After stacking, we flowed the cell-seeded collagen through both layers by pipetting into one inlet.

Under optimized collagen assembly conditions (5 mg/ml collagen with 5 min room temperature nucleation), we followed the development of capillary structures throughout the constructs daily for 1 week (Figures 3.11 and 3.12). We varied the density of human umbilical vein endothelial cells (HUVEC) (1 or 2 million cells/ml) to examine how initial cell density influenced the extent of capillary formation in the constructs with different channel sizes. In addition to endothelial cells, we coencapsulated adult adipose-derived mesenchymal stem cells (MSCs) within the patterned matrix. MSCs have been shown to increase the extent of anastomosis and to stabilize functional engineered capillaries *in vivo* [35]. For EC–MSC coculture experiments, we labeled cells with orange and green live cell fluorescent dyes so they could be tracked by fluorescence microscopy.

Each day, we placed the constructs in a microscope incubation chamber and took an array of image stacks using a motorized stage (35 *X*–*Y* sites by 25 *Z*-planes). The immobilization of the constructs on the PDMS supports greatly

◀ completion of gelling (note the images are from different constructs from the same condition). Collagen reflectance images are in the left column and are overlaid with brightfield images in the right column to highlight cell and channel (top of images) boundaries. Collagen concentration is 6 mg/ml. Scale bar is 20 μm.

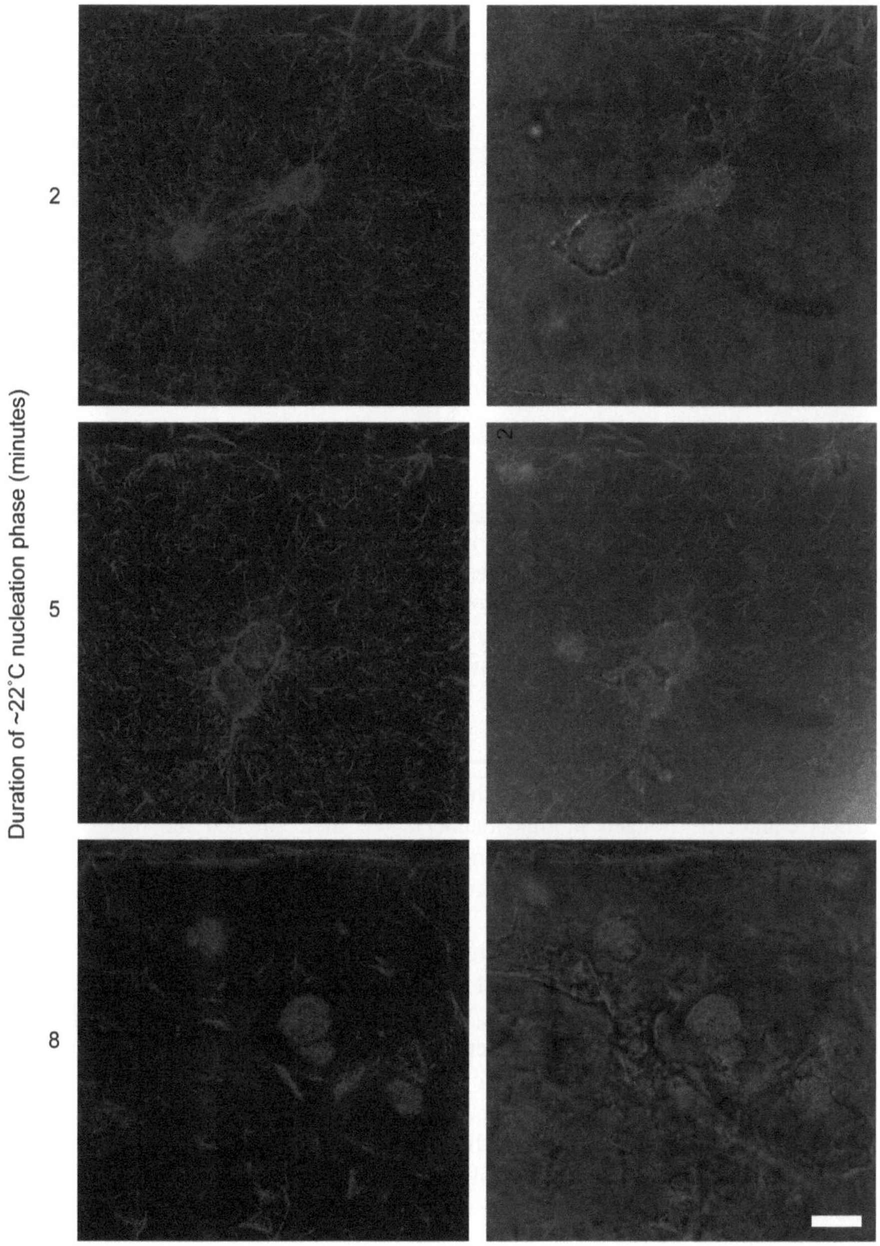

FIGURE 3.8

Two-step, two-temperature assembly [42] to engineer the structure of patterned collagen matrices. Collagen reflectance images are in the left column and are overlaid with

(*Continued*)

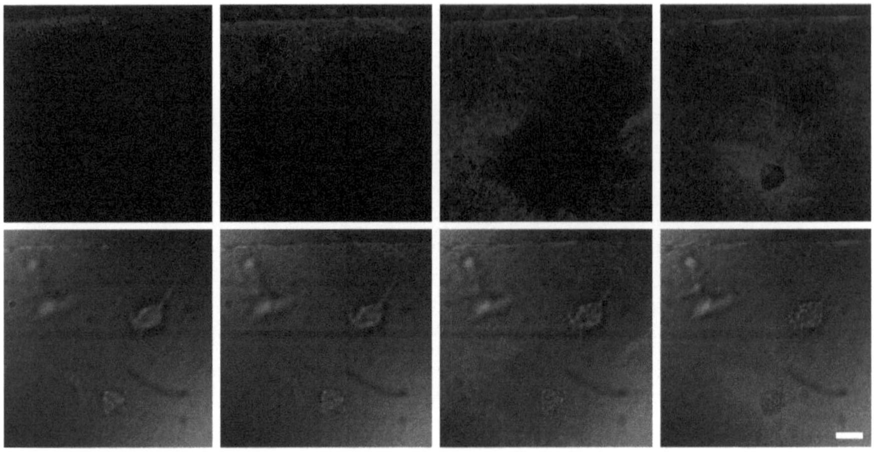

FIGURE 3.9

Images showing cell contraction of patterned collagen from the cover glass surface. Top row are collagen reflectance images and the bottom row are merged with brightfield images to show the cell and channel boundaries. In the bottom row, cells near the top of the channel are out of focus in the background. Images were acquired approximately 24 h after culture was started. Scale bar is 20 μm.

facilitated consistent imaging with the motorized stage (as freely floating constructs cannot be imaged in this manner). We then stitched the image stacks together to create a 3D image covering the 4 mm × 8 mm region of the construct inlets and channels (Figure 3.11).

In both EC only and EC–MSC cocultures, we were able to visualize the development of capillary tube structures throughout the inlets and channels in each layer of the construct (Figure 3.11). Higher cell density cultures qualitatively resulted in more uniform seeding and more extensive capillary formation, especially for the 100 × 100 cross-section channels (identified morphologically in brightfield images). We expected to see MSCs localizing to capillary walls, as has been observed in HUVEC–pericyte cocultures in 3D collagen matrices [44,45]. However, we did not qualitatively observe a biased localization of MSC to the EC ablumenal surfaces over 1 week of *in vitro* culture (Figure 3.12). Since we used growth medium optimized for EC capillary morphogenesis,

◀ brightfield images in the right column to highlight cell and channel (top of images) boundaries. Collagen concentration is 6 mg/ml. Images were taken approximately 1 h from completion of collagen gelling. Scale bar is 20 μm.

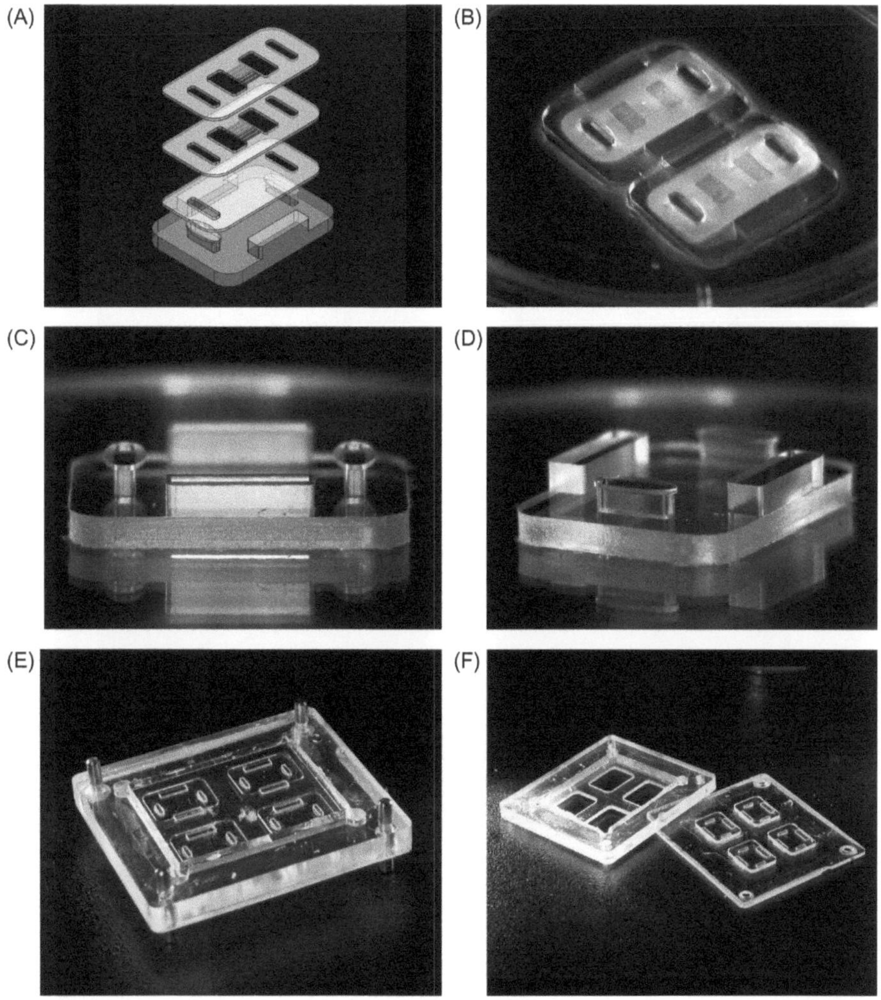

FIGURE 3.10

Fabrication of multilayer microfluidic hydrogels using PDMS supports. (A) Schematic diagram of multilayer assembly. (B) Macro image of constructs on supports after 4 days in culture. (C, D) Close up photographs of implant supports. The dimensions of the supports were 16 mm × 12 mm × 3 mm L × W × D. (E, F) Acrylic molds for the PDMS supports were fabricated by CNC milling. (E) The molds consisted of two layers that are aligned using steel pins. (F) The thin layer (right) defined the height and shape of the posts and guiderails while the thicker layer (left) defined the outer dimensions and thickness of the base layer. To mold the supports, PDMS was poured into the molds and degassed. A smooth piece of acrylic is used to seal the mold so that the bottom surface of the support is optically clear. The PDMS supports are still attached to the thinner mold layer in the image above.

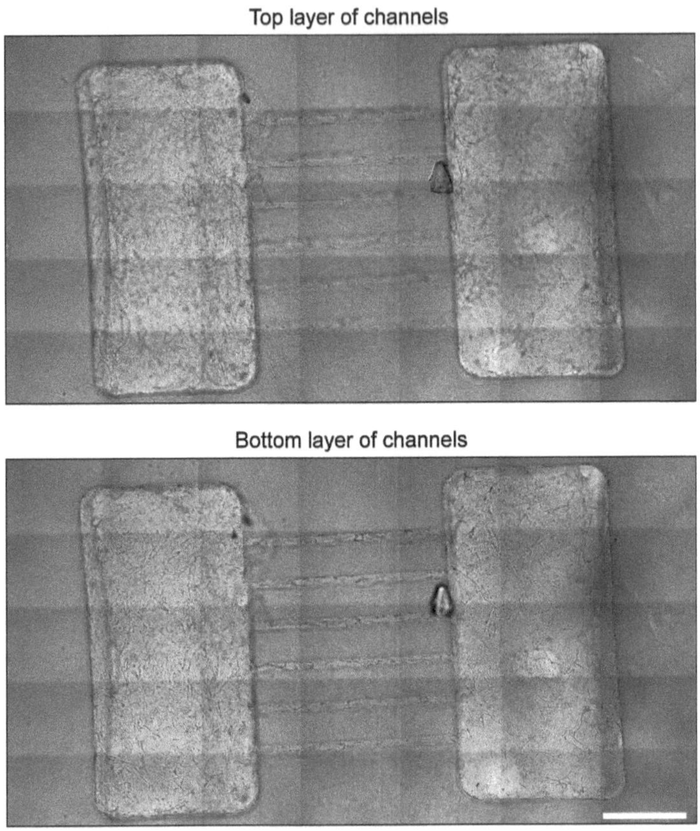

FIGURE 3.11

Tracking of vascular morphogenesis through *in vitro* culture. Shown are the top and bottom layers of a construct after 6 days in culture. This construct was seeded with 2 million HUVEC per ml and 0.5 million MSC per ml. The width of these channels is 100 μm and the length is 2.5 mm. Scale bar is 1 mm.

the soluble factor microenvironment may need to be further tuned to support perivascular localization of MSCs.

The goal of prevascularization is to accelerate capillary formation before grafting of an engineered tissue, in order to promote faster vascularization of the scaffold upon implantation, and hence improved viability of the tissue *in vivo*. Despite the intimate connection between network structure and function observed in native tissues, the relationship between geometry of the prevascularized network and effectiveness of the prevascularization strategy remains poorly explored. To address this important but open question in a systematic manner, it will be important to use a tissue engineering strategy that can both recapitulate the native-like microvascular structure (such as open lumens and cell polarization) that imparts to

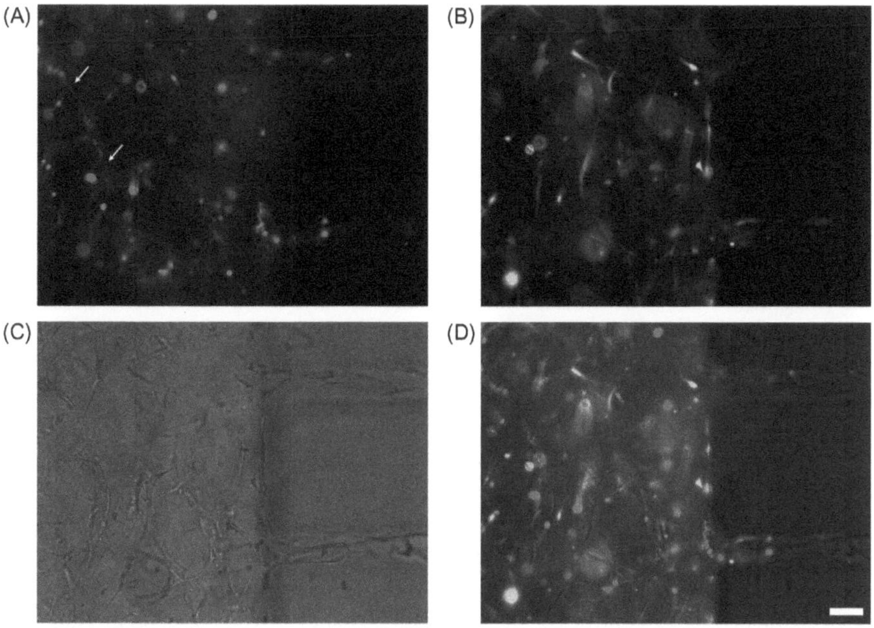

FIGURE 3.12

Interactions of EC and MSC coencapsulated in patterned collagen matrices: (A) ECs, some lumen structures are indicated with the arrows, (B) MSCs, (C) brightfield image, and (D) merged, a region from the same construct shown in Figure 3.9 at 3 days in culture. Scale bar is 100 μm.

the network its microvascular function, while allowing for systematic variations of network geometry. Toward this goal, this work has described the optimization of patterned collagen matrix structure for interface stability and vascular morphogenesis, and the formation of multilayer microfluidic scaffolds that support stable collagen-fiber interfacing on four walls. Our results show that EC can undergo 3D capillary morphogenesis within spatially constricted, micropatterned collagen matrices. In the next phase of this work, we plan to test the ability for these microfabricated vascular networks to integrate with host vasculature to become functional. Ultimately, we hope to investigate whether control over the structure of prevascularized networks can improve integration and long-term function of engineered grafts.

3.5 Conclusion and future directions

The merging of microfluidics with 3D cell culture methods has enabled new capabilities for precisely constructing and controlling cellular microenvironments and engineered tissues. However, while the capacity to better mimic native tissue environments has been improved as the sophistication of microfluidic 3D culture and fabrication methods has increased, challenges arise for analyzing and understanding such systems of intermediate complexity between 2D culture dishes and native tissue environments. Variation of multiple 3D microenvironment parameters (composition and structure of ECM, concentration and distribution of soluble and matrix-bound factors, fluid flow and mechanical forces) and tracking of cellular responses can quickly lead to large and complex data sets from experiments. Automated and semiautomated imaging and analysis techniques [46] that are integrated with microfluidic culture platforms could aid in building quantitative biological insights from large data sets.

Further development of microfluidic fabrication techniques for 3D engineered tissues is likely needed to adequately mimic certain native biological structures. For example, the 2.5D nature of most microfluidics fabrication techniques has limited ability to truly mimic some 3D structures found in native tissues (such as branching tree vascular networks). Alternative fabrication methods, such as 3D printing [19,47], can build more arbitrary 3D microenvironments than is capable of using microfluidics methods described in this chapter. Electrostatic discharge can be also used to create 3D fluid tree networks; however, this technique does not permit precision control of exact and consistent network structures and may be limited in terms of the biomaterials which can be used [48].

While the last decade has seen an explosion of methods using microfluidics for 3D cell culture and tissue engineering, researchers are now focusing on developing new insights and clinical utilities from such systems. Further investigation into how microfabricated tissues integrate and perform *in vivo* will be critical for determining the degree of control of tissue structure that is required to optimize graft function. Since cellular systems are their own architects, the degree to which engineers must control the environment to elicit desired biological response is largely unclear for most tissue systems. Microfluidic engineering of 3D microenvironments, along with other tissue engineering approaches, will further contribute to our knowledge of how to manipulate cellular behaviors to obtain clinically useful results.

References

[1] B.W. Lee, et al., Strongly binding cell-adhesive polypeptides of programmable valencies, Angew. Chem. Int. Ed. Engl. 49 (11) (2010) 1971–1975.

[2] S.-J. Wang, et al., Differential effects of EGF gradient profiles on MDA-MB-231 breast cancer cell chemotaxis, Exp. Cell Res. 300 (1) (2004) 180–189.

[3] C.D. Chin, K. Khanna, S.K. Sia, A microfabricated porous collagen-based scaffold as prototype for skin substitutes, Biomed. Microdevices 10 (3) (2008) 459–467.

[4] Y.K. Cheung, et al., Direct patterning of composite biocompatible microstructures using microfluidics, Lab Chip 7 (5) (2007) 574–579.

[5] Y. Shin, et al., In vitro 3D collective sprouting angiogenesis under orchestrated ANG-1 and VEGF gradients, Lab Chip 11 (13) (2011) 2175–2181.

[6] J. Lii, et al., Real-time microfluidic system for studying mammalian cells in 3D microenvironments, Anal. Chem. 80 (10) (2008) 3640–3647.

[7] M. Cabodi, et al., A microfluidic biomaterial, J. Am. Chem. Soc. 127 (40) (2005) 13788–13789.

[8] N.W. Choi, et al., Microfluidic scaffolds for tissue engineering, Nat. Mater. 6 (11) (2007) 908–915.

[9] A.P. Golden, J. Tien, Fabrication of microfluidic hydrogels using molded gelatin as a sacrificial element, Lab Chip 7 (6) (2007) 720–725.

[10] Y. Ling, et al., A cell-laden microfluidic hydrogel, Lab Chip 7 (6) (2007) 756–762.

[11] A. Paguirigan, D.J. Beebe, Gelatin based microfluidic devices for cell culture, Lab Chip 6 (3) (2006) 407–413.

[12] B.M. Gillette, et al., In situ collagen assembly for integrating microfabricated three-dimensional cell-seeded matrices, Nat. Mater. 7 (8) (2008) 636–640.

[13] B.M. Gillette, et al., Engineering extracellular matrix structure in 3D multiphase tissues, Biomaterials 32 (32) (2011) 8067–8076.

[14] W. Tan, T.A. Desai, Layer-by-layer microfluidics for biomimetic three-dimensional structures, Biomaterials 25 (7–8) (2004) 1355–1364.

[15] Y.K. Cheung, et al., Microscale control of stiffness in a cell-adhesive substrate using microfluidics-based lithography, Angew. Chem. Int. Ed. Engl. 48 (39) (2009) 7188–7192.

[16] Y.H. Jang, et al., An integrated microfluidic device for two-dimensional combinatorial dilution, Lab Chip 11 (19) (2011) 3277–3286.

[17] T. Thorsen, S.J. Maerkl, S.R. Quake, Microfluidic large-scale integration, Science 298 (5593) (2002) 580–584.

[18] M.D. Tang, A.P. Golden, J. Tien, Molding of three-dimensional microstructures of gels, J. Am. Chem. Soc. 125 (43) (2003) 12988–12989.

[19] W. Lee, et al., On-demand three-dimensional freeform fabrication of multi-layered hydrogel scaffold with fluidic channels, Biotechnol. Bioeng. 105 (6) (2010) 1178–1186.

[20] G.D. Pins, M. Toner, J.R. Morgan, Microfabrication of an analog of the basal lamina: biocompatible membranes with complex topographies, FASEB J. 14 (3) (2000) 593–602.

[21] M.S. Stosich, et al., Vascularized adipose tissue grafts from human mesenchymal stem cells with bioactive cues and microchannel conduits, Tissue Eng. 13 (12) (2007) 2881–2890.

[22] V.L. Cross, et al., Dense type I collagen matrices that support cellular remodeling and microfabrication for studies of tumor angiogenesis and vasculogenesis in vitro, Biomaterials 31 (33) (2010) 8596–8607.

[23] C.J. Bettinger, et al., Silk fibroin microfluidic devices, Adv. Mater. 19 (5) (2007) 2847–2850.

[24] S.-Y. Cheng, et al., A hydrogel-based microfluidic device for the studies of directed cell migration, Lab Chip 7 (6) (2007) 763–769.

[25] Y.S. Song, et al., Engineered 3D tissue models for cell-laden microfluidic channels, Anal. Bioanal. Chem. 395 (1) (2009) 185–193.

[26] B.M. Gillette, et al., Dynamic hydrogels: switching of 3D microenvironments using two-component naturally derived extracellular matrices, Adv. Mater. 22 (6) (2010) 686–691.

[27] C.J. Bettinger, et al., Three-dimensional microfluidic tissue-engineering scaffolds using a flexible biodegradable polymer, Adv. Mater. 18 (2) (2005) 165–169.
[28] C. Fidkowski, et al., Endothelialized microvasculature based on a biodegradable elastomer, Tissue Eng. 11 (1–2) (2005) 302–309.
[29] M. Shin, et al., Endothelialized networks with a vascular geometry in microfabricated poly(dimethyl siloxane), Biomed. Microdevices 6 (4) (2004) 269–278.
[30] G.M. Price, J. Tien, Methods for forming human microvascular tubes in vitro and measuring their macromolecular permeability, Methods Mol. Biol. 671 (2011) 281–293.
[31] K.M. Chrobak, D.R. Potter, J. Tien, Formation of perfused, functional microvascular tubes in vitro, Microvasc. Res. 71 (3) (2006) 185–196.
[32] M. Kamei, et al., Endothelial tubes assemble from intracellular vacuoles in vivo, Nature 442 (7101) (2006) 453–456.
[33] A. Alajati, et al., Spheroid-based engineering of a human vasculature in mice, Nat. Methods 5 (5) (2008) 439–445.
[34] X. Chen, et al., Prevascularization of a fibrin-based tissue construct accelerates the formation of functional anastomosis with host vasculature, Tissue Eng. A 15 (6) (2009) 1363–1371.
[35] J.M. Melero-Martin, et al., Engineering robust and functional vascular networks in vivo with human adult and cord blood-derived progenitor cells, Circ. Res. 103 (2) (2008) 194–202.
[36] P.L. Tremblay, et al., Inosculation of tissue-engineered capillaries with the host's vasculature in a reconstructed skin transplanted on mice, Am. J. Transplant. 5 (5) (2005) 1002–1010.
[37] J. Yang, et al., Telomerized human microvasculature is functional in vivo, Nat. Biotechnol. 19 (3) (2001) 219–224.
[38] B.M. Gillette, Microfabrication of Collagen-Based Extracellular Matrices for Tissue Engineering (Ph.D. dissertation), Columbia University, New York, NY, 2011.
[39] A.L. Sieminski, R.P. Hebbel, K.J. Gooch, The relative magnitudes of endothelial force generation and matrix stiffness modulate capillary morphogenesis in vitro, Exp. Cell. Res. 297 (2) (2004) 574–584.
[40] G.E. Davis, D.R. Senger, Endothelial extracellular matrix: biosynthesis, remodeling, and functions during vascular morphogenesis and neovessel stabilization, Circ. Res. 97 (11) (2005) 1093–1107.
[41] G.E. Davis, D.R. Senger, Extracellular matrix mediates a molecular balance between vascular morphogenesis and regression, Curr. Opin. Hematol. 15 (3) (2008) 197–203.
[42] Y.L. Yang, S. Motte, L.J. Kaufman, Pore size variable type I collagen gels and their interaction with glioma cells, Biomaterials 31 (21) (2010) 5678–5688.
[43] K.E. Sung, et al., Control of 3-dimensional collagen matrix polymerization for reproducible human mammary fibroblast cell culture in microfluidic devices, Biomaterials 30 (27) (2009) 4833–4841.
[44] A.N. Stratman, et al., Endothelial-derived PDGF-BB and HB-EGF coordinately regulate pericyte recruitment during vasculogenic tube assembly and stabilization, Blood 116 (22) (2010) 4720–4730.
[45] A.N. Stratman, et al., Pericyte recruitment during vasculogenic tube assembly stimulates endothelial basement membrane matrix formation, Blood 114 (24) (2009) 5091–5101.

[46] H. Parsa, R. Upadhyay, S.K. Sia, Uncovering the behaviors of individual cells within a multicellular microvascular community, Proc. Natl. Acad. Sci. U.S.A. 108 (12) (2011) 5133–5138.

[47] C. Norotte, et al., Scaffold-free vascular tissue engineering using bioprinting, Biomaterials 30 (30) (2009) 5910–5917.

[48] J.-H. Huang, et al., Rapid fabrication of bio-inspired 3D microfluidic vascular networks, Adv. Mater. 21 (35) (2009) 3567–3571.

CHAPTER

4 Fabrication of Advanced Microcontainer Arrays for Perfused 3D Cell Culture in Microfluidic Bioreactors

Roman Truckenmüller*, Stefan Giselbrecht*, Eric Gottwald and Arne Schleunitz

Institute for Biomedical Technology and Technical Medicine, University of Twente, The Netherlands

4.1 Introduction

The last 40 years have impressively shown that including additional dimensionality in model cell culture systems is an important component in designing model cell culture microenvironments [1]. Moving from 2D to 3D cell culture systems has generated a significant amount of basic science knowledge because of the ability to recapitulate native tissue microenvironments. The newfound capacity to fabricate multidimensional cell culture systems suggests that new considerations must be accommodated in defining appropriate *in vitro* cell culture systems for use in testing hypotheses. For example, we have learned from studies carried out by Sutherland [2] that the tumor microenvironment is an intrinsically diverse environment that influences the responsiveness to therapeutic agents. Resistance to radiation therapy can be conferred to cells cultured in 3D that is not present when the identical cells are cultured in 2D [3]. Engineering the 3D microenvironment can influence a wide range of cellular behavior including proliferation, differentiation, angiogenesis and immune response, and programmed cell death [4,5]. The diversity in phenotypes derived from different tissue sources requires highly specialized cell culture systems that can recapitulate the native microenvironment. Conversely, highly specialized cell culture environments restrict the potential scope of applications. This dichotomy leads to a rethinking process that should result in identifying the key definable parameters in 3D cell culture. In turn, this could result in a universal platform whose modular design allows for the piecewise integration of relevant microenvironmental properties, but

*These authors contributed equally to this chapter.

Microfluidic Cell Culture Systems. ISBN: 978-1-4377-3459-1

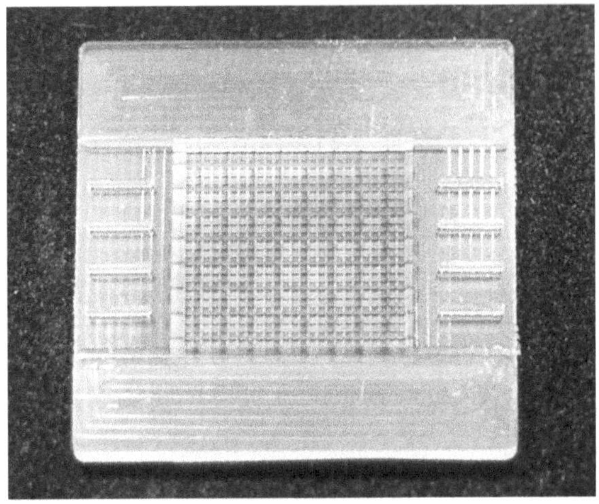

FIGURE 4.1

3^D-KITChip from PMMA comprising 625 cell containers (outer dimensions of the chip: $20 \times 20 \times 1\ mm^3$).

also is able to incorporate additional exogenous cues. With this in mind, a variety of 2D and 3D microfluidic cell culture platforms have previously been fabricated using a broad spectrum of materials [6–10]. One prominent example is a microfluidic 3D cell culture technology termed "3^D-KITChip". This device is a versatile platform that allows cells to be cultured in a variety of conditions including different active flow modes, flow regimes, fluid compositions, and gas supply possibilities. The simplicity of these devices allows many mammalian cell types to be cultured in precisely defined environmental factors. A modular design of the platform concept suggests that this system is suitable for the integration of additional characteristics to reproduce native organotypic cultures as accurately as possible. The 3^D-KITChip is a polymer chip in form of a roughly $10 \times 10\ mm^2$ sized array of up to 1156 microcontainers arranged in orthogonal rows and columns and housed in a $20 \times 20\ mm^2$ frame (Figure 4.1) [11–13]. The containers of the chip have porous bottoms or walls to allow for active perfusion of the cells and organoids. To support this strategy, the KITChip is integrated into single- or multichip bioreactors, where it can additionally be operated in a superfusion or in a mixed superfusion-perfusion mode. Guiding structures on the chip frame help to control the medium flow over the intermediate microcontainer array. The characteristic size of the cell containers mimic the typical spacing of capillaries in vascular networks found in many mammalian tissues. Reducing the characteristic dimensions of the device to this length scale ensures that cells located on the interior of multicellular aggregates are supplied with sufficient nutrients and waste removal [14–16]. Potential applications of the KITChip as such or in corresponding bioreactors range from fundamental research, such as stem cell research, toxicology and research in pharmaceutical industry including absorption, distribution, metabolism, excretion, and toxicity studies. Microfluidic devices may also be used for translational therapies including the use of biohybrid extracorporeal organ support systems [17,18].

The fabrication of the porous KITChip can be pursued using three distinct processes:

1. Micro injection molding or hot embossing of the chip, thinning down of the chip bottom by milling (after first removing the sprue and runner in case of injection molding, or the peripheral residual layer in case of hot embossing), and laser drilling or ablation of holes in the container bottoms.
2. Micro injection molding or hot embossing of the chip, completely removing of the chip bottom by milling, and solvent vapor bonding of a porous bottom in form of, for example, a commercial ion-track-etched membrane to the back of the chip.
3. Thermoforming of the chip from a heavy-ion-irradiated membrane, etching out of the ion tracks to achieve pores in the container bottoms and sidewalls, and cutting out the flexible chip.

The individual process steps are described in detail in the following two Sections 4.2 and 4.3. The design of the KITChip is adapted to the respective molding process. In case of the micro-injection-molded or hot-embossed chips, the cell containers have a cubic shape with an edge length of 300 μm and are arranged in a grid of 350 μm, i.e., the walls between the containers are 50 μm thin. In case of the thermoformed chip, the containers have a circular-cylindrical shape with round bottoms or with rounded transitions from container walls to bottoms, in each case with a diameter and depth of approximately 300 μm, and are arranged in a 400 μm grid. The functionalization of the cell containers as artificial cellular microenvironments is discussed in Section 4.4. Section 4.5 deals with the integration of the KITChip into the fluidic microbioreactors.

4.2 Micromolding of cell container arrays

The 3^D-KITChip is a prominent example of a microfluidic device for cell culture that can be fabricated by polymer micromolding or replication [19–21]. These processes are amenable for cell culture studies due to the compatibility with rapid prototyping techniques and the relatively low cost per piece. As with many microfluidic-based devices for biomedical applications [22,23], these systems are envisioned to be a one-time use component. Another important aspect in this context is the diversity of the processable materials and of their modification or functionalization. Microfluidic devices have been fabricated from a number of materials including amorphous or semicrystalline commodity, engineering, high-performance or biodegradable polymers. Examples include, but are not limited to, polymethylmethacrylate (PMMA), polycarbonate (PC), polystyrene (PS), cycloolefin polymers (COP) and copolymers (COC), polyglycolic acid, polylactic acid (PLA), and polycaprolactone (PCL). In contrast to many established microfluidic platforms, the KITChip can be fabricated by three primary categories of micromolding processes:

1. micro injection molding,
2. (micro) hot embossing,
3. microscale thermoforming.

Hot embossing and microthermoforming are partly similar in their general strategy, but vary in the specifics of the process. The raw starting materials in these and related processes are typically thermoplastic polymer films. The manufacturing equipment employed in many of these methods are hot presses and other tools for thermomechanical manipulation of materials [24].

4.2.1 Micro injection molding

In brief, in injection molding, granules of a thermoplastic polymer are plasticized and homogenized by heat, friction, and shear forces in a heated barrel containing a rotating and reciprocating screw, and injected by the forward moving screw in a tool cavity provided with the mold. Here, the polymer melt is cooled down and therefore solidifies before then the molded parts are ejected from the mold after opening the tool. After demolding, the parts are still connected to the sprue and the runners, see also Figure 4.2. These have to be removed first before the parts can be used directly or after further processing. For micro injection molding [25–27], the injection system and the thermal management of the tool and mold is adapted to the smaller shot weights and the smaller details of the molded parts, respectively. Concerning the thermal management, that means that micro injection molding is carried out as a "variotherm(al)" process with the mold or tool being cycled between molding and demolding temperature. During injection, the tool

FIGURE 4.2

Injection-molded 3^D-KITChips from PMMA still connected with the film gates and the sprue.

FIGURE 4.3

Section of an injection-molded KITChip from PMMA (scanning electron microscope (SEM) image).

Source: *Reproduced with permission from Knedlitschek et al. [13], copyright 1999, American Society of Mechanical Engineers.*

FIGURE 4.4

Section of the mold used to injection-mold the KITChip fabricated from brass by mechanical micromachining (SEM image; side length of the cubic protrusions: 300 μm; grid of the protrusions: 350 μm).

Source: *Reproduced with permission from Giselbrecht et al. [28], copyright 2004, the Institution of Engineering and Technology.*

cavity together with the micromold is evacuated. Injection molding of the KITChip (Figure 4.2 and 4.3) [12,28] is performed using dedicated equipment (Microsystem 50, Battenfeld). The corresponding mold or mold insert from brass (Figure 4.4) is fabricated by mechanical micromachining using diamond tools.

4.2.2 Hot embossing

In hot embossing [29,30], a cut sheet from a thermoplastic polymer is heated up in the evacuated space between a mold plate and a substrate plate via these plates, and, when softened or flowing, pressed against the mold by the substrate plate. Then, the polymer is cooled down via the plates, and the parts are demolded as the distance between plates is increased. After demolding, the parts are still connected to each other and or to the residual layer. The parts first have to be diced, or separated from the residual layer, before they can be used. Embossing of the KITChip [31], see also Figure 4.12, is performed in a professional hot embossing machine (WUM, Jenoptik). The mold from brass is fabricated by mechanical micromachining using diamond tools.

4.2.3 Microscale thermoforming

Micro injection molding and hot embossing are established fabrication techniques with well-characterized processing conditions. However, microscale thermoforming is a more novel-processing approach that has recently emerged. In microthermoforming [24], similar to macroscopic thermoforming [32], a thin film of a thermoplastic polymer is heated to a softened but not flowing state and forced into the cavities or over the protrusions of a micromold. The mold space is concurrently evacuated to facilitate complete replication. In case of "micro pressure forming", forming is carried out by using compressed gas as a source of pressure

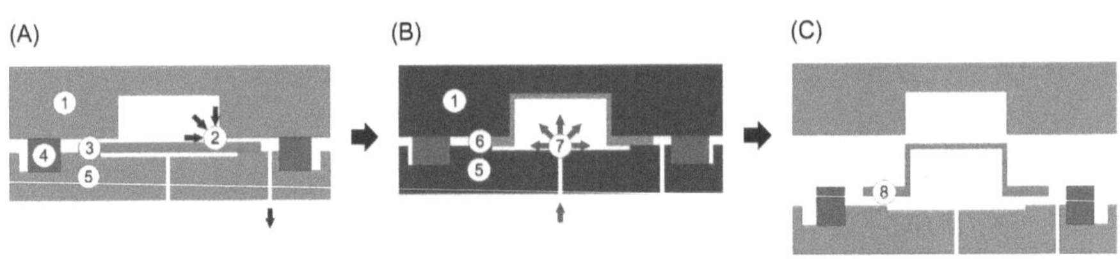

FIGURE 4.5

Process steps of micro pressure forming: (A) insertion of a thin thermoplastic film into a microthermoforming tool, evacuation of the tool, (B) heating up the tool, forming of the softened film by compressed nitrogen, (C) cooling down the tool, and demolding and unloading the solidified film microstructure (using the example of a negative forming [32] variant; simplified scheme with only a single mold cavity; (1) mold, (2) vacuum, (3) thermoplastic film to be thermoformed, (4) axial seal, (5) counter plate, (6) softened and formed film, (7) compressed nitrogen, (8) solidified thermoformed film microstructure; for (1), (5): blue or mid-gray and red or dark gray correspond to cold and heated). (For interpretation of the references to color in this figure legend, the reader is referred to the web version of this book.)

Source: *Reproduced with permission from Truckenmüller et al. [24], copyright 2011, Wiley-VCH.*

[33,34]. The device is finalized after molding by trimming away excess material and cutting the form to liberate individual devices from each other.

4.2.3.1 Micro pressure forming

In the micro pressure forming process (Figure 4.5), a cut sheet of a thin thermoplastic film is inserted into a microthermoforming tool (Figure 4.5A). The three-part tool consists of a plate-shaped micromold with mold cavities, a counter plate with openings for evacuation and gas pressurization, and an axial seal inbetween. The tool is mounted into a heated press. The press, and with it the tool, is closed to such an extent that vacuum sealing of the volume enclosed by the two tool plates and the seal is achieved, while the edges of the plastic sheet are not yet clamped between the plates. At this point, the entire tool is evacuated, fully closed so that the sheet is now clamped, and heated to increase the temperature of the polymer film around the softening temperature (Figure 4.5B). When the polymer becomes sufficiently soft, the film is then formed into the evacuated microcavities of the mold by applying compressed nitrogen. The temperature of the tool is reduced to below the softening temperature of the polymer (Figure 4.5C). When, sufficiently below the softening temperature of the polymer, the material is dimensionally stable again, the gas pressure is decreased. Then, the tool is opened and the thermoformed film microstructure is demolded and unloaded. Micro pressure forming of microfluidic devices such as the KITChip (Figure 4.6A, B) [28,35,36] is performed in a hot embossing machine (WUM, Jenoptik) supplemented with equipment for gas pressurization (Figure 4.7A, B). The corresponding mold from brass is fabricated by mechanical micromachining using diamond end mills (Figure 4.8).

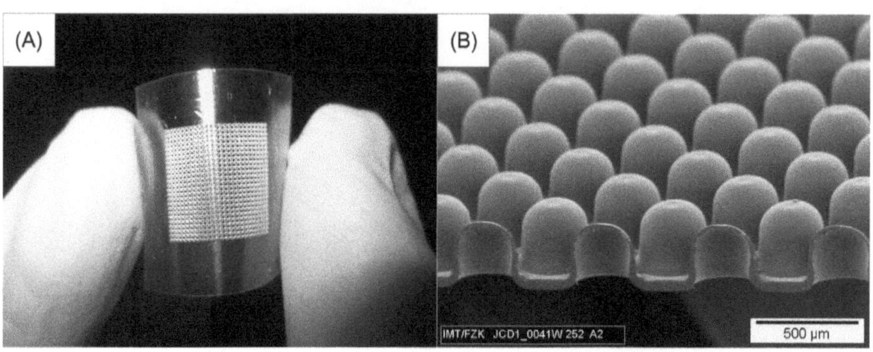

FIGURE 4.6

KITChip micro-pressure-formed from a 50 μm thin PC film: (A) flexible chip bent between the thumb and the forefinger of a hand, (B) section of a chip (SEM image; cross-sectional view, upside down; inner container diameter and depth: approximately 300 μm; grid of microcontainer array: 400 μm).

Source: *(A) Reproduced with permission from Truckenmüller et al. [35], copyright 2008, the Royal Society of Chemistry. (B) Reproduced with permission from Giselbrecht et al. [28], copyright 2004, the Institution of Engineering and Technology.*

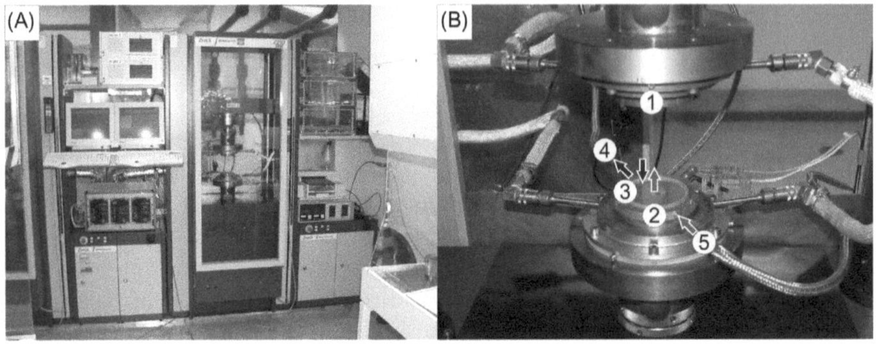

FIGURE 4.7

Microthermoforming machine and tool: (A) Machine for micro pressure forming on a laboratory scale based on a converted hot embossing press (WUM, Jenoptik) provided with (B) a corresponding tool ((1) micromold, (2) counter plate with evacuation and gas pressurization openings, (3) seal, (4) vacuum, (5) compressed nitrogen).

Source: *(A) Reproduced with permission from Truckenmüller et al. [24], copyright 2011, Wiley-VCH. (B) Reproduced with permission from Truckenmüller et al. [35], copyright 2008, the Royal Society of Chemistry.*

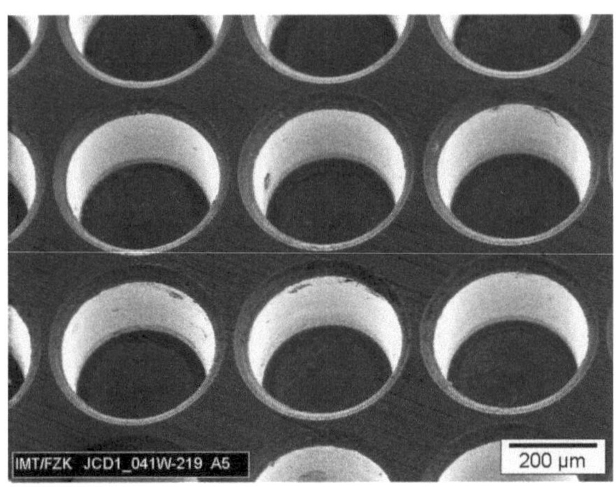

FIGURE 4.8

Section of the mold used to pressure-form the KITChip fabricated from brass by micromilling (SEM image; mean diameter of microcavities: 325 μm; cavity depth: 300 μm; grid of cavity array: 400 μm; draft angles: 5°; cavities additionally equipped with 20 μm wide 45°-bevels to further facilitate demolding).

Source: *Reproduced with permission from Truckenmüller et al. [35], copyright 2008, the Royal Society of Chemistry.*

4.2.3.2 Other microthermoforming processes

Besides by compressed gas, forming can be carried out by the softened or flowing polymer of a second film in case of "micro back molding" [37,38], into a paraffin bed within the "membrane microemboss (MeME)" process [39], by an elastomeric counter tool in "rubber-assisted hot embossing" [40] or between a tool and a matching counter tool [41]. The first and the third processes were successfully used to fabricate cell container arrays of a similar design as the KITChip [38,42].

4.3 Introducing porosity

As mentioned, three different process sequences can be applied to introduce porosity in the container bottoms of the 3^D-KITChip:

1. Laser drilling of holes in the thinned-down bottom of an injection-molded or hot-embossed chip.
2. Substituting the bottom of an injection-molded or hot-embossed chip with an ion-track-etched membrane.
3. Microscale thermoforming of a heavy-ion-irradiated membrane to a chip and etching out of the pores.

Pore diameters between 2 and 3 μm allow perfusion of the cell culture medium through the bottoms but not the passage of cells to the backside of the chip during seeding or by migration.

FIGURE 4.9

3^D-KITChips from PMMA fixed on a cooled mounting plate (with their fragile polymer microstructures frozen in water) for thinning down the chip bottoms by milling prior to their laser perforation.

4.3.1 Laser micromachining

As it is not yet possible to obtain KITChips with open pores in the bottoms of the chip containers directly by injection molding or hot embossing, the porosity must be introduced by a subsequent perforation of the chip bottom. For this, first the bottom is thinned down by milling to approximately 80 μm (Figure 4.9). The perforation is then carried out by a masked-based laser ablation process [11,12,28] (Figure 4.10A, B). Due to the beam guidance of the excimer laser

FIGURE 4.10

Sections of microcontainers of a KITChip from PMMA with laser-perforated bottoms (SEM images; pore size: 2–3 μm; pore density: 64 pores/microcontainer; white rectangle in the image on the top indicates the image section of the close-up view underneath).

Source: *Reproduced with permission from Eschbach et al. [11], copyright 2005, Wiley-Liss.*

used, 20–30 μm diameter orifices in the mask from stainless steel are reduced to 2–3 μm diameter spots on the backside of the chip bottom located 50 cm away from the mask.

4.3.2 Solvent vapor bonding of track-etched membranes

In a cheaper and faster process for introducing porosity into the bottom of the KITChip compared to the process presented in the last section, commercially available ion-track-etched membranes are integrated into the chip. For this, the bottom of the chip is not only thinned down but completely removed by mechanical machining using diamond mills. Then, the membrane is attached to the back of the chip in a solvent vapor bonding process. In this process, up to four machined KITChips and ion-track-etched membranes are inserted into a corresponding

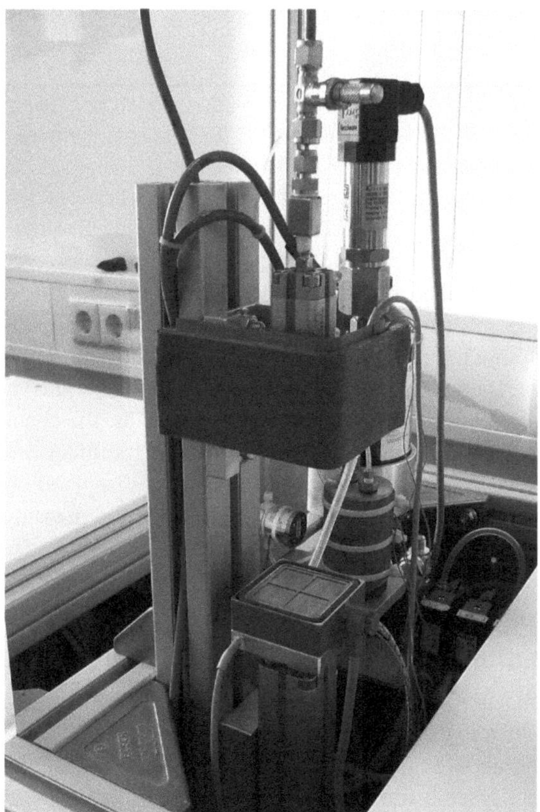

FIGURE 4.11

Apparatus for solvent vapor bonding of ion-track-etched membranes to KITChips with removed bottoms.

FIGURE 4.12

Section of a KITChip from PMMA with a solvent vapor-bonded porous bottom in form of a commercial ion-track-etched membrane from PC (SEM image; membrane thickness: 10 μm; pore size: 3 μm, pore density: 2×10^6 pores/cm^2).

Source: *Reproduced with permission from Gottwald et al. [12], copyright 2007, the Royal Society of Chemistry.*

apparatus (Figure 4.11). The apparatus consists of an upper plunger, a lower holder for the chips and membranes with openings for evacuation and letting in the vaporized solvent, and an axial seal ring inbetween. The chamber is closed and evacuated. Then, the parts to be bonded are exposed to the solvent vapor and pressed together. After a dwell time in the range of a few seconds, the chamber is evacuated again, thereby removing the solvent. Due to the short contact time between parts and solvent, only surface-near material of the parts is dissolved and a deformation of the bulk structure during pressing is avoided. Then, the pressure is released, the chamber is opened, and the bonded parts (Figure 4.12) are unloaded.

4.3.3 Combination of thermoforming and ion track technology

Comprehensive 3D perfusion design must be engineered in order to provide cell aggregates with sufficient supply of nutrients, gas exchange, and removal of potentially cytotoxic metabolites. Therefore, thermoformed microfluidic cell culture systems such as the KITChip must be designed to incorporate micro- or nanopores along all surfaces that contain viable cell populations. (Figure 4.13). This overlaid bulk modification is achieved by a "SMART" process, see Section 4.2, and that in form of a combination of microscale thermoforming and ion track etching [31,35,36]. Ion track technology [43,44] is an established, industrially applied

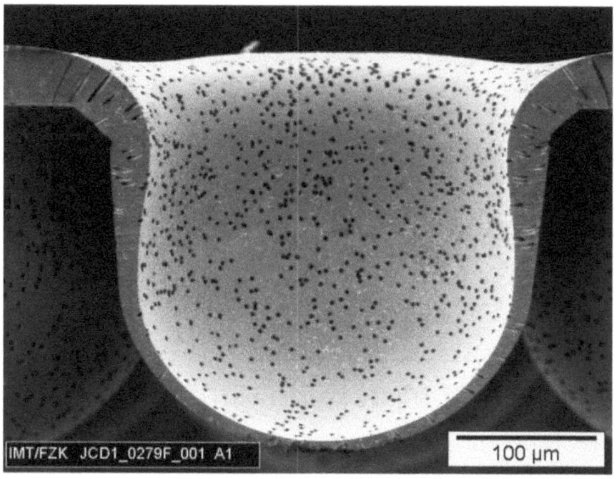

FIGURE 4.13

Microcontainer of a KITChip from PC with ion-track-etched pores perpendicular to the container walls (SEM image; cross-sectional view).

Source: *Reproduced with permission from Truckenmüller et al. [35], copyright 2008, the Royal Society of Chemistry.*

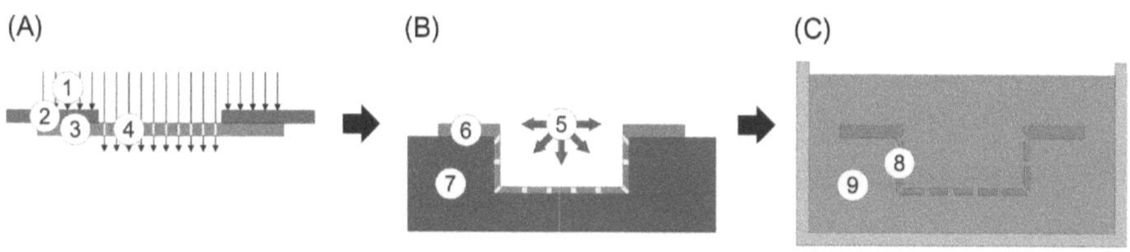

FIGURE 4.14

Process steps of the SMART process for the generation of micro- or nanopores in thermoformed films: (A) premodification, (B) microthermoforming, and (C) postmodification. Simplified scheme: (1) heavy ion radiation, (2) ion absorber mask, (3) PC film, (4) latent ion tracks, (5) compressed nitrogen, (6) thermoformed film, (7) heated thermoforming mold, (8) pores, (9) etch bath.

Source: *Reproduced with permission from Truckenmüller et al. [35], copyright 2008, the Royal Society of Chemistry.*

technology, mainly used to manufacture porous polymer membranes for micro- and ultrafiltration. In conventional track etching, a thin polymer film typically from PC, polyethylene terephthalate or polyimide (PI) and with thicknesses between 1 and 50 μm is irradiated with accelerated, energetic heavy ions from a corresponding ion source. The resulting process produces randomly distributed cylindrical tracks of physicochemically degraded polymer in the film. Then, the "latent" ion tracks are opened to cylindrical pores by wet-chemical etching. The pores typically have a diameter or size between 10 nm and 10 μm, and an areal density up to 5×10^9 pores/cm^2. Pore size and density can be adjusted independently. The density depends on the ion fluence during irradiation, while the size depends on the etching conditions such as etchant concentration, etching temperature, and duration. The SMART process represents a strategy for the generation of micropores in thermoformed films (Figure 4.14). In this process, the plane film is first heavy-ion-irradiated (Figure 4.14A) and then thermoformed into the desired geometry (Figure 4.14B). The pores are then etched into the three-dimensionally formed film (Figure 4.14C).

4.4 Functionalization of cell container arrays

Surface properties represent an additional parameter that must be carefully chosen when designing biomaterials for use in microfluidics. The intrinsic surface chemistry of the bulk material affords a certain degree of flexibility in achieving nominally hydrophilic or hydrophobic surfaces. However, coatings and surface modification techniques can greatly expand the functionality in this regard. Incorporating, for example, textured surfaces into microfluidic devices such as the 3^D-KITChip allow for an added component in controlling cell-materials interactions in perfused systems [15]. Surface modification of container arrays has important implications on cellular processes and cell fate including adhesion, migration, and proliferation to differentiation [45–48].

4.4.1 Protein coatings

To provide appropriate biochemical microenvironments via the surfaces of the containers of the KITChip, particularly concerning the balance between cell adhesion on the container surfaces and cell aggregation in the container volumes, the chip surfaces can be functionalized with extracellular matrix (ECM) components. Depending on the cell type used in a specific cell culture application and the application itself, ECM proteins such as different types or variants of collagen, also cross-linked, fibronectin, or laminin can be physisorbed on the chip surface in the course of incubating the chip in corresponding solutions prior to inoculating the chip with the cell suspension [12,13,49–51].

4.4.2 SMART processes for overlaid patterned surface and bulk modification of thermoformed cell container arrays

Due to the permanent material coherence of the polymer film during thermoforming, modifications of the film prior to its forming remain preserved beyond the forming operation. The "substrate modification and replication by thermoforming (SMART)" processes [12,35,36] take advantage of this fact. A SMART process is a sequence of a central microthermoforming process surrounded by pairs of associated pre- and postprocesses for micro- or nanopatterned modifications of the surface or bulk of the formed film (Figures 4.14, 4.16, and 4.19). The film modifications or functionalizations can be of a physical, chemical, or biological nature. In a SMART process, still on the unformed, plane film, premodifications of the film material define the locations where later, then already on the spatially formed film, the final local (post)modifications are generated. The definition of the modification sites before film forming is normally performed by highly anisotropic, directed lithographic processes. Due to the ability of the flexible thin film to adapt to planar masks or substrates, lithography-based premodification can be performed with maximum resolution by minimizing optical artifacts such as diffraction effects or effects of limited depth of field. The modification itself after film forming is usually performed by isotropic, undirected wet-chemical processes. The final isotropic modification provides access over the entire three-dimensionally formed film. This modification however is site-specific because of the premodification. Due to the task-sharing between the pre- and the postmodification, one can obtain highly resolved patterns of even temperature- and stress-sensitive modifications also on or in hardly accessible sidewalls and behind undercuts of thermoformed microstructures. In a simpler version of the SMART process, modifications are performed only prior to the forming step (Figure 4.19).

4.4.2.1 Cell adhesion micropatterns

For defined conditions within the 3D cell culture in the KITChip, it is critical to allow luminal attachment of cells to the interior surfaces of the containers as opposed to the barriers that separate them. For this, only the surfaces of the

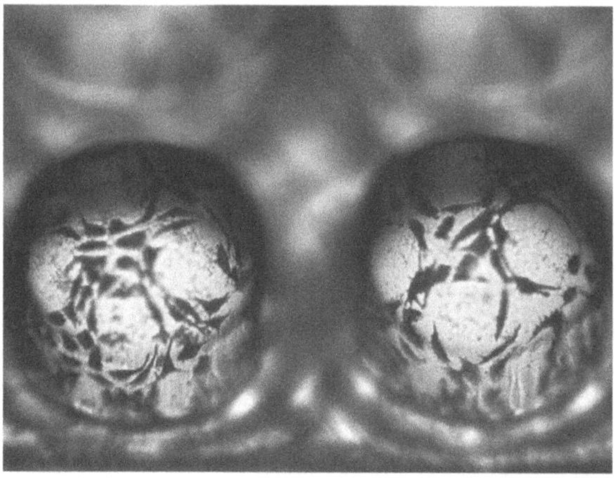

FIGURE 4.15

Microcontainers of a 3^D-KITChip from PS with fixed and crystal-violet-stained cells of the mouse fibroblast cell line L929 located only on the DUV-irradiated chessboard pattern (back view).

Source: *Reproduced with permission from Truckenmüller et al. [35], copyright 2008, the Royal Society of Chemistry.*

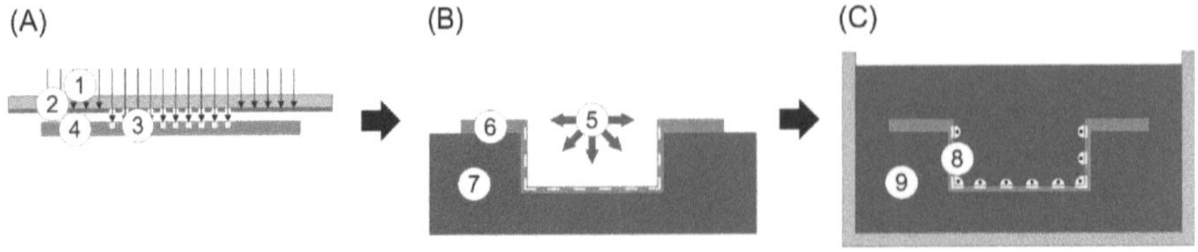

FIGURE 4.16

Process steps of the SMART processes for generation of cell adhesion micropatterns on thermoformed films: (A) premodification, (B) microthermoforming, and (C) postmodification. Simplified scheme: (1) DUV radiation, (2) photomask, (3) UV-modified domains, (4) PS film, (5) compressed nitrogen, (6) thermoformed film, (7) heated thermoforming mold, (8) adhering cells of the mouse fibroblast cell line L929, (9) cell culture medium.

Source: *Reproduced with permission from Truckenmüller et al. [35], copyright 2008, the Royal Society of Chemistry.*

curved container walls of a KITChip thermoformed from a film of an originally cell repellent polymer such as pristine PS can be adjusted to be cell attractive. The same strategy can be applied to control the spatial organization of cells within the containers (Figure 4.15), also when establishing 3D co-culture models on the basis of the chip. The corresponding overlaid surface modification of the KITChip can be achieved by a SMART process in the form of a combination of microscale thermoforming and micropatterned deep UV (DUV) irradiation [35,36]. Exposure of an appropriate polymer substrate to DUV radiation first results in a modified adsorption of plasma proteins on the substrate surface in cell culture because of the changed physicochemical properties of the polymer surface [52]. Subsequently, it leads to an increased cell adhesion due to the altered composition of the protein adsorbate on the substrate. This simple but effective method allows for highly resolved protein patterns [53,54] and the formation of protein gradients through nonbinary techniques such as gray-tone or moving masks [55,56]. The patterned DUV surface modification can also serve as a basis for one-step polymer grafting. The SMART process can generate cell adhesion micropatterns on thermoformed films (Figure 4.16). The film is first irradiated locally with DUV through a chrome-on-quartz photomask (Figure 4.16A) followed by thermoforming of the film (Figure 4.16B). In contrast to complex biomolecules, during forming, the robust surface chemistry of the photomodification of the polymer surface does not degrade or denature, only the areal density of the corresponding functional groups decreases in a precisely defined manner. Finally, a cell suspension is dispensed on the film scaffold with the cells adhering only to the irradiated domains in the course of cultivation (Figure 4.16C), possibly after a prior incubation of the scaffold in an appropriate protein solution. Starting with films from polymeric materials being cell-adherent in their native state, other SMART processes can pattern cells on three-dimensionally formed films by selectively establishing nonadherent domains (Figure 4.17) [57].

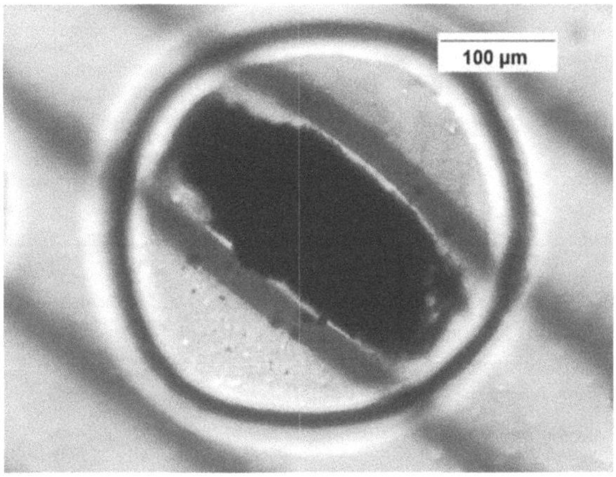

FIGURE 4.17

Microcontainer of a KITChip from N_2-plasma-treated PLA with fixed and crystal-violet-stained cells of the human hepatocellular carcinoma cell line HepG2 located only within the line pattern from a photo-cross-linked mixture of vinylated hyaluronic acid and ε-poly-L-lysine.

Source: *Reproduced with permission from Richter et al. [57], copyright 2010, Springer.*

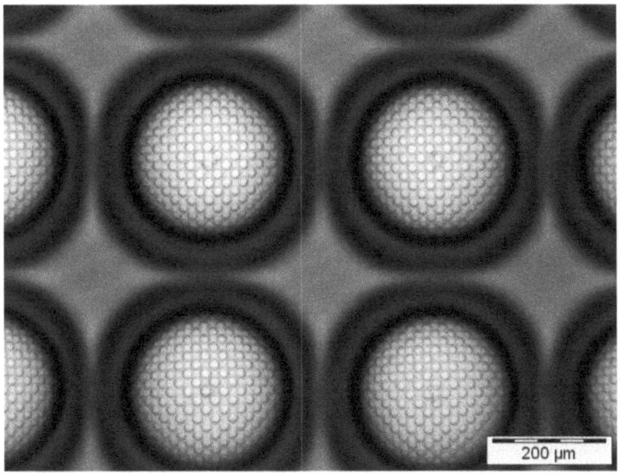

FIGURE 4.18

Microcontainers of a KITChip from PC with imprinted pillar pattern (back view).

Source: *Reproduced with permission from Truckenmüller et al. [24], copyright 2011, Wiley-VCH.*

4.4.2.2 Surface and bulk micro- and nano topographies

To provide defined topographical microenvironments within the containers of the thermoformed KITChip, the container walls can be provided with surface textures or bulk topographies (Figure 4.18) that can be engineered to recapitulate native protein structures in the ECM [7,58,59]. This overlaid modification can be achieved, for

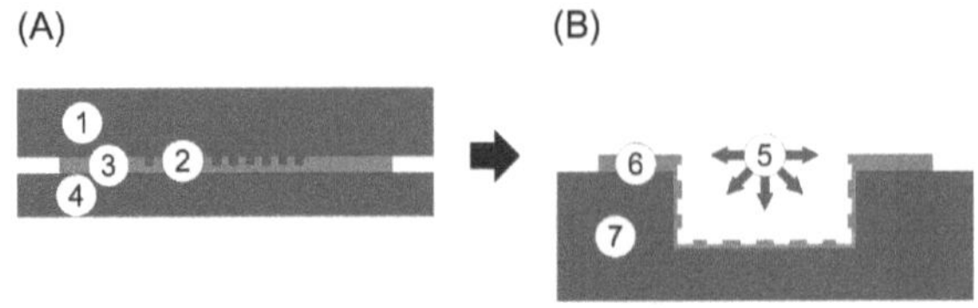

FIGURE 4.19

Process steps of the SMART process for the generation of surface or bulk micro- or nanotopographies on or in thermoformed films: (A) premodification and (B) microthermoforming. Simplified scheme: (1) heated imprint stamp, (2) topographies, (3) PC film, (4) heated substrate wafer, (5) compressed nitrogen, (6) thermoformed film, (7) heated thermoforming mold.

Source: *Reproduced with permission from Truckenmüller et al. [24], copyright 2011, Wiley-VCH.*

example, by a SMART process in the form of a combination of microscale thermoforming and hot embossing or thermal micro- or nanoimprint [37,38,60,61]. In this SMART process for the generation of surface or bulk micro- or nano topographies on or in thermoformed films (Figure 4.19), first, the film is thermally imprinted (Figure 4.19A). Then, the film is thermoformed (Figure 4.19B) with the thermoforming temperature preferably, but not necessarily, below the imprinting temperature. Alternatively, overlaid topographies can be generated by a combination of microthermoforming and DUV direct lithography or X-ray lithography [62]. Here, the film is first irradiated through a DUV or X-ray mask, then thermoformed, and finally developed.

4.4.2.3 Microelectrodes

For *in situ* characterization of cells cultivated in the containers of the thermoformed KITChip by impedance measurement or measurement of electrochemically accessible parameters, or for stimulation of electrically excitable cells, the curved container walls can be provided with 3D microelectrodes, even in highly stretched wall areas. This overlaid surface modification or coating can be achieved by a SMART process in the form of a combination of microscale thermoforming and micropatterned electroless plating [12,35]. In the SMART process for the generation of microelectrodes on thermoformed films, first, the film is conditioned or activated by local deposition of single metal seeds in the form of atom clusters, for example, by short-time sputter deposition through a slotted mask sheet. In contrast to closed metal layers, the disconnected seeds do not fail during forming. Rather their surface density increases. The film is then thermoformed and subjected to an electroless plating process in an autocatalytic chemically reductive electrolyte. The seeds then expand in dimension isometrically and eventually bridge with one another to form a continuous, coherent electrically conducting film. Further applications of metal thin films particularly from gold patterned on the chip surface are applications where biomolecules shall be covalently linked to specific surface domains via gold-thiol chemistry [63,64].

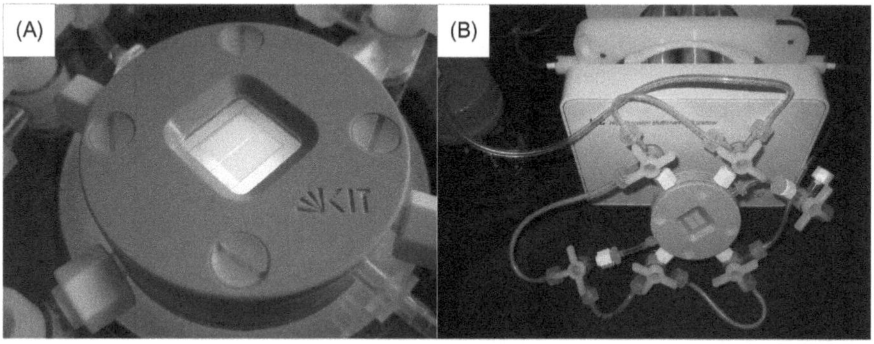

FIGURE 4.20

(A) Bioreactor, (B) connectors, tubings, three-way valves, medium reservoir, and cassette pump. The bioreactor housing consists of three disks from polyetheretherketone (PEEK), with two injection-molded 3^D-KITChips inserted. The upper and lower lid is screwed against the middle part housing the chips, and in each case sealed against a coverslip in its central opening by a silicone gasket. Similar gaskets are used to seal the top of the chips against the chip bottoms in the area of their frame.

4.5 Integration into microfluidic bioreactors

For an active supply with cell culture medium, microfluidic reactors can be inserted into modular bioreactor housings that can support the operation of multiple devices in parallel. In the case of 3^D-KITChip, up to 32 devices can be operated, but this maximum number may be increased depending upon the required complexity of the experiment. For fundamental basic science experiments, housing hardware can be expanded to host two microfluidic arrays (Figure 4.20A). The lids contain microscope coverslips that enable real-time microscopic analysis. The bioreactor is divided into three individually addressable microfluidic compartments; a central compartment which is located between the two chips and a compartment residing on the apical and basal surface of the central layer. It is possible to arrange different supply modes by simply adjusting the connectivity of the six inlet and outlet ports of the bioreactor via three-way valves. The superfusion mode can be characterized by a flow parallel to the top and bottom of each chip. This results in a low shear stress mode of operation and is therefore suited for cells sensitive for this type of interference. In the perfusion mode, the medium is entering the central compartment, passes the porous bottom of each chip, therefore passing the tissues inside its microcontainers, and leaves the bioreactor on top of each chip. By this type of operation, a gradient formation between the top and the bottom of the chips is effectively prevented. When interconnecting the three microfluidic compartments, even a mixed mode of operation is possible. The flow rates can be adjusted within a wide range due to the use of a cassette pump that can be equipped with different tubing. Because the whole setup consisting of a chip-containing bioreactor, a pump and a medium reservoir is a closed sterile loop (Figure 4.20B), it can be either placed in a conventional incubator or in an environment at 37°C with the support of additional gas lines. The used external gas mixing station is capable of a stepless variable mixing of all necessary gases. For monitoring purposes, fiber-optic O_2-, pH-, pCO_2-, or other sensors can be incorporated into the circulation.

4.6 Conclusion

Microfluidic cell culture devices have the potential to transform the traditional culture conditions by precisely engineering the cell culture microenvironment. Complex fabrication techniques form a critical cornerstone for the advancement of this novel paradigm. The 3^D-KITChip is a prominent example of this technology. Specifically, the 3^D-KITChip is a polymer chip for 3D cell culture in form of an array of microcontainers provided with porous bottoms or walls to allow an active perfusion of the cells therein with cell culture medium within dedicated microfluidic bioreactors. Depending on the application, these bioreactors can house 1–32 chips and can be operated in a superfusion, perfusion, or a mixed superfusion-perfusion mode. The KITChip is fabricated by micro injection molding, hot embossing, or microscale thermoforming. To introduce porosity in the chips, holes are laser-drilled in the container bottoms, commercial ion-track-etched membranes are bonded to chips with removed bottoms, or heavy-ion-irradiated membranes are thermoformed and the ion tracks are etched out. For tailoring the biophysicochemical microenvironment of the cells cultured in the containers of the KITChip, the container walls can be functionalized. In connection with introducing porosity as well as functionalizing the container walls, thermoforming allows not only to modify the bottoms of the containers but also their sidewalls, and that patterned with lithographic resolution. Modifications include cell adhesion micropatterns, surface and bulk micro- and nano topographies or microelectrodes. The advancements in microfabrication methods will continue to empower microfluidic cell culture technologies such as the KITChip and beyond.

References

[1] B. Altmann, A. Welle, S. Giselbrecht, R. Truckenmüller, E. Gottwald, The famous versus the inconvenient—or the dawn and the rise of 3D-culture systems, World J. Stem Cells 1 (2009) 1–6.

[2] R.M. Sutherland, Cell and environment interactions in tumor microregions: the multicell spheroid model, Science 240 (1988) 177–184.

[3] H. Dertinger, D. Hülser, Increased radioresistance of cells in cultured multicell spheroids. I. Dependence on cellular interactions, Radiat. Environ. Biophys. 19 (1981) 101–107.

[4] B. Altmann, S. Giselbrecht, K.F. Weibezahn, A. Welle, E. Gottwald, The three-dimensional cultivation of the carcinoma cell line HepG2 in a perfused chip system leads to a more differentiated phenotype of the cells compared to monolayer culture, Biomed. Mater. 3 (034120) (2008) 10.

[5] W. Mueller-Klieser, Three-dimensional cell cultures: from molecular mechanisms to clinical applications, Am. J. Physiol. Cell Physiol. 273 (1997) C1109–C1123.
[6] C. Fidkowski, M.R. Kaazempur-Mofrad, J.T. Borenstein, J.P. Vacanti, R. Langer, Y. Wang, Endothelialized microvasculature based on a biodegradable elastomer, Tissue Eng. 11 (1) (2005) 302–309.
[7] C.J. Bettinger, B. Orrick, A. Misra, R. Langer, J.T. Borenstein, Microfabrication of poly(glycerol-sebacate) for contact guidance applications, Biomaterials 27 (2006) 2558–2565.
[8] C.J. Bettinger, K.M. Cyr, A. Matsumoto, R. Langer, J.T. Borenstein, D.L. Kaplan, Silk fibroin microfluidic devices, Adv. Mater. 19 (19) (2007) 2847–2850.
[9] C.J. Bettinger, J.T. Borenstein, R. Langer, Microfabrication techniques in scaffold development, in: C.T. Laurencin, L. Nair (Eds.), Nanotechnology and Tissue Engineering: The Scaffold, CRC Press, Boca Raton, FL, 2008, pp. 87–119.
[10] C.J. Bettinger, J.T. Borenstein, R. Langer, Micro- and nanofabricated scaffolds, in: R. Lanza, R. Langer, J.P. Vacanti (Eds.), Principles of Tissue Engineering, third ed., Elsevier, Burlington, MA, 2007, pp. 341–358.
[11] E. Eschbach, S.S. Chatterjee, M. Nöldner, E. Gottwald, H. Dertinger, K.F. Weibezahn, et al., Microstructured scaffolds for liver tissue cultures of high cell density: morphological and biochemical characterization of tissue aggregates, J. Cell. Biochem. 95 (2005) 243–255.
[12] E. Gottwald, S. Giselbrecht, C. Augspurger, N. Dambrowsky, R. Truckenmüller, V. Piotter, et al., A chip-based platform for the generation of three-dimensional tissues in vitro, Lab Chip 7 (2007) 777–785.
[13] G. Knedlitschek, F. Schneider, E. Gottwald, T. Schaller, E. Eschbach, K.F. Weibezahn, A tissue-like culture system using microstructures: influence of extracellular matrix material on cell adhesion and aggregation, J. Biomech. Eng. Trans. ASME 121 (1999) 35–39.
[14] D.A. Bruzewicz, A.P. McGuigan, G.M. Whitesides, Fabrication of a modular tissue construct in a microfluidic chip, Lab Chip 8 (2008) 663–671.
[15] Y. Yang, K. Kulangara, J. Sia, L. Wang, K.W. Leong, Engineering of a microfluidic cell culture platform embedded with nanoscale features, Lab Chip 11 (9) (2011) 1638–1646.
[16] C.J. Bettinger, J.T. Borenstein, Biomaterials-based microfluidics for engineered tissue constructs, Soft Matter. 6 (20) (2010) 4999–5015.
[17] G.V. Mazariegos, D.J. Kramer, R.C. Lopez, A. Obaid Shakil, A.J. Rosenbloom, M. DeVera, et al., Safety observations in phase I clinical evaluation of the excorp medical bioartificial liver support system after the first four patients, ASAIO J. 47 (5) (2001) 471–475.
[18] J.F. Patzer, Oxygen consumption in a hollow fiber bioartificial liver—revisited, Artif. Organs 28 (1) (2004) 83–98.
[19] T. Hanemann, M. Heckele, V. Piotter, Current status of micromolding technology, Polym. News 25 (2000) 224–229.
[20] M. Heckele, W.K. Schomburg, Review on micro molding of thermoplastic polymers, J. Micromech. Microeng. 14 (2004) R1–14.
[21] V. Piotter, W. Bauer, T. Hanemann, M. Heckele, C. Müller, Replication technologies for HARM devices: status and perspectives, Microsyst. Technol. 14 (2008) 1599–1605.
[22] S.M. Kim, S.H. Lee, K.Y. Suh, Cell research with physically modified microfluidic channels: a review, Lab Chip 8 (7) (2008) 1015–1023.

[23] A.W. Martinez, S.T. Phillips, G.M. Whitesides, Three-dimensional microfluidic devices fabricated in layered paper and tape, Proc. Natl. Acad. Sci. U.S.A. 105 (50) (2008) 19606–19611.
[24] R. Truckenmüller, S. Giselbrecht, N. Rivron, E. Gottwald, V. Saile, A. van den Berg, et al., Thermoforming of film-based biomedical microdevices, Adv. Mater. 23 (2011) 1311–1329.
[25] J. Giboz, T. Copponnex, P. Mele, Microinjection molding of thermoplastic polymers: a review, J. Micromech. Microeng. 17 (2007) R96–109.
[26] V. Piotter, T. Hanemann, R. Ruprecht, J. Hausselt, Injection molding and related techniques for fabrication of microstructures, Microsyst. Technol. 3 (1997) 129–133.
[27] V. Piotter, K. Mueller, K. Plewa, R. Ruprecht, J. Hausselt, Performance and simulation of thermoplastic micro injection molding, Microsyst. Technol. 8 (2002) 387–390.
[28] S. Giselbrecht, T. Gietzelt, E. Gottwald, A.E. Guber, C. Trautmann, R. Truckenmüller, et al., Microthermoforming as a novel technique for manufacturing scaffolds in tissue engineering (CellChips®), IEEE Proc. Nanobiotechnol. 151 (2004) 151–157.
[29] M. Heckele, W. Bacher, K.D. Müller, Hot embossing—the molding technique for plastic microstructures, Microsyst. Technol. 4 (1998) 122–124.
[30] M. Worgull, *Hot Embossing—Theory and Technology of Microreplication*, William Andrew, Norwich, NY, USA, 2009.
[31] S. Giselbrecht, E. Gottwald, R. Truckenmüller, C. Trautmann, A. Welle, A.E. Guber, et al., Microfabrication of chip-sized scaffolds for three-dimensional cell cultivation, J. Vis. Exp. 15 (2008) http://www.jove.com/index/details.stp?ID=699, doi: 10.3791/699.
[32] J.L. Throne, *Technology of Thermoforming*, Hanser, Munich, Germany, 1996.
[33] R. Truckenmüller, Z. Rummler, T. Schaller, W.K. Schomburg, Low-cost thermoforming of micro fluidic analysis chips, J. Micromech. Microeng. 12 (2002) 375–379.
[34] R. Truckenmüller, S. Giselbrecht, Microthermoforming of flexible, not buried hollow microstructures for chip-based life sciences applications, IEEE Proc. Nanobiotechnol. 151 (2004) 163–166.
[35] R. Truckenmüller, S. Giselbrecht, C.A. van Blitterswijk, N. Dambrowsky, E. Gottwald, T. Mappes, et al., Flexible fluidic microchips based on thermoformed and locally modified thin polymer films, Lab Chip 8 (2008) 1570–1579.
[36] S. Giselbrecht, T. Gietzelt, E. Gottwald, C. Trautmann, R. Truckenmüller, K.F. Weibezahn, et al., 3D tissue culture substrates produced by microthermoforming of pre-processed polymer films, Biomed. Microdevices 8 (2006) 191–199.
[37] M. Reinhardt, A. Schleunitz, R. Truckenmüller, S. Bartels, E. Gottwald, V. Saile, et al., Mikrostrukturierte polymerfolien mit integrierter nanotopographie als zell-interface, Proc. 1. GMM Workshop Mikro-Nano-Integration, Seeheim, Germany, 2009.
[38] R. Truckenmüller, S. Giselbrecht, M. Escalante-Marun, M. Groenendijk, B. Papenburg, N. Rivron, et al., Fabrication of cell container arrays with overlaid surface topographies, Biomed. Microdevices 14 (1) (2012) 95–107.
[39] M. Ikeuchi, K. Ikuta, Artificial capillary network chip for in vitro 3D tissue culture, Proc. 14th International Conference on Solid-State Sensors, Actuators and Microsystems, *Transducers*, Lyon, France, 2007, pp. 1337–1340.
[40] P. Nagarajan, D.G. Yao, Rubber-assisted micro forming of polymer thin films, Microsyst. Technol. 15 (2009) 251–257.

[41] H. Dreuth, C. Heiden, Thermoplastic structuring of thin polymer films, Sens. Actuators A 78 (1999) 198–204.

[42] M. Buitinga, C.J.M. Loomans, M.A. Engelse, M. Karperien, E.J.P. de Koning, A.A. van Apeldoorn, et al., Nano-engineered microchips for extrahepatic islet of Langerhans transplantation, Proceedings of the Second Tissue Engineering and Regenerative Medicine International Society World Congress, TERMIS, Seoul, Republic of Korea, 2009.

[43] R.L. Fleischer, P.B. Price, R.M. Walker, Nuclear Tracks in Solids. Principles and Applications, University of California Press, Berkeley, CA, 1975.

[44] R. Spohr, *Ion Tracks and Microtechnology. Principles and Applications*, Vieweg, Braunschweig, Germany, 1990.

[45] S. Gerecht, C.J. Bettinger, Z. Zhang, J. Borenstein, G. Vunjak-Novakovic, R. Langer, The effect of actin disrupting agents on contact guidance of human embryonic stem cells, Biomaterials 28 (28) (2007) 4068–5077.

[46] M.J. Dalby, A. Hart, S.J. Yarwood, The effect of the RACK1 signalling protein on the regulation of cell adhesion and cell contact guidance on nanometric grooves, Biomaterials 29 (3) (2008) 282–289.

[47] C.J. Bettinger, Z. Zhang, S. Gerecht, J.T. Borenstein, R. Langer, Enhancement of *in vitro* capillary tube formation by substrate nanotopography, Adv. Mater. 20 (1) (2008) 99–103.

[48] C. Bettinger, R. Langer, J. Borenstein, Engineering substrate topography at the micro- and nanoscale to control cell function, Angew. Chem. 48 (30) (2009) 5406–5415.

[49] E. Gottwald, B. Lahni, D. Thiele, S. Giselbrecht, A. Welle, K.F. Weibezahn, Chip-based three-dimensional cell culture in perfused micro-bioreactors, J. Vis. Exp. 15 (2008) http://www.jove.com/index/details.stp?ID=564, doi:10.3791/564.

[50] C.J. Flaim, S. Chien, S.N. Bhatia, An extracellular matrix microarray for probing cellular differentiation, Nat. Methods 2 (2) (2005) 119–125.

[51] S.R. Khetani, S.N. Bhatia, Microscale culture of human liver cells for drug development, Nat. Biotechnol. 26 (2007) 120–126.

[52] G.K. Toworfe, R.J. Composto, C.S. Adams, I.M. Shapiro, P. Ducheyne, Fibronectin adsorption on surface-activated poly(dimethylsiloxane) and its effect on cellular function, J. Biomed. Mater. Res. 71A (2004) 449–461.

[53] M.E. Manwaring, J.F. Walsh, P.A. Tresco, Contact guidance induced organization of extracellular matrix, Biomaterials 25 (17) (2004) 3631–3638.

[54] Y. Ito, Surface micropatterning to regulate cell functions, Biomaterials 20 (23-24) (2000) 2333–2342.

[55] N. Zaari, P. Rajagopalan, S.K. Kim, A.J. Engler, J.Y. Wong, Photopolymerization in microfluidic gradient generators: microscale control of substrate compliance to manipulate cell response, Adv. Mater. 16 (23–24) (2004) 2133–2137.

[56] J.A. Burdick, A. Khademhosseini, R. Langer, Fabrication of gradient hydrogels using a microfluidics/photopolymerization process, Langmuir 20 (13) (2004) 5153–5156.

[57] C. Richter, M. Reinhardt, S. Giselbrecht, D. Leisen, R. Truckenmüller, A. Blau, et al., Spatially controlled cell adhesion on three-dimensional substrates, Biomed. Microdevices 12 (2010) 787–795.

[58] X.F. Walboomers, H. Croes, L.A. Gensel, J.A. Jansen, Contact guidance of rat fibroblasts on various implant materials, J. Biomed. Mater. Res. 47 (1999) 204–212.

[59] A.-S. Andersson, P. Olsson, U. Lidberg, D. Sutherland, The effects of continuous and discontinuous groove edges on cell shape and alignment, Exp. Cell Res. 288 (1) (2003) 177–188.

[60] S. Giselbrecht, M. Reinhardt, A. Schleunitz, E. Gottwald, R. Truckenmüller, Microstructured cell culture chips with integrated nanotopography, Proceedings of the Fifth International Conference on Microtechnologies in Medicine and Biology, MMB, Quebec City, QC, Canada, 2009, pp. 246–247.

[61] A. Schleunitz, R. Truckenmüller, M. Reinhardt, S. Giselbrecht, Fabrication of 3D cell containers with integrated topography by combined microscale thermoforming and thermal nanoimprint, Proceedings of the Eighth International Conference on Nanoimprint and Nanoprint Technology, NNT, San Jose, CA, 2009.

[62] S. Giselbrecht, M. Reinhardt, T. Mappes, M. Börner, E. Gottwald, C. van Blitterswijk, et al., Closer to nature—bio-inspired patterns by transforming latent lithographic images, Adv. Mater. 23 (11) (2011) 1311–1329.

[63] P. Harder, M. Grunze, R. Dahint, G.M. Whitesides, P.E. Laibinis, Molecular conformation in oligo(ethylene glycol)-terminated self-assembled monolayers on gold and silver surfaces determines their ability to resist protein adsorption, J. Phys. Chem. B 102 (2) (1998) 426–436.

[64] N.T. Flynn, T.N.T. Tran, M.J. Cima, R. Langer, Long-term stability of self-assembled monolayers in biological media, Langmuir 19 (26) (2003) 10909–10915.

CHAPTER

Mechanobiological Approaches for the Control of Cell Motility

5

Sang-Hee Yoon*, Young Kyun Kim and Mohammad R.K. Mofrad
Department of Bioengineering, University of California, Berkeley, CA, USA

5.1 Introduction

Cell movement is a multistep process that results from concerted biological queues such as environmental factors, signal transduction, and cytoskeletal rearrangement. The migratory behavior of an adherent cell is known as a sequential combination of the following processes: the morphological polarization of a cell into a teardrop shape; the extension of the cell membrane; the formation of cell–substrate adhesions at a leading edge; the translocation of the cell body via contractile force; the detachment of cell–substrate adhesions at a trailing edge. Directed cell motility is essential to the development and maintenance of multicellular organisms which require the orchestrated movement of adherent cells to specified sites in particular directions [1]. Any abnormality in cell motility is known to lead to serious pathophysiological consequences such as mental retardation, vascular disease, tumor formation, and metastasis [2]. A better control of cell motility therefore has a major impact on the development of new therapeutic strategies for the pathophysiological consequences. For example, a successful control of cell motility can lead to the development of new medical dressings for fast wound healing. In detail, fibroblasts and epithelial cells move intensively toward open wounds during wound healing process, especially proliferation phase [3,4]. If this chemotactic cell movement is controlled (i.e., expedited with controllability), then the wound healing process will be significantly promoted.

Due to the biological importance of cell motility in pathophysiological phenomena, many studies have been intensively conducted to understand the migratory behavior of adherent cells. Previous microelectromechanical systems (MEMS)-based approaches have primarily addressed the effect of chemically and

*Current address: Wyss Institute for Biologically Inspired Engineering, Harvard University, Cambridge MA, USA.

Microfluidic Cell Culture Systems. ISBN: 978-1-4377-3459-1

topographically patterned substrates on cell–substrate interactions that regulate cell motility [5]: microscale (isotropic and anisotropic) chemical patterns on a variety of substrates were used to characterize cell patterning and growth along the patterns [7] which were closely related to cell motility, and nanoscale (isotropic and anisotropic) chemical patterns were utilized to regulate collective cell functions such as migration, adhesion, proliferation, and differentiation [8,9]; (micro and nano) isotropic topographical patterns (e.g., evenly or randomly distributed pits or protrusions) were fabricated to examine the collective cell functions [10,11], and (micro and nano) anisotropic topographical patterns (e.g., parallel troughs and ridges) were studied to understand cell alignment along the patterns [12,13]. Although these attempts were successful in characterizing the effect of chemical and topographical patterns on the migratory behavior of adherent cells, their success was far from controlling cell motility (e.g., direction and rate of cell movement).

Recently, controlling the direction and rate of cell movement has been found to have a prominent association with new therapy developments for diseases such as cardiovascular diseases, asthma, and cancer. To meet the needs of the time, several passive methods using MEMS technology were developed to control the motility of adherent cells. At first, "wound healing" assays were developed to manipulate cell movement. The assays involved deliberately inflicting wounds on a confluent monolayer of adherent cells [14–16]. They induced the cells to spontaneously polarize and migrate toward the wounds. However, these assays showed very limited controllability of cell movement—no control for the movement path of adherent cells. Since Harrison [17] found cell responses to solid structures (i.e., contact guidance), a control for cell motility using a variety of topographical features has been extensively researched. For example, parallel grooves [18–21], spikes [22], three-dimensional post arrays [23], meshes [24], and porosities [25,26] were used to enhance cell movement along the topographical features and to suppress cell movement across them. However, these methods were still far from a full control of cell movement, i.e., moving an adherent cell to a specified site in a particular direction. Active methods using MEMS technology were also carried out to regulate cell movement. Chemotactic assays [27–32] were the mainstream in these approaches. The assays adjusted the concentration distribution of chemoattractants (or chemorepellents) using two microchambers connected to each other through a microvalve. For example, cells and chemoattractants (or chemorepellents) were placed in each of the chambers and the gradient of chemical signals around the cells was manipulated by regulating the microvalve. Although the chemotactic assays somewhat controlled cell movement, they were unable to control cell motility at a single-cell level. Moreover, the assays depended on the use of active microdevices (e.g., microvalves and microregulators) which placed restrictions on the extent of their biomedical applications. As a slightly different approach, various experimental techniques which manipulate cell detachment (or cell adhesion) were developed to control the motility of adherent cells. The techniques can be categorized into photolithography [33,34], e-beam lithography [35], dip-pen lithography [36], nanoimprint lithography [37], microcontact printing [38,39], elastomeric stencil [40], ink-jet printing [41], and optical tweezer [42]. Although these techniques successfully manipulated cell detachment (or cell adhesion), they failed to show a concrete evidence of controlled cell movement. Furthermore, these techniques applied unintended mechanical stimuli (e.g., cell denaturation [42]) to cells, which considerably deformed the cells before and during experiment.

In this chapter, we present two mechanobiological approaches using MEMS technology, each of which respectively manipulates cell motility in passive and active ways. As a passive approach, a biological assay carved with diverging micropatterns, termed "Rome platform," which makes all adherent cells lead to the Rome (target site) is developed. This method has the following specialties in controlling the motility of adherent cells. First of all, this controls the rate and direction of cell movement by imposing one-directional morphological polarity on adherent cells with diverging micropatterns, which is as similar as possible to *in vivo* cell movement. This method therefore makes it possible to manipulate the movement of *undisturbed* adherent cells. Additionally, this controls cell movement in a *passive* way, i.e., by adjusting the dimensions of diverging micropatterns. To control the motility of adherent cells in an active way, one part of an adherent cell is selectively detached from a substrate using a mechanobiological assay, nicknamed "biological breadboard (BBB)." This assay spatiotemporally manipulates cell detachment at a subcellular level using the reductive desorption of a gold–thiol self-assembled monolayer (SAM) at a negative activation potential of −0.9 to −1.65 V. This assay has the following features in manipulating cell movement. This controls the migratory behavior of *living and intact* cells. By employing an arginine–glycine–aspartic (RGD) peptide as a cell adhesion motif, the BBB provides adherent cells with a microenvironment that is as similar as possible to *in vivo* microenvironment. This is because a microenvironment is one of the most dominant factors in determining the motility of adherent cells.

5.2 Passive control of cell motility

5.2.1 Rome platform

A Rome platform consists of parallel and narrow troughs (zone I), diverging troughs (zone II), and parallel and wide troughs (zone III), as shown in Figure 5.1. The dimensions of the diverging micropatterns are chosen to be trough depth (or ridge height) of 3 μm, divergence angle of 0.5–5.0°, and width of trough/ridge of 3–10 μm. As the dimensions are within a range comparable with cell (especially, NIH 3T3 fibroblasts) size, the migratory behavior of the cells is significantly influenced by the diverging micropatterns.

The diverging micropatterns achieve one-directional cell movement by imposing one-directional morphological polarity on adherent cells. The working principle of the Rome platform, which passively controls the cell movement, can be explained by both cell biology and fluid mechanics. An adherent cell on the diverging micropatterns senses microscopic structural nonuniformities (i.e., diverging troughs and

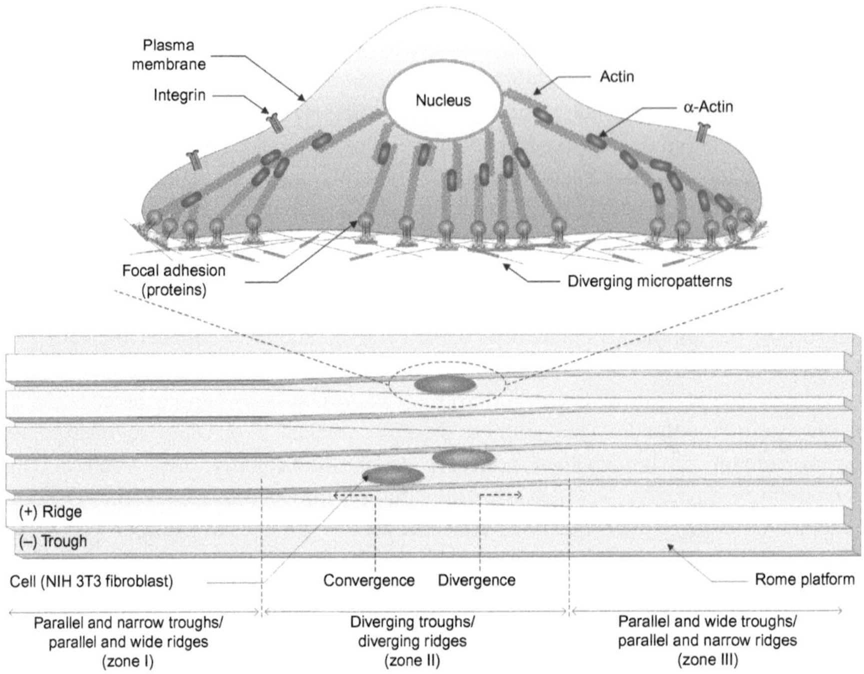

FIGURE 5.1

A biological assay carved with diverging micropatterns, termed "Rome platform," to control cell movement in a passive way. The Rome platform is composed of parallel and narrow troughs (zone I), diverging troughs (zone II), and parallel and wide troughs (zone III). The zone II has different angles of 0.5°, 1.0°, 2.0°, 4.0°, and 5.0° and two widths of trough/ridge of 3 μm (beginning of divergence) and 10 μm (end of divergence). This assay imposes one-directional morphological polarity on adherent cells using the zone II, thus manipulating the motility of adherent cells.

ridges), and starts to change its morphology into a teardrop shape (Figure 5.2A). This change in cell morphology makes the cell move into the diverging direction [1,43]. Even when a cell initially travels in the opposite direction (i.e., converging direction), the diverging micropatterns make the cell change its movement direction (Figure 5.2B, C). That is, the microscopic structural nonuniformities of the diverging micropatterns apply one-directional spatial stimuli to the cell through cell–substrate interactions, thus forcing the cell to move in a specified direction.

From a fluid mechanics viewpoint, cell movement in a diverging micropattern (Figure 5.2A, bottom) can be assumed as a viscous flow in a rectangular microchannel with a channel length of l_{ch}, a width of w_{ch}, and a height of h_{ch}. The flow resistance in the microchannel, Ω_{ch}, is expressed as

$$\Omega_{ch} = \frac{12\eta l_{ch}}{{h_{ch}}^3 w_{ch}} \left\{ 1 - \sum_{n=1,3,5,\ldots}^{\infty} \frac{1}{n^5} \frac{192}{\pi^5} \frac{h_{ch}}{w_{ch}} \tanh\left(\frac{n\pi w_{ch}}{2h_{ch}} \right) \right\}^{-1} \tag{5.1}$$

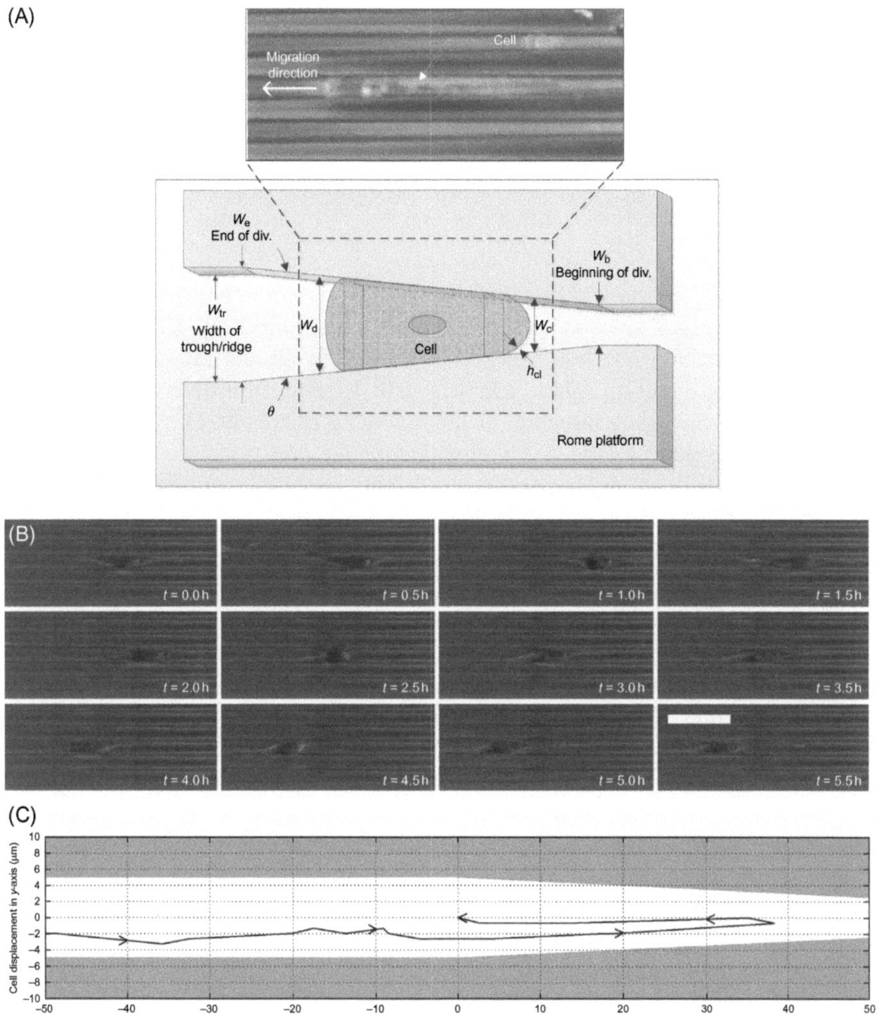

FIGURE 5.2

Passive control of cell (NIH 3T3 fibroblasts) movement using the Rome platform. (A) Cell movement guided by the Rome platform. An adherent cell changes its morphological shape along the troughs and ridges of the diverging micropatterns. The change in cell morphology on the Rome platform, similar to that of *in vivo* cell movement, makes the cell move in the diverging direction of the diverging micropatterns. (B) Sequential optical images showing the migratory behavior of NIH 3T3 fibroblasts on the diverging micropatterns. Even when an adherent cell starts to move in the opposite direction (converging direction of the diverging micropatterns), the diverging micropatterns compel the cell to move in the diverging direction. (C) Movement path of the cell obtained from the sequential optical images of (B). Scale bar of (B) is 50 μm.

where η is a fluid viscosity [44]. The ratio of fluidic resistance in a diverging direction to that in a converging direction, Ω_d/Ω_c, is written as

$$\frac{\Omega_d}{\Omega_c} = \frac{w_c}{w_d}\left\{\frac{1-\sum_{n=1,3,5,\ldots}^{\infty}(1/n^5)(192/\pi^5)(h_{cl}/w_c)\tanh(n\pi w_c/2h_{cl})}{1-\sum_{n=1,3,5,\ldots}^{\infty}(1/n^5)(192/\pi^5)(h_{cl}/w_d)\tanh(n\pi w_d/2h_{cl})}\right\} \approx \frac{w_c}{w_d} < 1 \tag{5.2}$$

where w_c and w_d are the widths of the leading and the trailing edges with a height of h_{cl}, respectively. The calculated ratio is less than 1, meaning cell movement in a diverging direction has less energy loss caused by fluidic resistance than that in a converging direction. Thus, adherent cells located on a diverging micropattern move in a diverging direction of the diverging micropattern, rather than in a converging direction.

5.2.2 Materials and methods

5.2.2.1 Microfabrication

The fabrication of the Rome platform started by making a reusable master mold on a silicon (Si) wafer. A 4 in (500 μm thick) Si wafer was cleaned with a piranha solution of 1:1 (v/v) 96% sulfuric acid (H_2SO_4) and 30% hydrogen peroxide (H_2O_2) for 10 min. To prepare a master mold, 3 μm thick photoresist (SU-8 2002, MicroChem Corp.) was spin-coated at 1000 rpm for 30 s, followed by soft baking at 95°C for 2 min, exposure at 135 mJ/cm^2, postexposure baking at 95°C for 2 min, and developing with an organic solvent solution (SU-8 developer, MicroChem Corp., Figure 5.3A). A polydimethylsiloxane (PDMS, MicroChem Corp.) stamp was fabricated by pouring PDMS over the SU-8 master, curing at 65°C for 2 h, and peeling it off from the SU-8 master mold (Figure 5.3B). A photocurable hybrid polymer substrate (ORMOCER, MicroChem Corp.) was prepared through spin-coating (2000 rpm, 30 s) and ultraviolet (UV) curing (300 mJ/cm^2 at 365 nm wavelength) of 2 ml of ORMOCER resin on a 150 μm thick microscope cover glass that was pretreated with an oxygen plasma chamber (PM-100 Plasma Treatment System, March Plasma Systems, Inc.) at 100 W for 30 s (Figure 5.3C). After additional spin-coating of 1 ml of ORMOCER resin and prebaking at 80°C for 2 min, the PDMS stamp was imprinted on the ORMOCER layer to transfer diverging micropatterns (Figure 5.3D). This ORMOCER substrate was UV cured (Figure 5.3E), followed by demolding and hard baking at 150°C for 1 h (Figure 5.3F).

The fabricated assay was transparent in the visual spectrum (Figure 5.3G). Our assay consisted of parallel and narrow troughs (zone I), diverging troughs (zone II), and parallel and wide troughs (zone III). Five kinds of 3 μm thick assays (Figure 5.3H) have different divergence angles, θ, of 0.5, 1.0, 2.0, 4.0, and 5.0° in the zone II, a narrow width of trough (beginning of divergence), w_b, of

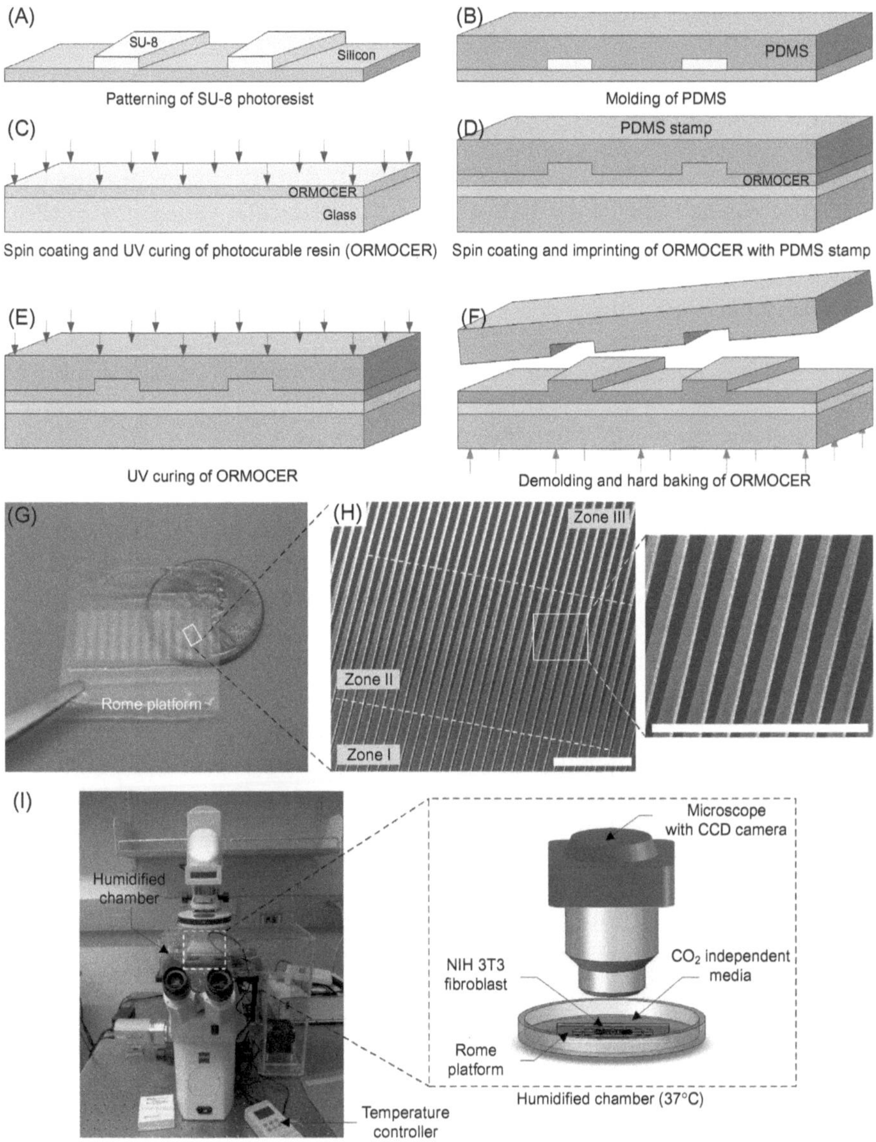

FIGURE 5.3

Microfabrication and experimental setup of the Rome platform. (A–F) Fabrication process: patterning of SU-8 photoresist on a 4 in silicon wafer to prepare a master for PDMS molding (A); PDMS molding to fabricate a reusable PDMS stamp (B); first spin-coating and UV curing of ORMOCER resin on a microscope cover glass to fabricate the substrate of the Rome platform (C); second spin coating of ORMOCER resin, followed by

(*Continued*)

3 μm in the zone I, and a wide width of trough (end of divergence), w_e, of 10 μm in the zone III.

5.2.2.2 Cell culture

NIH 3T3 mouse embryonic fibroblast cells (NIH 3T3 fibroblasts) were cultured in a Dulbecco's modified eagle medium (DMEM, GIBCO™) supplemented with 10% fetal bovine serum (FBS, GIBCO™) and 1% Penicillin–Streptomycin (Pen/Strep, GIBCO™) at 37°C in a humidified atmosphere of 5% CO_2. Cells were passaged every 5 days as follows. The cells were washed once in a 1× phosphate buffered saline (PBS, Sigma-Aldrich) solution and then trypsinized with a 0.5% Trypsin–EDTA solution (Sigma-Aldrich). After centrifuging the cells, they were inoculated into a new petri dish. NIH 3T3 fibroblasts with a passage number of 10–20 were used in our experiments. Before each experiment, the fabricated assay was sterilized with 70% ethanol, washed twice with a 1× PBS solution, and incubated with a CO_2 independent medium (GIBCO™) supplemented with 10% FBS and 1% Pen/Strep for 1 h at room temperature. The assay was placed in a petri dish containing a 5 ml CO_2 independent medium with a cell suspension of about 1×10^4 cells/ml. After 1 h of cell seeding, unadhered cells were removed by washing in a 1× PBS solution, followed by replacement with a fresh CO_2 independent medium (10% FBS and 1% Pen/Strep). All experiments were carried out after 24 h of cell seeding at 37°C in a humidified chamber.

5.2.2.3 Confocal immunofluorescence microscopy

The state of contact of NIH 3T3 fibroblasts to diverging micropatterns was assessed from immunofluorescence images (Figure 5.4F). Cells were fixed with a 4% (v/v) formaldehyde (Fisher Scientific) diluted with chilled 1× PBS solution for 15 min. The fixed cells were permeabilized with a 200 μl 0.5% (v/v) Triton X-100 (Sigma-Aldrich) diluted with 1× PBS solution at room temperature for 10 min and then were washed three times with a 1× PBS solution, followed by blocking nonspecific binding using a 3% (w/v) nonfat dry milk (Cell Signaling Technology, Inc.) diluted with 1× PBS solution at 4°C for 1 h and washing the cells once with a 1× PBS solution. A 10 μl methanolic stock solution of rhodamine phalloidin (Biotium, Inc.) was diluted with a 200 μl 1× PBS solution

◀ imprinting with the PDMS stamp to define micropatterns (D); second UV curing (E); demolding and hard baking (F). (G and H) Microfabricated assay. Optical photograph of the assay placed on top of a one-cent coin, showing its transparency (G); scanning electron microscopy (SEM) images of the assay (H). (I) Experimental setup. NIH 3T3 fibroblasts are loaded on the assay at 37°C in a CO_2 independent media supplemented with 10% FBS and 1% Pen/Strep. The migratory behavior of the cells is monitored with an inverted microscope equipped with a charge-coupled device (CCD) camera to obtain their sequential images. Scale bars of (H) are 100 μm.

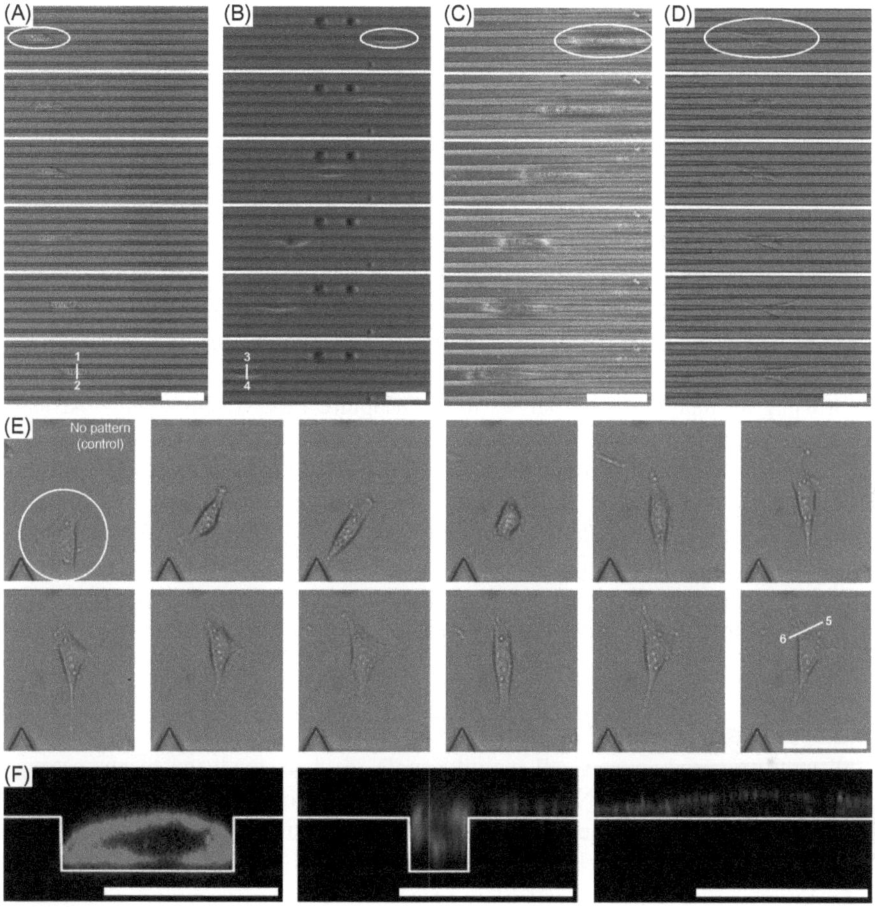

FIGURE 5.4

Optical and confocal images of NIH 3T3 fibroblasts moving on the Rome platform. (A) Guided cell movement along a single parallel and wide trough (zone III). (B) Guided cell movement along a single parallel and narrow trough (zone I). (C) Guided cell movement along a single diverging trough (zone II). (D) Guided cell movement along multiple (more than two) micropatterns. (E) Unguided cell movement on a flat substrate with no topographical feature (control group). All images of (A–E) are taken every 30 min 24 h after cell loading. (F) Confocal images of the cells placed on the zone III (left, along line 1–2 of (A)), on the zone I (center, along line 3–4 of (B)), and on the flat substrate (right, along line 5–6 of (E)). The actin filaments of the cells were stained with rhodamine phalloidin (red) for immunofluorescence imaging. The confocal images show the cells contact to both two sides of the trough and bottom substrate. Scale bars of (A–E) are 50 μm and those of (F) are 10 μm. (For interpretation of the references to color in this figure legend, the reader is referred to the web version of this book.)

supplemented with 1% Bovine Serum Albumin (Fisher Scientific) for each immunofluorescence staining. The actin filaments of the cells were stained by incubating the cells with this solution for 20 min at room temperature and then washing three times with a 1× PBS solution. Confocal immunofluorescence images were obtained with a confocal laser scanning microscope (LSM 510, Carl Zeiss MicroImaging, Inc.).

5.2.2.4 Time-lapse microscopy

The migratory behavior of NIH 3T3 fibroblasts on the Rome platform was monitored with an inverted microscope (Axio Observer A1, Carl Zeiss MicroImaging, Inc.) equipped with a CCD digital camera (Retiga-SRV, QImaging; Figure 5.3I). The location of each cell was automatically tracked every 30 min interval for 24 h using an imaging and analysis tool (QCapture PRO 5, QImaging). At least 10 migrating cells, isolated from each other, were selected for statistical analysis in each experiment, thus obtaining 48 images for each cell—total 480 images from each experiment. Our experiments were conducted at least 10 times for each assay. The movement path of each cell was tracked from the obtained time-sequential images. In detail, after identifying a cell of interest, the coordinates of the cell (especially, nucleus) were obtained by analyzing the time-sequential images and then the relative displacement of the cell at each time interval was calculated.

5.2.2.5 Analysis of cell motility

Cell motility speed, s, and cell motility angle, ϕ, were analyzed to characterize the migratory behavior of NIH 3T3 fibroblasts. Cell motility speed and cell motility angle at each time interval, Δt, were calculated from the relative displacement, δx and δy, of each cell.

$$s = \sqrt{\left(\frac{\delta x}{\Delta t}\right)^2 + \left(\frac{\delta y}{\Delta t}\right)^2} \tag{5.3}$$

$$\phi = \arctan\left(\frac{\delta y}{\delta x}\right) \tag{5.4}$$

Cell motility angle is between 0° and 90°, where 0° represents a direction along micropatterns (x-axis direction) and 90° means a direction across them (y-axis direction). Cell movement on a flat substrate with no topographic feature therefore has a cell motility angle of about 45° [24].

The migratory behavior of NIH 3T3 fibroblasts migrating on the Rome platform was also quantified by fitting the mean-squared displacement, $\langle d^2(t)\rangle$, of the cells to a persistent random walk model. This mean-squared displacement was obtained from the location data of the cells over entire experiment time, ultimately yielding the directional persistence time and random motility coefficient of each cell. When calculating the mean-squared displacement, we used an overlapping time interval sampling method (e.g., 0–1, 0.5–1.5, and 1–2 h) over a time interval of $i\Delta t$ larger than the smallest time interval (i.e., 30 min). All mean-squared displacements were averaged as follows:

$$\langle d^2(t)\rangle = \frac{1}{N-i+1}\sum_{k=0}^{N-i+1}[\{x((k+i)\Delta t)-x(k\Delta t)\}^2 + \{y((k+i)\Delta t)-y(k\Delta t)\}^2] \quad (5.5)$$

where N is the total number of time intervals over all experiment time and x and y are the coordinates of the cell at each time interval [45]. This equation yielded a series of $\langle d^2(t)\rangle$ for increasing time interval value, $i\Delta t$. Directional persistence time and random motility coefficient were found by fitting the mean-squared displacement calculated from experimental data to a theoretical mean-squared displacement [45–48].

$$\langle d^2(t)\rangle = 2n_d\mu\{t - P(1 - e^{-t/P})\} \quad (5.6)$$

where n_d is a number of dimensions tracked in the experiments (i.e., $n_d = 2$ in our analysis), μ is a random motility coefficient defined as $\mu = s^2P/n_d$ [46], and P is a directional persistence time. The random motility coefficient can be interpreted as a diffusional representation of moving cell distribution, while the cell motility speed indicates a simple rate of cell movement. The directional persistence time means an average length of time between significant changes in a direction of cell movement along micropatterns. The values of P and μ for cells moving on the Rome platform were obtained from experimental data corresponding to one-third of total cell movement because the true movement path of cells was unable to be monitored for enough time [49]. Some cells whose measured mean-squared displacement showed much discrepancy from theoretical mean-squared displacement were not considered in this analysis.

Statistical analysis was made using a commercial statistical tool (Minitab 14, Minitab, Inc.) to quantify the migratory behavior of adherent cells (e.g., cell motility speed, cell motility angle, directional persistence time, and random motility coefficient). Each measurement result was analyzed using one-way analysis of variance (ANOVA) with Bonferroni *post hoc* test. *P*-values less than or equal to 0.05 were considered statistically significant. All data were represented as mean ± standard error (of the mean) values.

5.2.3 Results and discussion

5.2.3.1 Effect of micropatterns on cell motility

The effect of the dimensions and shape of micropatterns on cell motility was characterized. The movement of NIH 3T3 fibroblasts on a variety of substrates with different micropatterns was compared to that of the cells on a flat substrate. The nucleus location of each NIH 3T3 fibroblast, seeded on each substrate, was recorded every 30 min for 24 h using time-lapse microscopy (Figure 5.4A–E).

Continuous movement paths of at least five cells of each substrate were obtained by analyzing the measured cell location (Figure 5.5A–F).

NIH 3T3 fibroblasts on a flat substrate (control group, Figures 5.4E and 5.5F) had randomly oriented protrusions and freely moved without any directionality in their movement. The percentage for cells which moved within sectors of $-22.5°$ to $22.5°$ and 157.5 to $202.5°$ was 24.1%, and the percentage for cells which traveled within sectors of 67.5–112.5° and 247.5–292.5° was 25.9% where 0° meant a positive x-axis direction.

The cells, seeded on parallel micropatterns with different widths of trough/ridge of 3 and 10 μm, moved along the micropatterns. In the parallel micropatterns with a width of trough/ridge of 10 μm (Figure 5.4A), 75.9% and 60.3% of the cells migrated along troughs and ridges, respectively, while 5.2% and 12.6% of the cells moved across troughs and ridges, respectively. Here, "along troughs and ridges" means sectors of $-22.5°$ to $22.5°$ and 157.5 to $202.5°$ and "across troughs and ridges" indicates sectors of 67.5–112.5° and 247.5–292.5° where the direction of the parallel micropatterns is set as 0°. In the parallel micropatterns

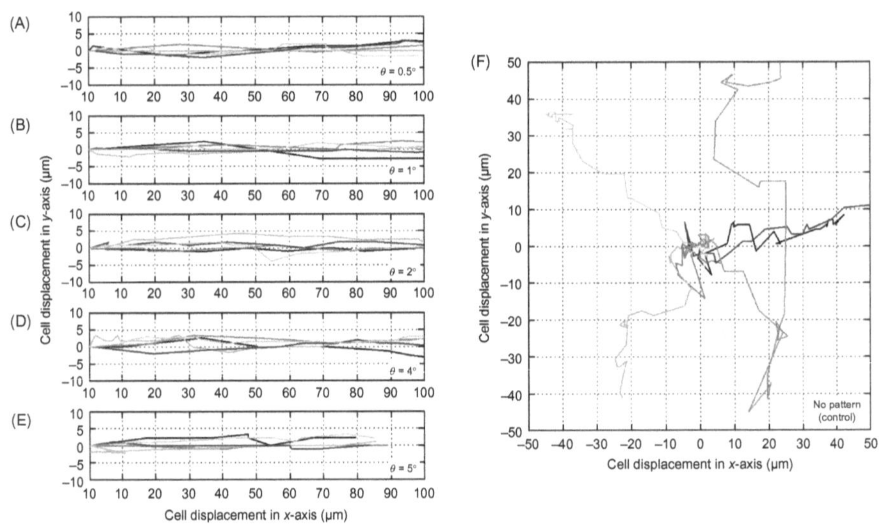

FIGURE 5.5

Movement paths of NIH 3T3 fibroblasts migrating on the Rome platform. The paths connect the positions of each cell (especially, nucleus) every 30 min, thus showing the trajectory of each cell which moves on the Rome platform. (A–E) Measured movement paths of the cells along diverging micropatterns with different divergence angles of 0.5° (A), 1.0° (B), 2.0° (C), 4.0° (D), and 5.0° (E). (F) Measured movement paths of the cells migrating on a flat substrate with no topographical feature (control group). The guided cell motility paths (A–E) are compared to the unguided ones (F).

with a width of trough/ridge of 3 μm (Figure 5.4B), average percentages for cells which traveled along troughs and ridges were 79.8% and 61.2%, respectively, whereas average percentages for cells which moved across troughs and ridges were only 4.4% and 11.8%, respectively. These results let us know the following biological facts. First of all, adherent cells tend to move along parallel micropatterns rather than across parallel micropatterns. When adherent cells sense the physical cues (e.g., troughs and ridges) of the parallel micropatterns, the cells start to stretch themselves along the micropatterns to maximize their contact area to the physical cues (Figure 5.4F). This change in cell morphology results in guided cell movement along the micropatterns. Secondly, this tendency is in inverse proportion to the width of trough/ridge, showing that the degree of change in cell morphology by micropatterns is correlated to the direction of cell movement.

In the diverging micropatterns with different divergence angles of 0.5°, 1.0°, 2.0°, 4.0°, and 5.0° (Figures 5.4C and 5.5A–E), the movement of NIH 3T3 fibroblasts was also well-guided along the micropatterns. On average, 76.4% and 72.1% of the cells traveled along the troughs and ridges of the diverging micropatterns, respectively, and 4.6% and 7.6% of the cells moved across the troughs and ridges, respectively. The percentage for cells which travel along troughs and ridges increased as the divergence angle of the diverging micropatterns increased. This suggests that the micropatterns with a larger divergence angle can lead to a relatively sudden change in cell morphology, compared to the micropatterns with a smaller divergence angle, which positively affects cell movement along the micropatterns.

5.2.3.2 Effect of micropatterns on migratory characteristics of adherent cells

We also characterized the effect of the dimensions and shape of diverging micropatterns on cell migratory characteristics such as cell motility speed, cell motility angle, directional persistence time, and random motility coefficient. The cell motility speed was used to quantify the rate at which cells migrate; the cell motility angle was measured to indicate the movement direction of adherent cells; the directional persistence time was calculated to gage how long the diverging micropatterns maintain cell movement without significant changes in the movement direction of NIH 3T3 fibroblasts; the random motility coefficient was computed to explore how the cells disperse on the diverging micropattern. The cell migratory characteristics were determined by modeling the measured movement paths of the cells (Figure 5.5) as a persistent random walk model [45–48]. As diverging micropatterns were fully described by three parameters (w_b, w_e, and θ, where w_b and w_e are two widths of trough/ridge at which the divergence of diverging micropatterns begins and ends, respectively, and θ is a divergence angle; see Figure 5.2A), the effects of both two widths of trough/ridge and divergence angle on the migratory characteristics of adherent cells were independently quantified.

Dependence of cell migratory characteristics on width of trough/ridge

First of all, the effect of the width of trough/ridge of micropatterns on cell motility speed was explored using our assay (Figure 5.6A). The cell motility speed, s, calculated from Eq. (5.2) had an inverse relation to the width of trough/ridge, w_{tr}, represented as s (μm/min) = $0.26 + 0.88/w_{tr}$ (μm). This can be explained by the following biological phenomena. Assuming no volume change in adherent cells during cell movement, the cells moving on the micropatterns with a narrow width of trough/ridge have a longer displacement per each movement cycle than those traveling on the micropatterns with a wide one. Moreover, the micropatterns with a narrow width of trough/ridge (3 μm in this study) are known to weaken the strength of cell–substrate adhesion slightly, thus maximizing cell motility speed [47]. The motility speed of the cell moving on the micropatterns with a width of trough/ridge of 45.5 ± 3.5 μm was identical to that of the cell migrating on a flat substrate (control group), indicating the micropatterns with a width of trough/ridge larger than 45.5 ± 3.5 μm were recognized as a flat substrate by the cells. This demonstrates that the ceiling and floor values for the width of trough/ridge of the micropatterns which are designed to passively control the motility of NIH 3T3 fibroblasts are 3 μm or less and about 42–49 μm, respectively. Remarkably, when NIH 3T3 fibroblasts were guided by multiple (more than one), identical micropatterns (denoted with an asterisk symbol in Figure 5.6), the motility speed of the cells was determined by the largest dimension of the micropatterns to which the circumferential marginal zone of the cells contacted (see Figure 5.4D). This suggests that cell mechanosensors which receive and respond to external physical cues (e.g., micropatterns in this study) are intensively activated at the circumferential marginal zone of the cell rather than at the central zone.

The dependence of cell motility angle on the width of trough/ridge was also explored by analyzing the measured movement paths of NIH 3T3 fibroblasts (Figure 5.6B). The cell motility angle, calculated using Eq. (5.3), was proportional to the width of trough/ridge, and the cell motility angle obtained from the micropatterns was much smaller than that of a control group. This shows that the micropatterns with a narrow width of trough/ridge are helpful to achieve directed cell movement.

In contrast to the cell motility speed, the directional persistence time got maximized at a width of trough/ridge of 10 μm (Figure 5.6C). That is, the directional persistence time evidently decreased on the micropatterns with narrow (3 μm) and wide (24.5 μm) widths of trough/ridge. The NIH 3T3 fibroblasts on the micropatterns with a narrow width of trough/ridge seem to easily escape from the micropatterns and to sense other topographical features around the micropatterns due to a relatively small contact area to the micropatterns, whereas the cells on the micropatterns with a wide width of trough/ridge are thought to incompletely sense the micropatterns. This indicates the width of trough/ridge of 10 μm, comparable to the size of NIH 3T3 fibroblasts, can prevent the cells from frequently

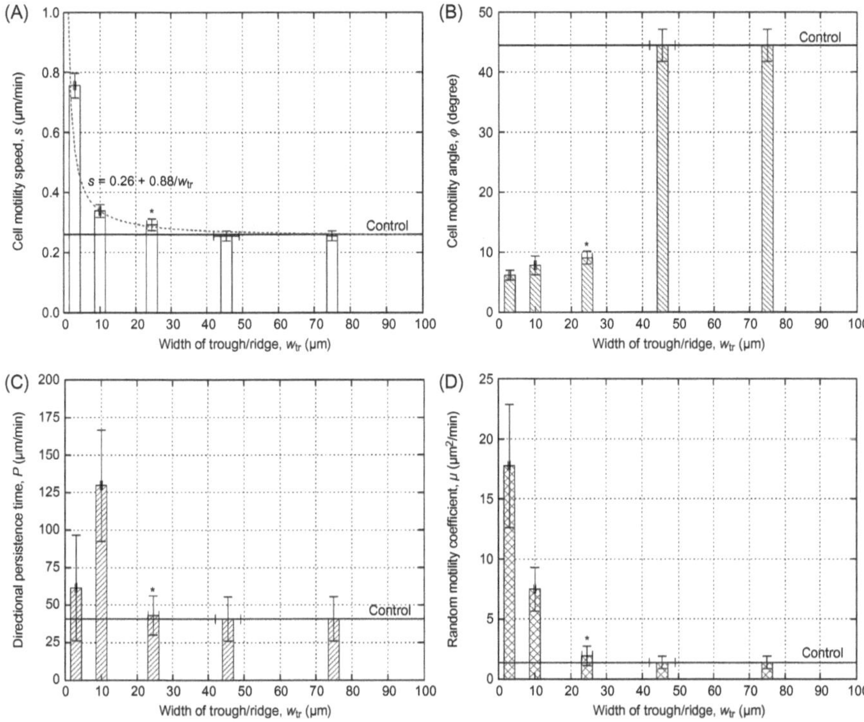

FIGURE 5.6

Quantification of the migratory behavior of NIH 3T3 fibroblasts in response to (parallel) micropatterns with different widths of trough/ridge. (A) Cell motility speed, s (μm/min), as a function of the width of trough/ridge of micropatterns, w_C (μm). The measured cell motility speed decreases as the width of trough/ridge increases, expressed as $s = 0.26 + 0.88/w_C$. This result suggests that there are both ceiling and floor values for the dimensions of micropatterns which can influence cell movement, and the ceiling and floor values of NIH 3T3 fibroblasts are 3 μm (or less) and about 42–49 μm, respectively. (B) Cell motility angle as a function of the width of trough/ridge of micropatterns. The measured cell motility angle is proportional to the width of trough/ridge, w_C, at $w_C \leq 45.5 \pm 3.5$ μm but is the same as that of a control group at $w_C > 45.5 \pm 3.5$ μm. (C) Directional persistence time as a function of the width of trough/ridge of micropatterns. The directional persistence time of NIH 3T3 fibroblasts is maximized when the width of trough/ridge is about 10 μm. (D) Random motility coefficient as a function of the width of trough/ridge of micropatterns. The measured random motility coefficient is in inverse proportion to the pitch of ridges, which is explained in Figure 5.7A. The symbol of asterisk denotes the cells moving on multiple, identical micropatterns. All results are compared with the migratory characteristics of a control group. Error bars of (A–D) are the standard errors of the means.

changing their movement direction, thus forcing the cells to move in a specified direction for a relatively long time.

The random motility coefficient was also calculated as a function of the width of trough/ridge (Figure 5.6D). The calculated random motility coefficient was inversely proportional to the width of trough/ridge of micropatterns, like the cell motility speed. In other words, the amount of the cells moving across a unit area through a unit structural nonuniformity in a unit time decreased as the width of trough/ridge of micropatterns increased. Although micropatterns with 3 μm width of trough/ridge did not have the longest directional persistence time, the random motility coefficient of the micropatterns was highest. This is because the random motility coefficient is also related to the cell motility speed (Figure 5.6A) that significantly increases as the width of trough/ridge of micropatterns decreases. The random motility coefficient of the cell moving on the micropatterns with a width of trough/ridge of 45.5 μm was also the same as that of a control group.

Together with the experimental results of cell motility speed and cell motility angle (Figure 5.6A, B), we conclude as follows. Micropatterns with a narrow width of trough/ridge are of assistance to achieve fast, one-directional, and large-amount cell movement although they sacrifice the directional persistence time. When adherent cells need to be guided for a long time, the width of trough/ridge of the micropatterns must be comparable to the size of the cells (e.g., about 10 μm for NIH 3T3 fibroblasts). Two widths of trough/ridge of the diverging micropatterns which control the motility of NIH 3T3 fibroblasts need to be in the range of 3–24.5 μm (less than 45.5 μm).

Dependence of cell migratory characteristics on divergence angle

The effect of the divergence angle of diverging micropatterns on the cell motility speed, cell motility angle, directional persistence time, and random motility coefficient of NIH 3T3 fibroblasts was investigated using the Rome platform. The cell motility speed was characterized as a function of the divergence angle of diverging micropatterns (Figure 5.7A). Although the diverging micropatterns prominently enhanced the cell motility speed, compared to that on a flat substrate (control group), the measured cell motility speed was almost insensitive to the divergence angle. Together with the dependence of cell motility speed on the width of trough/ridge of parallel micropatterns (see Figure 5.6A), the motility speed of adherent cells that contact to diverging micropatterns can be controlled by adjusting two widths of trough/ridge of diverging micropatterns at which divergence begins and ends, respectively. This result points out how to design the diverging micropatterns for the control of cell motility. Like cell movement on the parallel micropatterns, the cells on the troughs of the diverging micropatterns moved slightly faster than the cells on the ridges. The reason for this behavior seems that the troughs offer higher stability in maintaining cell movement than the ridges.

We also quantified the dependence of cell motility angle on the divergence angle of diverging micropatterns (Figure 5.7B). The cell motility angle was much smaller than that of a control group; the measured cell motility angle was in direct

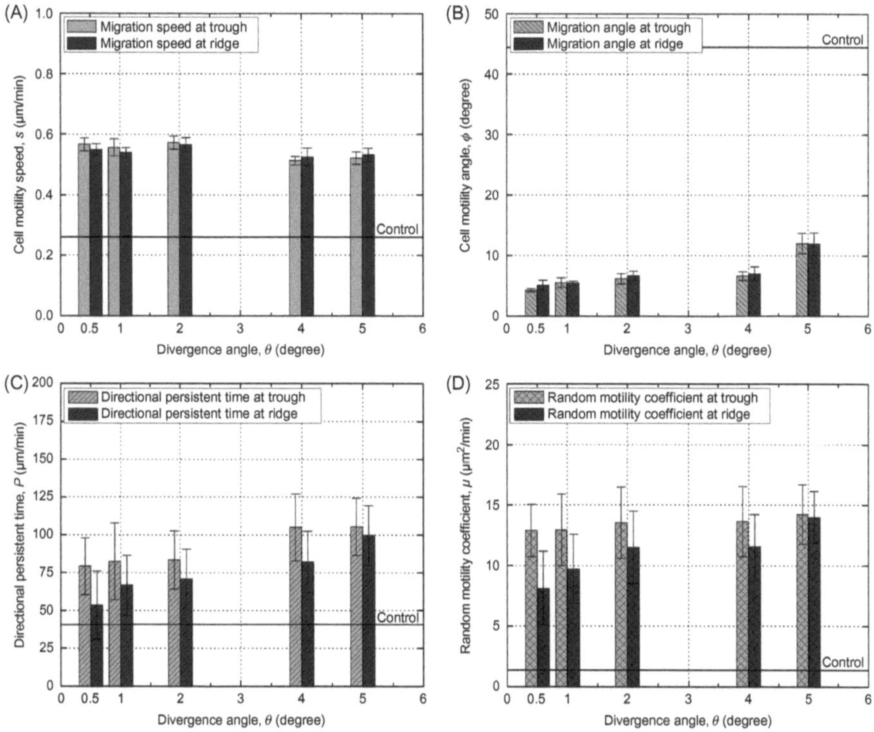

FIGURE 5.7

Quantification of the migratory behavior of NIH 3T3 fibroblasts using diverging micropatterns with different divergence angles. (A) Cell motility speed as a function of the divergence angle of diverging micropatterns. All diverging micropatterns are identical in their two widths of trough/ridge (w_w and w_n) at which divergence begins and ends, respectively. This shows cell motility speed is almost insensitive to the divergence angle. Compared to a flat substrate, the troughs of the diverging micropatterns expedite cell movement slightly (not prominently) more than the ridges. This indicates that the migratory behavior of adherent cells can be controlled by both troughs and ridges, but the troughs are more effective. (B) Cell motility angle as a function of the divergence angle of diverging micropatterns, showing the diverging micropatterns, have a capability to force adherent cells to move in one direction. The measured cell motility angle is in direct proportion to the divergence angle. (C) Directional persistence time as a function of the divergence angle of diverging micropatterns. The measured directional persistence time is also proportional to the divergence angle. The directional persistence time obtained from the cells moving on the troughs is larger than that obtained from the cells traveling on the ridges because the cells tend to move from the ridges to the troughs more than the opposite. (D) Random motility coefficient as a function of the divergence angle of diverging micropatterns. All quantification results show that the diverging micropatterns can control the direction and rate of cell movement in a passive way, and the troughs of the diverging micropatterns are more effective to control cell movement than the ridges. All results are compared with the migratory characteristics of a control group. Error bars of (A–D) are the standard errors of the means.

proportion to the divergence angle; the trough and ridges have no prominent difference in cell motility angle. These measurements show that the diverging micropatterns with a large-divergence angle impose a relatively high-level morphological polarity on adherent cells, compared to those with a small-divergence angle, which makes the cells deviate a little from their designated course within a reasonable range (not randomly). However, this behavior never affects the controllability of diverging micropatterns which determines the movement direction of adherent cells.

Next, the directional persistence time was measured as a function of the divergence angle (Figure 5.7C). The increase in the divergence angle contributed to increase the directional persistence time; the cell moving on the troughs had a longer directional persistence time than the cells moving on the ridges. The former is due to the relatively high morphological polarity caused by the diverging micropatterns with a large-divergence angle. Namely, the cells guided by diverging micropatterns with a large-divergence angle have blunter leading edge and sharper trailing edge than those guided by diverging micropatterns with a small-divergence angle, thus maintaining the direction of cell movement (without significant changes) for a longer time. The latter is because the troughs make the cells straightly move for a longer time (by using their upward sides) than the ridges. These results suggest that the diverging micropatterns with a large-divergence angle are preferable to control cell movement in a designated direction for a long time, although they have an acceptable loss in manipulating the direction of cell movement.

The random motility coefficient of the NIH 3T3 fibroblasts moving on the diverging micropatterns was plotted as a function of the divergence angle of diverging micropatterns (Figure 5.7D). The obtained random motility coefficient was proportional to the divergence angle; the cells moving on the troughs of the diverging micropatterns had a higher random motility coefficient than those moving on the ridges. The experimental results show that the level of morphological polarity, determined by the divergence angle of the diverging micropatterns, has an inevitable correlation with random motility coefficient.

All migratory characteristics (Figure 5.7A–D), obtained from NIH 3T3 fibroblasts moving on the diverging micropatterns, reveal the following facts: (i) in contrast to parallel micropatterns which guide adherent cells from right to left or vice versa along them, diverging micropatterns make adherent cells to move in a specified direction (i.e., diverging direction of the micropatterns), thus effectively controlling the movement direction of the cells; (ii) when adherent cells are guided using diverging micropatterns, the movement speed of the cells is determined not by the divergence angle of the diverging micropatterns but by the width of trough/ridge (i.e., two widths of trough/ridge where divergence begins and ends, respectively); (iii) the motility angle, directional persistence time, and random motility coefficient of adherent cells can be effectively controlled by adjusting the divergence angle of diverging micropatterns, which yields useful propositions for the design of diverging micropatterns which are intended to control cell motility.

5.3 Active control of cell motility

5.3.1 Biological breadboard

A BBB consists of identical gold electrodes patterned on a Pyrex substrate, which entitles this platform to a high degree-of-freedom in manipulating cell detachment at a subcellular level (Figure 5.8A). The Pyrex substrate and gold electrodes are modified with polyethylene glycol (PEG) and RGD-terminated thiol (RTT), respectively. The RTT modification is intended to make the gold electrodes cell-adhesive by tethering an RGD peptide to a gold surface via thiol compound.

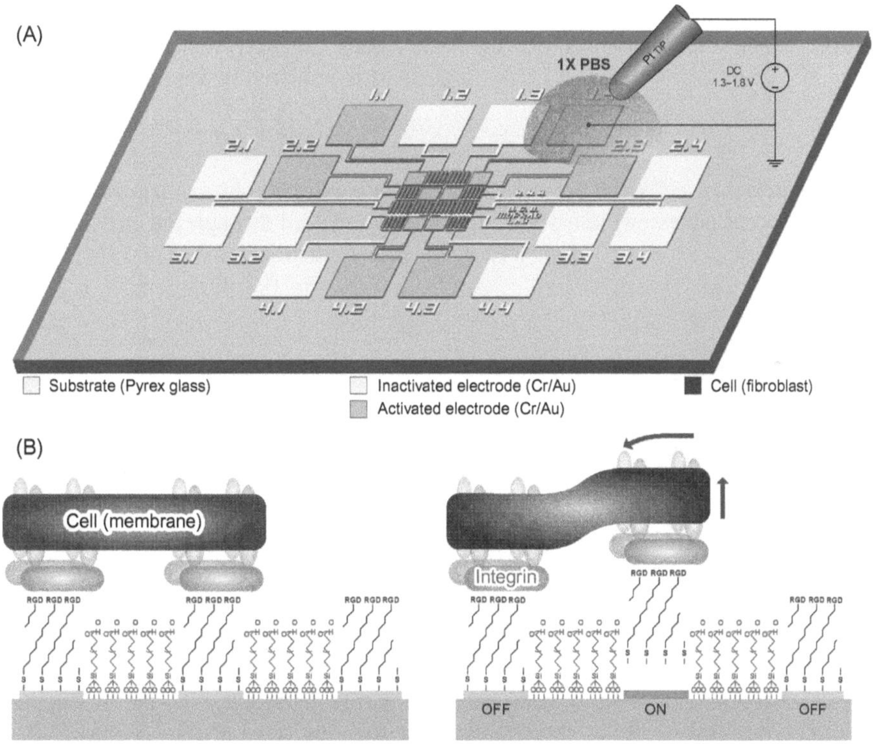

FIGURE 5.8

A BBB for the active control of cell motility. (A) Schematic of the BBB consisting of gold electrodes patterned on a Pyrex substrate. This platform has the addressability, multifunctionality, and reusability in manipulating cell detachment due to its structure and working principle. (B) Cell detachment manipulation using the BBB. A cell adheres to all gold electrodes functionalized with RTT (left). The cell (or part of the cell) is detached when a target electrode (second one from left) is activated by applying an activation potential which reductively desorbs a gold–thiol SAM.

The thiol compound creates a SAM on the gold electrodes following the spontaneous chemisorption

$$R - S - H + Au \rightarrow R - S - Au + \tfrac{1}{2}H_2 \qquad (5.7)$$

where R is a substituent [50]. The RTT therefore offered a cell (strictly speaking, integrin)-binding site which is almost same as *in vivo* microenvironment for cell adhesion. The PEG modification on the Pyrex substrate is designed to achieve a cell-resistive surface where hydrated neutral PEG chains sterically repulse cells.

In the spatiotemporal manipulation of cell detachment, adherent cells are loaded on the BBB modified with RTT and PEG and then start to stretch on the RGD peptide of the gold electrodes. On the manipulation of cell detachment, the entire cells (or part of the cell) are detached from the BBB by activating gold electrode(s) which breaks a chemical bonding between gold and thiol (Figure 5.8B). When the RTT-modified gold electrodes are activated with activation potential of −0.9 to −1.8 V, a gold–thiol SAM is reductively desorbed [51]

$$R - S - Au + H^+ + e^- \rightarrow R - S - H + Au \qquad (5.8)$$

The detached cells (or part of the cell) sense no external mechanical anchorage (i.e., focal adhesion) to a substrate and are liquefied through the gel–sol transition of their cytoskeleton, followed by cellular retraction.

5.3.2 Materials and methods

5.3.2.1 Microfabrication

The microfabrication of the BBB began with a 500 μm thick, 4 in, Corning 7740 Pyrex glass wafer. After cleaning the wafer in a piranha solution (H_2SO_4: H_2O_2 = 1:1), 1 μm thick LOR resist (LOR 10 A, MicroChem Corp.) and 2 μm thick positive photoresist (S1818, Rohm and Haas Corp.) were sequentially spin-coated to prepare a double-layer resist stack for lift-off process. This double-layer resist stack was patterned through lithography (Figure 5.9A, left), followed by the deposition of 5 nm thick chromium (Cr) adhesion layer and 100 nm thick gold (Au) layer through e-beam evaporation (Figure 5.9A, center). Next, the Cr/Au layer was patterned in an organic solvent mixture (BAKER PRS-3000 Stripper, Mallinckrodt Baker, Inc.), thus fabricating two kinds of BBBs (Figure 5.9A, right). One for the manipulation of cell detachment at a cellular level was composed of gold electrodes with a length of 500 μm and a width of 500 μm (Figure 5.9B), the other for the manipulation of cell detachment at a subcellular level consisted of gold lines with a width of 10 μm and a gap of 3 μm (Figure 5.9C). The fabricated BBB was wire-bonded in a chip carrier and then was assembled with a cell-culture-well (Figure 5.9D, left). For the BBB which was used with an inverted microscope, the deposition thickness of the gold layer was reduced to 30 nm to make the BBB transparent (Figure 5.9D, right).

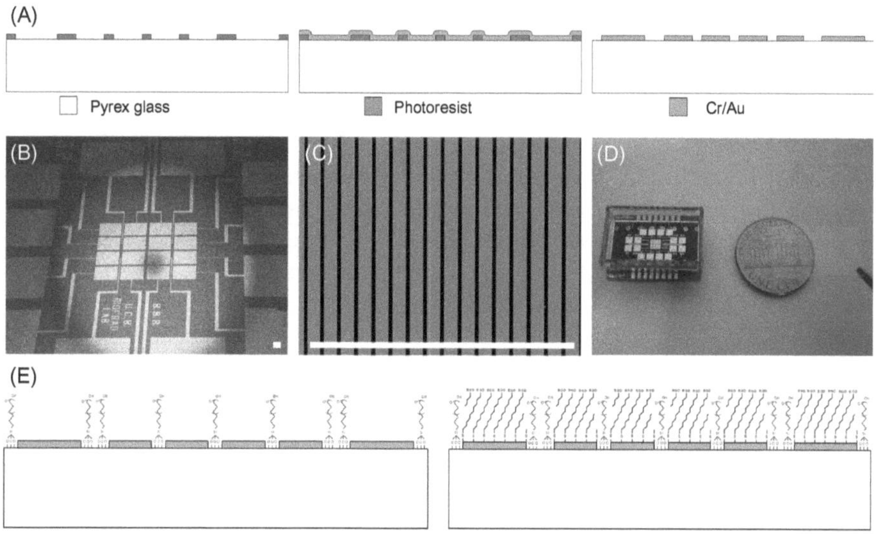

FIGURE 5.9

Microfabrication and surface modifications of the BBB. (A) The BBB is microfabricated by patterning gold electrodes on a Pyrex substrate: patterning of a photoresist layer through lithography (left); deposition of a Cr/Au layer through e-beam evaporation (center); patterning of the deposited Cr/Au layer through lift-off process (right). (B) Microfabricated BBB for the manipulation of cell detachment at a cellular level, each electrode of which is 500 μm in length and 500 μm in width. (C) Microfabricated BBB for the subcellular detachment, each electrode of which is 10 μm in width and 3 μm in gap. (D) Photograph of the BBB before (right) and after (left) assembly, showing its transparency. (E) Surface modification processes of the BBB. The BBB is incubated with a PEG solution to make a Pyrex substrate cell resistive (left) and then it is incubated with a synthesized RTT solution to make gold electrodes cell adhesive (right). Scale bars of (B) and (C) are 100 μm.

5.3.2.2 PEG modification on Pyrex substrate

The fabricated BBB was cleaned with an oxygen plasma chamber (PM-100 Plasma Treatment System, March Plasma Systems, Inc.) at 100 W for 30 s. The cleaned BBB was incubated with 2% (v/v) m-PEG silane (Gelest, Inc.) and 1% (v/v) hydrochloric acid (HCl, Fisher Scientific) dissolved in anhydrous toluene (Fisher Scientific) for 2 h (Figure 5.9E, left). This treatment was carried out in a glove box under a nitrogen purge. The incubated BBB was sequentially rinsed in fresh toluene and ethanol, followed by drying with nitrogen and curing at 120°C for 2 h.

5.3.2.3 RTT functionalization on gold electrodes

The gold electrodes were functionalized with an RTT solution which was synthesized by chemically combining *cyclo* (Arg-Gly-Asp-D-Phe-Lys) (Peptides

International, Inc.) with dithiobis(succinimidyl undecanoate) (Dojindo Molecular Technologies, Inc.) as follows. The *cyclo* (Arg-Gly-Asp-D-Phe-Lys) was dissolved in dimethoxysulfoxide (DMSO, Sigma-Aldrich) to get 1 mM RGD peptide aliquot and stored at −20°C. This reaction was made in a glove box under a nitrogen purge to protect the RGD peptide. The dithiobis(succinimidyl undecanoate) was also prepared in 1 mM aliquot in DMSO and stored at −20°C. This preparation was also made in moisture-free environment. When functionalizing the gold electrodes, two aliquots were warmed to room temperature in a desiccator. The RGD peptide aliquot was mixed with 1% (v/v) triethylamine (Fisher Scientific) for 5 min to make all primary amines of a lysine amino acid unprotonated. The same volume of the dithiobis(succinimidyl undecanoate) aliquot was added to the RGD peptide aliquot and then mixed using a vortex mixer for 4 h to synthesize an RTT solution. For the RTT functionalization of the gold electrodes (Figure 5.9E, right), the PEG-treated BBB was incubated with this solution for 1 h at room temperature to promote spontaneous chemisorption between thiol and gold, followed by sonificating in DMSO for 3 min and rinsing in ethanol and PBS (Sigma-Aldrich).

5.3.2.4 Contact angle measurement

The contact angles of two surfaces (i.e., PEG-treated Pyrex substrate and RTT-functionalized gold surface) were measured in a sessile drop mode using a goniometer (KRÜSS582, KRÜSS) to characterize the changes in wetting properties of the surfaces through surface modifications. The contact angle of each surface was measured 10 times and then was averaged. The contact angle of PEG-treated Pyrex substrate was compared to that of untreated Pyrex substrate, and the contact angle of RTT-functionalized gold surface was compared to those of untreated gold surface and thiol-treated gold surface.

5.3.2.5 X-ray photoelectron spectroscopy characterization

A sample for X-ray photoelectron spectroscopy (XPS) characterization was prepared by performing the RTT functionalization on a Cr/Au layer patterned on a silicon wafer. An untreated gold pattern was run as a control group. The XPS analysis was conducted using a customized electron spectroscopy for chemical analysis (ESCA, Omicron NanoTechnology) at 1×10^{-8} Torr. All measured spectra were referenced to the position of Au 4f peak. The scans were collected over a range of 20 eV around the peak of interest with a pass energy of 23.5 eV.

5.3.2.6 Electrochemical characterization of reductive desorption of gold–thiol SAM

A Cr/Au layer patterned on a silicon wafer was functionalized in the RTT solution to prepare the sample for cyclic voltammetry (CV) characterization. The potentiodynamic electrochemical characterization was performed using a three-electrode system composed of the gold surface of the sample (working electrode), platinum electrode (counter electrode), and Ag/AgCl electrode (reference electrode). The CV of the reductive desorption of a gold–thiol SAM was measured in Dubecco's phosphate buffered saline (DPBS, Sigma-Aldrich) using a scanning potentiostat (Model 362, EG&G Co.). A scan started cathodically from 0 to −2 V, then annodically back to 0 V at a scan rate of 50 mV/s.

5.3.2.7 Cell culture

NIH 3T3 fibroblasts were cultured in DMEM (GIBCO™) supplemented with 10% FBS (GIBCO™) and 1% Pen/Strep (GIBCO™) at 37°C in humidified 5% CO_2 atmosphere. The cell was passaged every 4 days. NIH 3T3 fibroblasts with a passage number of 5–20 were used in this study. For each experiment, the surface-modified BBB was sterilized and then was placed on a petri dish containing 5 ml cell-culture medium with a cell suspension of 1×10^6 cells/ml. For subcellular detachment, the cell concentration was changed into 1×10^4 cells/ml. Unadhered cells were removed by washing the BBB in PBS 1 h after cell loading. All experiments were carried out 24 h after cell loading.

5.3.3 Results and discussion

5.3.3.1 Surface modifications

The degree of PEG treatment on Pyrex substrate was verified by both contact angle measurement and cell loading test. The measured contact angle of the PEG-treated Pyrex substrate was $61.5 \pm 3.8°$ on average, while that of untreated Pyrex substrate was $25.7 \pm 1.5°$ (Figure 5.10A). Additionally, NIH 3T3 fibroblasts were seeded on both surfaces and then cell adhesion to each surface was visualized 24 h after cell loading (Figure 5.10B). These results show the PEG treatment makes the Pyrex substrate strongly hydrophobic, thus effectively preventing cell adhesion.

The RTT functionalization on gold surface was characterized by both contact angle measurement and XPS survey. The average contact angles measured from untreated gold surface, thiol-treated gold surface, and RTT-functionalized gold surface were $67.3 \pm 2.5°$, $53.3 \pm 1.3°$, and $24.6 \pm 2.8°$, respectively (Figure 5.10C). The existence of RGD peptides tethered to the gold electrodes was also investigated using XPS. The measured XPS survey spectrum shows the presence of gold, thiol, and RGD peptides (Figure 5.10D): the peaks of Au 4s, Au 4p, Au 4d, and Au 4f indicate the presence of gold (Au(111)); the peaks of S $2p_{1/2}$ and S $2p_{3/2}$ are corresponding to sulfur from the thiol (right inset); the peaks of C 1*s*, O 1*s*, O *KLL*, and N 1*s* (left inset) demonstrate there are carbon, oxygen, and nitrogen from the amine ($-NH_2$) and carboxylic (−COOH) groups of the RGD peptides. Together with the results of contact angle measurements, this confirms RGD peptides are well linked to the gold electrodes via thiol through RTT functionalization.

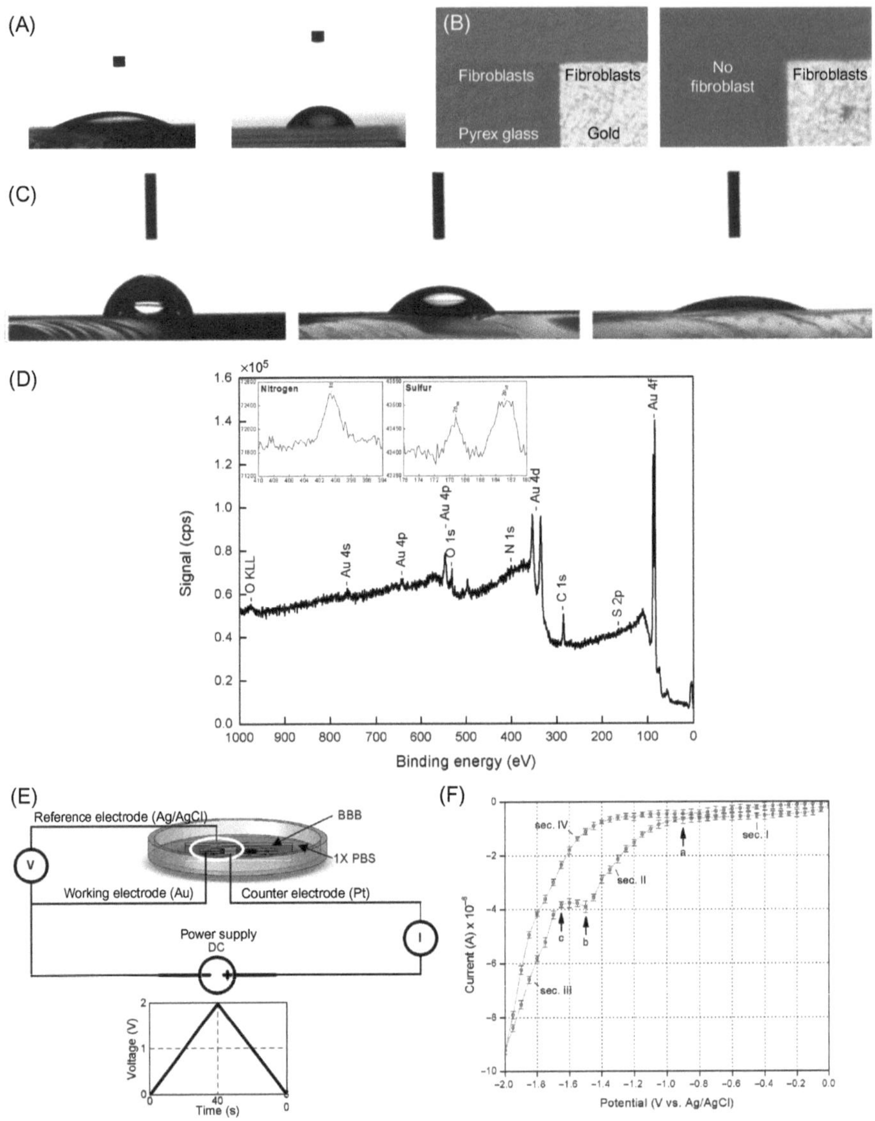

FIGURE 5.10

Characterizations of the surface modifications of the BBB and potentiodynamic electrochemical characterization of the reductive desorption of a gold–thiol SAM. (A) Contact angles measured from a Pyrex substrate before (left) and after (right) PEG modification. The contact angle is changed from $25.7 \pm 1.5°$ to $61.5 \pm 3.8°$ through PEG modification. (B) Cell (NIH 3T3 fibroblasts) loading on a Pyrex substrate before (left) and after (right) PEG modification, showing the Pyrex substrate is changed from cell adhesive

(*Continued*)

5.3.3.2 Reductive desorption of gold–thiol SAM

The reductive desorption of a gold–thiol SAM was verified by measuring CV curves using a three-electrode system (Figure 5.10E). The current at the working electrode (gold electrode of the BBB) was plotted versus the applied voltage with respective to a reference electrode (Figure 5.10F). The part of the CV (sec. I) ranging from 0.00 to −0.90 V was corresponding to no reductive desorption of the SAM; the potential range (sec. II) showed the deductive desorption began and ended at −0.90 V (point a) and −1.65 V (point c), respectively; this electrochemical reaction got maximized at −1.50 V (point b). This measurement shows that the activation potential must be larger than −0.90 V to break the chemical binding between gold and thiol. Considering the generation of hydrogen due to the electrolysis of cell-culture medium at around −2.00 V, the optimum potential for the reductive desorption of a gold–thiol SAM is thought to be −1.00 (larger than −0.90 V) to −1.80 V (less than −2.00 V) at which NIH 3T3 fibroblasts are successfully detached from the gold electrode of the BBB without viability.

5.3.3.3 Detachment of adherent cells for the control of cell motility

The spatiotemporal manipulation of cell detachment at a cellular level was characterized by disconnecting entire living cell(s) from the gold electrodes of the BBB. The biological characterization was conducted at −1.50 V for two NIH 3T3 fibroblasts (Figure 5.11A), subconfluent cells with 25% confluence (Figure 5.11B), subconfluent cells with 50% confluence, subconfluent cells with 75% confluence, and confluent cells with 100% confluence (Figure 5.11C). As an evaluation index for the speed of cell detachment at a cellular level, a time required to detach 95% of the adhered cells (CD-time) was measured. The CD-times of the above five cases (averaged from at least 10 measurements) were 45.2 ± 6.8, 36.7 ± 8.7, 31.1 ± 4.2, 24.8 ± 5.5, and 21.1 ± 3.5 s, respectively (Figure 5.11D), showing an inverse proportion to cell confluence. This is because the detachment of a cell applies additional vertical force to neighboring cells through cell–cell interactions at confluent conditions where all cells are connected to each other by the cell–cell interactions. This measurement indicates cell–cell interactions have a correlation to

◀ to cell resistive. (C) Contact angles measured from bare gold ($67.3 \pm 2.5°$, left), thiol-treated gold ($53.3 \pm 1.3°$, center), and RTT-functionalized gold ($24.6 \pm 2.8°$, right). (D) XPS survey spectrum of the gold surface modified with RTT. Detected are a gold peak from a gold surface, a sulfur peak from thiol, a nitrogen peak from the amine group of an RGD peptide, and carbon and oxygen peaks from the carboxylic acid group of an RGD peptide. (E) Experimental setup for potentiodynamic electrochemical characterization where the gold surface of the BBB, a platinum electrode, and a Ag/AgCl electrode work as working, counter, and reference electrodes, respectively. (F) CV measured from the gold surface modified with RTT, indicating the reductive desorption of the gold–thiol SAM starts and finishes at −0.90 and −1.65 V, respectively, and gets maximized at −1.50 V.

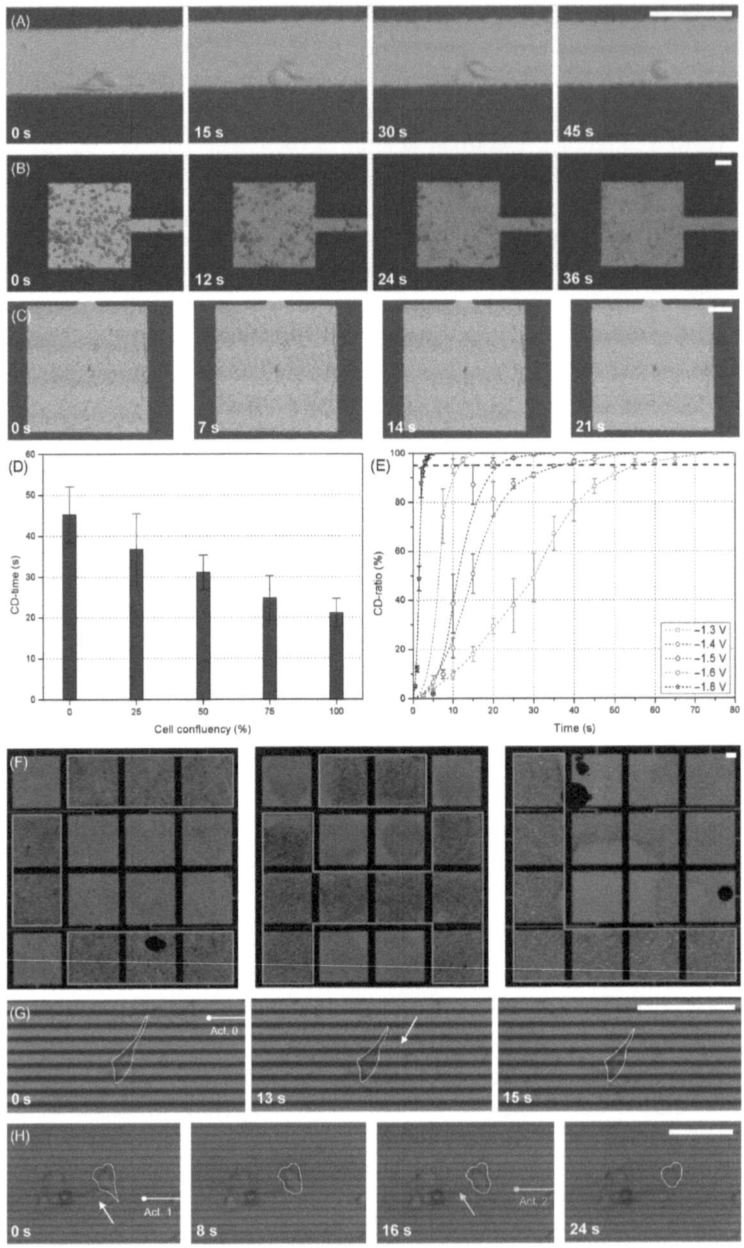

FIGURE 5.11

Spatiotemporal manipulation of cell detachment for the control of cell motility using the BBB. (A–C) Sequential optical images showing the spatiotemporal manipulation of cell

(Continued)

cell detachment behavior. The dependence of cell detachment on an activation potential was also investigated by counting the ratio of detached cells to total cells (CD-ratio) as a function of time and activation potential (Figure 5.11E). The CD-ratio increased as the activation potential increased because the reductive desorption of a gold–thiol SAM was in direct proportion to the activation potential (see Figure 5.10F). The measured CD-ratio was monotonically increasing with two inflection points (s-shape curves), demonstrating a large deviation in the strength of integrin binding to extracellular matrix proteins or another cell. A programmable cell patterning was also performed. A single four-by-four BBB was used to spatio-temporally manipulate the detachment of NIH 3T3 fibroblasts at a cellular level (Figure 5.11F). That is, the cells were sequentially patterned in a shape of "C," "A," or "L" using the BBB and then the BBB was reused for the next cell patterning after cleaning it. The BBB was cleaned through two-step cleaning processes (i.e., activation of gold electrodes and simplified organic cleaning of the device). The activation of gold electrodes was for yielding original gold through the reductive desorption of a gold–thiol SAM, and the organic cleaning was for removing biological remnants (e.g., dead cells during cell loading).

The cell detachment at a subcellular level was also achieved by disconnecting the part of a cell from the BBB having gold lines with a width of 10 μm and a gap of 3 μm (see Figure 5.9C). When we activated a gold line on which one end of the cytoplasm of the cell was placed, the part of the cell was successfully detached and then the detached cytoskeleton finished its retraction within 16 s

---

◀ detachment at a cellular level for two cells (0% confluent cells), 25% confluent cells, and 100% confluent cells. The two cells are detached from a gold electrode at an activation potential of −1.50 V and their CD-time is 45.2 ± 6.8 s (A). The average CD-time of the 25% confluent cells is 36.7 ± 8.7 s (B). The average CD-time of the 100% confluent cells is 21.1 ± 3.5 s (C). (D) Measured CD-time as a function of cell confluence. This shows the CD-time is in inverse proportion to cell confluence. (E) Measured CD-ratio as a function of time and activation potential for 100% confluent cells. The CD-ratio is monotonically increasing with two inflection points (s-shape curve) indicating there is a large deviation in the strength of integrin binding to extracellular matrix proteins or another cell. (F) Programmable cell patterning using the manipulation of cell detachment on a single four-by-four BBB. The cells are patterned in shapes of "C," "A," and "L," verifying the addressability, multifunctionality, and reusability of the BBB. (G and H) Spatiotemporal manipulation of cell detachment at a subcellular level using the BBB. The part of an NIH 3T3 fibroblast is detached by a single activation (Act. 0). This subcellular detachment is accompanied by the spontaneous retraction of the detached cytoskeleton (G). The part of a cell is sequentially detached by a series of activations (Act. 1 and Act. 2). Scale bars are 100 μm.

after subcellular detachment (Figure 5.11G). The retraction speed of the detached cytoskeleton at a subcellular level is faster than that of the detached cell at a cellular level because a single cell which has no constraint in stretching itself is under higher tension than confluent cells. Next, the part of NIH 3T3 fibroblasts was detached with a series of activations (Figure 5.11H). The second activation (Act. 2) was applied 16 s after the first one (Act. 1). This experiment reveals that repetitive activations within dozens of seconds do not sacrifice the viability of the cell and are able to gradually increase the amount of subcellular detachment. These successful manipulations of subcellular detachment open a possibility for the control of cell motility using the BBB. In detail, the movement direction of an adherent cell is controlled by selectively activating some of the gold electrodes of the BBB. When a cell is located on the origin and adheres to four gold electrodes that lie in each quadrant, the cell is thought to move in the first quadrant by activating the gold electrode in the third quadrant (or all gold electrodes except the one in the first quadrant). This is because the RGD peptides tethered on inactivated gold electrode(s) have much higher affinity for cell adhesion than bare gold of activated ones.

5.4 Summary

Two MEMS-based mechanobiological methods for the control of cell motility (i.e., direction and rate of cell movement) have been presented as preliminary attempts to develop new therapeutic solutions for pathophysiological phenomena related to cell motility. The first method was a Rome platform that passively controlled cell movement using diverging micropatterns. The Rome platform was composed of 3 μm thick diverging micropatterns with different divergence angles of 0.5–5.0° and two widths of trough/ridge of 3 and 10 μm where divergence begun and ended, respectively. This platform achieved one-directional cell movement (i.e., cell movement in a diverging direction of the diverging micropatterns) by imposing morphological polarity on adherent cells using the diverging micropatterns. The rate of cell movement was controlled by changing two widths of trough/ridge where divergence begun and ended, respectively. The experimental characterizations inform us of the following biological facts: (i) diverging micropatterns affect the migratory behavior of adherent cells, and the changes in the dimensions and shape of the diverging micropatterns resulted in the changes in the direction and rate of cell movement in a controllable manner; (ii) there are both ceiling and floor values for the dimensions of the diverging micropatterns which can influence cell migration (e.g., the ceiling and floor values for width of trough/ridge are respectively 3 μm (or less) and about 42–49 μm for NIH 3T3 fibroblasts); (iii) whether the morphological polarization of adherent cells is biological (like *in vivo* cell movement) or artificial (like cell movement on the Rome platform), the morphological polarization is sufficient to determine the movement direction of the cells; (iv) together with the results of cell movement on multiple, identical micropatterns, cell mechanosensors which receive and respond to external physical cues (e.g., micropatterns in this study) are intensively activated at the circumferential marginal zone of the cell rather than at the central zone. A BBB was also developed as an active method for the control of cell motility. The BBB, composed of multiple gold electrodes patterned on a Pyrex substrate, was surface modified with RTT and PEG to achieve a cell-adhesive surface on the gold electrodes and a cell-resistive surface on the Pyrex substrate, respectively. This assay controlled cell detachment at a subcellular level by reductively desorbing a gold–thiol SAM with an activation potential of −0.90 to −1.65 V. When the chemical binding between gold and thiol was broken by activating a gold electrode of interest, one part of a cell, located on the electrode, was detached from the activated gold electrode and then experienced its fast retraction. The detached (and then retracting) cytoskeleton experienced a 3 times decrease in its elastic modulus and also a 30 times increase in its damping coefficient within a few seconds. In the manipulation of cell movement using the BBB, one part of a cell which was relatively in the opposite direction of intended cell movement was deliberately detached. The detached cytoskeleton was spontaneously retracted, which lead to the successful control of cell movement.

The extrapolation of these MEMS-based mechanobiological methods to other cells might help us to better understand the migratory nature of adherent cells, thereby resulting in a development of new therapeutic strategies related to cell movement. Ongoing work is focusing on more in-depth control of cell motility by developing large-scale assays to shed light on the dynamics of cell motility. Combined with molecular dynamics models [52–54], the proposed devices for the programmable control of cell movement will offer an opportunity for studies of molecular cell biomechanics related to cell motility.

References

[1] D.A. Lauffenburger, A.F. Horwitz, Cell migration: a physically integrated molecular process, Cell 84 (1996) 359–369.

[2] A.J. Ridley, M.A. Schwartz, K. Burridge, R.A. Firtel, M.H. Ginsberg, G. Borisy, et al., Cell migration: integrating signals from front to back, Science 302 (2003) 1704–1709.

[3] V. Jones, K. Harding, J. Stechmiller, G. Schultz, Acute and chronic wound healing, in: S. Baranoskiand, E.A. Ayello (Eds.), Wound Care Essentials: Practice Principles, Lippincott Williams & Wilkins, Philadelphia, 2007, pp. 64–76.

[4] G.C. Gurtner, S. Werner, Y. Barrandon, M.T. Longaker, Wound repair and regeneration, Nature 453 (2008) 314–321.

[5] J.Y. Lim, H.J. Donahue, Cell sensing and response to micro- and nanostructured surfaces produced by chemical and topographic patterning, Tissue Eng. 13 (2007) 1879–1891.

[6] M.N. Yousaf, B.T. Houseman, M. Mrksich, Using electroactive substrates to pattern the attachment of two different cell types, Proc. Natl. Acad. Sci. U.S.A. 98 (2001) 5992–5996.
[7] A. Tourovskaia, T. Barber, B.T. Wickes, D. Hirdes, B. Grin, D.G. Castner, et al., Micropatterns of chemisorbed cell adhesion–repellent films using oxygen plasma etching and elastomeric masks, Langmuir 19 (2003) 4754–4764.
[8] S.P. Massia, J.A. Hubbell, An RGD spacing of 440 nm is sufficient for integrin $\alpha_v\beta_3$-mediated fibroblast spreading and 140 nm for focal contact and stress fiber formation, J. Cell Biol. 114 (1991) 1089–1100.
[9] K.Y. Lee, E. Alsberg, S. Hsiong, W. Comisar, J. Linderman, R. Ziff, et al., Nanoscale adhesion ligand organization regulates osteoblast proliferation and differentiation, Nano. Lett. 4 (2004) 1501–1506.
[10] A.M. Green, J.A. Jansen, J.P. van der Waerden, A.F. von Recum, Fibroblast response to microtextured silicone surfaces: texture orientation into or out of the surface, J. Biomed. Mater. Res. 28 (1994) 647–653.
[11] M.J. Dalby, D. Giannaras, M.O. Riehle, N. Gadegaard, S. Affrossman, A.S. Curtis, Rapid fibroblast adhesion to 27 nm high polymer demixed nano-topography, Biomaterials 25 (2004) 77–83.
[12] X.F. Walboomers, W. Monaghan, A.S. Curtis, J.A. Jansen, Attachment of fibroblasts on smooth and microgrooved polystyrene, J. Biomed. Mater. Res. 46 (1999) 212–220.
[13] F. Johansson, P. Carlberg, N. Danielsen, L. Montelius, M. Kanje, Axonal outgrowth on nano-imprinted patterns, Biomaterials 27 (2006) 1251–1258.
[14] C.D. Nobes, A. Hall, Rho GTPases control polarity, protrusion, and adhesion during cell movement, J. Cell Biol. 144 (1999) 1235–1244.
[15] S. Etienne-Manneville, A. Hall, Integrin-mediated activation of Cdc42 controls cell polarity in migrating astrocytes through PKCzeta, Cell 106 (2001) 489–498.
[16] A. Kodama, I. Karakesisoglou, E. Wong, A. Vaezi, E. Fuchs, ACF7: an essential integrator of microtubule dynamics, Cell 115 (2003) 343–354.
[17] R.G. Harrison, The reaction of embryonic cells to solid structures, J. Exp. Zool. A Ecol. Genet. Physiol. 17 (1914) 521–544.
[18] C. Oakley, D.M. Brunette, Response of single, pairs, and clusters of epithelial cells to substratum topography, Biochem. Cell Biol. 73 (1995) 473–489.
[19] A. Curtis, C. Wilkinson, Topographical control of cells, Biomaterials 18 (1997) 1573–1583.
[20] X.F. Walboomers, H.J. Croes, L.A. Ginsel, J.A. Jansen, Growth behaviour of fibroblasts on microgrooved polystyrene, Biomaterials 19 (1998) 1861–1868.
[21] D.M. Brunette, B. Chehroudi, The effects of the surface topography of micromachined titanium substrata on cell behavior *in vitro* and *in vivo*, J. Biomech. Eng. 121 (1999) 49–57.
[22] Y.A. Rovensky, V.I. Samoilov, Morphogenetic response of cultured normal and transformed fibroblasts, and epitheliocytes, to a cylindrical substratum surface. Possible role for the actin filament bundle pattern, J. Cell Sci. 107 (1994) 1255–1263.
[23] K. Kolind, A. Dolatshahi-Pirouz, J. Lovmand, F.S. Pedersen, M. Foss, F. Besenbacher, A combinatorial screening of human fibroblast responses on micro-structured surfaces, Biomaterials 31 (2010) 9182–9191.

[24] H. Jeon, H. Hidai, D.J. Hwang, K.E. Healy, C.P. Grigoropoulos, The effect of micronscale anisotropic cross patterns on fibroblast migration, Biomaterials 31 (2010) 4286–4295.

[25] S. Takeuchi, Wound healing in the cornea of the chick embryo. III. The influence of pore size of millipore filters on the migration of isolated epithelial sheets in culture, Dev. Biol. 51 (1976) 49–62.

[26] H.J. Fitton, B.A. Dalton, G. Beumer, G. Johnson, H.J. Griesser, J.G. Steele, Surface topography can interfere with epithelial tissue migration, J. Biomed. Mater. Res. A 42 (1998) 245–257.

[27] N.L. Jeon, H. Baskaran, S.K.W. Dertinger, G.M. Whitesides, L. Van de Water, M. Toner, Neutrophil chemotaxis in linear and complex gradients of interleukin-8 formed in a microfabricated device, Nat. Biotechnol. 20 (2002) 826–830.

[28] F. Lin, C. Nguyan, S. Wang, W. Saadi, S. Gross, N.L. Jeon, Neutrophil migration in opposing chemoattractant gradients using microfluidic chemotaxis devices, Ann. Biomed. Eng. 33 (2005) 475–482.

[29] J.P. Diao, L. Young, S. Kim, E.A. Fogarty, S.M. Heilman, P. Zhou, et al., A three-channel microfluidic device for generating static linear gradients and its application to the quantitative analysis of bacterial chemotaxis, Lab Chip 6 (2006) 381–388.

[30] C.W. Frevert, G. Boggy, T.M. Keenan, A. Folch, Measurement of cell migration in response to an evolving radial chemokine gradient triggered by a microvalve, Lab Chip 6 (2006) 849–856.

[31] S. Koyama, D. Amarie, H. Soini, M. Novotny, S.C. Jacobson, Chemotaxis of mouse sperm on microfluidic devices, Anal. Chem. 78 (2006) 3354–3359.

[32] N. Blow, Cell migration: our protruding knowledge, Nat. Methods 4 (2007) 589–594.

[33] P. Clark, P. Connolly, A.S. Curtis, J.A. Dow, C.D. Wilkinson, Topographical control of cell behavior. I. Simple step cues, Development 99 (1987) 439–448.

[34] B. Chehroudi, T.R. Gould, D.M. Brunette, Titanium-coated micromachined grooves of different dimensions affect epithelial and connective-tissue cells differently *in vivo*, J. Biomed. Mater. Res. 24 (1990) 1203–1219.

[35] J.W. Lussi, C. Tang, P.-A. Kuenzi, U. Staufer, G. Csucs, J. Vörös, et al., Selective molecular assembly patterning at the nanoscale: a novel platform for producing protein patterns by electron-beam lithography on SiO_2/indium tin oxide-coated glass substrates, Nanotechnology 16 (2005) 1781–1786.

[36] K.-B. Lee, S.J. Park, C.A. Mirkin, J.C. Smith, M. Mrksich, Protein nanoarrays generated by dip-pen nanolithography, Science 295 (2002) 1702–1705.

[37] J.D. Hoff, L.-J. Cheng, E. Meyhöfer, L.J. Guo, A.J. Hunt, Nanoscale protein patterning by imprint lithography, Nano. Lett. 4 (2004) 853–857.

[38] C.S. Chen, M. Mrksich, S. Huang, G.M. Whitesides, D.E. Ingber, Geometric control of cell life and death, Science 276 (1997) 1425–1428.

[39] N.Y. Lee, J.R. Lim, Y.S. Kim, Selective patterning and immobilization of biomolecules within precisely-defined micro-reservoirs, Biosens. Bioelectron. 21 (2006) 2188–2193.

[40] A. Folch, B.H. Jo, O. Hurtado, D.J. Beebe, M. Toner, Microfabricated elastomeric stencils for micropatterning cell cultures, J. Biomed. Mater. Res. 52 (2000) 346–353.

[41] E.A. Roth, T. Xu, M. Das, C. Gregory, J.J. Hickman, T. Boland, Inkjet printing for high-throughput cell patterning, Biomaterials 25 (2004) 3707–3715.

[42] A.L. Birkbeck, R.A. Flynn, M. Ozkan, D. Song, M. Gross, S.C. Esener, VCSEL arrays as micromanipulators in chip-based biosystems, Biomed. Microdevices 5 (2003) 47–54.

[43] B. Alberts, A. Johnson, J. Lewis, M. Raff, K. Roberts, P. Walter, Molecular Biology of the Cell, Garland Science, New York, 2002.

[44] S.-H. Yoon, Y.-H. Cho, High-precision digital microflow controllers using fluidic digital-to-analog converters composed of binary-weighted flow resistors, J. Microelectromech. Syst. 15 (2006) 967–975.

[45] D.I. Shreiber, V.H. Barocas, R.T. Tranquillo, Temporal variations in cell migration and traction during fibroblast-mediated gel compaction, Biophys. J. 84 (2003) 4102–4114.

[46] R.B. Dickinson, R.T. Tranquillo, Optimal estimation of cell movement indexes from the statistical analysis of cell tracking data, AIChE J. 39 (1993) 1995–2010.

[47] P.A. DiMilla, J.A. Stone, J.A. Quinn, S.M. Albelda, D.A. Lauffenburger, Maximal migration of human smooth muscle cells on fibronectin and type IV collagen occurs at an intermediate attachment strength, J. Cell Biol. 122 (1993) 729–737.

[48] J. Tan, W.M. Saltzman, Topographical control of human neutrophil motility on micropatterned materials with various surface chemistry, Biomaterials 23 (2002) 3215–3225.

[49] C.L. Stokes, D.A. Lauffenburger, S.K. Williams, Migration of individual microvessel endothelial cells: stochastic model and parameter measurement, J. Cell Sci. 99 (1991) 419–430.

[50] G. Karp, Cell and Molecular Biology: Concepts and Experiments, John Wiley & Sons, New York, 2004.

[51] B.A. Dalton, X.F. Walboomers, M. Dziegielewski, M.D. Evans, S. Taylor, J.A. Jansen, et al., Modulation of epithelial tissue and cell migration by microgrooves, J. Biomed. Mater. Res. 56 (2001) 195–207.

[52] S.E. Lee, S. Chunsrivirot, R.D. Kamm, M.R.K. Mofrad, Molecular dynamics study of talin–vinculin binding, Biophys. J. 95 (2008) 2027–2036.

[53] J. Golji, M.R.K. Mofrad, A molecular dynamics investigation of vinculin activation, Biophys. J. 99 (2010) 1073–1081.

[54] J. Golji, J. Lam, M.R.K. Mofrad, Vinculin activation is necessary for complete talin binding, Biophys. J. 100 (2011) 332–340.

CHAPTER

Transport Models for Three-Dimensional Cell Culture Systems

6

Niraj K. Inamdar[1], Linda G. Griffith[2] and Jeffrey T. Borenstein[3]

[1]*Department of Mechanical Engineering, Massachusetts Institute of Technology, Cambridge MA, USA*

[2]*Department of Biological Engineering, Massachusetts Institute of Technology, MA, USA*

[3]*Biomedical Engineering Center, Draper Laboratory, Cambridge MA, USA*

6.1 Introduction

A critical early step in the development of microfluidic cell culture systems is the establishment of a computational model capable of describing the transport of oxygen and other nutrients, as well as physical parameters such as fluid mechanical forces imparted to the cultured cells, in order to optimize its performance for specific applications. Due to the complex interactions between the physical and the biological processes taking place within typical microfluidic cell culture systems and the rapid advance of available computational tools, these platforms are particularly amenable to modeling-based approaches. In this chapter, we will introduce a framework that addresses several of the primary physical principles that govern flow and transport within typical microfluidic cell culture systems, discuss examples of transport models for specific applications, and review various approaches that have been developed to optimize solute transport and related fluid mechanical properties in these devices.

6.1.1 Critical design criteria for *in vitro* models

One of the principal objectives of *in vitro* cell culture models is to create a physiologically relevant microenvironment. Salient system parameters and interactions include oxygen and nutrient concentrations and gradients, cell–cell signaling, cell–surface interactions, and parameters related to fluid flow and shear stress. Transport models may be used to predict the behavior of existing device configurations or as a tool to obtain desired transport characteristics within a device during initial design stages. In either case, usefulness of the model will depend on the robustness and applicability of the approach, its physical relevance, and the validity of various assumptions that

Microfluidic Cell Culture Systems. ISBN: 978-1-4377-3459-1

are invoked. Here, we address several of the physiological parameters important to consider in the design and analysis of microfluidic tissue culture systems.

Oxygen delivery: For most cell types, controlling the delivery of molecular oxygen to the culture is of critical importance. Oxygen transport and consumption by a cultured population of cells within a microfluidic device is often the primary focus of investigation [1,2], especially for cultures of highly oxygen-sensitive cell types such as stem cells and hepatocytes. Consequently, in the design and analysis of a microfluidic tissue culture system, it is important to consider the following:

- The saturation concentration of oxygen in the cell culture medium at typical operating conditions for the device. Operating conditions depend on the particular study, for instance Standard Temperature and Pressure (STP) conditions (20°C, 1 atm) or incubator conditions (35°C, 5% CO_2).
- The physical nature and location of the oxygen reservoir. In some cell culture systems, the oxygen reservoir might be the ambient air, so that in operation, the cell culture system draws its oxygen from the medium–air interface; in other systems, the oxygen may be drawn by the tissue culture from oxygen-saturated culture medium. The location of the reservoir in relation to the cell culture is also important, since the effectiveness of oxygen delivery to the culture will depend on this distance as well as the dimensions of the device and its flow properties.
- The diffusivity of molecular oxygen within cell culture medium. If transport is diffusion-driven (rather than convection-dominated), the rate at which transport occurs is encoded in the diffusivity (or diffusion coefficient) D of oxygen within the medium. The diffusivity of any species within another can either be determined experimentally or derived from kinetic theory. One expression commonly used to estimate the diffusivity of a molecular species is the Einstein relation:

$$D = \frac{kT}{6\pi\mu R} \tag{6.1}$$

where k is the Boltzmann constant, T is the temperature, μ is the viscosity of the fluid, and R is the radius of the molecule in question. For molecular oxygen in water at 298 K, and assuming that R is given by twice the covalent radius of oxygen, a value of $D \approx 1.86 \times 10^{-5}$ cm^2/s is obtained, which compares favorably with the experimentally determined value of $D \approx 2.2 \times 10^{-5}$ cm^2/s [3].
- The oxygen consumption parameters for the particular cell type and culture configuration. Consumption-related parameters may typically be derived from the binding constants that define the equations of chemical equilibrium between oxygen and cells; these binding constants are typically derived by experiments such as surface plasmon resonance [4].

Nutrient and waste transport: In addition to oxygen, control over the delivery of nutrients to tissue culture and removal of metabolic wastes are also important considerations. In general, for waste removal to be sufficient, it may be required that convective transport dominates diffusive transport. This argument is based on the rationale that, if diffusive transport were dominant, metabolic waste could simply diffuse back toward the cell culture and occupy the extracellular environment. This criterion can be stated in terms of the Péclet number (see below):

$$Pe \gg 1 \tag{6.2}$$

Growth factors: The transport of soluble growth factors is another important consideration in tissue culture systems, as growth factors mediate intercellular communication and are capable of initiating cell proliferation, differentiation, and growth. Growth factors interact with cells through binding receptors on the cell surface that are specific to a particular growth factor. Models for the kinetics of binding between growth factors and receptors utilize the binding and dissociation coefficients for the growth factor, the receptor, and the growth factor–receptor complex as input parameters, which are derived experimentally [5,6]. Since growth factors are typically linked with other complex cellular behaviors, it is possible to develop models which correlate the binding of growth factors to cell receptors with processes such as the secretion of cytokines by cells. These models may eventually be incorporated into comprehensive computational transport models.

Perfusion and gradients: Perfusion-related phenomena govern the delivery of nutrients to the tissues. In nonsteady-state perfusion conditions, as well as in situations in which the binding of solutes occurs, concentration gradients will be developed. These gradients lead to diffusion-driven transport; by controlling both gradients as well as fluid flow, the transport of solute can be tailored to desired rates. Methods and challenges associated with this are discussed in Section 6.9.

Shear stresses: The effects that mechanical forces exert on cell behavior are the subject of intense investigation. Mesenchymal stem cells (MSCs), for instance, are known to differentiate variously into osteoblasts and chondrocytes on the basis of shear forces [7], an observation with significant implications for regenerative therapies based on engineering stem cells. Both physiologically and within tissue culture systems, mechanical forces are often imparted during the presence of viscous fluid flow. It is therefore important to control flow within microfluidic culture systems not only to achieve optimal molecular transport behavior but also to control shear stress.

6.2 Fluid flow in cell culture systems

The microfluidic environment within a cell culture system may comprise fluidic, mechanical, and biochemical factors. The ability to predict and control these various factors is necessary in order to realize robust microfluidic cell culture systems.

6.2.1 Equations of fluid motion

Fluid flow and mass transport are generally coupled by virtue of the fact that the presence of a solute alters the course of the fluid motion of the solvent; this is true, for instance, in polymer solutions [8]. However, it is often the case that solutions are sufficiently dilute such that the equations for fluid motion of the solvent may be solved independently from those for the chemical transport of solute. For the analysis that follows, it is assumed that all solutions are suitably dilute.

For situations in which the viscosity of the fluid can be regarded as a constant and the fluid itself as incompressible, the equations describing the time evolution of the velocity field $\vec{u}$ throughout the domain of interest—the Navier–Stokes equations—are [9]

$$\rho\left[\frac{\partial \vec{u}}{\partial t} + (\vec{u}\cdot\nabla)\,\vec{u}\right] = -\nabla p + \mu\nabla^2\,\vec{u} + \vec{F} \tag{6.3}$$

where ∇ is the gradient operator and ∇^2 is the vector Laplacian appropriate to the coordinate system. ρ is the fluid density, p is the hydrostatic pressure, μ is the dynamic viscosity of the fluid, and $\vec{F}$ is an external body force per unit volume. This may be a gravitational force (which is usually negligible in a microfluidic cell culture system), or, more commonly, an electric force due to, say, the presence of charge groups on the boundaries of the domain or an externally applied field. In order to study electric fields as they develop within a conducting fluid medium, the Navier–Stokes equations must be coupled with Maxwell's equations for continuous media [10].

The Navier–Stokes equations are a representation of Newton's second law as applied to a differential volume element of fluid. Therefore, the left-hand side of Eq. (6.3) describes the acceleration of the fluid element, and both terms are associated with the inertia of the fluid. The right-hand side, on the other hand, describes the forces acting on the fluid element, which, in addition to the external body force $\vec{F}$, includes contributions from the pressure p and frictional (or viscous) forces applied by adjacent fluid volume elements.

The relation for mass conservation in differential form is the continuity equation:

$$\frac{\partial \rho}{\partial t} + \nabla\cdot(\rho\,\vec{u}) = 0 \tag{6.4}$$

In the case of constant density ρ, this reduces to the familiar form

$$\nabla\cdot\vec{u} = 0 \tag{6.5}$$

Stresses within the fluid are represented by the stress tensor σ_{ij} $(i,j = x,y,z)$, which describes the stresses (force per unit area) acting on a surface element of the fluid element. Per notational convention, σ_{ij} denotes the stress acting on a surface of the fluid element perpendicular to the ith direction and parallel to the jth axis. The diagonal entries of the stress tensor (i.e., the case in which $i = j$) include contributions from the hydrostatic pressure p and from stresses arising from internal motion of the fluid; the other elements of the stress tensor give the local shear stress acting on the fluid element. In Cartesian coordinates, let us denote the components of the velocity vector $\vec{u}$ by u_i ($i = x,y,z$), then the stress tensor is given by [9]:

$$\sigma_{ij} = -p\delta_{ij} + \mu\left(\frac{\partial u_i}{\partial x_i} + \frac{\partial u_j}{\partial x_i}\right) \tag{6.6}$$

where $\delta_{ij} = 1$, if $i = j$ and 0 otherwise. Expressions for the stress tensor in other coordinate systems are given in Appendix A.

6.2.2 Boundary conditions

The boundary condition for fluids most applicable to microfluidic systems is the "no-slip" condition, which stipulates that, due to molecular interactions between a fluid and an immovable boundary, the velocity of the fluid at the boundary equals that of the boundary. Hence,

$$\vec{u} = 0 \text{ on a static boundary (no-slip condition)} \tag{6.7}$$

If, on the other hand, the boundary is itself moving with a velocity $\vec{U}$, then we have

$$\vec{u} = \vec{U} \text{ on a moving boundary (no-slip condition)} \tag{6.8}$$

A more complex situation involves the case in which a fluid shares a boundary with another fluid. Examples include the flowing of two immiscible fluids parallel to one another or an air–liquid interface, the study of which is important in understanding the behavior of cells in lung or tracheal tissue [11]. To establish the boundary condition for fluid–fluid interfaces, we make use of Newton's third law and apply it to the stresses exerted by fluid elements on their respective sides of their common boundary. We label the two fluids "Fluid 1" and "Fluid 2." Let $n_j^{(k)}$ ($k = 1$ or 2) be the jth component of a unit vector normal to the boundary surface, the positive direction being reckoned as pointing outward from fluid k, toward the other fluid, and let $\sigma_{ij}^{(k)}$ be the corresponding stress tensor for fluid k. The forces exerted by the fluids on one another must be equal in magnitude and opposite in direction. Generally, the expression $\sum_{j=1}^{3}\left\{n_j^{(1)}\sigma_{ij}^{(1)} + n_j^{(2)}\sigma_{ij}^{(2)}\right\} = 0$ must hold; but since the normal vectors $n_j^{(k)}$ are in opposite directions, we may drop the (k) superscript and write the boundary condition as

$$\sum_{j=1}^{3} n_j\sigma_{ij}^{(1)} = \sum_{j=1}^{3} n_j\sigma_{ij}^{(2)} \tag{6.9}$$

A special case of interest occurs when one of the fluids (e.g., Fluid 2) is inviscid, so that the viscous contributions to Fluid 2's stress tensor (the second term on the right-hand side of Eq. (6.6)) vanish. Physically, this gives the case of a free surface and could be taken to represent, for instance, a situation in which Fluid 2 is air, whose ability to withstand the shear imparted on it by a fluid is negligible.

6.3 The theory of mass transport

6.3.1 Equation of mass transport

For a medium in which there are dissolved N different solutes in which the mixture is assumed dilute, the equation of mass transport for the ith species of solute in terms of its concentration C_i is

$$\frac{\partial C_i}{\partial t} + (\vec{u} \cdot \nabla)C_i = \nabla \cdot (D_i \cdot \nabla C_i) + R_i, \quad i = 1, \ldots, N \tag{6.10}$$

where $\vec{u}$ is the velocity field given by the solution to the Navier–Stokes equations and R_i is a reaction term indicating the presence of a source or sink of solute in the domain of interest. This is especially applicable if cells interact with the solute throughout the domain or if there are chemical receptors distributed throughout (see Section 6.8.1). D_i is the diffusivity, or diffusion coefficient, characteristic to the solute in the medium; it is in general a tensor, but in most cases, it may be regarded as both isotropic and constant throughout space and time, so that the right-hand side of Eq. (6.10) may be written as

$$D_i \nabla^2 C_i + R_i \tag{6.11}$$

The second term on the left-hand side of Eq. (6.10) is the contribution to transport from convection; the first term on the right-hand side of Eq. (6.10) is that from diffusion. Each species present within the medium will have its own transport equation like Eq. (6.10), and any interaction between species within the domain of interest is encoded within the reaction term R. The flux of species i (i.e., the amount of solute passing through a surface of unit area per unit time) at a given point and along the direction specified by a unit vector $\hat{n}$ is given by

$$(-D_i \nabla C_i + \vec{u}\, C_i) \cdot \hat{n} \tag{6.12}$$

The first term gives the flux due to diffusive transport and the second due to convective transport. We see from the diffusive flux term that diffusion occurs locally in the direction from highest to lowest solute concentrations.

6.3.2 Boundary conditions

Boundary conditions for the transport equation are specified in terms of either fixed concentrations or flux conditions. For instance, if one location in a cell culture device is exposed to a reservoir of solute A large enough that the concentration of A at the interface $C_{A,0}$ can be taken constant over time, we obtain

$$C_A(t)\big|_{\text{interface}} = C_{A,0} \tag{6.13}$$

Equation (6.13) is valid even if $C_{A,0}$ is time-dependent. Flux boundary conditions specify the nature of the flux normal to a boundary surface. Commonly, it is the case that the boundary is impenetrable to the solute. In this case, we write most generally for species A

$$(-D_A \nabla C_A + \vec{u}\, C_A) \cdot \hat{n}\big|_{\text{boundary}} = 0 \tag{6.14}$$

In other cases, the boundary may be permeable to the solute; this is the case if the cell culture system is fabricated from, say, polydimethylsiloxane (PDMS), and the solute of interest is oxygen to which PDMS is quite permeable [12]. Then, if k is a transfer coefficient for oxygen in PDMS and C_∞ is a characteristic oxygen concentration in the PDMS, we would write the appropriate boundary condition as

$$(-D_{O_2} \nabla C_{O_2} + \vec{u}\, C_{O_2}) \cdot \hat{n}\big|_{\text{boundary}} = k(C_{O_2}\big|_{\text{boundary}} - C_\infty) \tag{6.15}$$

In this case, D_{O_2} is the diffusion coefficient for oxygen in the fluid domain of interest. In cell culture systems, it is often the case that one of the boundaries is populated by cells that consume one or more of the chemical species. In this case, again, it is necessary to choose a suitable model describing the flux. For example, we may specify at the cell-populated boundary Michaelis–Menten kinetics, a binding model which is often applicable to enzymatic reactions that can occur at the cell membrane. If a tissue culture which consumes species A is cultured along a planar surface, we would write

$$(-D_A \nabla C_A + \vec{u}\, C_A) \cdot \hat{n}\big|_{\text{boundary}} = \left.\frac{K_1 C_A}{K_2 + C_A}\right|_{\text{boundary}} \tag{6.16}$$

where K_1 and K_2 are parameters specific to the problem at hand (see Sections 6.4.4 and 6.8.2).

6.4 Binding kinetics

This section will describe some models for chemical reactions that may be implemented as the reaction term within the transport equation [13]. Under the assumption of local chemical equilibrium (the "principle of detailed balance"), the equations given can then be directly incorporated into the transport equation to account for diffusive and convective transport.

6.4.1 Irreversible, first-order single-molecule reactions

Here, we consider the simple case in which a single chemical species A is converted irreversibly into another species B. Hence, the reaction reads

$$A \xrightarrow{k} B \tag{6.17}$$

Here, k is the rate coefficient of the reaction. By virtue of the law of mass action, we can write the evolution of the concentration species A, C_A, with time as

$$\frac{\mathrm{d}C_A}{\mathrm{d}t} = -kC_A \tag{6.18}$$

Again, this describes the time evolution of A at a point in space neglecting other transport mechanisms. The equation may be immediately incorporated into the transport equation as

$$\frac{\partial C_A}{\partial t} + (\vec{u} \cdot \nabla) C_A = \nabla \cdot (D_A \cdot \nabla C_A) - k C_A \tag{6.19}$$

6.4.2 Reversible, first-order single-molecule reactions

A reaction that may proceed in both directions between species A and B is written as

$$A \underset{k_R}{\overset{k_F}{\longleftrightarrow}} B \tag{6.20}$$

where k_F is the rate constant for the forward A to B reaction and k_R is the constant for the reverse reaction. The evolution of species A and B in terms of their concentrations C_A and C_B is

$$\left.\begin{aligned} \frac{dC_A}{dt} &= -k_F C_A + k_R C_B \\ \frac{dC_B}{dt} &= k_F C_A - k_R C_B \end{aligned}\right\} \tag{6.21}$$

The extension to the transport equation is straightforward:

$$\left.\begin{aligned} \frac{\partial C_A}{\partial t} + (\vec{u} \cdot \nabla) C_A &= \nabla \cdot (D_A \cdot \nabla C_A) - k_F C_A + k_R C_B \\ \frac{\partial C_B}{\partial t} + (\vec{u} \cdot \nabla) C_B &= \nabla \cdot (D_B \cdot \nabla C_B) + k_F C_A - k_R C_B \end{aligned}\right\} \tag{6.22}$$

6.4.3 Reversible, first-order bimolecular reactions

A first-order, bimolecular reaction is described by the following expression:

$$A \underset{k_R}{\overset{k_F}{\longleftrightarrow}} B + C \tag{6.23}$$

This model is applicable, for instance, to reactions in which A represents a complex between ligand and receptor, and B and C, the individual components of the complex. The chemical evolution of each species is given by:

$$\left.\begin{aligned} \frac{dC_A}{dt} &= -k_F C_A + k_R C_B C_C \\ \frac{dC_B}{dt} &= k_F C_A - k_R C_B C_C \\ \frac{dC_C}{dt} &= k_F C_A - k_R C_B C_C \end{aligned}\right\} \tag{6.24}$$

and the extension to the transport equation is

$$\left.\begin{aligned}\frac{\partial C_A}{\partial t}+(\vec{u}\cdot\nabla)C_A&=\nabla\cdot(D_A\cdot\nabla C_A)-k_F C_A+k_R C_B C_C\\ \frac{\partial C_B}{\partial t}+(\vec{u}\cdot\nabla)C_B&=\nabla\cdot(D_B\cdot\nabla C_B)+k_F C_A-k_R C_B C_C\\ \frac{\partial C_C}{\partial t}+(\vec{u}\cdot\nabla)C_B&=\nabla\cdot(D_B\cdot\nabla C_B)+k_F C_A-k_R C_B C_C\end{aligned}\right\}\tag{6.25}$$

6.4.4 Enzyme kinetics (Michaelis–Menten kinetics)

The kinetics governing enzymatic reactions is typically given by Michaelis–Menten kinetics [14] and is also applicable to a range of other biochemical processes such as oxygen uptake by cells. In order to derive an expression for the reaction rate in such processes, suppose that F denotes a free chemical species (taken to be an enzyme in the model problem), B a binding substrate, C the complex which forms from F and B, and P the final product of the resulting from the dissociation of the enzyme F from the complex C. The reaction can be described as follows:

$$F+B\underset{k_R}{\overset{k_F}{\longleftrightarrow}}C\overset{k_{F,C}}{\longrightarrow}F+P\tag{6.26}$$

Here, k_F is the forward rate coefficient for the $F+B$ to C reaction, k_R is the reverse rate coefficient for the same reaction, and $k_{F,C}$ is the rate coefficient for the reaction (assumed only forward) for the dissociation of the complex into the enzyme and the final product.

The time evolution equations for each species read then as

$$\left.\begin{aligned}\frac{dC_F}{dt}&=-k_F C_F C_B+k_R C_C+k_{F,C}C_C\\ \frac{dC_B}{dt}&=-k_F C_F C_B+k_R C_C\\ \frac{dC_C}{dt}&=k_F C_F C_B-k_R C_C-k_{F,C}C_C\\ \frac{dC_P}{dt}&=k_{F,C}C_C\end{aligned}\right\}\tag{6.27}$$

Adding the first and second of Eq. (6.27), we find that

$$d(C_F+C_C)/dt=0 \text{ or } C_F+C_C=\text{constant}\equiv C_0\tag{6.28}$$

During an enzymatic reaction, the quantity of complex C_C will typically be substantially less than C_B and C_P; hence, if we average over a time period greater than the very first stages of the reaction process (itself generally a very short time period), we can conclude that $dC_C/dt \ll dC_B/dt$ and dC_P/dt. Hence, we set the third of Eq. (6.27) to 0 to get

$$k_F C_F C_B - k_R C_C - k_{F,C} C_C = 0 \tag{6.29}$$

Now, substituting Eq. (6.28) in Eq. (6.29), we get

$$k_F(C_0 - C_C)C_B = (k_R + k_{F,C})C_C \tag{6.30}$$

or, rearranging,

$$\begin{aligned} C_0 C_B &= \left(\frac{k_R + k_{F,C}}{k_F} + C_B\right) C_C \\ \therefore C_C &= \frac{C_0 C_B}{(k_R + k_{F,C}/k_F) + C_B} \end{aligned} \tag{6.31}$$

Substituting this in the last of Eq. (6.27) gives

$$\frac{dC_P}{dt} = \frac{k_{F,C} C_0 C_B}{(k_R + k_{F,C}/k_F) + C_B} \tag{6.32}$$

We may compare this result with Eq. (6.16) with $K_1 \propto k_{F,C} C_0$ and $K_2 = (k_R + k_{F,C})/k_F$. In the literature, $k_{F,C} C_0$ is often denoted V_{max}, indicating that it is the maximum value which dC_P/dt can take. $(k_R + k_{F,C})/k_F$ is often called the Michaelis–Menten parameter and denoted K_M. The expression given here for K_1 from Eq. (6.16) is not exact since Eq. (6.16) is an expression for flux, not time rate of change. To make the correspondence exact, we would have to multiply $k_{F,C} C_0$ by a parameter expressing the local per unit area density of the receptors (cf. ρ_{cells} in Section 6.8.2).

6.5 Nondimensionalization

Nondimensionalization entails rescaling the variables in a given equation so as to simplify the equation and enable manipulation of geometrical and physical parameters of the problem by varying only one or two dimensionless expressions, as described in the following example.

Let x range from 0 to L and y from 0 to H for the following simplified transport equation, which may be arrived at from order of magnitude considerations (see Section 6.6):

$$U\frac{\partial C}{\partial x} = D\frac{\partial^2 C}{\partial y^2} \tag{6.33}$$

This equation may model for instance the transport of a solute dissolved in a medium flowing at constant velocity U in a high aspect ratio channel where transport occurs primarily by convection in the x-direction and by diffusion in the narrower y-direction; such a situation is in fact typical of many microfluidic systems. We prescribe the following boundary conditions:

$$\left.\begin{aligned} C(x=0) &= C_0 \\ D\left.\frac{\partial C}{\partial y}\right|_{y=0} &= KC(y=0) \\ \left.\frac{\partial C}{\partial y}\right|_{y=H} &= 0 \end{aligned}\right\} \tag{6.34}$$

where K is a constant that may describe, for instance, the consumption of solute by a culture of cells at the bottom surface of the channel consistent with first-order, irreversible kinetics. To proceed with nondimensionalization, we must consider all the variables in the transport equation and the boundary conditions and determine a "characteristic" scaling factor for each of them, the scaling factors usually supplied by the geometry and physical parameters of the problem. For example, for the coordinate variables x and y, either L or H may be chosen as a characteristic scaling factor L or H. In this case we choose H, as the primary focus may be the diffusion of solute downward in the y-direction to the cell culture. For the concentration C, C_0 is the logical choice for a scaling factor. Proceeding to scale all the variables, we define the following dimensionless variables:

$$\left.\begin{aligned} \hat{x} &\equiv x/H \\ \hat{y} &\equiv y/H \\ \hat{C} &\equiv C/C_0 \end{aligned}\right\} \tag{6.35}$$

By these definitions, we note that $\hat{x}$ ranges from 0 to L/H, while $\hat{y}$ ranges from 0 to 1. Moreover, we note that $\mathrm{d}\hat{x}/\mathrm{d}x = \mathrm{d}\hat{y}/\mathrm{d}y = 1/H$. Then, by the chain rule, Eq. (6.33) becomes

$$\begin{aligned} UC_0\frac{\partial \hat{C}}{\partial \hat{x}}\frac{\mathrm{d}\hat{x}}{\mathrm{d}x} &= DC_0\frac{\partial^2 \hat{C}}{\partial \hat{y}^2}\left(\frac{\mathrm{d}\hat{y}}{\mathrm{d}y}\right)^2 \\ \therefore \frac{U}{H}\frac{\partial \hat{C}}{\partial \hat{x}} &= D\frac{\partial^2 \hat{C}}{\partial \hat{y}^2}\frac{1}{H^2} \\ \therefore \frac{\partial \hat{C}}{\partial \hat{x}} &= \frac{D}{UH}\frac{\partial^2 \hat{C}}{\partial \hat{y}^2} \end{aligned} \tag{6.36}$$

The combination of parameters D/UH is the reciprocal of the so-called Péclet number, often denoted Pe, so that

$$Pe \equiv \frac{UH}{D} \tag{6.37}$$

Pe characterizes the relative importance of convective transport to diffusive transport in the system. Now, we must nondimensionalize the boundary conditions; we have

$$\left.\begin{array}{l} \hat{C}(\hat{x}=0)=1 \\ \left.\dfrac{\partial \hat{C}}{\partial \hat{y}}\right|_{\hat{y}=0}=\dfrac{KH}{D}\hat{C}(\hat{y}=0) \\ \left.\dfrac{\partial \hat{C}}{\partial \hat{y}}\right|_{\hat{y}=1}=0 \end{array}\right\} \tag{6.38}$$

The combination of parameters KH/D is known as the Damköhler number [13]; it is often written Da, and it characterizes the relative importance of flux due to chemical reactions and flux due to diffusion, in this case, at the $y=0$ boundary of the system. Equations (6.36) and (6.38) completely describe the original system of Eqs. (6.33) and (6.34) in nondimensional form. Note that by solving Eqs. (6.36) and (6.38), several parameters of interest are lumped into the two constants Pe and Da. Any solution to the nondimensional equations in terms of given Pe and Da will furnish solutions to the original *dimensional* equations for any combination of U, H, K, and L/H that gives those values of Pe and Da. Finally, in order to recover the solution to Eqs. (6.33) and (6.34), we simply rearrange Eq. (6.35), so that $x=\hat{x}H$, $y=\hat{y}H$, and $C=\hat{C}C_0$.

We may also nondimensionalize the Navier–Stokes equations. Suppose the fluid is confined to a flow channel defined by the rectangular parallelepiped $x\in[0,L]$, $y\in[0,H]$, and $z\in[0,W]$ and let flow be introduced at $x=0$ at a uniform velocity of $u=U$. The Navier–Stokes equations read as

$$\rho\left[\frac{\partial \vec{u}}{\partial t}+\left(\vec{u}\cdot\nabla\right)\vec{u}\right]=-\nabla p+\mu\nabla^2\vec{u}+\vec{F} \tag{6.39}$$

We now choose a critical length dimension, whose selection again depends on the specific problem at hand. We choose H for this dimension, and let us choose U as a characteristic flow speed. By dimensional analysis, we scale the pressure by $\mu U/H$, t by $\rho H^2/\mu$, and the external force by $\rho H^2/(\mu U)$. Summarizing, we have

$$\left.\begin{array}{l} \hat{x}\equiv x/H, \hat{y}\equiv y/H, \quad \text{and} \quad \hat{z}\equiv z/H \\ \hat{\vec{u}}\equiv\vec{u}/U \\ \hat{p}\equiv p\left(\dfrac{H}{\mu U}\right) \\ \hat{t}\equiv t\left(\dfrac{\mu}{\rho H^2}\right) \\ \hat{\vec{F}}\equiv\vec{F}\left(\dfrac{\mu U}{\rho H^2}\right) \end{array}\right\} \tag{6.40}$$

Then, the dimensionless Navier–Stokes equations read as

$$\frac{\partial\hat{\vec{u}}}{\partial\hat{t}}+\frac{\rho UH}{\mu}\left(\hat{\vec{u}}\cdot\nabla\right)\hat{\vec{u}}=-\nabla\hat{p}+\nabla^2\hat{\vec{u}}+\hat{\vec{F}} \tag{6.41}$$

where all gradients are now taken with respect to the dimensionless length variables. The combination of parameters $\rho UH/\mu$ is called the Reynolds number, often written *Re* or simply *R*. It describes the relative importance of inertia to viscosity in the fluid. Indeed, in the limit of high Reynolds number, we may neglect the viscous terms in the Navier–Stokes equation and obtain the Euler equations:

$$\frac{\partial \hat{\vec{u}}}{\partial t} + Re\left(\hat{\vec{u}} \cdot \nabla\right)\hat{\vec{u}} = -\nabla\hat{p} + \hat{\vec{F}} \tag{6.42}$$

which describes the so-called perfect, or inviscid, fluids (such as gases), for which viscosity is negligible. The limit of low Reynolds number, on the other hand, is very often relevant to flows in tissue culture systems. In this case, the inertial terms on the left-hand side of the Navier–Stokes equations are negligible, and we have the unsteady Stokes equations:

$$\frac{\partial \hat{\vec{u}}}{\partial t} = -\nabla\hat{p} + \nabla^2\hat{\vec{u}} + \hat{\vec{F}} \tag{6.43}$$

6.6 Order of magnitude analysis

Order of magnitude analysis is a useful method to obtain the order of magnitude estimates of solutions and to determine physically negligible terms in the equations of fluid motion and transport. This method proceeds by approximating variables and derivatives by using physical parameters directly obtainable from the statement of the problem at hand. Again, we proceed by example.

We consider the steady-state ($\partial/\partial t \to 0$) Navier–Stokes equations in the limit of low Reynolds number. Suppose the fluid is confined to a flow channel defined by the rectangular parallelepiped $x \in [0,L]$, $y \in [0,H]$, and $z \in [0,W]$, with $L \gg W \gg H$; again, let flow be introduced at $x = 0$ at a uniform velocity of $u = U$. Such dimensions are typical of long, high aspect ratio channels found in microfluidic cell culture systems. Let $\vec{u} \equiv (u, v, w)$. The Navier–Stokes equations and the continuity equation then read as

$$\left.\begin{aligned}
-\frac{\partial p}{\partial x} + \mu\left(\frac{\partial^2 u}{\partial x^2} + \frac{\partial^2 u}{\partial y^2} + \frac{\partial^2 u}{\partial z^2}\right) &= 0\\
-\frac{\partial p}{\partial y} + \mu\left(\frac{\partial^2 v}{\partial x^2} + \frac{\partial^2 v}{\partial y^2} + \frac{\partial^2 v}{\partial z^2}\right) &= 0\\
-\frac{\partial p}{\partial z} + \mu\left(\frac{\partial^2 w}{\partial x^2} + \frac{\partial^2 w}{\partial y^2} + \frac{\partial^2 w}{\partial z^2}\right) &= 0\\
\frac{\partial u}{\partial x} + \frac{\partial v}{\partial y} + \frac{\partial w}{\partial z} &= 0
\end{aligned}\right\} \tag{6.44}$$

We approximate the partial derivatives as follows:

$$\left.\begin{aligned} \frac{\partial}{\partial x} &\sim \frac{1}{L} \\ \frac{\partial}{\partial y} &\sim \frac{1}{H} \\ \frac{\partial}{\partial z} &\sim \frac{1}{W} \end{aligned}\right\} \tag{6.45}$$

We approximate u by U. The pressure, for which we do not have a characteristic value, will simply be denoted by P. We can use the continuity equation to estimate v and w, as well:

$$\begin{aligned} \frac{\partial u}{\partial x} + \frac{\partial v}{\partial y} + \frac{\partial w}{\partial z} = 0 \Rightarrow \frac{\partial u}{\partial x} \sim \frac{\partial v}{\partial y} \sim \frac{\partial w}{\partial z} \\ \therefore \frac{U}{L} \sim \frac{v}{H} \sim \frac{w}{W} \\ \therefore v \sim U\frac{H}{L} \text{ and } w \sim U\frac{W}{L} \end{aligned} \tag{6.46}$$

We may now substitute these relations into the Navier–Stokes equations; we get

$$\left.\begin{aligned} -\frac{P}{L} + \mu\left(\frac{U}{L^2} + \frac{U}{H^2} + \frac{U}{W^2}\right) \sim 0 \\ -\frac{P}{H} + \mu\left(\frac{H}{L}\frac{U}{L^2} + \frac{H}{L}\frac{U}{H^2} + \frac{H}{L}\frac{U}{W^2}\right) \sim 0 \\ -\frac{P}{W} + \mu\left(\frac{W}{L}\frac{U}{L^2} + \frac{W}{L}\frac{U}{H^2} + \frac{W}{L}\frac{U}{W^2}\right) \sim 0 \end{aligned}\right\} \tag{6.47}$$

Using the fact that $L \gg W \gg H$, we may now compare terms within each of the equations and subsequently conclude

$$\left.\begin{aligned} -\frac{P}{L} + \mu\frac{U}{H^2} \sim 0 \\ -\frac{P}{H} \sim 0 \\ -\frac{P}{W} \sim 0 \end{aligned}\right\} \tag{6.48}$$

P has been left in since we have no immediate means of estimating its order of magnitude. Substituting the corresponding original expressions into Eq. (6.41), we have

$$\left.\begin{aligned} -\frac{\partial p}{\partial x} + \mu\frac{\partial^2 u}{\partial y^2} &= 0 \\ -\frac{\partial p}{\partial y} &= 0 \\ -\frac{\partial p}{\partial z} &= 0 \end{aligned}\right\} \tag{6.49}$$

This reduced form of the Navier–Stokes equations gives rise to the well-known Poiseuille equation, which can be arrived at by noticing that obtaining a consistent solution to the equations requires the pressure gradient $\partial p/\partial x$ to be a constant and by applying no-slip boundary conditions to the top and the bottom of the channel in the y-direction:

$$u(y) = \frac{H^2}{2\mu}\left(\frac{\Delta p}{L}\right)\frac{y}{H}\left(1 - \frac{y}{H}\right) \tag{6.50}$$

$\Delta p/L$ has been written for $\partial p/\partial x$, Δp being the pressure drop along the length L of the channel.

6.7 Bulk parameter models

In many cases, and especially in tissue culture microfluidic networks comprising many channels, it is possible to use "bulk" (or "lumped") parameter models to determine quantities of interest such as flow rates and concentrations, rather than solving the full differential equations for the entire system. Let us consider one such network in Figure 6.1.

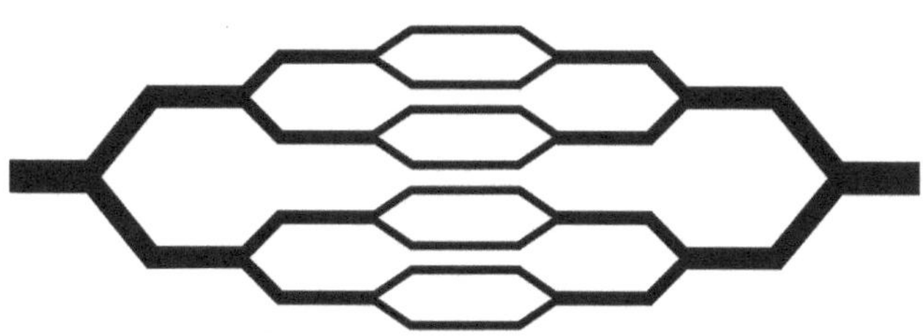

FIGURE 6.1

Schematic of an example of a microfluidic network for tissue culture. Using standard lithographic processes, channels of rectangular cross section can be arranged as shown. This network represents an artificial vasculature. In this device, fluid may be flowed through the left inlet (by, say, a syringe pump), with the outflow located at the right.

The network shown in Figure 6.1 represents an artificial vasculature, with each bifurcation suggesting that of a blood vessel or capillary. We suppose that the flow throughout the network is low Reynolds number, and we ignore for now the effect of losses such as those found at sharp corners and in fluidic connections. We also suppose that each of the channels in the network is of the type described in the previous section, although the methods we will use can easily be applied to arbitrary cross sections, as well. Let us calculate the volumetric flow rate Q associated with the flow profile given by Eq. (6.50):

$$\begin{aligned} Q &= W \int_{y=0}^{H} u(y) dy \\ &= \frac{H^3 W}{12\mu L} \Delta p \end{aligned} \tag{6.51}$$

For a channel whose width W is of dimension comparable to H but which is, along with H, still much less than L, the following approximate expression holds:

$$Q = \frac{H^3 W}{12\mu L}\left(1 - 0.630\frac{H}{W}\right)\Delta p \tag{6.52}$$

We notice that the fluidic network shown in Figure 6.1 resembles that of an electrical circuit. In fact, for low Reynolds flows, we may make a formal analogy between electrical and fluidic networks in which we simulate potential difference with pressure differences Δp along a fluidic channel element and electrical currents with flow rates Q. Then, by analogy, we identify the ratio $\Delta p/Q$ with the fluidic resistance R_{fluid} of the element. In this way, we can set up a combination of channels with given flow rates or driving pressures and solve for the system much as one would solve for an electrical circuit. In our example, we have

$$R_{\text{fluid}} = \frac{12\mu L}{H^3 W} \tag{6.53}$$

Consequently, we can build up a network of channels, with the resistance of each channel given by the above expression. Rules analogous to Kirchhoff's can also be applied; indeed, instead of the law of conservation of current at circuit nodes, we have the conservation of mass (Figure 6.2) with the sum of all incoming flow equal to that of all outgoing flow.

Mathematically, the law of conservation of mass for the fluid at the ith node of the network can be stated as

$$Q_{0,i} = Q_{1,i} + Q_{2,i} \tag{6.54}$$

In order to develop bulk parameter models for mass transfer, it is necessary to make several simplifying assumptions, as well. We demonstrate one such strategy. Suppose we have a microfluidic device in which convective transport

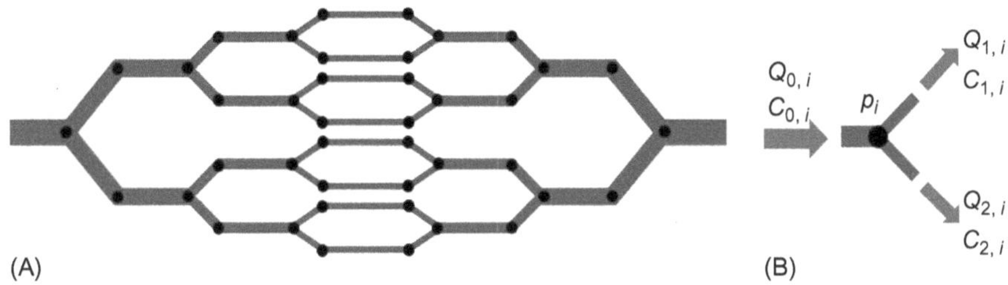

FIGURE 6.2

Schematic of microfluidic network with nodes and conservation of mass at nodes indicated. (A) Microfluidic network as shown in Figure 6.1 with nodes indicated. (B) Conservation of mass for fluid flow and solute at a given node i of a microfluidic network. $Q_{0,i}$ and $C_{0,i}$ represent the fluid flow rate and solute concentration, respectively, incoming at node i. $Q_{1,i}$ and $C_{1,i}$, and $Q_{2,i}$ and $C_{2,i}$ are the flow rates and solute concentrations leaving the node. p_i is the pressure at node i.

dominates and that solute is consumed throughout the device in a manner consistent with a first-order, irreversible model, then we may write the equation of transport as

$$u\frac{\mathrm{d}\overline{C}}{\mathrm{d}x} = k(\overline{C} - C_\infty) \tag{6.55}$$

where $\overline{C}$ is the average cross-sectional concentration of solute through the channel, C_∞ is some characteristic concentration, and k a mass transfer coefficient. If we integrate the above equation over the channel height H, then we have

$$\frac{\mathrm{d}\overline{C}}{\mathrm{d}x} = \frac{HWk}{Q}(\overline{C} - C_\infty) \tag{6.56}$$

If C_0 is the solute concentration at the inlet of the channel in question, we have

$$\overline{C}(x) = (C_0 - C_\infty)\exp\left(\frac{HWk}{Q}x\right) + C_\infty \tag{6.57}$$

The average outlet concentration is

$$\overline{C}(L) = (C_0 - C_\infty)\exp\left(\frac{HWk}{Q}L\right) + C_\infty \tag{6.58}$$

Then, by enforcing mass conservation for the solute species at the ith node,

$$C_{0,i} = C_{1,i} + C_{2,i} \tag{6.59}$$

it is possible to use the solutions for Q to now develop a system of equations by which we can solve for $\overline{C}$ over all elements of the network. An approach similar to this was taken by Cabodi et al. [15] in their investigation examining the possibility of controlled convective mass transfer in wound dressings. Instead of the model given by Eq. (6.58), the authors correlated the mass transfer coefficient to the flow rate by using existing experimental data. Numerous other possibilities exist, of course, and the model employed depends on the accuracy desired by the modeler.

It is worthwhile for us to examine an example from the literature in which a bulk parameter model has been applied to a tissue engineering system. Lee et al. [16] developed a microfluidic device whose unique geometry mimics that of the liver sinusoid. As shown in Figure 6.3A, the device configuration is such that primary, nutrient-enriched fluid flows through a horseshoe-like flow channel. Smaller quantities of fluid flows through a number of small, narrow channels (the "endothelial-like barrier") into a cell culture area, and back into the flow channel, transporting with it nutrients and waste. The cell culture area is presumably set to an outlet which is open to atmospheric pressure, p_{atm}. The bulk parameter model that the authors have used is shown in Figure 6.3B.

In comparing the authors' expression for the hydraulic resistance of a channel element, we notice that instead of considering the cross-sectional dimensions of the channel, they have considered instead the hydraulic diameter d_h of the channel, a parameter often used to approximate results derived for circular channels for rectangular ones by assigning an "effective diameter" to the noncircular channel (in this case, the authors have used the expression

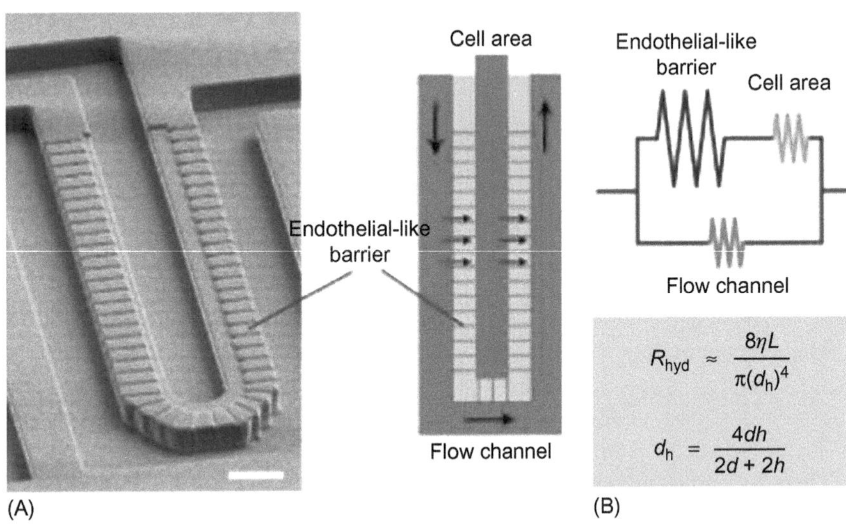

FIGURE 6.3

Geometry and bulk parameter network for artifical liver sinusoid microfluidic cell culture device. (A) Geometry of artificial liver sinusoid microfluidic cell culture device. (B) Bulk parameter equivalent resistance circuit given for geometry by Lee et al. [16].

Source: *Reproduced from Ref. [16] with permission from Wiley Periodicals, Inc.*

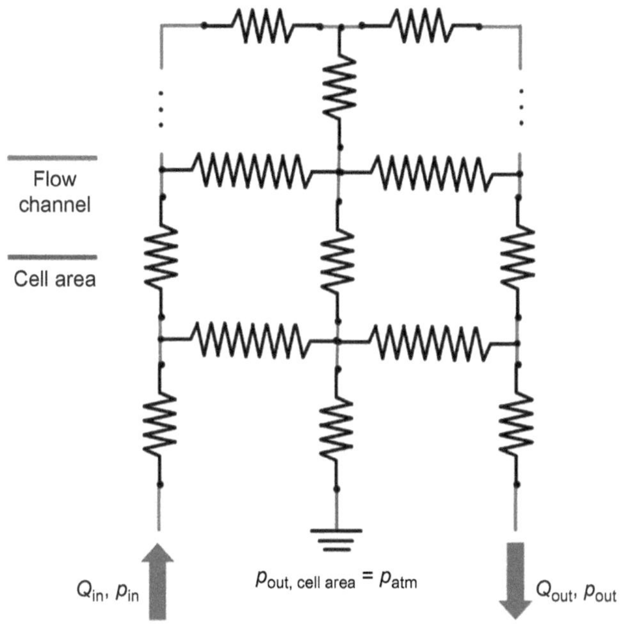

FIGURE 6.4

Bulk parameter microfluidic network for the geometry shown in Figure 6.3A. The pressure at the end of the cell area has been taken to be "ground," and set to atmospheric pressure p_{atm}. Inlet and outlet values for the flow rate and pressure are shown accordingly.

for the hydraulic resistance equivalent to Eq. (6.53) for circular channels). The most general expression of the hydraulic diameter is given by:

$$d_h = 4 \times \frac{\text{channel area}}{\text{channel perimeter}} \tag{6.60}$$

For circular channels, the usual expressions give the diameter of the channel itself, while for rectangular channels, the expression evaluates to that given by Lee et al. [16].

The authors [16] have assumed that all flow through the narrow channels can be accounted for through an effective resistance into the cell area. However, if an equivalent circuit model is constructed from the original layout (Figure 6.4), an easy simplification into equivalent resistances is not obvious, and therefore it may be more accurate to furnish a solution to the full network as depicted later.

6.8 Examples

In this section, we consider two examples of transport whose solutions will shed some light on the nature of transport processes in tissue culture systems.

6.8.1 Ligand–receptor binding model

The first case to be considered involves a free species diffusing through a matrix; throughout the matrix are distributed binding receptors specific to the free species. Such a situation may describe the diffusion of a ligand through a tissue matrix with receptors distributed throughout the tissue. Then, we may model the concentrations of the free species, the binding receptors, and the free species–binding receptor complex as a function of time by incorporating first-order binding kinetics into the transport equation. This type of analysis has been applied to the kinetics of drug delivery toward the development of an implantable system for treating hearing loss [17].

Let F, B, and C denote the free species, the receptor, and the complex, respectively. The width and thickness of the matrix are supposed long enough that the problem can be sufficiently considered to be one-dimensional in x, with x ranging from 0 to L. Let k_b be the binding coefficient between F and B, and let k_d be the dissociation constant for C. Since R and C are immobile (bound to the matrix), we write the transport equations for F, B, and C as follows:

$$\left.\begin{aligned} \frac{\partial C_F}{\partial t} &= D\frac{\partial^2 C_F}{\partial x^2} - k_b C_F C_B + k_d C_C \\ \frac{\partial C_B}{\partial t} &= -k_b C_F C_B + k_d C_C \\ \frac{\partial C_C}{\partial t} &= k_b C_F C_B - k_d C_C \end{aligned}\right\} \tag{6.61}$$

We must specify initial conditions for F, B, and C and two boundary conditions for F. Let the initial concentration of F be zero throughout, with the initial concentration of B uniform throughout the matrix and given by $C_{B,0}$; the initial concentration $C_{C,0}$ of complex is 0. Let us introduce a steady source of F at the $x=0$ left-hand boundary at a concentration of C_L, with $x=L$ side subject to the no-flux condition $\partial C_F/\partial x|_{x=L}=0$. Summarizing, we have

$$\left.\begin{aligned} C_F(x,t=0) = C_C(x,t=0) = 0 \\ C_B(x,t=0) = C_{B,0} \end{aligned}\right\}\text{Initial conditions} \tag{6.62}$$

$$\left.\begin{aligned} C_F(x=0,t) = C_L \\ \left.\frac{\partial C_F}{\partial x}\right|_{x=L} = 0 \end{aligned}\right\}\text{Boundary conditions} \tag{6.63}$$

The three coupled, nonlinear transport equations (6.61) along with the initial and boundary conditions (6.62) and (6.63) must be solved numerically, and a number of packages exist by which to accomplish this. The results shown in Figure 6.5 were generated using the PDEPE package within MATLAB (MathWorks, Natick, MA), and the physical parameters used are summarized in Table 6.1.

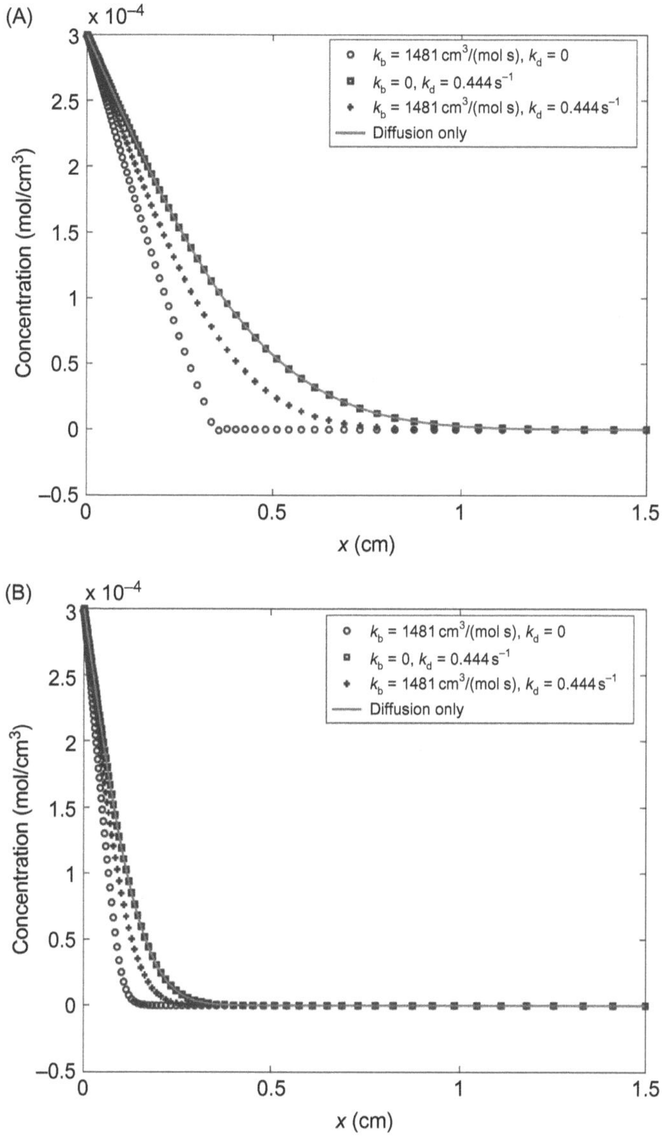

FIGURE 6.5

Concentration profiles for free species C_F as a function of distance x at $t = 2$ h for two different cases. The diffusion coefficient D in (A) is 10 times greater than that in (B). (A) Case 1: Concentration profiles for free species C_F as a function of distance x at $t = 2$ h for cases with binding only, dissociation only, and both binding and dissociation. $D = 1 \times 10^{-5}$ cm^2/s. (B) Case 2: Concentration profiles for free species C_F as a function of distance x at $t = 2$ h for cases with binding only, dissociation only, and both binding and dissociation. $D = 1 \times 10^{-6}$ cm^2/s. Note the depression in the profile due to the reduction in the diffusion coefficient.

Table 6.1 Summary of Physical Parameters Used in Example 6.8.1

	Case 1	Case 2
D	1×10^{-5} cm^2/s	1×10^{-6} cm^2/s
C_L	300×10^{-6} mol/cm^3	
$C_{B,0}$	270×10^{-6} mol/cm^3	
L	1.5 cm	
k_b	1481 cm^3/(mol s)	
k_d	0.444 s^{-1}	

Cases 1 and 2 are distinguished by a 10-fold decrease in D in Case 2.

In order to render Eqs. (6.61)–(6.63) more easily solvable, they were first nondimensionalized in accordance with the procedure outlined in Section 6.5; the reader is encouraged to go through this process as an exercise. After the solutions were calculated, they were redimensionalized and plotted (Figure 6.5). Figure 6.5 provides an understanding of the relative effect of binding and dissociation on C_F as reflected by varying k_b and k_d, as well as that of decreasing the diffusion coefficient D. In Figure 6.5B, we see the effect of decreasing D by a factor of 10 is generally to depress the concentration profile in space. This is simply a reflection of the fact that chemical species with lower diffusion coefficients in a given medium have a lower ability to penetrate the medium in the absence of convection.

With regards to the binding and dissociation constants, Figure 6.5 presents three further cases for both Case 1 and Case 2: (1) only binding (○), (2) only dissociation (□), and (3) both binding and dissociation (+). Comparison with the case in which only diffusion occurs (solid red line) is also given. We note first that since $C_{C,0} = 0$, the concentration profile for the case in which there is only dissociation matches with the case in which there is only diffusion, since there is no F to dissociate from any complex C in the matrix. If there is only binding, the effect is to diminish the concentration of F deeper into the matrix, as less F is present to diffuse. If there is both binding and dissociation, we naturally see a profile between those for just binding and just dissociation.

6.8.2 Mass transport in a membrane bilayer device

One strategy by which to deliver a uniform concentration gradient to a culture of cells independently of fluidic shear is the membrane bilayer device [18,19]. This device configuration comprises two channels or compartments separated by a porous membrane (in this case, assumed to be made from a polymer such as PDMS), with flow established both compartments (Figure 6.6A). Transport of solute to a cell population can be achieved by flowing solute-containing medium through the proximate channel, effectively shielding cultured cells from direct shear forces.

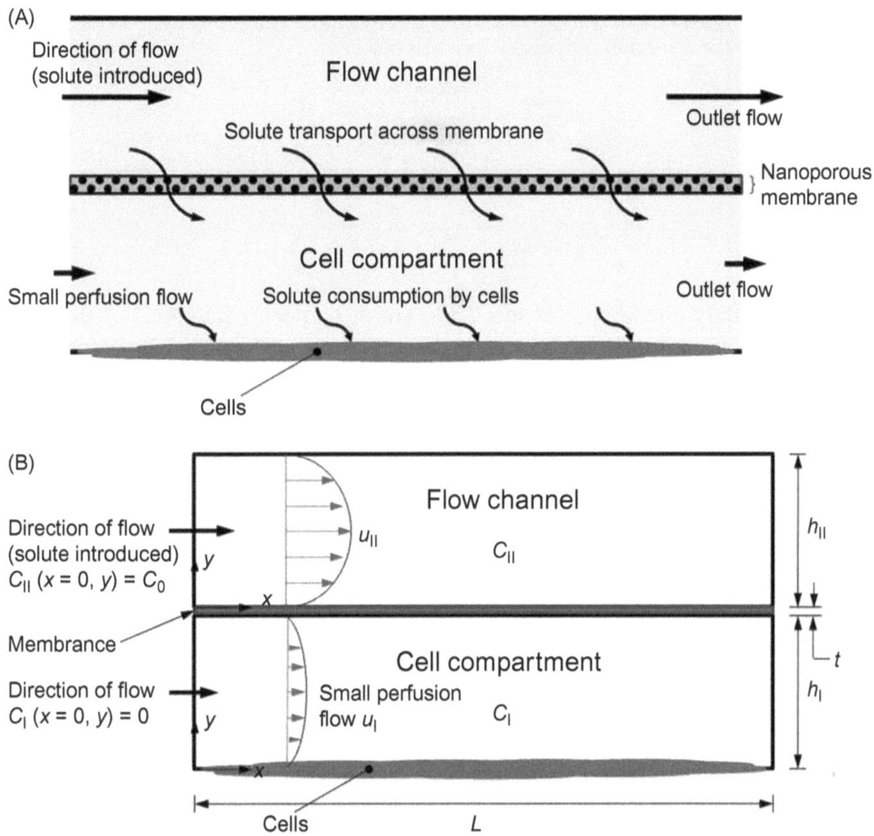

FIGURE 6.6

Schematic and geometry of bilayer microfluidic configuration. (A) Schematic of bilayer microfluidic device. (B) Defined geometrical parameters for bilayer microfluidic device.

In this example, we will consider the case of oxygen transport with a culture of MSCs within the device. A culture of MSCs is ideal for study within a microfluidic construct, since stem cells are known to differentiate within specific applied shear ranges [7]. Moreover, MSCs are sensitive to oxygen and should be expected to be cultured in low oxygen environments akin to that experienced *in vivo* for such cells. Let us set up the equations of the problem; definitions of geometrical parameters are given in Figure 6.6B.

We assume that the channels through which flow occurs are long and of high aspect ratio, so that the flow profile given by Eq. (6.50) is appropriate, and that all physical processes have reached steady state. We also assume that the Péclet number along the channel length satisfies Eq. (6.2), so that transport in the x-direction is convection-dominated while it is diffusion-dominated in the y-direction.

Denoting the upper, flow channel "II," and the lower, cell compartment channel "I," we may write the transport equations for each as

$$u_i \frac{\partial C_i}{\partial x} = D_i \frac{\partial^2 C_i}{\partial y^2}, \quad i = \mathrm{I}, \mathrm{II} \tag{6.64}$$

where the velocity profile is given by:

$$u_i(y) = \frac{H_i^2}{2\mu_i}\left(\frac{\Delta p_i}{L}\right)\frac{y}{H_i}\left(1 - \frac{y}{H_i}\right) \tag{6.65}$$

The boundary conditions for this setup are of particular interest. For the flow channel, we assume that we introduce the solute at the left-hand boundary at a concentration of C_0 uniform across the channel height

$$C_{\mathrm{II}}(x = 0,\ y) = C_0 \tag{6.66}$$

while we suppose a no-flux condition at the upper wall ($y = H_{\mathrm{II}}$)

$$D_{\mathrm{II}} \left.\frac{\partial C_{\mathrm{II}}}{\partial y}\right|_{y=H_{\mathrm{II}}} = 0 \tag{6.67}$$

For the condition at the membrane, we must take into account the transport within the membrane. However, within the membrane, there are no convection and reaction terms; therefore, we have to consider only diffusion, and, again by employing arguments from Section 6.6, we simply have to integrate the equation $D_{\mathrm{membrane}} \mathrm{d}^2C/\mathrm{d}y^2 = 0$, valid within the membrane and for a membrane such that its thickness $t \ll L$. The solution is a linear function with two integration constants to be determined. In order to determine these constants, we must use mass conservation to stipulate that the mass flux diffusing down from the flow channel equals that which enters the membrane and that the mass flux that leaves the membrane equals that entering the cell compartment. It is easily shown that these conditions give

$$\left.\begin{aligned} D_{\mathrm{II}} \left.\frac{\partial C_{\mathrm{II}}}{\partial y}\right|_{y=0} &= \frac{D_{\mathrm{membrane}}}{t}\left[C_{\mathrm{I}}(x, y = H_{\mathrm{I}}) - C_{\mathrm{II}}(x, y = 0)\right] \\ D_{\mathrm{I}} \left.\frac{\partial C_{\mathrm{I}}}{\partial y}\right|_{y=H_{\mathrm{I}}} &= D_{\mathrm{II}} \left.\frac{\partial C_{\mathrm{II}}}{\partial y}\right|_{y=0} \end{aligned}\right\} \tag{6.68}$$

In the cell compartment, for the purpose of this example, we assume that no solute is present at the left-hand boundary,

$$C_{\mathrm{I}}(x = 0, y) = 0 \tag{6.69}$$

while solute is consumed at the cell cultured surface at a rate consistent with Michaelis–Menten kinetics:

$$D_{\mathrm{I}}\frac{\partial C_{\mathrm{I}}}{\partial y}\bigg|_{y=0} = V_{\max}\rho_{\mathrm{cells}}\frac{C_{\mathrm{I}}(x, y=0)}{K_M + C_{\mathrm{I}}(x, y=0)} \tag{6.70}$$

If we suppose the extracellular environment to be hypoxic (low in oxygen), on the basis of the parameters C_0 and K_M given in Table 6.2, we may expand the right-hand side of Eq. (6.70) into a Taylor series and make the important simplification as

$$D_{\mathrm{I}}\frac{\partial C_{\mathrm{I}}}{\partial y}\bigg|_{y=0} \approx \frac{V_{\max}\rho_{\mathrm{cells}}}{K_M}C_{\mathrm{I}}(x, y=0) \tag{6.71}$$

The rest of the geometrical and physical parameters are also summarized in Tables 6.2 and 6.3. For example, we will examine three separate cases, 1–3, each representing increasing levels of cell seeding confluency and uptake rates for the cell culture. The shear stress experienced by the cell culture was set by the shear value at which osteogenic differentiation has been observed to begin in MSCs, while C_0 was set at 1% the saturation concentration of oxygen in the given medium.

Equations (6.64)–(6.69), and (6.71) may be solved in several ways; the solutions given below were obtained analytically by power series methods. In Figure 6.7,

Table 6.2 Summary of Physical Parameters used in Example 6.8.2 [18]

Parameter	Value	Description
D_{II}, D_{I}	3×10^{-5} cm²/s	Diffusion coefficient for oxygen in medium
D_{membrane}	1×10^{-5} cm²/s	Diffusion coefficient for oxygen in polymeric membrane
T	10 μm	Thickness of membrane
μ	1×10^{-2} dyne s/cm²	Dynamic Viscosity of water
H_{I}, H_{II}	50 μm	Height of channel
W	200 μm	Width of channel
L	1.5 cm	Length of channel
K_M	1×10^{-7} mol/cm³	Michaelis–Menten parameter for oxygen
C_{Sat}	2.15×10^{-7} mol/cm³	Oxygen concentration at saturation
C_0	2.15×10^{-9} mol/cm³	$0.01\times C_{\mathrm{Sat}}$, 1% saturation
τ	$\tau = 0.3$ dyn/cm²	Shear, determined by initiation of osteogenic differentiation in MSCs. Corresponds to a flow rate of 2.5 nl/s

Table 6.3 Values of $V_{\max}/K_M$ and ρ_{cells} for Three Separate Cases for Increasing Cell Uptake Rates and Cell Seeding Densities [18]

	Case 1	Case 2	Case 3
$V_{\max}/K_M$	5×10^{-5} cm³/10⁶ cells/s	1×10^{-4} cm³/10⁶ cells/s	2×10^{-4} cm³/10⁶ cells/s
ρ_{cells}	1×10^{3} cells/cm²	1×10^{5} cells/cm²	1×10^{6} cells/cm²

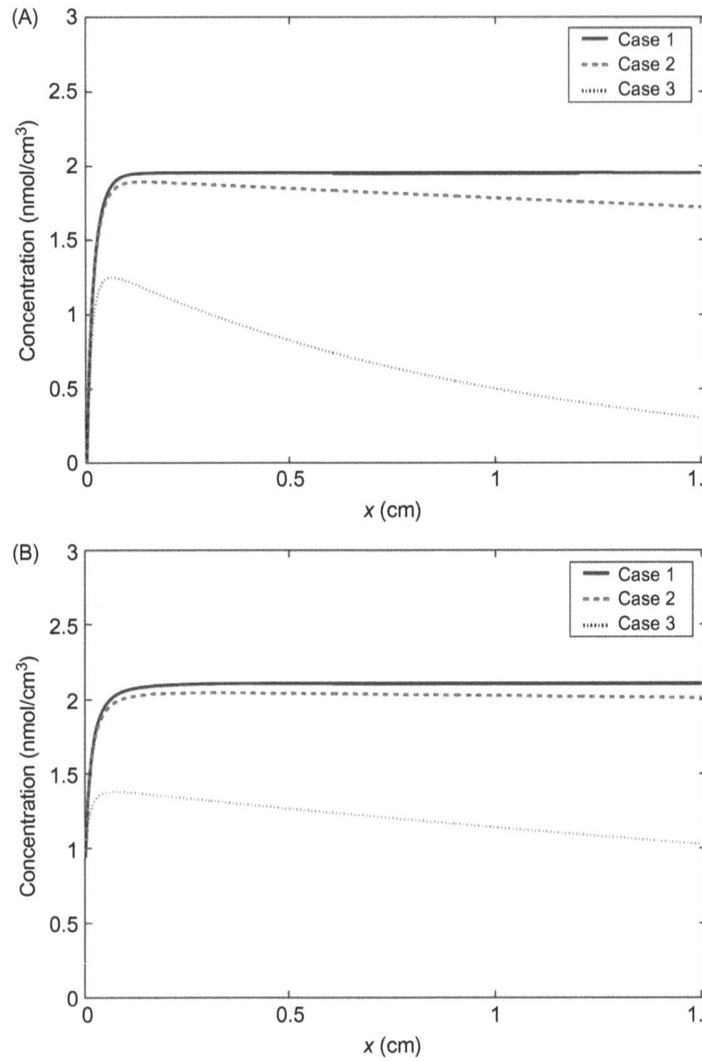

FIGURE 6.7

Concentration of oxygen at cell surface as a function of *x* for Cases 1, 2, and 3. (A) Concentration of oxygen at cell surface as a function of *x* for Cases 1, 2, and 3 with the flow rate in the cell compartment equal to 2.5 nl/s and that in the flow channel equal to 25 nl/s. (B) Concentration of oxygen at cell surface as a function of *x* for Cases 1, 2, and 3 with the flow rate in the cell compartment equal to 2.5 nl/s and that in the flow channel equal to 125 nl/s. This increase in the flow channel flow rate increases the concentration of solute along the length of the channel considerably.

we see the concentration of oxygen at the cell surface for Cases 1–3. In Figure 6.7A and B, the cell compartment has the same flow rate, 2.5 nl/s, set by the acceptable applied level of fluid shear stress. In Figure 6.7A, the flow rate in the flow channel is 25 nl/s but that in Figure 6.7B is 125 nl/s. As expected, we see that Case 3 has the greatest depletion along the length of the channel, owing to the relatively high uptake rate and confluency of the culture. In Figure 6.7B, we see the effect of increasing the flow rate in the flow channel fivefold is to flatten out the concentration profiles for Case 2, and especially for Case 3, with little change between the amount of oxygen delivered to the culture at the beginning of the channel and that delivered to the end of the channel. This increase was brought about without changing the flow rate (or the wall shear stress) in the cell compartment.

Finally, in Figure 6.8, we compare the concentration of oxygen at the cell surface for the membrane bilayer (uptake and seeding parameters equal to those

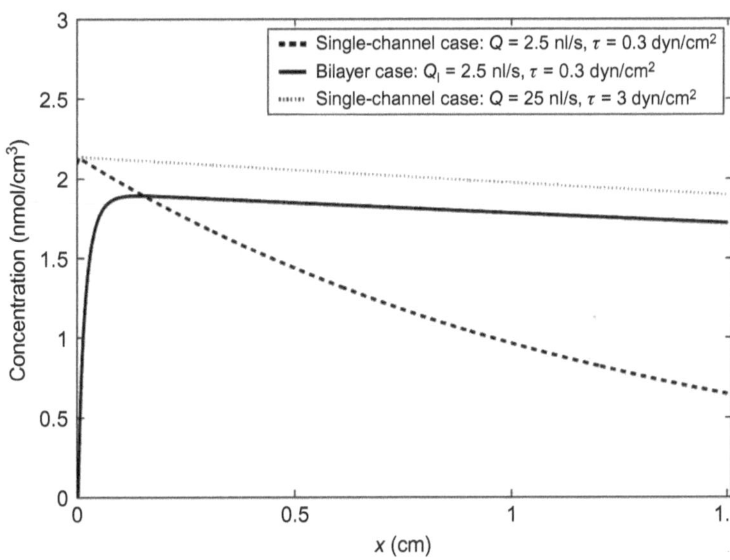

FIGURE 6.8

Comparison of concentration of oxygen at cell surface for bilayer and single channel. Concentration at the cell surface as a function of *x* for single channel (dotted and dashed lines) and bilayer (solid black line). For the bilayer, the cell compartment flow rate is given by 2.5 nl/s (corresponding to $\tau = 0.3$ dyn/cm^2), while the flow channel flow rate is 25 nl/s. For the single-channel configuration, we have flow rates given by 2.5 nl/s (dashed black line) and 25 nl/s (dotted blue line, corresponding to $\tau = 3.0$ dyn/cm^2). For the single-channel case corresponding to $\tau = 0.3$ dyn/cm^2, note the decaying consumption profile, while the bilayer consumption profile is nearly uniform. Other parameters are those for Case 2. (For interpretation of the references to color in this figure legend, the reader is referred to the web version of this book.)

for Case 2) to that of a simple, single-channel configuration with different flow rates. For the single-channel configuration, we see that, if the flow rate in the cell channel is set by $\tau = 0.3$ dyn/cm^2, there is considerable depletion along the channel length. It is only if this flow rate is increased by a factor of 10 that the concentration at the cell surface for the single-channel case can match the uniformity afforded by the bilayer at the lower flow rate.

6.9 Microfluidic approaches for flow and transport control

One of the greatest challenges in microfluidic cell culture systems is the need to exercise control over both the local fluidic and the chemical environment of a tissue construct. With the emergence of flow as an important part of a physiologically relevant culture system, the evolution from static culture dishes requires the additional understanding of how fluid flow and convective transport affect the transport qualities of a particular microfluidic tissue culture platform. With the presence of flow, transport is no longer isotropic throughout the fluid (as it typically is with diffusion), so the requirements laid out in Section 6.1 with regard to sufficient nutrient and oxygen delivery, waste removal, and concentration gradient management must be met within such a context.

In order to overcome the challenges associated with controlling flow while managing fluid mechanical properties, a number of innovative ideas have been proposed in the literature over the past decade. In the following, the general characteristics of several of these approaches and their potential benefits are briefly described.

6.9.1 Membrane bilayer

As demonstrated in Section 6.8.2, the bilayer membrane microfluidic configuration is potentially capable of delivering nearly uniform concentrations of solute to a culture of cells. This is in contrast to the inherent depletion of solute that occurs in a single-channel configuration, and it can be accomplished by independently controlling the fluidic shear exerted on a cell culture. Moreover, the bilayer geometry can be implemented using assembly methods similar to the soft lithography techniques typically utilized in the construct of microfluidic devices [18,19] (Figure 6.9).

Effective implementation of the bilayer construct naturally depends strongly on the transport characteristics of the membrane. Culture systems with two chambers separated by a membrane with low porosity can be modeled by neglecting convective transport, as described earlier. If the porosity (the fraction of the membrane covered by pores) of the membrane is sufficiently high, convective transport will be appreciable, and the effective transport of solute into the cell compartment will either be increased or decreased depending on which channel convection favors. Of course, the presence of transmembrane fluid can perturb the flow field within each channel, and it seems at first glance that it is preferable to minimize such transmembrane fluid flux in order to maintain as great control as possible over transport within each channel.

FIGURE 6.9

Implementation of bilayer microfluidic device.

Source: *Reproduced from Ref. [19] with permission from Springer.*

6.9.1.1 Managing fluid flow across membranes

In Section 6.8.2, Culture systems with two chambers separated by a membrane with low porosity can be modeled by neglecting convective transport, as described earlier. If the porosity (the fraction of the membrane volume covered by pores) of the membrane is sufficiently high, convective transport will be appreciable, and the effective transport of solute into the cell compartment will either be increased or decreased depending on which channel convection favors. The law that governs transport in porous materials is Darcy's law:

$$Q = -\frac{\kappa}{\mu}\nabla p \tag{6.72}$$

Here, Q is the local volumetric flow rate in the porous material, p is the hydrostatic pressure, μ is the viscosity of the fluid, and κ is a proportionality constant, sometimes called the permeability of the material. We see that fluid will flow locally in the direction from higher to lower pressure. For a membrane with cylindrical pores oriented along the thin dimension of the membrane, Eq. (6.72) can be rewritten in terms of the local velocity V of the transmembrane fluid flow as

$$V = -\frac{d_{\text{pore}}^2 \varpi}{32\mu}\frac{\Delta p}{t} \tag{6.73}$$

where d_{pore} is the diameter of each pore, ϖ is the porosity of the membrane, t is the thickness of the membrane, and Δp is the pressure difference across the membrane. Since one of the benefits of the bilayer configuration is the ability to saturate the flow channel with solute—accomplished by increasing the velocity in the upper channel—while minimizing flow in the lower channel, we may regard Δp as set. The most efficient way to decrease V would then be to decrease d_{pore}, so long as doing so does not prevent any desired macromolecule transport from taking place across the membrane.

6.9.2 Valves

Recently, Mosadegh et al. [20] have attempted to solve the problem of nonuniform concentration profiles by incorporating computer-actuated valves into their devices

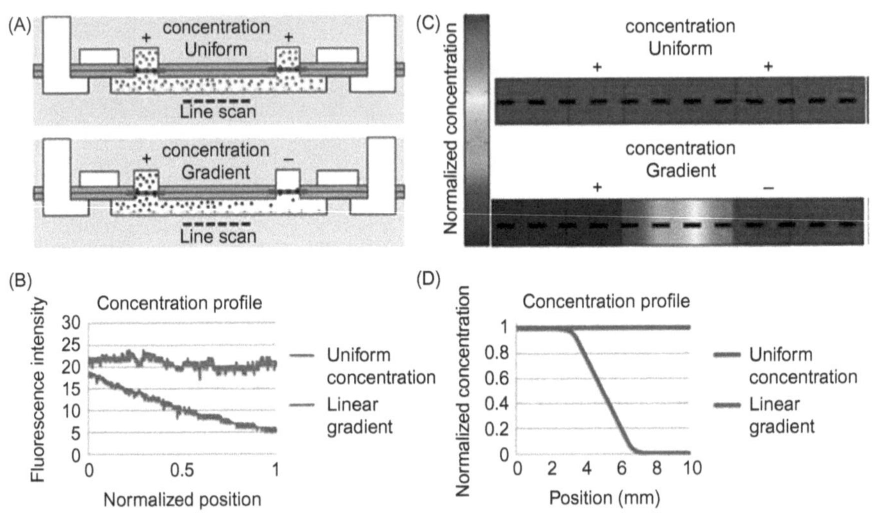

FIGURE 6.10

Concept and modeling results for actuated valve system, after Mosadegh [20]. (A) Distribution of diffusing species when present in both channels (top) or only in bottom channel (bottom), (B) Fluorescence Intensity Map for line scans for each case in (A), (C) Simulated concentration map for each case in (A), (D) Simulated concentration profile for each case in (A).

Source: *Reproduced from Ref. [20] with permission from the Royal Society of Chemistry Publishing.*

(Figure 6.10). The valve concept is similar to the bilayer in that solute is introduced from a reservoir by means of the valves and allowed to diffuse downward into the lower compartment. Precise control of solute introduction along the length of the chamber is enabled by using valves, but cross-sectional solute distribution may not be completely uniform. Also, since solute is not introduced continuously along the channel, depletion will occur in the intervals between the valves. The use of computer-controlled valves increases both the cost of the system and the manufacturing complexity. Nevertheless, if implemented, the control offered by the valve system could enable the investigation of the effect on tissue culture of small changes in solute concentrations.

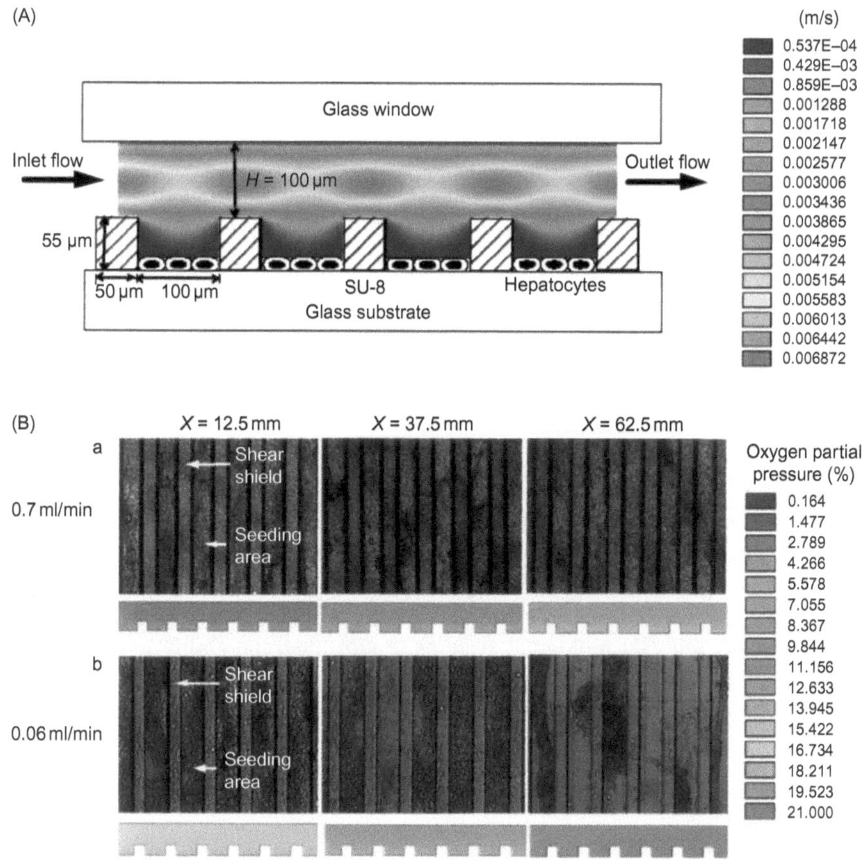

FIGURE 6.11

Modeling and implementation of microgroove-based microfluidic cell culture device. (A) Modeling results for flow field within microgroove device geometry. (B) Modeling results for concentration profile for microgroove tissue culture device with two different flow rates. Results are tabulated at three different locations along the channel length.

Source: *Reproduced from Park et al. [22] with permission from Wiley Periodicals, Inc.*

6.9.3 Grooves

Sandstrom et al. [21]and Park et al. [22] developed a method of decoupling fluidic shear from transport by culturing cells within microfabricated grooves set along the length of a channel in regular intervals. The greatest benefit of this device configuration is that, as shown in Figure 6.11A, the cells in question are exposed to little shear, since the grooves essentially shield the cells from the effect of direct fluid flow. Of course, for cell populations and studies for which exposure to shear stress is essential, the groove approach may not be preferable, and if a grooved geometry is still desired, the groove height may have to be modified. By Figure 6.11B, it is also clear that some depletion will occur along the length of the channel, since the device is still essentially a single-channel device. It also seems possible that cell–cell communication throughout the device will be limited due to the presence of the grooves. Nevertheless, the use of microfabricated grooves represents a unique and interesting method by which to isolate the effect of shear and transport for a tissue culture.

6.10 Conclusion

In this chapter, we have considered several of the principal mechanisms and computational tools relevant to the analysis of microfluidic cell culture systems. Mechanisms of fluid flow, mass transport, and binding kinetics are central to the behavior of microfluidic cell culture systems, and understanding their interactions is crucial to developing robust microfluidic cell culture platforms.

Several topics are not discussed in this chapter due to space limitations include magnetohydrodynamics, electrophoresis, and colloid and electrolyte theory, which are covered elsewhere in several excellent resources [9,12,23].

Microfluidic cell culture systems are highly complex; they seek to recreate physiologically relevant biological systems *in vitro*, and the ability to model such systems has powerful practical implications as new technologies are developed for drug delivery systems and platforms for drug discovery, regenerative medicine, and many other applications.

Appendix A Navier–Stokes equations in other coordinate systems [9]

A.1 Navier–Stokes equations in radial coordinates

Let v_r, v_θ, and v_z be the velocity in the radial, polar, and z-directions, respectively. Then for the equations of fluid motion, we have

$$\rho\left(\frac{\partial v_r}{\partial t}+v_r\frac{\partial v_r}{\partial r}+\frac{v_\theta}{r}\frac{\partial v_r}{\partial \theta}+v_z\frac{\partial v_r}{\partial z}-\frac{v_\theta^2}{r}\right) = -\frac{\partial p}{\partial r}+\mu\left[\frac{1}{r}\frac{\partial}{\partial r}\left(r\frac{\partial v_r}{\partial r}\right)+\frac{1}{r^2}\frac{\partial^2 v_r}{\partial \theta^2}+\frac{\partial^2 v_r}{\partial z^2}-\frac{2}{r^2}\frac{\partial v_\theta}{\partial \theta}-\frac{v_r^2}{r^2}\right] \quad \text{(A1)}$$

$$\rho\left(\frac{\partial v_\theta}{\partial t}+v_r\frac{\partial v_\theta}{\partial r}+\frac{v_\theta}{r}\frac{\partial v_\theta}{\partial \theta}+v_z\frac{\partial v_\theta}{\partial z}+\frac{v_r v_\theta}{r}\right) = -\frac{1}{r}\frac{\partial p}{\partial \theta}+\mu\left[\frac{1}{r}\frac{\partial}{\partial r}\left(r\frac{\partial v_\theta}{\partial r}\right)+\frac{1}{r^2}\frac{\partial^2 v_\theta}{\partial \theta^2}+\frac{\partial^2 v_\theta}{\partial z^2}+\frac{2}{r^2}\frac{\partial v_r}{\partial \theta}-\frac{v_\theta^2}{r^2}\right] \quad \text{(A2)}$$

$$\rho\left(\frac{\partial v_z}{\partial t}+v_r\frac{\partial v_z}{\partial r}+\frac{v_\theta}{r}\frac{\partial v_z}{\partial \theta}+v_z\frac{\partial v_z}{\partial z}\right) = -\frac{\partial p}{\partial r}+\mu\left[\frac{1}{r}\frac{\partial}{\partial r}\left(r\frac{\partial v_z}{\partial r}\right)+\frac{1}{r^2}\frac{\partial^2 v_z}{\partial \theta^2}+\frac{\partial^2 v_z}{\partial z^2}\right] \quad \text{(A3)}$$

The continuity equation is given by:

$$\frac{1}{r}\frac{\partial (r v_r)}{\partial r}+\frac{1}{r}\frac{\partial v_\theta}{\partial \theta}+\frac{\partial v_z}{\partial z}=0 \quad \text{(A4)}$$

The components of the stress tensor are given by:

$$\begin{aligned}
\sigma_{rr} &= -p+2\mu\frac{\partial v_r}{\partial r} & \sigma_{r\theta} &= \mu\left(\frac{1}{r}\frac{\partial v_r}{\partial \theta}+\frac{\partial v_\theta}{\partial r}-\frac{v_\theta}{r}\right)\\
\sigma_{\theta\theta} &= -p+2\mu\left(\frac{1}{r}\frac{\partial v_\theta}{\partial \theta}+\frac{v_r}{r}\right) & \sigma_{\theta z} &= \mu\left(\frac{\partial v_\theta}{\partial z}+\frac{1}{r}\frac{\partial v_z}{\partial \theta}\right)\\
\sigma_{zz} &= -p+2\mu\frac{\partial v_z}{\partial z} & \sigma_{rz} &= \mu\left(\frac{\partial v_r}{\partial z}+\frac{\partial v_z}{\partial r}\right)
\end{aligned} \quad \text{(A5)}$$

A.2 Navier–Stokes equations in polar coordinates

Let v_r, v_θ, and v_φ be the velocity in the radial, polar, and azimuthal directions, respectively. The Navier–Stokes equations are

$$\rho\left(\frac{\partial v_r}{\partial t}+v_r\frac{\partial v_r}{\partial r}+\frac{v_\theta}{r}\frac{\partial v_r}{\partial \theta}+\frac{v_\phi}{r\sin\theta}\frac{\partial v_r}{\partial \phi}-\frac{v_\theta^2+v_\phi^2}{r}\right) = -\frac{\partial p}{\partial r}+\mu\left[\frac{1}{r}\frac{\partial}{\partial r}\left(r^2\frac{\partial v_r}{\partial r}\right)+\frac{1}{r^2\sin\theta}\frac{\partial}{\partial \theta}\left(\sin\theta\frac{\partial v_r}{\partial \theta}\right)+\frac{1}{r^2\sin^2\theta}\frac{\partial^2 v_r}{\partial \phi^2} - \frac{2}{r^2\sin^2\theta}\frac{\partial (v_\theta\sin\theta)}{\partial \theta}-\frac{2}{r^2\sin\theta}\frac{\partial v_\phi}{\partial \phi}-\frac{2v_r}{r^2}\right] \quad \text{(A6)}$$

$$\rho\left(\frac{\partial v_\theta}{\partial t}+v_r\frac{\partial v_\theta}{\partial r}+\frac{v_\theta}{r}\frac{\partial v_\theta}{\partial \theta}+\frac{v_\phi}{r\sin\theta}\frac{\partial v_\theta}{\partial \phi}+\frac{v_r v_\phi}{r}-\frac{v_\phi^2\cot\theta}{r}\right)$$
$$=-\frac{1}{r}\frac{\partial p}{\partial \theta}+\mu\left[\frac{1}{r}\frac{\partial}{\partial r}\left(r^2\frac{\partial v_\theta}{\partial r}\right)+\frac{1}{r^2\sin\theta}\frac{\partial}{\partial\theta}\left(\sin\theta\frac{\partial v_\theta}{\partial\theta}\right)\right.$$
$$\left.+\frac{1}{r^2\sin^2\theta}\frac{\partial^2 v_\theta}{\partial\phi^2}-\frac{2\cos\theta}{r^2\sin^2\theta}\frac{\partial v_\phi}{\partial\phi}+\frac{2}{r^2}\frac{\partial v_r}{\partial\theta}-\frac{v_\theta}{r^2\sin^2\theta}\right] \tag{A7}$$

$$\rho\left(\frac{\partial v_\phi}{\partial t}+v_r\frac{\partial v_\phi}{\partial r}+\frac{v_\theta}{r}\frac{\partial v_\phi}{\partial \theta}+\frac{v_\phi}{r\sin\theta}\frac{\partial v_\phi}{\partial \phi}+\frac{v_r v_\phi}{r}+\frac{v_\phi v_\theta\cot\theta}{r}\right)$$
$$=-\frac{1}{r\sin\theta}\frac{\partial p}{\partial \phi}+\mu\left[\frac{1}{r}\frac{\partial}{\partial r}\left(r^2\frac{\partial v_\phi}{\partial r}\right)+\frac{1}{r^2\sin\theta}\frac{\partial}{\partial\theta}\left(\sin\theta\frac{\partial v_\phi}{\partial\theta}\right)\right.$$
$$\left.+\frac{1}{r^2\sin^2\theta}\frac{\partial^2 v_\phi}{\partial\phi^2}-\frac{2}{r^2\sin\theta}\frac{\partial v_r}{\partial\phi}+\frac{2\cos\theta}{r^2\sin^2\theta}\frac{\partial v_\theta}{\partial\phi}-\frac{v_\phi}{r^2\sin^2\theta}\right] \tag{A8}$$

The continuity equation is given by:

$$\frac{1}{r^2}\frac{\partial(r^2 v_r)}{\partial r}+\frac{1}{r\sin\theta}\frac{\partial(v_\theta\sin\theta)}{\partial\theta}+\frac{1}{r\sin\theta}\frac{\partial v_\phi}{\partial\phi}=0 \tag{A9}$$

The components of the stress tensor are given by:

$$\sigma_{rr}=-p+2\mu\frac{\partial v_r}{\partial r} \qquad \sigma_{r\theta}=\mu\left(\frac{1}{r}\frac{\partial v_r}{\partial\theta}+\frac{\partial v_\theta}{\partial r}-\frac{v_\theta}{r}\right)$$
$$\sigma_{\phi\phi}=-p+2\mu\left(\frac{1}{r\sin\theta}\frac{\partial v_\phi}{\partial\phi}+\frac{v_r}{r}+\frac{v_\theta\cot\theta}{r}\right) \qquad \sigma_{\theta\phi}=\mu\left(\frac{1}{r\sin\theta}\frac{\partial v_\theta}{\partial\phi}+\frac{1}{r}\frac{\partial v_\phi}{\partial\theta}-\frac{v_\phi\cot\theta}{r}\right)$$
$$\sigma_{\theta\theta}=-p+2\mu\left(\frac{1}{r}\frac{\partial v_\theta}{\partial\theta}+\frac{v_r}{r}\right) \qquad \sigma_{\phi r}=\mu\left(\frac{\partial v_\phi}{\partial r}+\frac{1}{r\sin\theta}\frac{\partial v_r}{\partial\phi}-\frac{v_\phi}{r}\right) \tag{A10}$$

Appendix B Transport equation in other coordinate systems

We consider only the case in which the diffusion coefficient is constant. The transport equation in cylindrical coordinates for species i is

$$\frac{\partial C_i}{\partial t}+v_r\frac{\partial C_i}{\partial r}+\frac{v_\theta}{r}\frac{\partial C_i}{\partial\theta}+v_z\frac{\partial C_i}{\partial z}=D_i\left[\frac{1}{r}\frac{\partial}{\partial r}\left(r\frac{\partial C_i}{\partial r}\right)+\frac{1}{r^2}\frac{\partial^2 C_i}{\partial\theta^2}+\frac{\partial^2 C_i}{\partial z^2}\right]+R_i \tag{A11}$$

where v_r, v_θ, and v_z are the solutions to the Navier–Stokes equations in cylindrical coordinates. The transport equation in spherical coordinates for species i is

$$\begin{aligned}\frac{\partial C_i}{\partial t} + v_r\frac{\partial C_i}{\partial r} + \frac{v_\theta}{r}\frac{\partial C_i}{\partial \theta} + \frac{v_\varphi}{r\sin\theta}\frac{\partial C_i}{\partial \phi} \\ = D_i\left[\frac{1}{r}\frac{\partial}{\partial r}\left(r^2\frac{\partial C_i}{\partial r}\right) + \frac{1}{r^2\sin\theta}\frac{\partial}{\partial\theta}\left(\sin\theta\frac{\partial C_i}{\partial\theta}\right) + \frac{1}{r^2\sin^2\theta}\frac{\partial^2 C_i}{\partial\phi^2}\right] + R_i\end{aligned} \tag{A12}$$

where v_r, v_θ, and v_φ are the solutions to the Navier–Stokes equations in cylindrical coordinates.

References

[1] C. Fehrer, R. Brunauer, G. Laschober, H. Unterluggauer, S. Reitinger, F. Kloss, et al., Reduced oxygen tension attenuates differentiation capacity of human mesenchymal stem cells and prolongs their lifespan, Aging Cell 6 (2007) 745–757.

[2] K. Domansky, W. Inman, J. Serdy, A. Dash, M.H.M. Lim, L.G. Griffith, Perfused multiwell plate for 3D liver tissue engineering, Lab Chip 10 (2010) 51–58.

[3] R.T. Ferrell, D.M. Himmelblau, Diffusion coefficients of nitrogen and oxygen in water, J. Chem. Eng. Data 12 (1967) 111–115.

[4] R. Green, R.A. Frazier, K.M. Shakesheff, M.C. Davies, C.J. Roberts, S.J.B. Tendler, Surface plasmon resonance of dynamic biological interactions with biomaterials, Biomaterials 21 (2000) 1823–1835.

[5] J.M. Taylor, W.M. Mitchell, S. Cohen, Characterization of the binding protein for epidermal growth factor, J. Biol. Chem. 249 (1974) 2188–2194.

[6] W. Weber, P.J. Bertics, G.N. Gill, Immunoaffinity purification of the epidermal growth factor receptor, J. Biol. Chem. 259 (1984) 14631–14636.

[7] J.R. Glossop, S.H. Cartmell, Effect of fluid flow-induced shear stress on human mesenchymal stem cells: differential gene expression of IL1B and MAP3K8 in MAPK signaling, Gene Exp. Patterns 9 (2009) 381–388.

[8] S.J. Hawarda, Z. Li, D. Lighter, B. Thomas, J.A. Odella, X.-F. Yuan, Flow of dilute to semi-dilute polystyrene solutions through a benchmark 8:1 planar abrupt micro-contraction, J. Non-Newtonian Fluid Mech. 165 (2010) 1654–1669.

[9] L.D. Landau, E.M. Lifshitz, Fluid Mechanics, Pergamon Books, Oxford, 1987.

[10] L.D. Landau, E.M. Lifshitz, L.P. Pitaevskii, Electrodynamics of Continuous Media, Pergamon Books, Oxford, 1984.

[11] D. Huh, H. Fujioka, Y.-C. Tung, N. Futai, R. Paine, J.B. Grotberg, et al., Acoustically detectable cellular-level lung injury induced by fluid mechanical stresses in microfluidic airway systems, Proc. Natl. Acad. Sci. U.S.A. 104 (2007) 18886–18891.

[12] G. Mehta, K. Mehta, D. Sud, J.W. Song, T. Bersano-Begey, N. Futai, et al., Quantitative measurement and control of oxygen levels in microfluidic poly(dimethylsiloxane) bioreactors during cell culture, Biomed. Microdevices 9 (2007) 123–134.

[13] A. Grodzinsky, Fields, Forces, and Flows in Biological Systems, Garland Science, New York, NY, 2011.

[14] G.E. Briggs, J.B.S. Haldane, A note on the kinematics of enzyme action, Biochem. J. 19 (1925) 338–339.

[15] M. Cabodi, V.L. Cross, Z. Qu, K.L. Havenstrite, S. Schwartz, A.D. Stroock, An active wound dressing for controlled convective mass transfer with the wound bed, J. Biomed. Mater. Res. B 82B (2007) 210–222.

[16] P.J. Lee, P.J. Hung, L.P. Lee, An artificial liver sinusoid with a microfluidic endothelial-like barrier for primary hepatocyte culture, Biotech. Bioeng. 97 (2007) 1340–1346.

[17] E.E. Leary Pararas, Z. Chen, J. Fiering, M.J. Mescher, E.S. Kim, M.J. McKenna, et al., Kinetics of reciprocating drug delivery to the inner ear, J. Control. Release 152 (2011) 270–277.

[18] N.K. Inamdar, L.G. Griffith, J.T. Borenstein, Transport and shear in a microfluidic membrane bilayer device for cell culture, Biomicrofluidics 5 (2011) 022213.

[19] A. Carraro, W.-M. Hsu, K.M. Kulig, W.S. Cheung, M.L. Miller, E.J. Weinberg, et al., *In vitro* analysis of a hepatic device with intrinsic microvascular-based channels, Biomed. Microdevices 10 (2008) 795–805.

[20] B. Mosadegh, M. Agarwal, H. Tavana, T. Bersano-Begey, Y.-s Torisawa, M. Morell, et al., Uniform cell seeding and generation of overlapping gradient profiles in a multiplexed microchamber device with normally-closed valves, Lab Chip 10 (2010) 2959–2964.

[21] C.E. Sandstrom, J.G. Bender, W.M. Miller, E.T. Papoutsakis, Development of novel perfusion chamber to retain nonadherent cells and its use for comparison of human "mobilized" peripheral blood mononuclear cell cultures with and without irradiated bone marrow stroma, Biotech. Bioeng. 50 (1996) 493–504.

[22] J. Park, F. Berthiaume, M. Toner, M.L. Yarmush, A.W. Tilles, Microfabricated grooved substrates as platforms for bioartificial liver reactors, Biotech. Bioeng. 90 (2005) 632–644.

[23] V. Levich, Physicochemical Hydrodynamics, Prentice Hall, Englewood Cliffs, NJ, 1962.

PART

2

Tissue Engineering Strategies

CHAPTER

7 Microfluidic Systems for Controlling Stem Cells Microenvironments

Šeila Selimović[1], Hirokazu Kaji[2], Hojae Bae[1] and Ali Khademhosseini[1]

[1]*Center for Biomedical Engineering, Department of Medicine, Brigham & Women's Hospital, Harvard Medical School, Boston MA, USA*

[2]*Department of Bioengineering and Robotics, Tohoku University, Sandai, Japan*

7.1 Introduction

Over the last few decades, laboratories in the natural and life sciences, as well as in the engineering fields, have seen a continuous reduction in the size of experimental platforms, leading from benchtop systems to lab-on-a-chip devices. This transition has been accompanied by automation via robotic instruments for speedy and precise sample handling, as well as by a steady increase in throughput. Only a few years ago researchers were manually conducting cumbersome protein crystallization trials using hundreds of microliters of expensive samples, one multiwell plate at a time. Today one can rely on a robotic setup, which offers a few orders of magnitude reduction in sample volume, or reduce the experiment down to nanoliter scale by using microfluidic devices [1]. Similarly, life scientists and bioengineers have been constrained to culturing cells in milliliter-sized flasks, but now microfluidics offer new approaches for cell culture, analysis (PCR [2], enzyme-linked immunosorbent assay [3]), electrophoresis [4]), and screening [5] that vastly reduce the amount of cells and expensive reagents and enable simultaneous preparation and testing of large sample arrays in short time periods with high levels of precision and resolution.

In the context of biology and biomedical engineering, stem cells are of particular interest due to their ability to develop into many different cell types [6,7], e.g., cardiomyocytes [8,9], hepatocytes [10–12], osteoblasts [13,14], neural cells [15–17], and endothelial cells [18] (Figure 7.1). Stem cells are sensitive to a variety of microenvironmental factors that regulate all aspects of cell behavior, including self-renewal and differentiation [19]. Hence, they are characterized by their ability to morph or differentiate into specific cell lineages as a response to spatiotemporally controlled intrinsic and extrinsic stimuli. The use of stem cells in tissue engineering

Microfluidic Cell Culture Systems. ISBN: 978-1-4377-3459-1

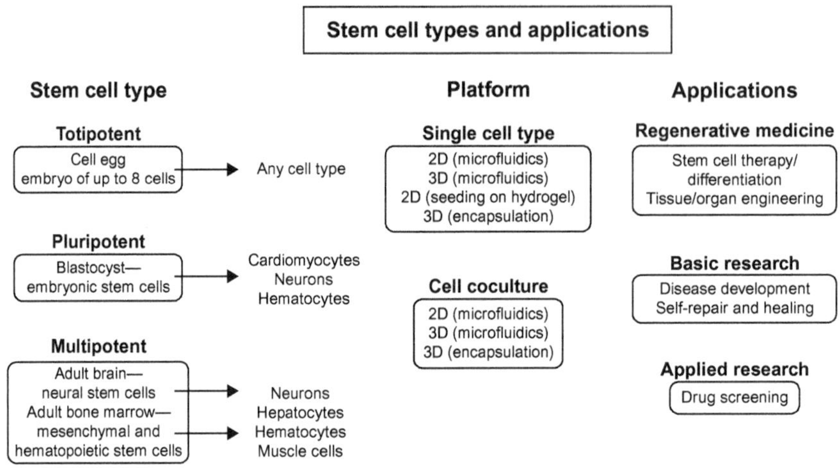

FIGURE 7.1

An overview of stem cell types, their sources and differentiation products; the most common research platforms and applications.

specifically, e.g., in organ repair and development, could potentially revolutionize medical treatments for life-threatening diseases like heart [6,7] or kidney failure [20], where current approaches cannot induce organ regeneration and too few organs are available for transplantation. They do, however, allow for the regulation of cell–microenvironment interactions, such as those between cells, between cells and extracellular matrix (ECM), cells and soluble factors, and cells and mechanical stimuli [21]. As a result, cell differentiation can be regulated via a controlled set of microenvironmental factors not just on the level of cell colonies, but also in individual cells *in vitro* [22]. Furthermore, it has been shown that different types of stem cells are specialized to survive and differentiate into different tissues and organs in the body. Modern technologies can be applied to such three-dimensional (3D) structures, which can then serve as biomimetic tissue models [23–25]. A current example is combining biomaterials (e.g. hydrogels) and microfabrication techniques such as microfluidics in order to form *in vitro* tissue structures [26].

Certain embryonic stem cell (ESC) lines, derived from both mouse and human cells [6,27], could potentially help introduce novel cell therapies, as they can be propagated indefinitely and so provide an unlimited number of cells. The use of human embryos for extraction of stem cells and generation for ESC lines is ethically controversial [28]. A novel paradigm to the problem of finding ESC sources, however, was established by Takahashi and Yamanaka [29], who succeeded in reprogramming murine fibroblasts into induced pluripotent stem (iPS) cells *in vitro*. It was shown that the iPS cells were in many ways similar to ESCs [30,31] in both molecular structure and function, which opened up a new realm of research opportunities in regenerative medicine [32–34]. One important question is how various microenvironmental factors contribute to stem cell self-renewal and differentiation [35].

An efficient use of the currently available cell cultures requires a vast reduction in sample volume and cost, as well as a precise and high-throughput experimental platform that enables the simultaneous study of hundreds or thousands of stem cell samples. Microfluidic technologies are the obvious choice for such experiments, in part because they offer excellent control over the cell microenvironment [36–38]. This chapter highlights the current microfluidic techniques for manipulating stem cells, investigating their behavior, and directing differentiation in response to controlled microenvironmental factors. This application of microscale engineering may be of great benefit to regenerative medicine.

7.2 Microfluidic elements for cell culture

Lab-on-a-chip systems are often referred to as micro total analysis systems or μTAS, as they often integrate multiple functions, from sample injection, mixing, and storage, via filtering and sorting, to optical analysis, incubation, and sample treatment, and finally extraction, including techniques specific to biological samples, such as cell culture and perfusion, cell lysis, polymerase chain reaction (PCR), and screening assays [39]. In this section, we introduce microfluidic elements and techniques used in the most common cell manipulation applications (Figure 7.2).

7.2.1 Cell sorting and filtering

Often test samples, such as whole blood, include different types of cells and pollutants. The sample components and cells of interests can be sorted by different mechanisms, most of which rely on gravitational [40,41], hydrodynamic [42,43], electric [44] (FACS and μFACS) [46]/magnetic [47,48], or dielectrophoretic [49,50] forces, or a combination thereof [41]. In essence, cells can be filtered by size [51,52], flow-line (used in laminar flow experiments), optically [53–57], acoustically [58–61], or by activating piezoelectric microfluidic elements [62]. Excellent reviews of most common cell sorting techniques have been given by Chen et al. [63] and Tsutsui and Ho [64].

Electrophoresis describes the effect of suspended charged particles and their counterions moving under the influence of an applied electric field. In contrast, dielectrophoresis relates to uncharged particles, which become polarized due to an applied nonuniform electric field and move along a particular direction. Dielectrophoresis was shown to be an efficient cell separation mechanism in microfluidic devices: neural stem/progenitor cells and neurons [65], human adipose cells [66], and human osteoblast-like cells [67] were all successfully enriched using this method.

In the case of magnetic sorting, often paramagnetic beads on the scale of 10–100 nm are attached to cells [39]. The beads then respond to the gradients in

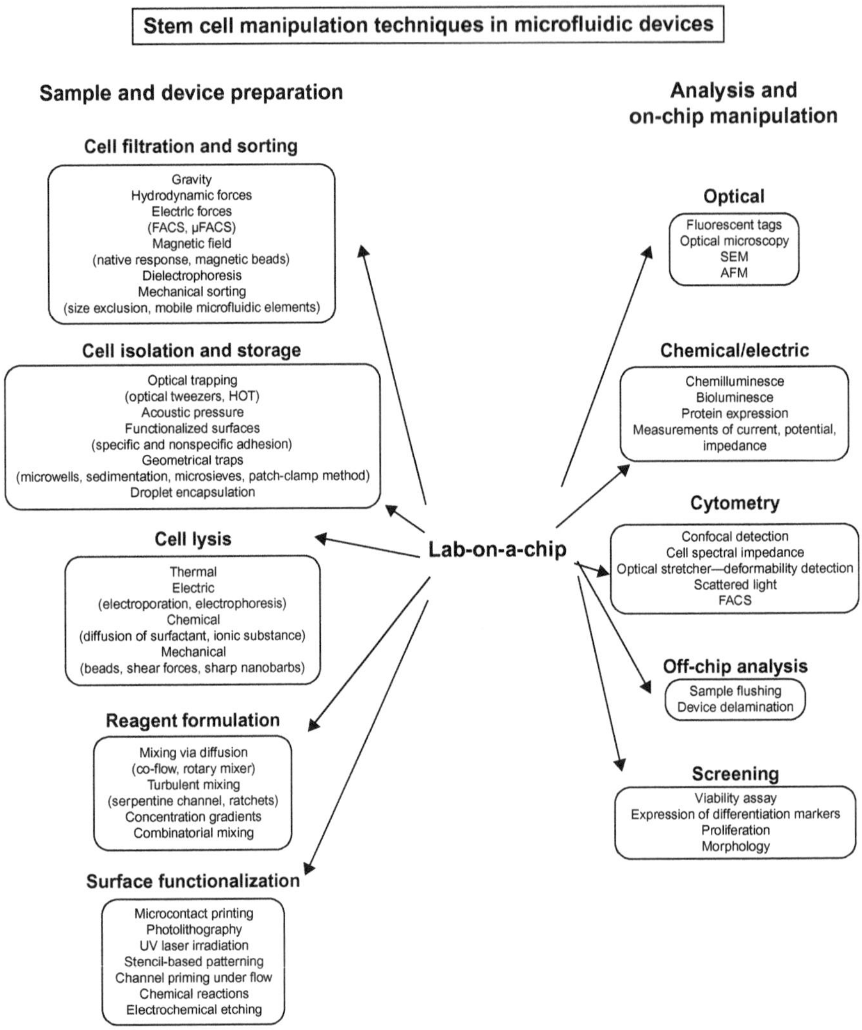

FIGURE 7.2

An overview of the techniques applied in microfluidic or lab-on-a-chip devices for manipulation of stem cells and testing reagents as well as methods of sample analysis.

applied magnetic fields. In other applications, ferromagnetic wires are used below the flow channels to form a magnetic field and affect hydrodynamic focusing [47]. Certain cell types, such as leukocytes and red blood cells, have a natural magnetic response—in that case, magnetic beads need not be used, rather, the cells can directly respond to an external applied magnetic field [68]. The removal of the magnetic tagging and washing steps is advantageous, as it streamlines the experiments, although generally tagged cells can be washed and both cells and magnetic beads can be reused.

Optical manipulation of cells is advantageous for the same reason, namely the reduction in contamination sources. Here, optical tweezers are the method of choice. Optical tweezers use a focused laser beam which exerts a force on dielectric particles. If the particle is displaced from the center of the beam, a net force will act to pull the particle toward the beam center, both laterally and axially, such that the particle remains close to the focus plane. Cells can be tagged with dielectric molecules that respond to this force, enabling single-cell observation [56,69]. For 2D trapping of multiple particles with different indices of refraction, a dual-beam setup is used in conjunction with a beam-splitter, such that an interference pattern is created (interferometric optical trapping) [70]. Yet, another level of complexity is reached with holographic optical trapping (HOT), pioneered in the laboratory of Grier [71]. This method allows for 3D trapping, by using holograms to split a single beam into *n* distinct beams, each of which is focused by a separate lens and so represents a single set of optical tweezers [55]. Optical tweezers and HOT have been used in conjunction with microfluidics and to trap stem cells [57,72]. Combining the two applications promises to be the next step in optical manipulation of stem cells using microfluidics. Disadvantages of optical trapping methods include heating and photodamage of cells. Often short laser wavelength or long irradiation times [73,74] can affect cell metabolism, division, and growth, while the local temperature increase can lead to heat shock and expression of corresponding signaling molecules, thereby affecting the measurements of interest. Thus, a careful choice of the laser type and exposure time, as well as trapping the dielectric tag molecules rather than whole cells, can help reduce damage to cells.

Another noncontact trapping method, acoustic trapping, is accomplished by exposing the cells inside a microfluidic device to a standing ultrasonic acoustic wave, with frequencies on the order of megahertz, which is applied via an external piezoelectric transducer. The acoustic wave is in fact a pressure wave, such that acoustic forces affect the hydrodynamics in the channel and so push the cells toward the pressure nodes. Such trapping of yeast cells [61] and neural stem cells [60] in the pressure nodes was demonstrated by Evander et al. and others. High cell viability as well as cell growth was observed at continuous transducer operation at 12.4 MHz and continuous perfusion. For 2D cell trapping and patterning, so-called acoustic tweezers have been developed [75]. Beyond trapping cells in clusters, the method can be used for particle separation.

7.2.2 Cell isolation and storage

A cell isolation device that has recently been the object of much attention is the circulating tumor cell (CTC) chip [76]. In this approach, support posts coated with specific antibodies are used as a filtering unit in a microchannel filled with whole blood. Only CTCs stick to the posts via specific adhesion, but not healthy red blood cells or leukocytes. In this case, the antibodies are the actual cell filtering and isolation element, while the presence and geometrical positioning of the posts increases the sensitivity of the device—over 90% of all CTCs in the sample can be captured in this fashion, at an initial concentration of only a few cells per milliliter of whole blood. It has been shown that the shape of the flow lines (in the case of the CTC chip, Hele-Shaw flow [76]), controlled by the orientation and geometry of the support posts, can strongly affect the efficiency of the cell filtering and isolation device, by redirecting cells along modified flow lines to particular capture sites.

Cells can be captured and stored in microfluidic devices using a number of other methods, such as via trapping in microsieves [78,79], encapsulation in droplets [26,27], sedimentation in storage wells [80], well-plating [81], or patch-clamp arrays [82]. Cell trapping in droplets is a high-throughput technique, as droplet production, e.g., via flow focusing, can operate in the kilohertz range [83] and allows for generating assays of thousands of samples in a short amount of time; however, this method is commonly utilized for single-cell experiments [84]. Additionally, nutrient exchange between the droplets and the reservoir is difficult to establish, and the picoliter- or nanoliter-sized droplets do not contain enough medium for long-term culture. Similarly, microsieves are also better suited to single-cell experiments than to cell culture and aggregate formation, and cells constantly experience flow and shear stress, which can affect cell behavior, e.g., by inducing mechanotransduction. Another method that applies mechanical force on cells is the patch-clamp and is therefore restricted to studies of the mechanical and electrical properties of cells. Single cells can also be trapped in microwells, as recently shown by Han et al. [85], but this method has been specifically designed for experiments on large cells, such as oocytes (~100 um).

In contrast to these techniques developed for single-cell storage and analysis, microwell platforms are preferentially used in cell culture applications [80,86,87] (Figure 7.3). The microwells are usually formed using a two-step photolithographic process and are placed adjacent to or underneath flow channels. Multiple cells are stored in the same microwell and stimulated in order to aggregate, spread on the substrate, and divide. Each microwell thus becomes a lab-on-a-chip representation of a standard milliliter-sized cell culture flask. By engineering multiple microwells on the same substrate and applying the same culture conditions to all of them, the statistics of the study can be enhanced. Alternatively, adding different growth factors to each well or set of wells can transform such a microscale platform into a high-throughput screening device. This transformation of a simple multimicrowell plate into a functionally complex microfluidic device with individually accessible and autonomous storage compartments is not straightforward, however, as it requires the incorporation of a medium perfusion system for enabling long-term cell survival as well as a multiplexer or concentration gradient generator for simultaneous testing of several combinations of growth factors (see section 7.2.5). Most previous devices have either enabled long term culture

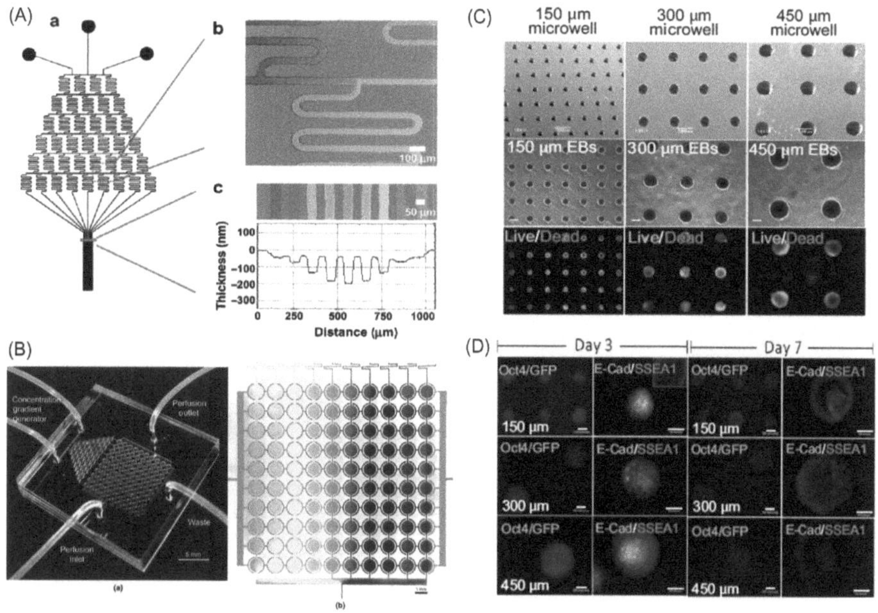

FIGURE 7.3

Microfluidic devices for stem cell experiments: the tree-like concentration gradient generator [88] is used to mix two or three input solutions via diffusion in a short amount of time (A); photograph of a high-throughput microfluidic cell culture array of 10 × 10 microwells and a concentration gradient applied across 10 columns of the device [89] (B); arrays of microwells for culturing ESCs (top), phase (middle), and fluorescence images (bottom row) of embryonic bodies grown in the microwells [90] (C); and expression of pluripotency markers in those embryonic bodies [90] (D).

Source: *All figures are reprinted with permission.*

or were useful in performing high throughput experiments (e.g., the device developed by Hattori et al. [91] allows for long-term perfusion experiments but is currently only capable of screening 16 different microenvironmental conditions).

Another challenge is achieving a uniform distribution of cells across all microwells, thereby ensuring statistically significant analytical results, and shielding the stored cells from a perfusion stream, such that fresh medium is delivered to the storage chamber and toxins are removed, but the cells are not exposed to strong shear stresses. Progress in this matter has been demonstrated by Hung et al. [89,92], who developed a promising design for perfusion cultures (Figure 7.3), with the caveat that compartmentalization valves are difficult to incorporate, thus limiting the device to low- and medium-throughput applications. Cioffi et al. [45] characterized the optimal design strategies for stable cell docking in microwells and minimizing shear stresses in perfusion application, and Jang et al. [80] advanced that work by adding cell localization valves. Closing the valves interrupts the flow in the microfluidic device and aids in positioning cells above the microwells, allowing them to sink to the bottom of these deep wells, where they were shielded from new flow of particles and media. This particular device has been shown to reduce variations in cell distribution across a large array of microwells and to allow for cell perfusion but has not yet been tested in screening applications.

7.2.3 Cell lysis

After sample filtration and cell sorting, cells often have to be disrupted or lysed on-chip prior to analysis. This can be accomplished via thermal [93,94], chemical [95], mechanical [96,97], or electrical [98,99] manipulation. A simple lysis chip was developed by Schilling et al. [95], in which cells and a lytic agent co-flow in a microchannel and interact via diffusion. The lytic agent can be any chemical that decreases the ionic strength of the solution, and the material released from the lysed cells then follows laminar flow lines into a detection channel. Alternatively, electric forces can also be used to dissolve the cell membrane, e.g., by electroporation [98] or electrophoresis [99]. In this case, the on-chip electrodes can have multiple applications, e.g., cell separation and lysis. More often, however, cells are lysed [100] on-chip in distinct, static lysis reactors, especially when several cell manipulation steps are integrated on the same platform, such as mixing, lysis, cell culture, or PCR. Alternatively, strong shear forces could be applied to the cells, e.g., by using high-pressure geometries, such as long and narrow microchannels [101], or cells can be lysed mechanically as they pass through a filter region lined with sharp nanoscopic barbs. [96] Finally, microbeads can be injected into the cell sample on-chip and sonicated, such that the beads shear off the cell membranes [102]. It should be noted that in all cases a high level of control over the functional elements of the microfluidic device is crucial, as only the cell membrane is to be affected by the lysis mechanism, but not the lysis products. As thermal lysis can easily lead to protein denaturation and mechanical techniques usually have efficiencies well below 50%, currently chemical and electrical lysis protocols are most widely integrated with microfluidic devices [39].

7.2.4 Surface patterning

In microfluidic-based cell culture platforms, one of the major challenges is efficient, quantifiable, and reproducible immobilization of cells in a defined area or compartment, where the cells are used for drug testing, observation or to create cocultures. The immobilization of cells on a substrate surface can be simply realized by locating suspended cells onto a surface treated for enhanced cell adhesion, since the adhesion and proliferation of living cells on a substrate surface are well recognized to depend on many surface characteristics, such as the surface charge, wettability, chemistry, and surface roughness [103,104]. Various techniques such as microcontact printing [105], microfluidic patterning [106], photolithography [107], UV laser irradiation [108], and stencil-based patterning [109] have been reported to enable cell micropatterning on a substrate surface. However, microchannels are less amenable to the patterning methods employed for planer surfaces due to the 3D geometries involved in the channel structures. Also, conventional surface modification methods which require close proximity to the substrate are restrictive when applied to microchannels due to the requirement to seal the channels after patterning. Nonetheless, several surface pattering methods of bonded microchannels have been reported. For example, Zhao et al. [110] utilized laminar flow surface patterning by wet chemistry to prepare parallel features along microchannels, in which streams of octadecyltrichlorosilane (OTS) in an organic solvent and OTS-free streams were used to pattern glass microchannels with hydrophobic stripes. Moreover, they employed SAMs (self-assembled monolayers) of silane with the photocleavable *o*-nitrobenzyl functionality to pattern bonded microchannels [111]. In this approach, UV irradiation was applied through various masks to cleave the nitrobenzyl-oxygen bond, yielding regions of hydrophilic carboxylic acid groups at the channel surface. The complexity of these patterns is only limited by the mask design and the resolution of the lithography technique. The use of an electrochemical method is another approach to pattern surfaces of bonded microchannels. For example, Kaji et al. [112] individually addressed specific regions of a bonded microchannel for protein immobilization. Hypobromous acid, which is electrochemically generated at a microelectrode in the flow channel, was used for oxidative removal of a heparin layer from the adjacent wall, and subsequent addition of proteins was locally adsorbed on the treated area. By repeating the patterning process, even multiple types of proteins can be site-specifically immobilized in a single channel. Moreover, they combined this technique with dielectrophoresis to efficiently position cells on cell-adhesive protein immobilized area [113]. Since the microelectrodes fabricated on the channel surface can be used for both electrogenerating the oxidant and producing dielectrophoretic forces, the electrode configuration is simple and does not sacrifice further miniaturization of microfluidic systems.

7.2.5 Microfluidic mixing, concentration gradients, and combinatorial solutions

Mixing of solutions in microfluidic channels is commonly governed by diffusion, as the flow conditions are usually confined to the low Reynolds number region ($Re < 1$) and thus to laminar flow. When operating in this region, miscible solutions are often co-flowed side-by-side into the same channels, such that (especially at low flow rates) there is a discernible interface between the liquids parallel to the direction of flow. Molecules, ions, and particles can diffuse across this interface, thereby being mixed. By adding another level of structural complexity to the microchannel, we can utilize this diffusion-based mixing to generate several mixtures with different ratios of the input solutions, leading to discrete or continuous concentration gradients. A well-known example is the tree-like gradient generator [40] (Figure 7.3), which takes n miscible aqueous solutions as inputs, combines them, mixes them via diffusion, and splits them into $(n + 1)$ outputs. Usually, this gradient generator takes only two inputs and produces up to 12 or more outputs. When the output solutions are directed to separate channels, we speak of a discrete concentration gradient; however, when they are combined in a single wide flow channel, we speak of a continuous concentration gradient, as further mixing occurs at the interfaces of the output flows, thereby blurring the discrete concentration values. Depending on the flow rates and channel dimensions in the mixing region, this structure can be used to produce linear and exponential concentration gradients. Other channel structures relying on diffusion as a passive mixing mechanism have been developed for logarithmic [131], exponential [41], and sigmoidal [114] gradients. In another instance, several miscible liquids are sequentially being pumped inside a toroidal channel or rotary pump [115], thus the mixing is active. Each liquid follows a parabolic Poiseuille flow profile, and a parabolic diffusion and mixing interface is generated perpendicular to the flow. The centers of these interfaces rotate faster than the edges, such that with each push the interfaces stretch until the solutions form thin strips that are wrapped around each other. Increasing the interfacial surface in this manner actively speeds up the mixing via diffusion across the interface.

Aside from diffusion, microfluidic designs can give rise to turbulent and chaotic mixing [48], even when the channel dimensions and flow rates are in the laminar flow region. This is especially advantageous in cases where the convective transfer of material is faster than diffusion mixing. A standard example is the serpentine structure developed by Liu et al. [116], in which chaotic mixing occurs as a result of eddies at points of largest curvature in the channel. The ratchet design developed by Stroock et al. [117] utilizes a similar principle by engineering a series of ratchets on the channel floor that are oriented at an angle with respect to the flow. When the liquid encounters a ratchet, it experiences the largest resistance in the direction of flow. An axial pressure gradient emerges that forces the solution to flow back across the top of the channel and effectively fold upon itself.

Any of the mixing elements listed above can be used for generating combinations of two or more miscible solutions. In contrast to gradient generators, which are usually confined to two or three on-chip solutions, combinatorial mixers can join many input liquids (usually eight or more) in different ratios and produce any desired number of output solutions [46–49]. A standard example is the multiplexer device developed by Thorsen et al. [118]. The multiplexer uses $2 \log_2 n$ actuation channels to control n flow channels and to produce and store 1000 different combinations on-chip. This programmable device represents efforts to integrate and execute in parallel physical, chemical, and biological applications, from metering and mixing different solutes, to directing these mixtures to their respective storage chambers and *in situ* high-throughput analysis.

7.2.6 On-chip cell culture

The miniaturization of bioreactors for cell culture is advantageous given the large surface-area-to-volume ratio associated with microfluidic systems. For instance, cell culture systems based on microfludics can be used to integrate more efficient mass exchange networks to facilitate the perfusion of oxygen and nutrients to the cells as well as the removal of waste. Moreover, the small volumes involved necessitate only small amounts of biological factors and produce less waste, which are particularly attractive for expensive bioassays or cell culture studies.

PDMS (poly(dimethylsiloxane)) is a common material for the fabrication of microfluidic cell culture platforms since it is relatively cheap, easy to mold, and highly permeable to gases. For example, Nishikawa et al. [38] demonstrated that a spheroid culture of rat hepatocytes can be stably attached to collagen-coated PDMS surfaces with sufficient oxygen permeation through the bottom PDMS surface. Also, multiple layers of micromachined PDMS films can be assembled into 3D microstructures for cell culture [1]. In addition, most microfluidic cell culture systems (fabricated from PDMS, glass, polyester, or optical glue) are optically transparent and compatible with conventional imaging techniques.

Long-term cell culture can be performed in microfluidic perfusion systems in which cells are constantly supplied with oxygen and nutrients while metabolic waste products are removed. Hung et al. [89] fabricated an array of cell culture chambers where the fresh medium is perfused continuously from ports uniformly across the array [89] (Figure 7.3B). The chamber arrays were also integrated with an upstream concentration gradient generator to enable cell assay studies to be performed with different reagent concentrations in each array column. Perfusion cell culture systems allow a constant cellular microenvironment to be maintained, which is of importance for the functionality of the cells in a long-term culture. Lee et al. [120] reported a microfluidic-based artificial liver sinusoid incorporating a microfabricated endothelial-like barrier, with mass transport properties similar to the liver acinus. By using this system, rat and human hepatocytes were sustained for a week. Also, Kane et al. [121] developed an array of 8×8 microfluidic wells, which maintained hepatocyte functions for up to a month.

Automatization of a series of cell culture procedures is an important aspect of cell culture systems based on microfluidics. Electrowetting-based digital microfluidics (DMF), an alternative to the conventional format of enclosed microchannels, is an attractive platform for implementing such purpose [122]. DMF devices are formed from an array of electrodes which are used to manipulate discrete fluidic droplets by applying a series of electrical potentials to those electrodes. Recently, Barbulovic-Nad et al. [123] reported a DMF device capable of implementing all of the steps required for mammalian cell culture including cell seeding, growth, detachment, and re-seeding on a fresh surface. In this approach, they demonstrated cell growth characteristics comparable to those found in conventional tissue culture.

The development of 3D cell culture systems is also important since they offer a more biologically relevant model to perform cell-based research and development, whereas cells cultured in 2D systems have been known to lose their functions or differentiation capability [124]. To date there have been many attempts to develop 3D culture systems. For instance, Tan and Desai [125] fabricated 3D hierarchical tissue-like microstructure through sequential deposition of cells and biopolymer matrix on specific regions within microchannels. Toh et al. [126] used micropillar arrays fabricated within a microchannel to implement 3D cell culture. In this approach, the array is located in the center of the microchannel, which divides the channel into a central compartment for cell culture and two side channels for medium perfusion.

7.2.7 Cell analysis on-chip

The general advantage of microfluidic cell analysis is its perfusion capability and transparency coupled with small volumes, fast reaction rates, and high throughput. Perfusion allows infusion of assay reagents and transparency enables imaging and analysis of cellular responses. Gradient generating microfluidic devices have been used for real-time monitoring of cell behavior including migration, proliferation, differentiation, and apoptosis [127]. For example, Jeon et al. [128] demonstrated the use of the gradient generator in studying neutrophil chemotaxis. They tested the neutrophilic chemotactic responses to different configurations of chemokine IL-8 gradient in the microfluidic device. In another example, Pihl et al. [129] developed a microfluidic device with a gradient generation component for pharmacological gradient profiling. Using this device, drug streams were held at different concentrations and voltage-gate K^+ ion channels were screened using scanning probe measurements.

The combination of cell microarrays and microfluidics has led to the creation of screening methods capable of systematically varying one or more parameters across the cell microarray in time and space. For example, Thompson et al. [130] developed a microfluidic platform for continuous monitoring of gene expression in a living cell array. Using this platform, they profiled the activation of the transcription factor NF-κB in HeLa S3 cells in response to varying doses of the inflammatory cytokine TNF (tumor necrosis factor)-α. Also, Kim et al. [131] developed a strategy to simultaneously create different dynamic soluble microenvironments across a cell microarray. Microfluidic channels upstream of the cell array were used to generate different temporal profiles of soluble factors, the effects of which were monitored by fluorescence in cell lines with specific gene reporters.

7.3 Controlling cellular microenvironments

The ability to control the environment of cell culture systems is crucial for *in vitro* cell function studies and for optimum design of tissue constructs that mimic the organizational complexity of *in vivo* tissue architectures [21]. *In vivo* cells integrate and interact with a microenvironment comprised of a milieu of

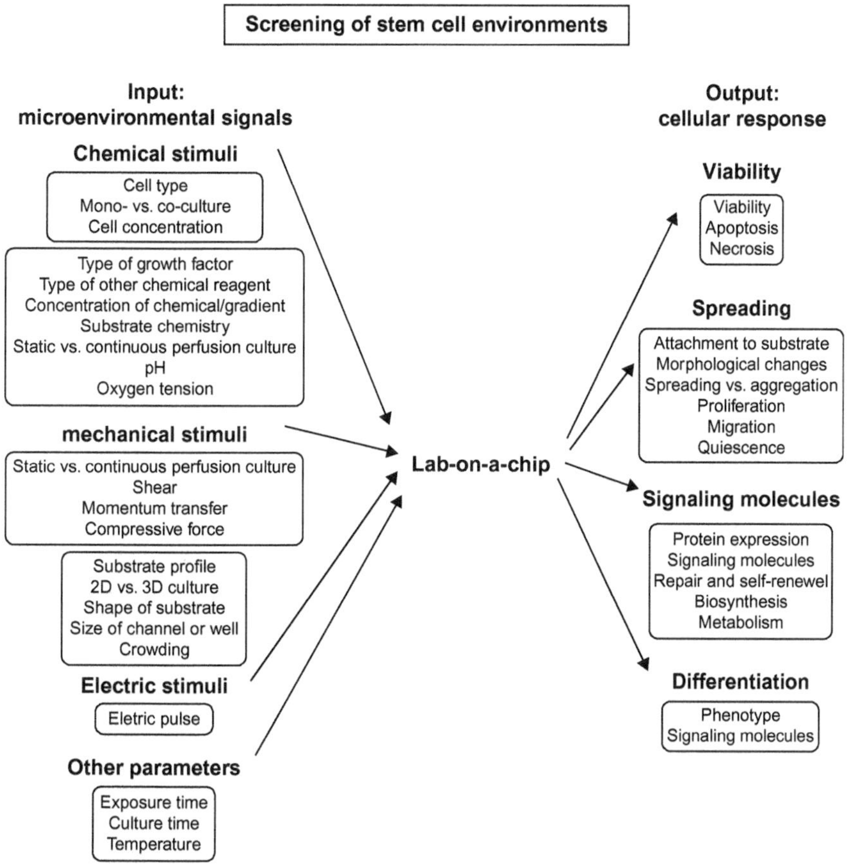

FIGURE 7.4

Microenvironmental factors affecting the stem cell behavior in microfluidic or lab-on-a-chip devices (input) and output variables (cellular response).

biochemical, biomechanical, and bioelectrical signals derived from surrounding cells, ECM, and soluble factors. These components vary in both time and space and are integral to the regulation of cellular behaviors.

Microscale technologies, some of which have been developed in microelectronics and MEMS (microelectromechanical systems), can control material features from nanometers to centimeters, thus providing unprecedented control over the interface between cells and the surrounding environment [132]. In addition, the biological, chemical, and mechanical properties of biomaterials can also be tuned to further control the cellular microenvironment. In the following sections, we review how the microscale technologies, especially microfluidics, can be used to control the cell–microenvironment interactions (Figure 7.4).

7.3.1 Soluble factors and chemical stimuli

As pointed out before, microfluidic devices offer a fast and affordable alternative to standard bench-scale experiments and can scan hundreds and even thousands of different conditions simultaneously using only microliter amounts of cells and reagents. In regard to chemical stimuli, the screening conditions include cell type, mono- versus coculture, cell concentration, type and concentration of a growth factor or other reagent, static versus continuous perfusion culture, and culture or exposure time. Often the experimental output or information that is being collected refers to attachment to substrate, viability versus apoptosis and necrosis, growth and multiplication, expression of proteins and other signaling molecules, and last, not least, cell differentiation. We have already listed several types of devices that contain generators of combinatorial mixtures and concentration gradients on one hand, and arrays of storage chambers on the other hand, which enable researchers to study the effects of variations in these experimental parameters on stem cell behavior. Many microfluidic applications for screening of cellular microenvironments now combine these two device elements and allow for testing of many soluble factors on cells and cell aggregates that are stored on-chip [91,114].

Mei et al. [37] and Warrick et al. [36] list in their review papers several high-throughput stem cell screening approaches. Beyond the screening experiments mentioned therein, researchers have studied the effect of the epidermal growth factor (EGF) on breast cancer cell chemotaxis [133], ligand binding [134], and, of course, cell attachment to a substrate, proliferation, and viability [135]. Further studies have focused on the behavior of ESCs in response to growth factors responsible for cardiogenesis, such as expressing brachyury (GFP-brachyury) (GFP-Bra) [52–54] and goosecoid (Gsc) (GFP-Gsc) [56,57], in response to combinations of soluble factors such as wnt3a, activin A, bone morphogenetic protein-4, and fibroblast growth factor-4 (FGF-4) [136]. Researchers have also explored the growth and differentiation potential of human neural cells [137], the effect of combinatorial signaling inputs on cell fate [136], and the optimal culture media composition to direct ESC differentiation [138,139].

The device developed by Chung et al. [137] was used to apply a nonlinear concentration gradient of growth factors to human neural stem cells. The device, utilizing the tree-like concentration gradient generator, was fabricated from PDMS, with a glass slide forming the bottom of the channels and the cell culture area. All features were coated with poly-L-lysine and mouse laminin to encourage cell adhesion to the substrate and to allow for continuous perfusion (i.e., constant flow of the growth factors) without the danger of disturbing the cells. The three gradient generator inputs contained a mixture of medium without a growth factor and a line filled with different growth factors in different experiments (EGF, FGF-2, and PDGF, platelet-derived growth factor). As expected, the cells exposed to the growth factors showed a marked increase in proliferation over the course of 7 days, which was proportional to the increase in growth factor concentration.

In contrast, the cells lacking the growth factors exhibited differentiation markers specific to astrocytes, when stained with antibodies.

A high-throughput application of microfluidics to controlling the stem cell microenvironment is described by Lecault et al. [140]. The experiment aimed to distinguish between the natural cell fate and the cell microenvironment as chief factors for the cell proliferation behavior. This was accomplished by keeping the culture parameters constant in one (controlled) experiment and letting them change naturally due to the characteristics of the PDMS device, e.g., the medium became more concentrated with time, as the water diffused into the device (uncontrolled experiment). On-chip, 1600 wells were precisely controlled via on-chip valves and filled with single mouse hematopoietic stem cells (HSCs), at a loading efficiency of up to 30%. The cells were first cultured under continuous perfusion and later the medium was refreshed every 2 h, and the cell multiplication frequency was measured. The cell viability and multiplication were much lower in the uncontrolled experiment as opposed to the case where the microenvironment was kept constant. Additionally, the withdrawal effect of steel factor and its reversibility were examined as a function of time on a device with over 6000 individual cell culture wells. The observations indicated that primary mouse HSCs could tolerate steel factor withdrawal of up to 16 h, but could not repair the damage after a longer withdrawal. The device design and high-throughput nature of the experiments were instrumental in tracking individual cells and reducing the measurement error, compared to standard 24- and 96-well plates, and in achieving statistically significant results.

7.3.2 Mechanical stimuli

7.3.2.1 Substrate properties

Another set of factors in (stem) cell behavior, specifically differentiation, includes the surface chemistry [4], shape (2D or 3D cylindrical [142], concave or convex microwells [143,144]), and rigidity [145] of the growth substrate. The effect of the shape and stiffness of culture substrates fabricated from the standard microfluidic device materials such as PDMS, polyester, glass, and polystyrene on stem cell morphology [146], aggregation [141,147,148], and differentiation [149] has been studied intensely. These parameters are of great importance for *in vitro* stem cell studies, as cells are known to behave differently in culture in comparison to *in vivo*.

Liu et al. [149] found that mouse ESCs (cell line R1) proliferated most on tissue culture petri dishes, slightly less on untreated glass, and least on PDMS. Similarly, the stem cell differentiation into neurons was preferred on the petri dish surface and was weak on glass and PDMS. It was hypothesized that the substrate stiffness was a factor in this as much as the substrate chemistry.

Fu et al. [150] studied the behavior of human mesenchymal stem cells (MSCs) in response to the rigidity of micromolded PDMS substrates. The authors varied the height of the posts to control the platform rigidity. While the platform was not strictly an autonomous microfluidic device, findings from this work can easily be applied to design a more complex lab-on-a-chip device for stem cell studies. It was observed that cells differentiated preferentially into osteoblasts on rigid microposts, while adipogenic differentiation was strongest on soft microposts. These histological findings and optical observations matched gene expression data.

Additionally, in studies that focus on growth of embryonic bodies inside storage microwells, the microwell shape is relevant. Commonly, microwells in PDMS or hydrogel-based devices are cylindrical or box-shaped, as this requires only the simplest lithography steps. It has been shown, however, that rounded and tapered wells allow for more efficient exchange of medium and enable the growth of a single spherical embryonic body [141], while cells tend to form several aggregates in cylindrical wells [143,144]. When only one aggregate is formed per well, the embryonic bodies then often adopt a cylindrical shape. Because this changes the surface-to-volume ratio, different cells receive different amounts of fresh nutrients and oxygen, such that some cells may not be as well-nourished and start dying.

7.3.2.2 Shear stress and other mechanical effects

A microenvironmental factor that is easily incorporated in a microfluidic platform is shear, as loading of cells and exchange of medium, in addition to delivery of growth factors, requires flow. Fluid shear stress can have a profound influence on the function of cells, ranging from altered metabolism to cell lysis. For example, ESC-derived endothelial cells can be induced to elongate and align in the direction of the flow by modulating gene expressions, in the similar transcriptional response to shear stress found in primary cells [151]. For microfluidic perfusion culture in 2D Poiseuille flow systems, shear stress at the channel wall can be simply estimated from the parabolic flow profile [152]. Typically, the shear stress can be controlled by changing the flow rate of medium or the dimensions of the microchannel, where the effect of shear stress on cell growth, differentiation, and migration can be studied. However, since the changes in the flow rate and the channel dimensions are accompanied by the variation of the content of the soluble microenvironment, the effects on nutrient delivery and cell secreted factors must also be considered.

Most experiments are designed to operate in the region of physiological shear. Voldman and coworkers [131,153], however, have explored a wide range of shears and its effect on mouse ESCs. They applied a set of flow rates varying logarithmically from 10^{-3} to 1 μl/min in cell culture channels. By conducting separate experiments in which they modified the concentration of nutrients in the culture medium, the researchers were able to decouple the effects of shear stress and nutrient concentration on stem cell growth. They concluded that the stem cells proliferated more at the higher flow rates, especially when compared to more robust cell types, such as 3T3 fibroblasts.

Aside from exploiting the device geometry and materials, as well as the flow properties inside the microfluidic device to affect stem cell behavior, researchers have also employed momentum transfer to act on the cells. Song et al. [154] used electromagnetic pulses to guide magnetic beads to collide in large microfluidic chambers with osteoblasts during particular phases of the cell cycle. The cell growth rates were shown to depend on bead size (4.5 μm beads caused a higher growth rate than 8.4 μm beads), impact frequency (the growth rate increased with frequency), and the timing of the cell cycle (largest growth was observed in the G1 phase and smallest in the G2 phase). Although high frequencies (up to 1 MHz) were used, up to 95% of all cells were viable.

Several groups have studied the impact of pneumatic force on differentiation of human MSCs cultured in microfluidic devices. One microfluidic platform [155] incorporated PDMS membranes of different thicknesses, which, when activated, exerted a force on the stored cells. It was determined that cells exposed to pressures above 2 psi in this chip could not survive for more than a few minutes. Sim et al. [156] were interested in osteogenesis in human MSCs in response to a cyclic compressive force. They observed an increase in proliferation in the stimulated cells as compared to the control group, which did not experience a compression, but also identified an optimal pressure that led to peak proliferation. It was also shown that the stimulated cells entered the early stage of differentiation to osteoblasts, while the control group remained undifferentiated. This was evident in the increased expression levels of alkaline phosphatase and calcium and decreased levels of CD90 in the mechanically stimulated group. In contrast, Wan et al. [157] developed a hybrid PDMS-hydrogel microfluidic device to apply a uniaxial cyclic strain to ESC-derived embryonic bodies seeded in collagen gels. When differentiation into cardiomyocytes was initiated with growth factors, application of a 10% strain at 1 Hz late in the culture was observed to inhibit the differentiation process.

7.3.3 Electric stimuli

In certain cases, stem cell behavior, e.g., differentiation, can be affected by creating an artificial link between the inside of a cell and its microenvironment. This can be done by electroporating individual cells, which increases their electrical conductivity and permeability of the membrane to outside molecules [158]. Valero et al. [159] have accomplished just that in their microfluidic device, with 75% efficiency. The researchers first transfected single human MSCs with DNA that coded for green fluorescent-erk1 fusion protein. Then, they electroporated the cells one at a time in an environment of FGF, which resulted after a few minutes in nuclear translocation of the fusion protein. This method could potentially be applied to studies of cellular processes at the single cell level, by increasing the strength of the interaction between cells and their chemical microenvironment.

Direct application of electric fields to certain cell types is also known to affect their behavior. In one experiment, 4-day-old embryoid bodies formed from mouse embryonic fibroblasts were exposed to a 90 s DC pulse varied between 100 and 500 V/m. The highest electric field increased cell differentiation into cardiomyocytes [160]. In another experiment, human ESCs were subjected to a square-wave

DC pulse of 1000 V/m [161]. The amount of reactive oxygen generated by the cells as a result of this treatment increased proportionally with the duration of the electric pulse. Finally, Radisic et al. [162] treated neonatal rat myocytes with AC pulses over the course of 5 days and observed a sevenfold increase in cell contraction, as well as cell alignment.

7.3.4 Cell–cell contact and coculture

Cell–cell interactions that occur through immediate cell contact, intracellular protein transfer, or paracrine soluble factors are crucial for maintaining stem cell homeostasis in the stem cell niche. For example, MSCs have been shown to differentiate into multiple lineages in a cell-type contact-dependent manner [163]. For this reason, significant research has attempted to investigate the mechanisms by which cells signal to and influence each other. However, designing experiments with unsophisticated coculture systems is difficult and limiting in scope. Modern technologies such as microfabrication and microfluidics have allowed researchers to conduct heterotypic coculture experiments in a more precise and controlled manner.

Microfluidics offers several approaches to pattern various cell types. For example, Chiu et al. [164] used 3D microfluidic systems to pattern two different cell types in complex, discontinuous structures. Since many isolated channels can be contained in the multilayered stamp, multiple cell types can be patterned more easily than through microcontact printing, although the placement of the cell populations in contact with each other is not possible because of the presence of stamp walls separating the compartments. Recently, Torisawa et al. [165] reported a microfluidic method to form coculture spheroids of various geometries and compositions. They used a two-layered microfluidic device that sandwiches a semiporous membrane so that flow occurs from the top channel through the membrane to the bottom channel. Arbitrary cellular arrangement was possible by regulating the geometric features of the bottom channel so that as culture media drained, the flow hydrodynamically focused cells onto the membrane only over the regions of the bottom channel. When the top channel had multiple inlets, cells could be seeded in adjacent laminar streams, allowing different cell types to be patterned simultaneously in well-defined spatial arrangements.

Recent advances in the ability to engineer surface properties of substrates have allowed researchers to dynamically modulate the interactions between cells and the substrate surface in real time using external trigger such as light, voltage, heat, and microelectrodes [74]. These techniques can be used for the sequential patterning of multiple cell types and control over the adhesion and motility of individual cell types. Also, dielectrophoresis, a phenomenon in which particles are manipulated based upon the interactions between a nonuniform electric field and charge polarizations induced in the particles, have been used for patterning different cell types. For example, Suzuki et al. [166] fabricated periodic and alternate cell patterns incorporating two types of adhesive cells using negative dielectrophoresis. An interdigitated array electrode with four independent micro-electrode subunits was used as a template to form cellular micropatterns. In this system, the dielectrophoretic force was induced by applying an AC voltage (typically 12 V_{pp}, 1 MHz) to direct cells toward a weaker region of electric field strength. After removing excess cells from the device, a second cell type was introduced into the device and, by changing the AC voltage mode, these cells were guided to other areas to form a different pattern.

Another approach to create coculture system is to use mechanically configurable devices. Hui and Bhatia [119] developed a technique for the dynamic control of cell–cell adhesion that could affect cellular phenotype. In this set up, a microfabricated silicon substrate consisting of two interlocking parts was manually manipulated to bring cells in close proximity to each other. The two parts could be joined in discrete configurations such that different types of cells are adjacent to one another or are separated by a micron-scale gap. Kaji et al. [77] monitored cell movements in a coculture system in which two complementary substrates, which could be mechanically assembled, were used to generate the coculture. Moreover, they integrated a microfluidic device with the controlled coculture system, allowing signaling factors secreted from one cell type to be directed via the culture medium to other cells.

7.3.5 Development and mimicking of ECMs

Interactions between the cells and the ECM are also relevant for cell function. In particular, the process of cell binding to molecules in the matrix, as well as mechanical properties of ECM such as topography and rigidity serve as a blueprint for many aspects of cell behavior, including growth, spreading, differentiation, and on a larger scale tissue formation [169,170]. For example, it has been shown that MSCs tend to differentiate into neurons when adhering to soft, brain-like substrates. In contrast, rigid matrices which have a modulus similar to collagenous bone, provide cues for osteogenic differentiation [171].

ECM building blocks include proteins (collagen, laminin, etc.) and polysaccharides (e.g., hyaluronic acid) that form fibers or sheets. These topographic features cover a wide size range from nanometers to centimeters. The response of stem cells to variations in the ECM topography consists of activating different signaling pathways and expressing particular proteins. This in turn stimulates cytoskeletal, biochemical, and biomechanical modifications in the cells. Experimentally, these mechanical cues are imitated by employing different biomaterials and different microfluidic substrates, including coating of flow channels and storage chambers [172].

Microarray technologies, which consist of spotting ECM molecules, have become popular to study the role of ECM components on stem cell phenotypes in a high-throughput manner. A broad range of ECM molecule combinations can be printed on glass slides [173,174]. For example, the response of human ESCs to various extracellular signals have been investigated using the nanoliter-scale synthesis technique [87]. In another study, combinations of ECM proteins were microarrayed and used for ESC differentiation studies [175]. Recently, technologies are being developed for creating 3D hydrogel-based microarrays which have more similarity to *in vivo* counterparts than 2D monolayers [176].

7.4 Challenges and outlook

The advance in microscale technologies has led to dramatic enhancements in studying biological systems. In particular, there has been much success in controlling the cellular microenvironment and investigating cellular responses to the changes of the cellular microenvironment. However, much research and development has yet to be carried out at the intersection of microfabrication and cell biology. Challenges include developing more robust systems to allow long-term cell culture that can control the mechanical factors that influence stem cell behavior in addition to the existing biochemical perturbation systems. Also, developing microfluidic cell analysis systems that can accurately and reproducibly handle biological samples, involving protein extraction and analysis from whole cell lysates or blood plasma, are crucial since most of the reported systems have been limited to DNA/RNA analysis.

One of the major challenges in investigating the cell microenvironment is the enormous number of possible combinations from different factors and interactions. Many cues and biological signaling events during development are context dependent, and basic cell behaviors, such as migration differ drastically between 2D and 3D environments. Thorough investigation of cell behaviors *in vitro* including the ability to discover mechanisms of the observed cell behaviors requires microscale technologies with fine control over the engineered cellular microenvironment. For example, techniques for creating arrays of 3D microgel combinations and its integration into lab-on-a-chip devices will be extremely useful in investigating cell–microenvironment interactions. Although such systems face a number of technical challenges including the creation of combinatorial experiments with cell-laden microgels and the high-throughput screening of its outcomes in 3D, it is promising for systematic investigation of cues and developmental signals that direct stem cell differentiation in an organized space.

More specifically, current limitations of microfluidic devices for stem cell studies include uniform loading of cells across the microwell array, dehydration of the cell samples and reagents, temporal localization of nonadherent cells, and extraction of individually stored cells and embryoid bodies. Solutions have been made to address some of these existing challenges: cells can be uniformly seeded in storage wells by using directed flow (e.g., via microvalves) or by introducing a high cell seeding density into the chip. Dehydration can be avoided by introducing a water reservoir into a porous polymeric device; alternatively, nonporous materials such as glass can be used [140]. Nonadherent cells can be *de facto* immobilized by taking the advantage of very deep wells, where a recirculation region is present. This prevents cells from being flushed out under the flow. Last, not least, individual cells (in single cell experiments) and cell aggregates can be extracted by employing flow directing microvalves and pumps.

The microscale technologies contain great potential for the advancement of the field of regenerative medicine. The challenges and improvements mentioned earlier will enable researchers to efficiently recreate cell microenvironments. This will provide base knowledge for developing laboratory grown tissues and will bring much closer to reality the generation of engineered organs as well as regeneration of defective tissues, thereby providing hope for millions of current and future patients with no other alternatives for cure.

Acknowledgments

This work was partially supported by the National Institutes of Health (EB008392 HL092836, EB009196, DE019024), National Science Foundation (DMR0847287), Institute for Soldier Nanotechnology, Office of Naval Research, and US Army Corps of Engineers. Hirokazu Kaji acknowledges the support from JSPS Fellowships for Research Abroad.

References

[1] E. Leclerc, Y. Sakai, T. Fujii, Cell culture in 3-dimensional microfluidic structure of PDMS (polydimethylsiloxane), Biomed. Microdev. 5 (2003) 109–114.

[2] J. Liu, M. Enzelberger, S. Quake, A nanoliter rotary device for polymerase chain reaction, Electrophoresis 23 (10) (2002) 1531–1536.

[3] C.M. Vajdic, et al., Increased incidence of squamous cell carcinoma of eye after kidney transplantation, J. Natl. Cancer Inst. 99 (17) (2007) 1340–1342.

[4] S.P. Forry, et al., Cellular immobilization within microfluidic microenvironments: dielectrophoresis with polyelectrolyte multilayers, J. Am. Chem. Soc. 128 (42) (2006) 13678–13679.

[5] A. Khademhosseini, et al., Cell docking inside microwells within reversibly sealed microfluidic channels for fabricating multiphenotype cell arrays, Lab Chip 5 (2005) 1–8.

[6] J.A. Thomson, et al., Embryonic stem cell lines derived from human blastocysts, Science 282 (5391) (1998) 1145–1147.

[7] F.M. Watt, B.L. Hogan, Out of Eden: stem cells and their niches, Science 287 (5457) (2000) 1427–1430.

[8] I. Kehat, et al., Human embryonic stem cells can differentiate into myocytes with structural and functional properties of cardiomyocytes, J. Clin. Invest. 108 (3) (2001) 407–414.

[9] C. Mummery, et al., Differentiation of human embryonic stem cells to cardiomyocytes: role of coculture with visceral endoderm-like cells, Circulation 107 (21) (2003) 2733–2740.

[10] T. Ishii, et al., In vitro differentiation and maturation of mouse embryonic stem cells into hepatocytes, Exp. Cell Res. 309 (1) (2005) 68–77.
[11] N. Lavon, O. Yanuka, N. Benvenisty, Differentiation and isolation of hepatic-like cells from human embryonic stem cells, Differentiation 72 (5) (2004) 230–238.
[12] T. Hamazaki, et al., Hepatic maturation in differentiating embryonic stem cells in vitro, FEBS Lett. 497 (1) (2001) 15–19.
[13] R.C. Bielby, et al., In vitro differentiation and in vivo mineralization of osteogenic cells derived from human embryonic stem cells, Tissue Eng. 10 (9–10) (2004) 1518–1525.
[14] J.M. Karp, et al., Cultivation of human embryonic stem cells without the embryoid body step enhances osteogenesis in vitro, Stem Cells 24 (4) (2006) 835–843.
[15] L.M. Bjorklund, et al., Embryonic stem cells develop into functional dopaminergic neurons after transplantation in a Parkinson rat model, Proc. Natl. Acad. Sci. U.S.A. 99 (4) (2002) 2344–2349.
[16] J.H. Kim, et al., Dopamine neurons derived from embryonic stem cells function in an animal model of Parkinson's disease, Nature 418 (6893) (2002) 50–56.
[17] M. Schuldiner, et al., Induced neuronal differentiation of human embryonic stem cells, Brain Res. 913 (2) (2001) 201–205.
[18] S. Levenberg, et al., Endothelial cells derived from human embryonic stem cells, Proc. Natl. Acad. Sci. U.S.A. 99 (7) (2002) 4391–4396.
[19] J.A. Burdick, G. Vunjak-Novakovic, Engineered microenvironments for controlled stem cell differentiation, Tissue Eng. A 15 (2) (2009) 205–219.
[20] J.J. Heit, S.K. Kim, Embryonic stem cells and islet replacement in diabetes mellitus, Pediatr. Diabetes 5 (Suppl. 2) (2004) 5–15.
[21] A. Khademhosseini, et al., Microscale technologies for tissue engineering and biology, Proc. Natl. Acad. Sci. U.S.A. 103 (8) (2006) 2480–2487.
[22] B. Murtuza, J.W. Nichol, A. Khademhosseini, Micro- and nanoscale control of the cardiac stem cell niche for tissue fabrication, Tissue Eng. B Rev. 15 (4) (2009) 443–454.
[23] A. Atala, Tissue engineering of artificial organs, J. Endourol. 14 (1) (2000) 49–57.
[24] D.F. Ren, et al., Evaluation of RGD modification on collagen matrix, Artif. Cells Blood Substit. Biotechnol. 34 (3) (2006) 293–303.
[25] L.J. Bonassar, C.A. Vacanti, Tissue engineering: the first decade and beyond, J. Cell Biochem. Suppl. 30–31 (1998) 297–303.
[26] H. Shin, Fabrication methods of an engineered microenvironment for analysis of cell–biomaterial interactions, Biomaterials 28 (2) (2007) 126–133.
[27] G.R. Martin, Isolation of a pluripotent cell line from early mouse embryos cultured in medium conditioned by teratocarcinoma stem cells, Proc. Natl. Acad. Sci. U.S.A. 78 (12) (1981) 7634–7638.
[28] A. Rolletschek, A.M. Wobus, Induced human pluripotent stem cells: promises and open questions, Biol. Chem. 390 (9) (2009) 845–849.
[29] K. Takahashi, S. Yamanaka, Induction of pluripotent stem cells from mouse embryonic and adult fibroblast cultures by defined factors, Cell 126 (4) (2006) 663–676.
[30] G. Amabile, A. Meissner, Induced pluripotent stem cells: current progress and potential for regenerative medicine, Trends Mol. Med. 15 (2) (2009) 59–68.
[31] K. Hochedlinger, K. Plath, Epigenetic reprogramming and induced pluripotency, Development 136 (4) (2009) 509–523.

[32] A.D. Ebert, et al., Induced pluripotent stem cells from a spinal muscular atrophy patient, Nature 457 (7227) (2009) 277–280.

[33] K. Tateishi, et al., Generation of insulin-secreting islet-like clusters from human skin fibroblasts, J. Biol. Chem. 283 (46) (2008) 31601–31607.

[34] I.H. Park, et al., Disease-specific induced pluripotent stem cells, Cell 134 (5) (2008) 877–886.

[35] A.M. Wobus, K.R. Boheler, Embryonic stem cells: prospects for developmental biology and cell therapy, Physiol. Rev. 85 (2) (2005) 635–678.

[36] J.W. Warrick, W.L. Murphy, D.J. Beebe, Screening the cellular microenvironment: a role for microfluidics, IEEE Rev. Biomed. Eng. 1 (1) (2008) 75–93.

[37] Y. Mei, M. Goldberg, D. Anderson, The development of high-throughput screening approaches for stem cell engineering, Curr. Opin. Cell Biol. 11 (4) (2007) 388–393.

[38] M. Nishikawa, et al., Stable immobilization of rat hepatocytes as hemispheroids onto collagen-conjugated poly-dimethylsiloxane (PDMS) surfaces: importance of direct oxygenation through PDMS for both formation and function, Biotechnol. Bioeng. 99 (6) (2008) 1472–1481.

[39] C. Yi, et al., Microfluidics technology for manipulation and analysis of biological cells, Anal. Chim. Acta 560 (2006) 1–23.

[40] D. Huh, et al., Gravity-driven microhydrodynamics-based cell sorter (microHYCS) for rapid, inexpensive, and efficient cell separation and size-profiling, in: 2nd Annual International IEEE-EMBS Special Topic Conference on Microtechnologies in Medicine & Biology, Madison, WI, 2002, pp. 466–469.

[41] B. Yao, et al., A microfluidic device based on gravity and electric force driving for flow cytometry and fluorescence activated cell sorting, Lab Chip 4 (6) (2004) 603–607.

[42] M. Yamada, M. Seki, Microfluidic particle sorter employing flow splitting and recombining, Anal. Chem. 78 (4) (2006) 1357–1362.

[43] M. Chabert, J.-L. Viovy, Microfluidic high-throughput encapsulation and hydrodynamic self-sorting of single cells, Proc. Natl. Acad. Sci. USA 105 (9) (2008) 3191–3196.

[44] Y. Sun, et al., Design, simulation and experiment of electroosmotic microfluidic chip for cell sorting, Sens. Actuators A 133 (2) (2007) 340–348.

[45] M. Cioffi, et al., A computational and experimental study inside microfluidic systems: the role of shear stress and flow recirculation in cell docking, Biomed. Microdev. 12 (4) (2010) 619–626.

[46] A.Y. Fu, et al., A microfabricated fluorescence-activated cell sorter, Nat. Biotechnol. 17 (1999) 1109–1111.

[47] M. Berger, et al., Design of a microfabricated magnetic cell separator, Electrophoresis 22 (2001) 3883–3892.

[48] H. Suzuki, C.-M. Ho, N. Kasagi, A chaotic mixer for magnetic bead-based micro cell sorter, J. Microelectromech. Syst. 13 (5) (2004) 779–790.

[49] X. Hu, et al., Marker-specific sorting of rare cells using dielectrophoresis, Proc. Natl. Acad. Sci. U.S.A. 102 (44) (2005) 15757–15761.

[50] Y. Li, et al., Continuous dielectrophoretic cell separation microfluidic device, Lab Chip 7 (2007) 239–248.

[51] S.K. Murthy, et al., Size-based microfluidic enrichment of neonatal rat cardiac cell populations, Biomed. Microdev. 8 (3) (2006) 231–237.

[52] J. Moorthy, D.J. Beebe, In situ fabricated porous filters for microsystems, Lab Chip 3 (2003) 62–66.

[53] M.M. Wang, et al., Microfluidic sorting of mammalian cells by optical force switching, Nat. Biotechnol. 23 (2005) 83–87.
[54] T.D. Perroud, et al., Microfluidic-based cell sorting of *Francisella tularensis* infected macrophages using optical forces, Anal. Chem. 80 (16) (2008) 6365–6372.
[55] E. Dufresne, D.G. Grier, Optical tweezer arrays and optical substrates created with diffractive optical elements, Rev. Sci. Instrum. 69 (1998) 1974–1977.
[56] E. Eriksson, et al., A microfluidic system in combination with optical tweezers for analyzing rapid and reversible cytological alterations in single cells upon environmental changes, Lab Chip 7 (2007) 71–76.
[57] D. Howard, et al., The manipulation of live mouse embryonic stem cells using holographic optical tweezers, Eur. Cell. Mater. 16 (3) (2008) 61.
[58] J. Shi, et al., Focusing microparticles in a microfluidic channel with standing surface acoustic waves (SSAW), Lab Chip 8 (2008) 221–223.
[59] T. Franke, et al., Surface acoustic wave actuated cell sorting (SAWACS), Lab Chip 10 (2010) 789–794.
[60] M. Evander, et al., Noninvasive acoustic cell trapping in a microfluidic perfusion system for online bioassays, Anal. Chem. 79 (7) (2007) 2984–2991.
[61] J.F. Spengler, W.T. Coakley, Ultrasonic trap to monitor morphology and stability of developing microparticle aggregates, Langmuir 19 (2003) 3635–3642.
[62] C.H. Chen, et al., Microfluidic cell sorter with integrated piezoelectric actuator, Biomed. Microdev. 11 (6) (2009) 1223–1231.
[63] P. Chen, et al., Microfluidic chips for cell sorting, Front. Biosci. 13 (2008) 2464–2483.
[64] H. Tsutsui, C.-M. Ho, Cell separation by non-inertial force fields in microfluidic systems, Mech. Res. Commun. 36 (1) (2009) 92–103.
[65] J.L. Prieto, et al., Dielectrophoretic separation of heterogeneous stem cell populations, in: 14th International Conference on Miniaturized Systems for Chemistry and Life Sciences, Groningen, The Netherlands, 2010, pp. 890–892.
[66] J. Vykoukal, et al., Enrichment of putative stem cells from adipose tissue using dielectrophoretic field-frow fractionation, Lab Chip 8 (8) (2008) 1386–1393.
[67] R.S.W. Thomas, et al., Trapping single human osteoblast-like cells from a heterogeneous population using a dielectrophoretic microfluidic device, Biomicrofluidics 4 (2) (2010) 022806.
[68] K.-H. Han, A.B. Frazier, Continuous magnetophoretic separation of blood cells in microdevice format, J. Appl. Phys. 96 (2004) 5803–5807.
[69] H. Zhang, K.-K. Liu, Optical tweezers for single cells, J. R. Soc. Interface 5 (24) (2008) 671–690.
[70] A.E. Chiou, et al., Interferometric optical tweezers, Opt. Commun. 133 (1997) 7–10.
[71] B. Sun, Y. Roichman, D.G. Grier, Theory of holographic optical trapping, Opt. Exp. 16 (2008) 15765–15776.
[72] H. Zhang, K.-K. Liu, A.E. Haj, Optical manipulation of stem cells, Eur. Cell. Mater. 16 (3) (2008) 98.
[73] G. Leitz, et al., Stress response in *Caenorhabditis elegans* caused by optical tweezers: wavelength, power, and time dependence, Biophys. J. 82 (4) (2002) 2224–2231.
[74] K.C. Neuman, et al., Characterization of photodamage to *Escherichia coli* in optical traps, Biophys. J. 77 (5) (1999) 2856–2863.
[75] J. Shi, et al., Acoustic tweezers: patterning cells and microparticles using standing surface acoustic waves (SSAW), Lab Chip 9 (2009) 2890–2895.

[76] S. Nagrath, et al., Isolation of rare circulating tumour cells in cancer patients by microchip technology, Nature 450 (2007) 1235–1239.

[77] S. Kaji, et al., Directing the flow of medium in controlled cocultures of HeLa cells and human umbilical vein endothelial cells with a microfluidic device, Lab Chip. 10 (2) (2010) 2374–2379.

[78] Z. Wang, et al., High-density microfluidic arrays for cell cytotoxicity analysis, Lab Chip 7 (2007) 740–745.

[79] M. Yang, C. Li, J. Yang, Cell docking and on-chip monitoring of cellular reactions with a controlled concentration gradient on a microfluidic device, Anal. Chem. 74 (2002) 3991–4001.

[80] Y.-H. Jang, et al., Deep wells integrated with microfluidic valves for stable docking and storage of cells, Biotechnol. J. 6 (2) (2011) 156–164.

[81] C.G. Conant, M.A. Schwartz, C. Ionescu-Zanetti, Well plate-coupled microfluidic devices designed for facile image-based cell adhesion and transmigration assays, J. Biomol. Screen. 15 (2010) 102–106.

[82] A.Y. Lau, et al., Open-access microfluidic patch-clamp array with raised lateral cell trapping sites, Lab Chip 6 (2006) 1510–1515.

[83] P. Garstecki, et al., Formation of monodisperse bubbles in a microfluidic flow-focusing device, Appl. Phys. Lett. 85 (13) (2004) 2649–2651.

[84] J.F. Edd, et al., Controlled encapsulation of single-cells into monodisperse picolitre drops, Lab Chip 8 (2008) 1262–1264.

[85] C. Han, et al., Integration of single oocyte trapping, in vitro fertilization and embryo culture in a microwell-structured microfluidic device, Lab Chip 10 (2010) 2848–2854.

[86] A. Khademhosseini, et al., Molded polyethylene glycol microstructures for capturing cells within microfluidic channels, Lab Chip 4 (5) (2004) 425–430.

[87] D.G. Anderson, S. Levenberg, R. Langer, Nanoliter-scale synthesis of arrayed biomaterials and application to human embryonic stem cells, Nat. Biotechnol. 22 (7) (2004) 863–866.

[88] N.L. Jeon, et al., Generation of solution and surface gradients using microfluidic systems, Langmuir 16 (22) (2000) 8311–8316.

[89] P.J. Hung, et al., Continuous perfusion microfluidic cell culture array for high-throughput cell-based assays, Biotech. Bioeng. 89 (1) (2005) 1–8.

[90] Y.-S. Hwang, et al., Microwell mediated control of embryoid body size regulates embryonic stem cell fate via differential expression of WNT5a and WNT11, Proc. Natl. Acad. Sci. U.S.A. 106 (40) (2009) 16978–16983.

[91] K. Hattori, S. Sugiura, T. Kanamori, Microenvironment array chip for cell culture environment screening, Lab Chip 11 (2011) 212–214.

[92] P.J. Hung, et al., A novel high aspect ratio microfluidic design to provide a stable and uniform microenvironment for cell growth in a high throughput mammalian cell culture array, Lab Chip 5 (2005) 44–48.

[93] T. Smekal, et al., Design, fabrication and testing of thermal components and their integration into a microfluidic device, in: 2002 Inter Society Conference on Thermal Phenomena, San Diego, CA, 2002, pp. 1039–1045.

[94] C. Ke, et al., Single step cell lysis/PCR detection of *Escherichia coli* in an independently controllable silicon microreactor, Sens. Actuators B 120 (2) (2006) 538–544.

[95] E.A. Schilling, A.E. Kamholz, P. Yager, Cell lysis and protein extraction in a microfluidic device with detection by a fluorogenic enzyme assay, Anal. Chem. 74 (8) (2002) 1798–1804.

[96] D.D. Carlo, K.-H. Jeong, L.P. Lee, Reagentless mechanical cell lysis by nanoscale barbs in microchannels for sample preparation, Lab Chip 3 (2003) 287–291.

[97] S.-S. Yun, et al., Mechanical cell lysis chip with ultra-sharp nano-blade array fabricated by crystalline wet etching of (110) silicon, in: 2010 IEEE 23rd International Conference on Micro Electro Mechanical Systems (MEMS), Wanchai, Hong Kong, 2010, pp. 356–359.

[98] H. Lu, M.A. Schmidt, K.F. Jensen, A microfluidic electroporation device for cell lysis, Lab Chip 5 (2005) 23–29.

[99] Q. Ramadan, et al., Simultaneos cell lysis and bead trapping in a continuous flow microfluidic device, Sens. Actuators B 113 (2006) 944–955.

[100] C.-Y. Lee, et al., Integrated microfluidic systems for cell lysis, mixing/pumping and DNA amplification, J. Micromech. Microeng. 15 (2005) 6.

[101] H. Kido, et al., A novel, compact disk-like centrifugal microfluidics system for cell lysis and sample homogenization, Colloids Surf., B 58 (1) (2007) 44–51.

[102] M.T. Taylor, et al., Lysing bacterial spores by sonication through a flexible interface in a microfluidic system, Anal. Chem. 73 (2001) 492–496.

[103] D. Falconnet, et al., Surface engineering approaches to micropattern surfaces for cell-based assays, Biomaterials 27 (16) (2006) 3044–3063.

[104] J.Y. Lim, H.J. Donahue, Cell sensing and response to micro- and nanostructured surfaces produced by chemical and topographic patterning, Tissue Eng. 13 (8) (2007) 1879–1891.

[105] P. Thiebaud, et al., PDMS device for patterned applicaiton of microfluids to neuronal cells arranged by microcontact printing, Biosens. Bioelectron. 17 (1–2) (2002) 87–93.

[106] D.A. Bruzewicz, A.P. McGuigan, G.M. Whitesides, Fabrication of a modular tissue construct in a microfluidic chip, Lab Chip 8 (2008) 663–671.

[107] V. Dhir, et al., Patterning of diverse mammalian cell types in serum free medium with photoablation, Biotechnol. Prog. 25 (2) (2009) 594–603.

[108] W. Pfleging, et al., Laser-assisted modification of polystyrene surfaces for cell culture applications, Appl. Surf. Sci. 253 (23) (2007) 9177–9184.

[109] S.R. Khetani, S.N. Bhatia, Microscale culture of human liver cells for drug development, Nat. Biotechnol. 26 (1) (2008) 120–126.

[110] B. Zhao, J.S. Moore, D.J. Beebe, Surface-directed liquid flow inside microchannels, Science 291 (5506) (2001) 1023–1026.

[111] B. Zhao, J.S. Moore, D.J. Beebe, Principles of surface-directed liquid flow in microfluidic channels, Anal. Chem. 74 (16) (2002) 4259–4268.

[112] H. Kaji, M. Hashimoto, M. Nishizawa, On-demand patterning of protein matrixes inside a microfluidic device, Anal. Chem. 78 (15) (2006) 5469–5473.

[113] H. Kaji, et al., Patterning adherent cells within microchannels by combination of electrochemical biolithography technique and repulsive dielectrophoretic force, Electrochemistry 76 (2008) 555–558.

[114] S. Selimovic, et al., Generating nonlinear concentration gradients in microfluidic devices for cell studies, Anal. Chem. 83 (2011) 2020–2028.

[115] H.P. Chou, M.A. Unger, S.R. Quake, A microfabricated rotary pump, Biomed. Microdev. 3 (2001) 323–330.

[116] R.H. Liu, et al., Passive mixing in a three-dimensional serpentine microchannel, J. Microelectromech. Syst. 9 (2000) 190–197.

[117] A.D. Stroock, et al., Chaotic mixer for microchannels, Science 25 (2002) 647–651.

[118] T. Thorsen, S.J. Maerkl, S.R. Quake, Microfluidic large-scale integration, Science 298 (5593) (2002) 580–584.

[119] E.E. Hui, S.N. Bhatia, Micromechanical control of cell–cell interactions, Proc. Natl. Acad. Sci. U.S.A 104 (14) (2007) 5722–5726.

[120] P.J. Lee, P.J. Hung, L.P. Lee, An artificial liver sinusoid with a microfluidic endothelial-like barrier for primary hepatocyte culture, Biotech. Bioeng. 97 (5) (2007) 1340–1346.

[121] B.J. Kane, et al., Liver-specific functional studies in a microfluidic array of primary mammalian hepatocytes, Anal. Chem. 78 (13) (2006) 4291–4298.

[122] M.J. Jebrail, A.R. Wheeler, Let's get digital: digitizing chemical biology with microfluidics, Curr. Opin. Chem. Biol. 14 (5) (2010) 574–581.

[123] I. Barbulovic-Nad, S.H. Au, A.R. Wheeler, A microfluidic platform for complete mammalian cell culture, Lab Chip 10 (12) (2010) 1536–1542.

[124] J. El-Ali, P.K. Sorger, K.F. Jensen, Cells on chips, Nature 442 (7101) (2006) 403–411.

[125] W. Tan, T.A. Desai, Layer-by-layer microfluidics for biomimetic three-dimensional structures, Biomaterials 25 (7–8) (2004) 1355–1364.

[126] Y.C. Toh, et al., A novel 3D mammalian cell perfusion-culture system in microfluidic channels, Lab Chip 7 (3) (2007) 302–309.

[127] M. Singh, C. Berkland, M.S. Detamore, Strategies and applications for incorporating physical and chemical signal gradients in tissue engineering, Tissue Eng. B Rev. 14 (4) (2008) 341–366.

[128] N. Li Jeon, et al., Neutrophil chemotaxis in linear and complex gradients of interleukin-8 formed in a microfabricated device, Nat. Biotechnol. 20 (8) (2002) 826–830.

[129] J. Pihl, et al., Microfluidic gradient-generating device for pharmacological profiling, Anal. Chem. 77 (13) (2005) 3897–3903.

[130] D.M. Thompson, et al., Dynamic gene expression profiling using a microfabricated living cell array, Anal. Chem. 76 (14) (2004) 4098–4103.

[131] L. Kim, et al., Microfluidic arrarys for logarithmically perfused embryonic stem cell culture, Lab Chip 6 (2006) 394–406.

[132] G.M. Whitesides, et al., Soft lithography in biology and biochemistry, Annu. Rev. Biomed. Eng. 3 (2001) 335–373.

[133] S.-J. Wang, et al., Differential effects of EGF gradient profiles on MDA-MB-231 breast cancer cell chemotaxis, Exp. Cell Res. 300 (2004) 180–189.

[134] D.B. Mountcastle, E. Freire, R.L. Biltonen, Generation of continuous ligand-macromolecule binding isotherms. Use of exponential concentration gradients, Biopolymers 15 (2) (1976) 355–371.

[135] K. Norrby, L. Franzen, A tissue model for the study of cell proliferation in vitro, Dev. Biol. Plant. 16 (1) (1980) 31–37.

[136] C.J. Flaim, et al., Combinatorial signaling microenvironments for studying stem cell fate, Stem. Cells Dev. 17 (1) (2008) 29–39.

[137] B.G. Chung, et al., Human neural stem cell growth and differentiation in a gradient-generating microfluidic device, Lab Chip 5 (2005) 401–406.

[138] N.S. Hwang, S. Varghese, J. Elisseeff, Controlled differentiation of stem cells, Adv. Drug. Deliv. Rev. 60 (2) (2008) 199–214.

[139] V.V. Abhyankar, et al., Human embryonic stem cell culture in microfluidic channels, in: Seventh International Conference on Miniaturized Chemical and Biochemical Analysis Systems, Squaw Valley, CA, 2003, pp. 17–20.
[140] V. Lecault, et al., High-throughput analysis of single hematopoietic stem cell proliferation in microfluidic cell culture arrays, Nat. Methods 8 (2011) 581–586.
[141] Š. Selimović, et al., Microfabricated polyester conical microwells for cell culture applications, Lab Chip 11 (2011) 2325–2332.
[142] L. Kang, et al., Cell confinement in patterned nanoliter droplets in a microwell array by wiping, J. Biomed. Mater. Res. A 93A (2010) 547–557.
[143] E. Kang, et al., Development of a multi-layer microfluidic array chip to culture and replate uniform-sized embryoid bodies without manual cell retrieval, Lab Chip 10 (2010) 2651–2654.
[144] Y.Y. Choi, et al., Controlled-size embryoid body formation in concave microwell arrays, Biomaterials 31 (2010) 4296–4303.
[145] C.-M. Lo, et al., Cell movement is guided by the rigidity of the substrate, Biophys. J. 79 (2000) 144–152.
[146] T. Yeung, et al., Effects of substrate stiffness on cell morphology, cytoskeletal structure, and adhesion, Cell Motil. Cytoskeleton 60 (2005) 24–34.
[147] H.-C. Moeller, et al., A microwell array system for stem cell culture, Biomaterials 29 (2008) 752–763.
[148] Y. Sakai, Y. Yoshiura, K. Nakazawa, Embryoid body culture of mouse embryonic stem cells using microwell and micropatterned chips, J. Biosci. Bioeng. 111 (1) (2011) 85–91.
[149] L. Liu, et al., A micro-channel-well system for culture and differentiation of embryonic stem cells on different types of substrate, Biomed. Microdev. 12 (3) (2010) 505–511.
[150] J. Fu, et al., Mechanical regulation of cell function with geometrically modulated elastomeric substrates, Nat. Methods 7 (9) (2010) 733–736.
[151] C.M. Metallo, et al., The response of human embryonic stem cell-derived endothelial cells to shear stress, Biotech. Bioeng. 100 (4) (2008) 830–837.
[152] L. Kim, et al., A practical guide to microfluidic perfusion culture of adherent mammalian cells, Lab Chip 7 (6) (2007) 681–694.
[153] Y.C. Toh, J. Voldman, Multiplex microfluidic perfusion identifies shear stress mechanosensing mediators in mouse embryonic stem cells, in: 14th International Conference on Miniaturized Systems for Chemistry and Life Sciences, Groningen, The Netherlands, 2010, pp. 10–12.
[154] S.-H. Song, J. Choi, H.-I. Jung, A microfluidic magnetic bead impact generator for physical stimulation of osteoblast cells, Electrophoresis 31 (16) (2010) 2762–2770.
[155] H.W. Wu, et al., A microfluidic device for chemical and mechanical stimulation of mesenchymal stem cells, in: 14th International Conference on Miniaturized Systems for Chemistry and Life Sciences, Groningen, The Netherlands, 2010, pp. 13–15.
[156] W.Y. Sim, et al., A pneumatic micro cell chip for the differentiation of human mesenchymal stem cells under mechanical stimulation, Lab Chip 7 (2007) 1775–1782.
[157] C.-R. Wan, S. Chung, R.D. Kamm, Differentiation of embryonic stem cells into cardiomyocytes in a compliant microfluidic system, Annals. Biomed. Eng. 39 (6) (2011) 1840–1847.

[158] R. Ziv, et al., Micro-electroporation of mesenchymal stem cells with alternating electrical current pulses, Biomed. Microdev. 11 (1) (2009) 95–101.
[159] A. Valero, et al., Gene transfer and protein dynamics in stem cells using single cell electroporation in a microfluidic device, Lab Chip 8 (1) (2008) 62–67.
[160] H. Sauer, et al., Effects of electrical fields on cardiomyocyte differentiation of embryonic stem cells, J. Cell Biochem. 75 (4) (1999) 710–723.
[161] E. Serena, et al., Electrical stimulation of human embryonic stem cells: cardiac differentiation and the generation of reactive oxygen species, Exp. Cell Res. 315 (20) (2009) 3611–3619.
[162] M. Radisic, et al., Functional assembly of engineered myocardium by electrical stimulation of cardiac myocytes cultured on scaffolds, Proc. Natl. Acad. Sci. U.S.A. 101 (52) (2004) 18129–18134.
[163] L. Wang, et al., Characterization and expression of amphioxus ApoD gene encoding an archetype of vertebrate ApoD proteins, Cell Biol. Int. 31 (1) (2007) 74–81.
[164] D.T. Chiu, et al., Patterned deposition of cells and proteins onto surfaces by using three-dimensional microfluidic systems, Proc. Natl. Acad. Sci. U.S.A 97 (6) (2000) 2408–2413.
[165] Y.S. Torisawa, et al., Microfluidic hydrodynamic cellular patterning for systematic formation of co-culture spheroids, Integr. Biol. 1 (11–12) (2009) 649–654.
[166] M. Suzuki, et al., Negative dielectrophoretic patterning with different cell types, Biosens. Bioelectron. 24 (4) (2008) 1043–1047.
[167] D. Xu, J.W. Chow, Y.T. Wang, Effects of turn angle and pivot foot on lower extremity kinetics during walk and turn actions, J. Appl. Biomech. 22 (1) (2006) 74–79.
[168] C. Chen, et al., Simultaneous control of microorganism, disinfection by-products and bio-stability by sequential chlorination disinfection, Huan Jing Ke Xue 27 (1) (2006) 74–79.
[169] R.C. Dutta, A.K. Dutta, Cell-interactive 3D-scaffold; advances and applications, Biotechnol. Adv. 27 (4) (2009) 334–339.
[170] E. Weisenberg, Pocket companion to Robbins pathologic basis of disease, Arch. Pathol. Lab. Med. 124 (10) (2000) 1566.
[171] A.J. Engler, et al., Matrix elasticity directs stem cell lineage specification, Cell 126 (4) (2006) 677–689.
[172] S. Levenberg, et al., Differentiation of human embryonic stem cells on three-dimensional polymer scaffolds, Proc. Natl. Acad. Sci. U.S.A. 100 (22) (2003) 12741–12746.
[173] D.N. Howbrook, et al., Developments in microarray technologies, Drug. Discov. Today 8 (14) (2003) 642–651.
[174] D.G. Anderson, et al., Biomaterial microarrays: rapid, microscale screening of polymer–cell interaction, Biomaterials 26 (23) (2005) 4892–4897.
[175] C.J. Flaim, et al., Combinatorial signaling microenvironments for studying stem cell fate, Stem Cells Dev. 17 (1) (2008) 29–39.
[176] T.G. Fernandes, et al., Three-dimensional cell culture microarray for high-throughput studies of stem cell fate, Biotechnol. Bioeng. 106 (1) (2010) 106–118.

CHAPTER

Vascularization of Microfluidic Hydrogels

8

Joe Tien, Keith H.K. Wong and James G. Truslow
Department of Biomedical Engineering, Boston University, Boston, MA, USA

8.1 Introduction

Current methods for engineering vascularized tissues rely primarily on the self-organization of vascular cells (endothelial cells, pericytes, and smooth muscle cells) or their progenitors in a scaffold and/or on the recruitment of new vessels via release of angiogenic factors [1,2]. While these approaches certainly accelerate the process of graft vascularization, they still require several days to achieve perfusion of the graft. It is difficult to see how thick, densely cellularized grafts (e.g., a myocardial patch) could survive the stagnant transition period that separates *in vitro* culture from *in vivo* perfusion.

In clinical tissue transplantation, grafted tissues are almost always (re)perfused upon completion of the surgical procedure. For instance, successful transfer of a "free flap" between distant regions of the human body requires microvascular anastomoses between the small vessels that feed the tissue and those that reside in the recipient bed [3]. Insetting the graft without establishing such anastomoses usually leads to necrosis of the graft (with the exception of thin tissues such as epidermal grafts).

These issues have prompted us and others to investigate how to create tissues that can be immediately perfused. Our approach is to form scaffolds that contain microfluidic networks that serve both as channels for perfusion and as templates for vascular growth [4–6]. We emphasize that such an approach differs from the "prevascularization" that often results in organization of vascular cells into short segments that have not been shown to sustain flow across the entire scaffold [7,8]. This chapter describes methods to form, vascularize, and optimize microfluidic scaffolds for perfusion of engineered tissues, and proposes a computational algorithm to simplify the design of microfluidic scaffolds for this application.

Microfluidic Cell Culture Systems. ISBN: 978-1-4377-3459-1

8.2 Design criteria for microfluidic scaffolds

What basic features should microfluidic scaffolds possess? First, the channels in these scaffolds must span the entire dimension of the scaffold. In this way, application of a pressure difference across the scaffold can lead to flow within the channels. Although flow is possible even in the absence of channels (since scaffolds are generally porous), such interstitial flows appear to be limited to the *in vitro* setting; to maintain nutritive flows, they require large pressure gradients and generate large interstitial shear stresses [9]. Second, the channels should have dimensions appropriate for the vascular system. Given that we envision eventually perfusing these scaffolds with blood or with a suspension of red blood cells, it would not make sense to design channels narrower than ~5 μm. Third, the scaffolds must promote adhesion and spreading of vascular cells, as we intend to form vessels within the channels. As noted in subsequent sections, the adhesion strength between cells and scaffold channels will affect the design of the microfluidic geometry. Fourth, the scaffold must be able to support passive diffusion of oxygen and solutes from the channel to cells embedded within the scaffold, as well as resist the mechanical stresses induced by perfusion. In particular, the scaffold must not have such a small pore size that it restricts the transport of serum proteins.

We have chosen to make microfluidic scaffolds from hydrogels of extracellular matrix (ECM) proteins, chiefly, type I collagen and fibrinogen. These materials are inherently adhesive to vascular cells, promote the formation of endothelium, and can have pore sizes greater than the sizes of most proteins. By learning how to build microfluidic networks in these materials and how to vascularize the networks, we hope to uncover broad principles that can be applied to the vascularization of non-ECM-based materials.

8.3 Forming and vascularizing microfluidic gels

8.3.1 Forming microfluidic gels

We have taken two approaches to form microfluidic ECM gels. The first one is subtractive, in which a channel forms when a sacrificial material is removed from the gel. Examples include the removal of a cylindrical template to form a single channel, and the dissolution of a micromolded gelatin mesh to form a network [5,10]. The second approach is additive, in which the irreversible adhesion of two gels into a monolithic whole can result in the formation of channels [6]. Both approaches are applicable to a wide variety of materials, although we have used them primarily in ECM gels.

8.3.1.1 Subtractive methods

To form ECM gels that contain single channels, we have used thin stainless steel needles as removable templates (Figure 8.1A) [10,11]. The needles are typically ~120 μm in diameter (we use 02 Japanese gauge acupuncture needles); to obtain

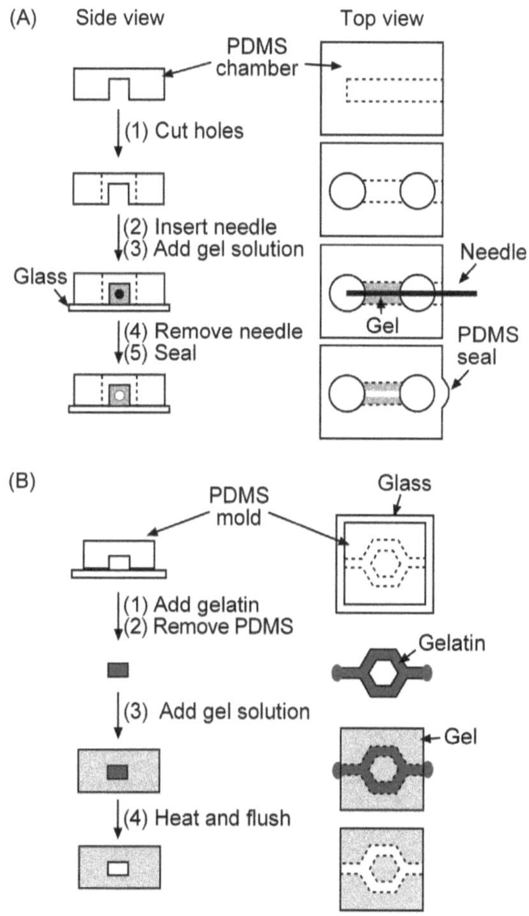

FIGURE 8.1

(A) Needle-based method to form a single channel. (B) Gelatin-based method to form a network of channels.

Source: *Modified from Refs. [5,10] with permission from Elsevier and the Royal Society of Chemistry.*

narrower needles, we etch them for a few minutes in a nickel etchant. We then adsorb a layer of bovine serum albumin (BSA) onto the needles to prevent gels from adhering to them [12]. We finally introduce liquid gel precursor around the BSA-coated needles, taking care that both ends of a needle are left uncovered. Gelation, followed by removal of a needle, yields a gel that contains a channel of the same diameter as the needle.

In practice, various external aids are required to obtain a well-formed channel. The ECM gels that we use can be quite soft and hence can be easy to deform irreversibly. We typically restrain the gels within a mechanical support, so that the gels do not shift during removal of a needle. Likewise, we use a guide to eliminate vibrations and lateral movement of the needle during its removal. With these precautions, this procedure can routinely make open channels as narrow as ~50 μm. Although it is possible to form narrower channels by using thinner needles, we have found that vascularizing such channels is difficult (see Section 8.3.2).

By arranging needles in arrays (either side-by-side or atop each other), it is possible to form gels that contain parallel arrays of channels. Such a configuration mimics the hexagonal vascular geometry analyzed by the physiologist Krogh [13], and its regularity allows for computational models and/or analytical expressions that describe its transport properties (see Section 8.4). These arrays appear to be well suited for perfusing constructs *in vitro*. Because they contain numerous inlets and outlets (one pair per channel), however, it is currently impractical to use them as vascular templates for microsurgical grafting.

To obtain greater versatility in the arrangement of channels within microfluidic gels, we have developed a subtractive lithographic method to replicate patterns inside ECM gels (Figure 8.1B) [5]. The subtractive element in this method is a molded gelatin mesh. We typically generate these meshes by thermosetting concentrated gelatin solutions within standard silicone polydimethylsiloxane (PDMS)-based microfluidic networks. Other methods for making gelatin microstructures (e.g., molding within hollow fibers) are possible, although they lack the versatility of a lithographic approach in the types of patterns generated (Golden and Tien, unpublished data). As with the needles described earlier, we encapsulate the gelatin meshes in ECM gels (taking care to leave the ends free). Since gelatin gels melt at 37°C, heating the encapsulated gels and imposing a pressure gradient across the ECM gel removes the molded structure, leaving behind microfluidic channels that replicate the geometry of the mesh. We have found that gelatin swells noticeably (by 20–40%, depending on preparation conditions), so it is important to design narrower meshes to obtain channels of the desired widths.

The strength and flexibility of gelatin meshes provide interesting opportunities to build thick microfluidic scaffolds. Meshes can be stacked or folded to form overlapping networks that span a thicker volume [5]. Because the lithographic procedure generates two-dimensional (2D) networks, stacked or folded meshes cannot reproduce the highly interconnected three-dimensional (3D) geometries characteristic of vascular beds *in vivo*. Nevertheless, using meshes that possess a treelike network geometry can greatly reduce the number of inlets and outlets in a microfluidic scaffold, which makes microsurgical implantation more plausible.

8.3.1.2 Additive methods

We have also developed an additive procedure to form microfluidic scaffolds [6]. We based this method on the procedures previously developed for forming 3D networks in PDMS devices, in which stacking and bonding of planar PDMS layers led to layer-by-layer construction of a 3D structure [14]. Unlike PDMS, however, hydrogels do not seal tightly to each other; in fact, even the modest forces exerted by living cells are sufficient to pry apart passively adherent gels. Thus, we identified compounds that could be used to bond gels irreversibly.

These compounds were "perturbants" that, if added at sufficiently high concentration, would dissolve the gels. At a subthreshold concentration, perturbants partially solubilize gels and generate locally high concentrations of gel monomer. Upon removal of the perturbant, the gel monomers repolymerize to "glue" passively adherent gels together. Examples of gel-bonding solutes include classical chaotropes such as urea and guanidine, as well as ECM-specific ones such as glycerol (for type I collagen). In practice, we mold gels so that two of them would form a microfluidic network when placed in contact. Addition and removal of perturbant occur by perfusion through these networks.

In principle, this bonding procedure can be applied iteratively to form 3D networks in gels, much as is currently done in PDMS. The practical difficulties in aligning patterned soft gels across large areas, however, have limited its application to date to 2D networks [6].

8.3.2 Vascularizing gels

8.3.2.1 Basic concepts

To use the barren channels in a microfluidic scaffold as templates for the formation of vessels, we introduce suspensions of vascular cells (endothelial cells in most studies, although we have examined cocultures of endothelial and mesenchymal cells) by perfusion [10,15]. As long as the cells do not clog the channels, they will adhere, spread, and proliferate within the channels. Channels of diameter $\geq$100 μm can be seeded routinely; with practice, those of diameter $\geq$50 μm can also be seeded. In channels narrower than 50 μm, clogs invariably form, and these channels can only be vascularized by promoting migration from adjacent larger vessels (Price and Tien, unpublished results). Once channels are seeded, we place them under a pressure difference to induce flow. To maintain the pressure head, we find that manually recirculating the perfusion media is sufficient.

Cells from large vessels (e.g., umbilical vein) and from microvessels are both suitable for vascularization of microfluidic scaffolds, although subtle differences exist in the final structures. In all cases, endothelial cells settle and adhere preferentially to the lower surface of channels. Over the next 1–2 days, the cells migrate along the channel and proliferate to form confluent lumens. The lumens never occlude over time. In vessels made from human umbilical vein endothelial cells, the endothelium initially aligns parallel to the direction of flow; in vessels of human dermal microvascular endothelial cells, the endothelium does not align in any direction under flow.

Vascularization of cylindrical channels leads, not surprisingly, to vessels that have cylindrical lumens [10]. Vascularization of channels that have rectangular cross sections (e.g., scaffolds made by gelatin- or perturbant-based lithographic methods) yield vessels that have rounded lumens, at least in the soft ECM gels we have tested. We believe that the inherent contractility of the seeded cells is partly responsible for generating a surface tension that leads to stress concentration at the sharp edges of the channels and subsequent rounding of the channel shape.

8.3.2.2 Promoting vascular stability and function

Although the open structure of vascularized channels supports initial perfusion, we have found that long-term perfusion is not guaranteed. That is, the microfluidic nature of a scaffold is necessary, but not sufficient, to allow immediate and sustained perfusion. The endothelium is responsible for changes in perfusion, since barren channels can be perfused essentially without limit. The many ways in which endothelium can disrupt the channel profile and thereby alter the flow resistance include: contracting the scaffold, invading the scaffold, and delaminating from the scaffold [10,16,17]. Contraction and invasion typically occur within a few days postseeding, whereas delamination can present over several days.

These changes can be limited, in part, by altering the bulk physical properties of the scaffold [10]. Scaffold contraction results from competition between the cellular contractility and the gel stiffness. We have found that scaffolds of shear modulus of at least several hundred Pascal are rigid enough to resist deformation. Invasion of the scaffold, however, correlates with scaffold porosity. Scaffolds with Darcy permeabilities of at most ~0.1 μm^2 are nonporous enough to resist invasion. In practice, both of these requirements are met in 6–10 mg/ml type I collagen gels that are formed at pH 7–7.5 and ≤25°C.

To understand how to limit endothelial delamination (and thus enable stable long-term perfusion), we have considered the forces and flows within vascularized microfluidic scaffolds in the presence of perfusion [16,17]. In particular, we have proposed that a balance of forces exists at the interface between the endothelium and the channel surface [16]; for stable vessels, the following inequality must hold:

$$\sigma_{\text{adh}} + P_t > \frac{\gamma_{\text{EC}}}{R} \tag{8.1}$$

Here, the interfacial adhesion stress σ_{adh} promotes vascular stability, as does transmural pressure P_t (defined as perfusion pressure in the lumen minus interstitial pressure in the scaffold). Endothelial contractility γ_{EC} induces a destabilizing stress that is inversely proportional to channel radius R. This hypothesis assumes that the delamination that we have observed is purely physical in origin (e.g., is independent of proteases). It predicts that modulation of endothelial contractility and/or perfusion stresses can enhance stability, as described below; we have not investigated altering the adhesivity of the scaffold surface.

8.3.2.3 Modulation of intravascular cyclic adenosine monophosphate (AMP)

It is well known from studies of endothelial cells that endothelial contractility results from activation of the actomyosin filament [18]. Here, the second messenger cyclic AMP (cAMP) serves to "relax" endothelial cells by activating protein kinase A, which phosphorylates the GTPase RhoA, and thus inhibits generation of tension [19]; nonclassical relaxation mechanisms that bypass protein kinase A also exist [20,21]. We have thus investigated to what extent elevation of cAMP levels could stabilize vessels in microfluidic scaffolds [16]. Since the cell membrane is impermeable to cAMP, we used an esterified analog (dibutyryl cAMP) and a phosphodiesterase inhibitor (Ro-20-1724); these compounds were added directly to the perfusate. As expected, addition of cAMP-elevating compounds led both to decreased contractility (as measured by radial strain of the lumen) (Figure 8.2A) and increased stability (Figure 8.2B). At the highest concentrations of dibutyryl cAMP (400 μM) with Ro-20-1724 (20 μM), vessels did not delaminate over the span of 2 weeks; the longest we have kept vessels under these conditions is ~2 months.

Strangely, the dose-dependence of reduced contractility and increased vascular stability did not match. Whereas the loss of contractility required the highest concentrations of cAMP-elevating compounds, stabilization was already apparent at lower concentrations of dibutyryl cAMP (80 μM). Thus, loss of contractility is not necessary for stabilization.

To determine if other effects of elevated cAMP levels might stabilize vessels, we examined the barrier function and turnover rates. Time-lapse videos of the efflux of fluorescent solutes from the perfusate into the scaffold revealed that the vessels were extremely leaky in the absence of cAMP-elevating agents (Figure 8.2C). Addition of moderate concentrations of dibutyryl cAMP (80 μM) was sufficient to largely eliminate these leaks. Turnover rates in the absence of cAMP-elevating compounds were small and consistent with the nonzero proliferation rates observed in postconfluent endothelium in culture. Only in the presence of high concentrations of cAMP-elevating compounds did the turnover rates decrease to essentially zero. These dose-dependences suggest that leaks may destabilize vessels in microfluidic scaffolds, whereas cell turnover (which itself can induce leaks [22,23]) is not a major factor. Indeed, computational models of the pressure profile within the scaffold near a leak showed that a leak allowed lumenal pressure to "diffuse" into the scaffold, thereby elevating scaffold pressure and reducing transmural pressure. These models indicated that the large numerous leaks observed in the absence of cAMP-elevating compounds resulted in a loss of 1–2 cm H_2O of stabilizing transmural pressure, which may be sufficient to induce delamination of the endothelium from the scaffold wall.

8.3.2.4 Modulation of perfusion stresses

We have also investigated to what extent changes in perfusion stresses affect vascular stability [17]. Our original motivation for these studies was to understand an early observation that vessels under higher flow invariably sustained perfusion for a longer period. As seen with cAMP-mediated stabilization, flow-induced stabilization was accompanied by elimination of leaks from the vessel wall (Figure 8.2D). The same mechanism of stabilization (loss of leaks to increase transmural pressure) appears to apply whether flow or cAMP levels are increased.

To elucidate the detailed mechanism behind flow-induced stabilization, we considered the stresses induced by flow. We calculated that increases in flow rate resulted in simultaneous increases in average lumenal pressure, shear stress, and transmural pressure. By altering lumenal pressures without changing flow rate,

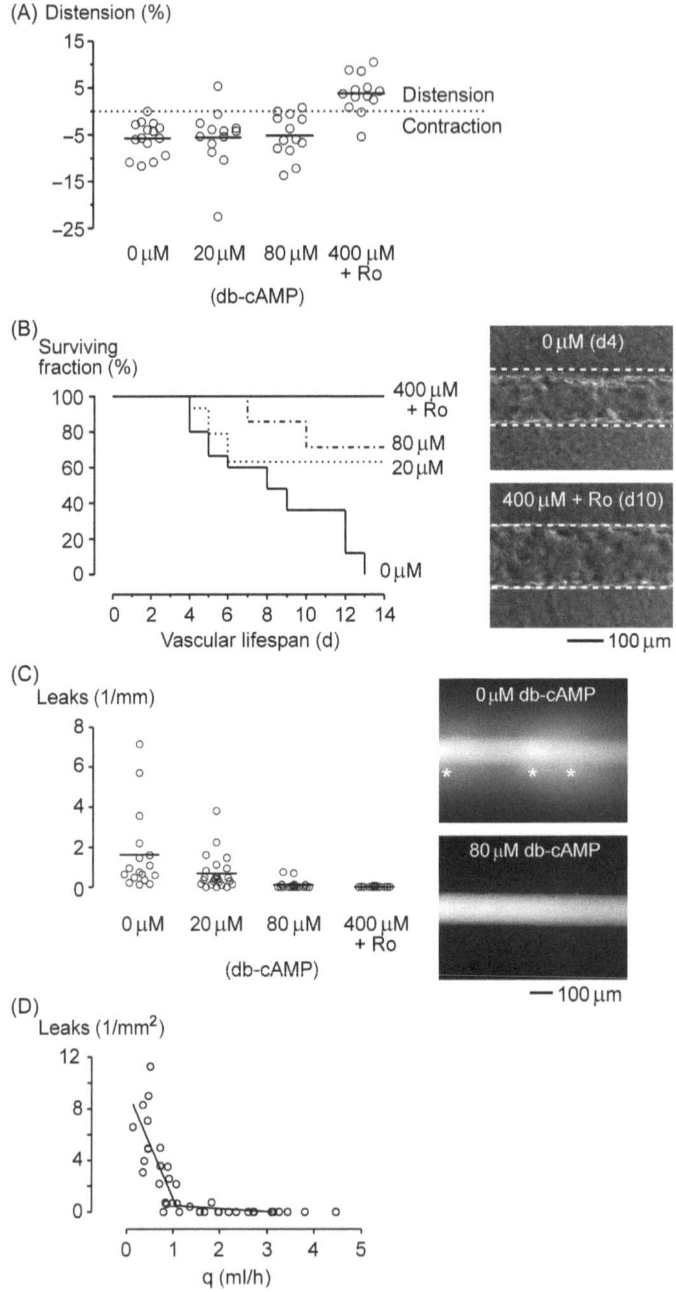

FIGURE 8.2

(A) Increases in cAMP levels reduce or reverse contractility of the vessel wall.
(B) Increases in cAMP levels prevent endothelial delamination and prolong lifespan of the

(Continued)

we found that the vascular phenotype did not depend on hydrostatic pressure. To disentangle the effects of changes in shear stress and transmural pressure, we devised scaffolds that resulted in regions of different shear, but similar transmural pressure, in the same vessel. In some cases, we placed nonvascularized empty channels near a vessel to reduce transmural pressure to zero. Our data indicated that shear stress was the main controller of barrier function; vessels that were exposed to shears above 10–15 dyn/cm^2 were leakproof. Transmural pressure, however, was the main direct controller of vascular stability, and we found that transmural pressure needed to be positive to obtain stable vessels.

These findings imply that high flow affects stability in two complementary ways: First, by inducing high transmural pressure, high flow directly stabilizes vessels. Second, by inducing high shear, high flow eliminates leaks, and thereby allows a positive transmural pressure to be maintained.

These mechanistic insights, along with those obtained from cAMP-elevation studies, are summarized in Figure 8.3. Note that leaky vessels tend to be unstable, and we believe this correlation holds in general (see Section 8.6.1 for another example). Long-term perfusion in vascularized microfluidic scaffolds is aided by good vascular barrier function.

8.4 Design considerations

To complement the experimental studies described earlier, we have initiated computational modeling of the transport within vascularized microfluidic scaffolds [24,25]. Here, the goal is to provide straightforward algorithms to design microfluidic scaffolds for a desired outcome. In particular, we would like to design these scaffolds so that perfusion of the vessels can maintain the oxygen concentration within the scaffold and transmural pressure at the vessel surface above their respective thresholds. These thresholds can be measured experimentally; for transmural pressure, the threshold will depend on the contractility of the endothelium and the adhesive strength between endothelium and scaffold, as shown by Eq. (8.1).

8.4.1 Perfusion

When analyzing the perfusion properties of a specific microfluidic scaffold, we not only wish to know whether the network geometry is sufficient for oxygen

◀ vessels. *Right*, representative images of vessels under long-term perfusion. Dotted lines indicate walls of the collagen channels. (C) Increases in cAMP levels reduce the number of leaks observed in the vessel wall. *Right*, fluorescence images of vessels perfused with 0 or 80 μM db-cAMP. Asterisks indicate locations of leaks. (D) Increases in lumenal flow rate reduce the leakiness of the vessel wall.

Source: *Modified from Refs. [16,17] with permission from Elsevier.*

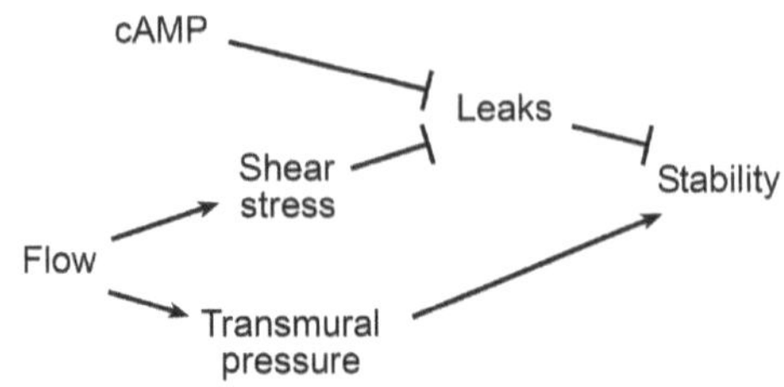

FIGURE 8.3

Mechanisms that affect the mechanical stability of the vessel wall. Arrows (→) indicate positive interactions. Perpendicular segments (⊣) indicate inhibitory interactions.

transport, but also to determine which designs are "better" than others. For instance, inflating the diameter of a vessel (e.g., 1 mm) would certainly lead to extremely large fluid flows and large oxygen transfer coefficients. Such a vascular configuration, however, seems artificial to us. It is important to choose an optimizing function that does not lead to extreme vascular designs. Among the various optimizing functions that we have considered include vascular volume fraction (i.e., the fraction of the tissue that consists of vessel lumen), efficiency of oxygen extraction (i.e., ratio of outflow to inflow oxygen concentration), and perfusion power. A vascular design is considered better if it leads to lower vascular volume fraction, higher oxygen extraction efficiency, or lower perfusion power. In all cases, we constrain the driving perfusion pressures at the inlets and outlets of the scaffold.

As a first step, we have analyzed the effectiveness of oxygen transport in a parallel hexagonal array, since the many symmetry planes in this model enable reduction of the computational geometry [25]. In this work, we used vascular volume fraction as the optimizing function. Fluid flow within the lumens was governed by the steady-state Navier–Stokes equations:

$$\rho(\mathbf{v}_{\text{vessel}} \cdot \nabla)\mathbf{v}_{\text{vessel}} = \eta \nabla^2 \mathbf{v}_{\text{vessel}} - \nabla P_{\text{vessel}} \tag{8.2}$$

Fluid flow within the scaffold obeyed Darcy's law:

$$\mathbf{v}_{\text{scaffold}} = -K \nabla P_{\text{scaffold}} \tag{8.3}$$

The two flows were coupled at the vascular surface by Starling's law for the filtration velocity (neglecting oncotic terms):

$$v_n = L_P (P_{\text{vessel}} - P_{\text{scaffold}}) \tag{8.4}$$

Finally, transport of oxygen within the lumen and in the scaffold was governed by steady-state reaction–convection–diffusion equations:

$$\mathbf{v}_{\text{scaffold}} \cdot \nabla c_{\text{scaffold}} = D_{O_2} \nabla^2 c_{\text{scaffold}} - q_{O_2} \tag{8.5}$$

$$\mathbf{v}_{\text{vessel}} \cdot \nabla c_{\text{vessel}} = D_{O_2} \nabla^2 c_{\text{vessel}} \tag{8.6}$$

Here, ρ and η are the density and viscosity of the perfusate; $\mathbf{v}_{\text{vessel}}$ and $\mathbf{v}_{\text{scaffold}}$, P_{vessel} and P_{scaffold}, and c_{vessel} and c_{scaffold} are the fluid velocities, fluid pressures, and oxygen concentrations in the vessels and scaffold; K is the hydraulic conductivity of the scaffold; L_P is the endothelial hydraulic conductivity; D_{O2} is the oxygen diffusivity, and q_{O2} is the oxygen consumption rate per volume of scaffold.

Each combination of vessel diameter and spacing yielded a minimum oxygen concentration in the scaffold. Using a parametric sweep, we systematically found combinations for which the minimum oxygen concentration equaled a set threshold (5% O_2). Of these combinations, the one that had the lowest vascular volume fraction was designated the optimal design. By altering the other set points in the model (scaffold thickness, oxygen consumption rate, etc.), we found optimal designs as a function of various parameters (Figure 8.4).

Surprisingly, the optimal vessel diameters were on the order of ~100 μm, which is much larger than the diameters of capillaries *in vivo*. Thus, for perfusion of engineered tissues, it may not be necessary to replicate the exact geometry and scale of vascular beds *in vivo*. We note that ~100-μm diameter vessels can be formed using needle-, gelatin-, or perturbant-based microfluidic scaffolds, so experimental tests of these models are possible.

The unusually symmetric vascular geometry suggested that approximate analytical solution may be possible. When we neglected all axial diffusion of oxygen, filtration and interstitial flow, analytical optimization yielded the following implicit expressions for the optimal diameter D and spacing H:

$$\frac{\alpha}{2} - \frac{2\beta L}{\alpha}\left[\left(\frac{H}{D}\right)^2 - 1\right]\left\{\beta\gamma\left[\left(\frac{H}{D}\right)^2 - 1\right] + \delta\left[\left(\frac{H}{D}\right)^2 \ln\left(\frac{H}{D}\right)^2 - \left(\frac{H}{D}\right)^2 + 1\right]\right\} = 0 \tag{8.7}$$

$$D^2 = \frac{8\beta L}{\alpha}\left[\left(\frac{H}{D}\right)^2 - 1\right] \tag{8.8}$$

Here, L is the scaffold thickness, whereas α, β, γ, and δ are functions of the remaining parameters (perfusion pressures, inlet oxygen concentration, etc.) (see Ref. [25] for more details). Comparisons of the approximate optima with the computationally determined ones showed agreement to within 5% in nearly all cases. We propose to use these expressions in the rational design of microfluidic scaffolds for perfusion of tissues (see Section 8.5).

8.4.2 Drainage

Our experimental work has shown that transmural pressure must exceed a threshold value for vessels to be stable against delamination from the scaffold. To promote vascular stability, we must find ways to control transmural pressure in a

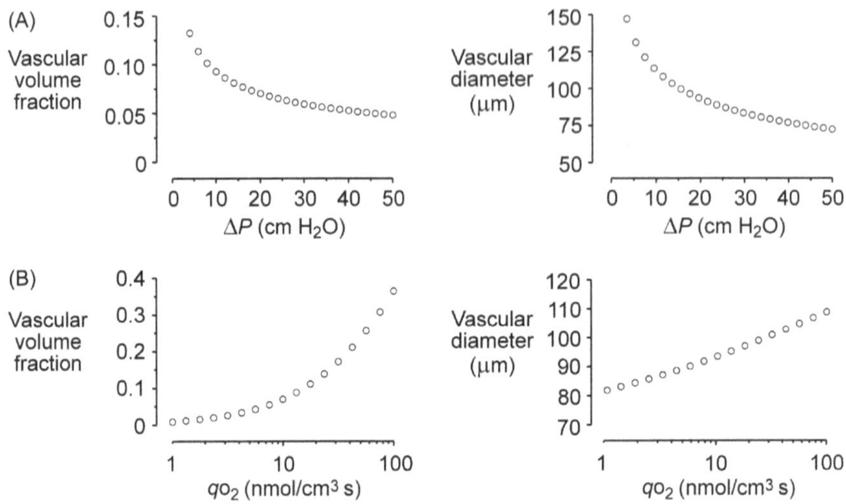

FIGURE 8.4

(A) The effect of perfusion pressure on optimal volume fraction and vessel diameter. (B) The effect of oxygen consumption rate per volume on optimal volume fraction and vessel diameter.

Source: *Modified from Ref. [25] with permission from the American Institute of Physics.*

vascularized scaffold. Since the pressures in the vessels are largely preset by perfusion pressures, the primary way to alter transmural pressure is to control the pressure within the scaffold ($P_{scaffold}$). We have analyzed the ability of nonvascularized empty channels ("drains") within the scaffold to modulate scaffold pressure and thereby maintain transmural pressure [24].

As with the models of perfusion, we considered a hexagonal array of vessels. Here, however, we computationally switched a subset of vessels to drains by deleting the endothelium and by setting the inlet and outlet pressures to a drainage pressure (lower than the perfusing pressures) (Figure 8.5A). Although such a vascularized, drained scaffold would be difficult to form experimentally—it would be easier if the drains were all perpendicular to the vessels—this model of parallel vessels and drains has the advantage of computational tractability.

Since we did not consider oxygen transport in these models, the equations to be solved were Eqs. (8.2)–(8.4), with the hydraulic conductivity L_P made infinite at the walls of the drainage channel. For each set of parameters (perfusion pressures, drainage pressure, etc.), we calculated the lowest transmural pressure in the model (Figure 8.5B). Instead of optimizing the drainage design (e.g., by finding the lowest drainage volume fraction that maintained transmural pressure above a threshold), we wanted to determine if simplifying design rules existed. In particular, would it be possible to treat a drain as if it had a "radius-of-action", within which all vessels would have transmural pressure above threshold and thus be stable?

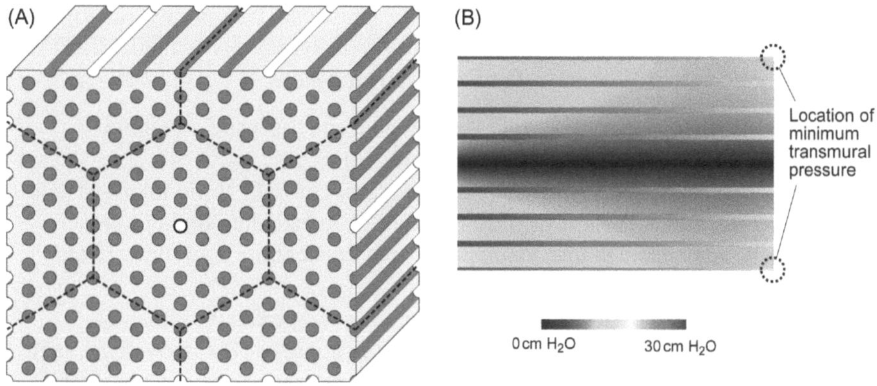

FIGURE 8.5

(A) Cutaway view of a perfused tissue construct with drainage channels. Vessels are gray, while drainage channels are white. Dashed lines represent planes of symmetry that can be exploited during computational analysis. (B) Cross-sectional map of interstitial pressure and lumenal pressure in a representative tissue construct. The drainage channel is in the middle, with four vessels on either side. Flow is from left to right. The circle denotes the point of minimum transmural pressure (8.7 cm H_2O in this model).

Source: *Modified from Ref. [24] with permission from Elsevier.*

We found that it is always possible (e.g., by lowering the drainage pressure) to cause transmural pressure to meet a threshold value for any given vascular and drainage configuration. Moreover, for large ratios of numbers of vessels to drains, the transmural pressure profile becomes a function of the distance from the drain only. For small ratios of numbers of vessels to drains (i.e., a high density of drainage channels), however, interactions between drains raises the overall transmural pressure; in a sense, having more drains increases their radii-of-action.

We found that the effectiveness of a drain in increasing transmural pressure depends largely on the vascular surface area per volume of scaffold, rather than on the individual vascular geometric parameters *per se*. This result allows one to decouple the design of perfusion and drainage systems, in that a given perfusion network (which may have a complex interconnected geometry) can be analyzed for drainage by considering an equivalent parallel vascular array with the same surface area per volume. This result also implies that the relative orientation of drains and vessels is not critical in determining whether drainage is sufficient.

8.5 Design algorithm

We wish to unify these experimental and computational results to answer the following question: Given a particular scaffold and desired seeding density (and, thus, oxygen consumption rate), how does one design the microfluidic geometry of the scaffold to enable long-term perfusion upon vascularization? Our proposed design algorithm consists of the following steps:

1. Determine or select the threshold oxygen concentration and threshold transmural pressure desired in the scaffold.
2. Choose reasonable perfusion pressures (e.g., 40 cm H_2O at the inlet and 20 cm H_2O at the outlet) and use Eqs. (8.7) and (8.8) to calculate optimal vessel diameters and spacings that maintain oxygen concentration in the scaffold above threshold.
3. From the optimized vascular geometry and chosen perfusion pressures, calculate the shear stress in the vessels. If it is above ~10–15 dyn/cm^2, then the scaffolds can be perfused without leaks. If the shear stress is below this threshold, then special care must be taken to ensure that the vessels do not leak under perfusion (e.g., by supplementing the perfusate with large amounts of cAMP-elevating compounds).
4. Use computational models to determine the maximum drainage spacing for a chosen drainage pressure (e.g., 0 cm H_2O) so that transmural pressure is above threshold. We have found computationally that drainage diameter is only a weak determinant of drainage capacity.

It is important to note that this algorithm yields designs with no built-in safety margin. In general, using more vessels will always improve the perfusion capacity, at the cost of increasing vascular volume fraction; for a given vascular design, using more drains will always improve the drainage capacity. Likewise, this algorithm offers a specific class of microfluidic designs, in which the vessels are parallel and evenly spaced, as are the drains. If more complex geometries (e.g., highly interconnected networks) are desired, then the algorithm can be used to generate starting points for the vascular and drainage design.

8.6 Summary

This chapter summarizes our work to date on forming, vascularizing, and optimizing microfluidic hydrogels for use as tissue scaffolds. Microfluidic scaffolds represent a promising new class of biomaterials that has the potential to directly address the issue of tissue vascularization. We have developed methods to form microfluidic ECM gels, and some of these methods should be suitable for polysaccharide or synthetic polymer gels.

By studying the properties of single vessels (made from scaffolds that contain a single channel), we have deduced several guiding principles for stable vascularization. First, leaks can cause eventual delamination of the vessel wall from the scaffold. Second, delamination is consistent with a physical process governed by an excess of destabilizing over stabilizing stresses. Third, signals that eliminate leaks (high cAMP levels, high shear stress) favor stable perfusion. Fourth, transmural pressure must exceed a threshold value to maintain stability.

By programming these principles into computational models, we have developed an algorithm for microfluidic design. It is important to ensure not only that perfusion is sufficient to maintain tissue viability, but also that drainage is sufficient to maintain vascular stability. We have obtained approximate analytical expressions that yield optimal vascular geometries for perfusion. For drainage, at best we can reduce the computational geometry to a simpler one that contains parallel channels.

8.6.1 Future directions

What obstacles remain toward application of microfluidic scaffolds in tissue reconstruction? Below, we list those we feel are most critical:

First, these experimental methods should be extended to other materials of biomedical interest. We think fibrin gels, alginates, and polyethylene glycol gels represent good places to explore, especially if analyzed in clinically approved formulations.

Second, the studies should make use of more than just endothelial cells. Although endothelial tubes and networks may provide the bare minimum for perfusion with blood, it would be beneficial to incorporate pericytes or smooth muscle cells into the vessel wall. Likewise, the presence of oxygen-consuming parenchymal cells may alter the sensitivity of the vessel wall to cAMP, shear stress, or transmural pressure, so measurements of these thresholds in scaffolds that contain both vascular and nonvascular cells is essential.

Third, the network design should be limited to a small number of inlets and outlets. This requirement will help to make surgical anastomosis practical; in clinical reconstructive surgery, the grafted tissue typically has just a pair of feeding vessels. Probably the simplest way to organize such designs is to make them modular, by coupling treelike portions to an exchange region that consists of parallel arrays of vessels. Computational models should help greatly in predicting which microfluidic designs are appropriate.

New ways to stabilize vessels in microfluidic scaffolds are also welcome. Perfusion with whole blood may reveal cell-borne factors that aid in perfusion. We have identified the macromolecular content of the perfusate as an additional stabilizing factor [26]. It would be helpful to obtain a broad set of stabilizing signals that would provide the experimentalist with some freedom in selecting the most practical way to perfuse a vascularized microfluidic scaffold.

Acknowledgments

This work was supported by the National Institute of Biomedical Imaging and Bioengineering under grant EB005792 and the National Heart, Lung, and Blood Institute under grant HL092335.

References

[1] Z. Lokmic, G.M. Mitchell, Engineering the microcirculation, Tissue Eng. B 14 (2008) 87–103.

[2] M. Lovett, K. Lee, A. Edwards, D.L. Kaplan, Vascularization strategies for tissue engineering, Tissue Eng. B 15 (2009) 353–370.

[3] R.K. Khouri, Avoiding free flap failure, Clin. Plast. Surg. 19 (1992) 773–781.

[4] C.M. Nelson, J. Tien, Microstructured extracellular matrices in tissue engineering and development, Curr. Opin. Biotechnol. 17 (2006) 518–523.

[5] A.P. Golden, J. Tien, Fabrication of microfluidic hydrogels using molded gelatin as a sacrificial element, Lab Chip 7 (2007) 720–725.

[6] G.M. Price, K.K. Chu, J.G. Truslow, M.D. Tang-Schomer, A.P. Golden, J. Mertz, et al., Bonding of macromolecular hydrogels using perturbants, J. Am. Chem. Soc. 130 (2008) 6664–6665.

[7] R. Montesano, L. Orci, P. Vassalli, *In vitro* rapid organization of endothelial cells into capillary-like networks is promoted by collagen matrices, J. Cell Biol. 97 (1983) 1648–1652.

[8] A.F. Black, F. Berthod, N. L'Heureux, L. Germain, F.A. Auger, *In vitro* reconstruction of a human capillary-like network in a tissue-engineered skin equivalent, FASEB J. 12 (1998) 1331–1340.

[9] M. Radisic, L. Yang, J. Boublik, R.J. Cohen, R. Langer, L.E. Freed, et al., Medium perfusion enables engineering of compact and contractile cardiac tissue, Am. J. Physiol. Heart Circ. Physiol. 286 (2004) H507–H516.

[10] K.M. Chrobak, D.R. Potter, J. Tien, Formation of perfused, functional microvascular tubes *in vitro*, Microvasc. Res. 71 (2006) 185–196.

[11] G.M. Price, J. Tien, Subtractive methods for forming microfluidic gels of extracellular matrix proteins, in: S.N. Bhatia, Y. Nahmias (Eds.), Microdevices in Biology and Engineering, Artech House, Boston, MA, 2009, pp. 235–248.

[12] M.D. Tang, A.P. Golden, J. Tien, Molding of three-dimensional microstructures of gels, J. Am. Chem. Soc. 125 (2003) 12988–12989.

[13] A. Krogh, The number and distribution of capillaries in muscles with calculations of the oxygen pressure head necessary for supplying the tissue, J. Physiol. 52 (1919) 409–415.

[14] D.T. Chiu, N.L. Jeon, S. Huang, R.S. Kane, C.J. Wargo, I.S. Choi, et al., Patterned deposition of cells and proteins onto surfaces by using three-dimensional microfluidic systems, Proc. Natl. Acad. Sci. U.S.A. 97 (2000) 2408–2413.

[15] G.M. Price, J. Tien, Methods for forming human microvascular tubes in vitro and measuring their macromolecular permeability, in: A. Khademhosseini, K.-Y. Suh, M. Zourob (Eds.), Biological Microarrays (Methods in Molecular Biology), vol. 671, Humana Press, Totowa, NJ, 2011, pp. 281–293.

[16] K.H.K. Wong, J.G. Truslow, J. Tien, The role of cyclic AMP in normalizing the function of engineered human blood microvessels in microfluidic collagen gels, Biomaterials 31 (2010) 4706–4714.

[17] G.M. Price, K.H.K. Wong, J.G. Truslow, A.D. Leung, C. Acharya, J. Tien, Effect of mechanical factors on the function of engineered human blood microvessels in microfluidic collagen gels, Biomaterials 31 (2010) 6182–6189.

[18] Z.M. Goeckeler, R.B. Wysolmerski, Myosin light chain kinase-regulated endothelial cell contraction: the relationship between isometric tension, actin polymerization, and myosin phosphorylation, J. Cell Biol. 130 (1995) 613–627.
[19] Z.M. Goeckeler, R.B. Wysolmerski, Myosin phosphatase and cofilin mediate cAMP/cAMP-dependent protein kinase-induced decline in endothelial cell isometric tension and myosin II regulatory light chain phosphorylation, J. Biol. Chem. 280 (2005) 33083–33095.
[20] X. Cullere, S.K. Shaw, L. Andersson, J. Hirahashi, F.W. Luscinskas, T.N. Mayadas, Regulation of vascular endothelial barrier function by Epac, a cAMP-activated exchange factor for Rap GTPase, Blood 105 (2005) 1950–1955.
[21] M.J. Lorenowicz, M. Fernandez-Borja, M.R.H. Kooistra, J.L. Bos, P.L. Hordijk, PKA and Epac1 regulate endothelial integrity and migration through parallel and independent pathways, Eur. J. Cell Biol. 87 (2008) 779–792.
[22] S.-J. Lin, K.-M. Jan, G. Schuessler, S. Weinbaum, S. Chien, Enhanced macromolecular permeability of aortic endothelial cells in association with mitosis, Atherosclerosis 73 (1988) 223–232.
[23] S.-J. Lin, K.-M. Jan, S. Chien, Role of dying endothelial cells in transendothelial macromolecular transport, Arteriosclerosis 10 (1990) 703–709.
[24] J.G. Truslow, G.M. Price, J. Tien, Computational design of drainage systems for vascularized scaffolds, Biomaterials 30 (2009) 4435–4443.
[25] J.G. Truslow, J. Tien, Perfusion systems that minimize vascular volume fraction in engineered tissues, Biomicrofluidics 5 (2011) 022201.
[26] A.D. Leung, K.H.K Wong, J. Tien, Plasma expanders stabilize human microvessels in microfluidic scaffolds, J. Biomed. Mater. Res. A 100 (2012) 1815–1822.

CHAPTER

Microfluidic Vascular Networks for Engineered Tissues

9

Leon M. Bellan, Holly Chamberlain, Diana Wu and Robert Langer
Koch Institute for Integrative Cancer Research, Massachusetts Institute of Technology, MA, USA

9.1 Introduction

The field of tissue engineering is devoted to forming materials that are able to functionally mimic natural tissue and organs [1,2]. These engineered materials in many cases are not exact replicas of natural tissue (which could be produced by decellularizing natural tissue followed by seeding of the desired cell types) but are fabricated so as to provide enough functionality (structural, biochemical, etc.) to serve as a clinically useful replacements for damaged tissue or as *in vitro* models for scientific study or drug development. The ability to manufacture "off-the-shelf" tissue and even organ replacements that can support cell growth is extremely attractive but far from trivial. Cell survival and behavior is particularly dependent on a wide variety of environmental parameters, including the mechanical properties of the surrounding materials, the presence or absence of biomolecules and other cells, local pH and temperature, and the presence of various ions and dissolved gasses. Moreover, cells must be able to exchange biochemicals and gasses with their surrounding environment in order to survive; however, a cell's ability to access these necessary materials is limited by diffusion of these materials through the extracellular environment. The diffusion-limited distance between a cell and a source of nutrients and gasses is typically on the scale of 100 μm [3]. Cells further than this limit from an appropriate nutrient and gas source will die. To overcome this limit, vascular networks provide advection of materials to actively transport biochemicals and gasses throughout the body.

In order to introduce vasculature into the engineered tissues, several strategies have been adopted [3,4]. Growth factors have been used to encourage new vessels to sprout from preexisting vessels in native tissue, a process known as angiogenesis [4–7]. Alternatively, cells that can differentiate into vessel forming cells can be transplanted to the area where vascularization is desired. Depending on the application, either method can be used. However, for many applications,

Microfluidic Cell Culture Systems. ISBN: 978-1-4377-3459-1

particularly those requiring tissue thicker than roughly 100 μm, the vasculature must be present within the scaffold (i.e., a "prevascularized scaffold") for viable cell growth. Such a scaffold can then be used to support cells that are either pre-seeded within the scaffold or that invade from surrounding native tissue, as it would provide a roughly uniform distribution of oxygen and nutrients throughout the scaffold thickness. This is likely to decrease the failure rate of implantation and increase the integration and function of the engineered tissue.

There are several challenges that must be addressed when fabricating a prevascularized artificial tissue construct. Of these, perhaps the two most important are scaling the microchannel network fabrication into the third dimension to produce a complex three-dimensional (3D) fluidic system that mimics a natural microvascular system and producing a vascular network within materials that are appropriate for artificial tissue.

9.2 3D Microfluidics fabrication techniques

Several studies have shown *in vitro* that cells cultured in 3D environments behave differently from those grown on surfaces in 2D culture [8–12]. Cells take local cues from mechanical forces due to interaction with the surrounding environment (whether natural or engineered) in addition to biochemical signals; thus growing cells in a 3D matrix will give a different result from cells growing on a flat surface. Not only has scientific study demonstrated the importance of 3D cell culture, but clinical work indicates the need for better tissue replacements that are either more easily vascularized by host ingrowth or prevascularized. For example, engineered skin replacements for patients with extensive burns often fail because of a delay or lack of vascularization in the scaffold [13,14]. The ability to engineer other, thicker organs that contain a higher level of functional 3D complexity is particularly dependent on having a robust set of processes to produce fluidic channels in a 3D architecture.

It is necessary to understand the growth of cells in engineered scaffolds not only for regenerative purposes but also to understand various pathological mechanisms, such as the development and propagation of tumors in controlled biomimetic systems. For example, researchers have found that using a 3D model for mesothelioma tumors allowed for better evaluation of cellular apoptotic resistance. Tumor cell apoptotic resistance is usually studied in 2D culture, but it was found that in 3D structures, cells have increased resistance [15]. In general, the fact that tumor development is profoundly linked to angiogenesis strongly supports the use of 3D culture techniques in which these processes can be better replicated and manipulated [16].

In order to create thick tissues and organs that can be used for both clinical applications and scientific study, engineers must create scaffolds containing a vascular network that penetrates the scaffold volume. Thus, there has been a push for the development of 3D microfluidic systems as the needs to understand tissue-scale biology and to develop organ assist and replacement devices grow. By transitioning from 2D to 3D fluidic systems, it becomes possible to dramatically scale up the volume of cell supporting artificial tissues. In order to create accurate 3D tissues, both traditional, such as microfabrication and solid freeform fabrication (SFF), and nontraditional fabrication techniques have been investigated.

9.2.1 Microfabrication of vasculature

Traditional microfabrication techniques, derived from the semiconductor industry, were some of the earliest processes used to produce microfluidic devices, including artificial vascular systems. These techniques are well suited for this purpose because the feature sizes and the extent of the patterns produced are in the range required for prototypical artificial vasculature. Moreover, the scalability (in 2D) and reproducibility of these techniques have been validated by extensive usage in the microelectronics industry.

Photolithography, the process used to create integrated circuits, can also be used to create patterned microchannels. A photomask is used to control the light used to induce chemical reactions in a photosensitive thin film ("photoresist"). The exposed thin films are chemically treated to either remove the exposed areas (positive tone) or the unexposed areas (negative tone). The pattern from the photomask can then be etched into the substrate (typically a silicon wafer) using the photoresist as a protective layer. The etched substrate can be used either as the channel system itself or as a template for another material (which will have the inverse structure of the substrate). Etched substrates may also be used for soft lithography processes, wherein elastomeric stamps or molds are used to create structures. This process is illustrated in Figure 9.1. The most common material used to create soft lithography molds is polydimethylsiloxane (PDMS), introduced by Whitesides and coworkers [17–19]. Flexible and robust PDMS templates are quickly and easily produced from silicon master molds, a process that has effectively enabled the dramatic growth of the field of microfluidics. PDMS can be used either as a template to form patterns in other materials or as the matrix itself, depending on the application. Channels patterned with photolithography have been used to form a wide range of devices, including artificial vascular systems [20–23].

Traditional microfabrication techniques inherently create 2D vascular patterns; the channel system exists in a single plane and there is no complexity to the channel architecture in the third dimension. To resolve this problem, multiple slabs of patterned material can be stacked, creating a 3D structure. This process has been used to produce 3D knots, basket weave, and other complex 3D fluidic structures [24] as well as to solve mathematical problems using fluidic devices [25]. This process has been exploited to produce 3D microfluidic systems that mimic vasculature as well. Leclerc et al. [26,27] used several stacked layers of microfluidic channels in PDMS to perfuse cell cultures of HepG2 and 3T3–L1 cells. Borenstein et al. [28] have also exploited the multilayer stacking technique and

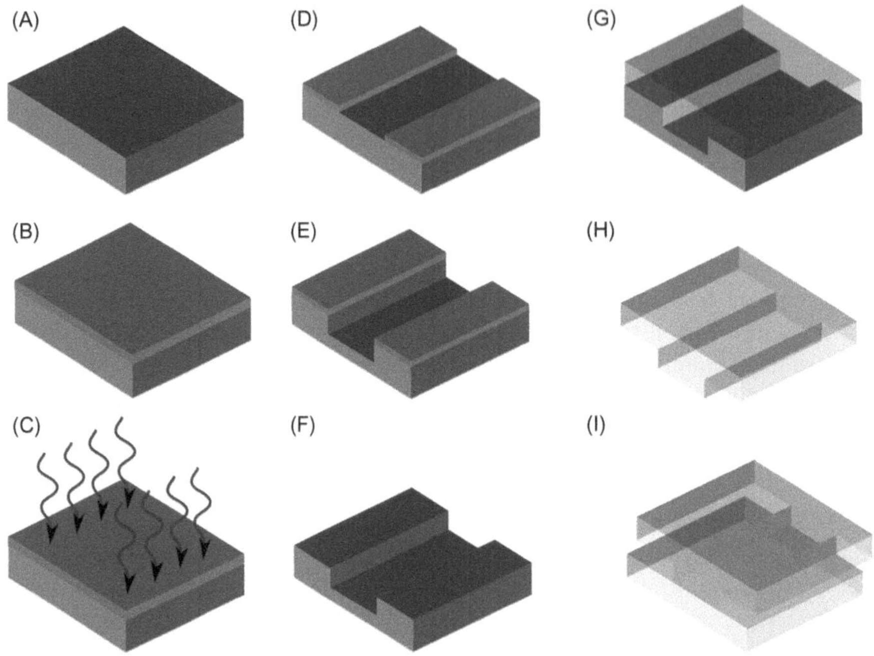

FIGURE 9.1

Illustration showing standard steps for photolithographic patterning of a silicon wafer to produce a fluidic system in PDMS. The silicon wafer (A) is coated with a layer of photoresist (B) which is then selectively exposed to light using a photomask (C). Upon development, the exposed areas of photoresist remain whereas the unexposed areas are dissolved away, assuming a negative tone process (D). The photoresist is used as a mask to selectively etch the silicon wafer (E) and is then removed (F), leaving an etched pattern in the silicon. To mold a separate material, this material (e.g. PDMS) is then poured on the silicon wafer (G) and allowed to solidify. The material is then removed from the patterned wafer (H) and subsequently attached to another slab (I).

successfully created a multilayer fluidic device consisting of 35 stacked PDMS layers to act as a renal dialysis system (Figure 9.2).

Traditional microfabrication techniques based on photolithography can be used to produce surface patterns to control cell adhesion, orientation, migration, etc., in addition to channel patterns. The fluidic channel configuration is inherently easy to manipulate because the pattern can be transferred from computer design to photomask to substrate with extreme accuracy. The results are precisely defined channels and other structures that can be stacked to create a 3D structure. Traditional lithography is a straightforward, well-established method of making microstructures. However, when transitioning to 3D fluidic devices, difficulties can result from the stacking method such as misalignment and an increased risk of leakage due to the

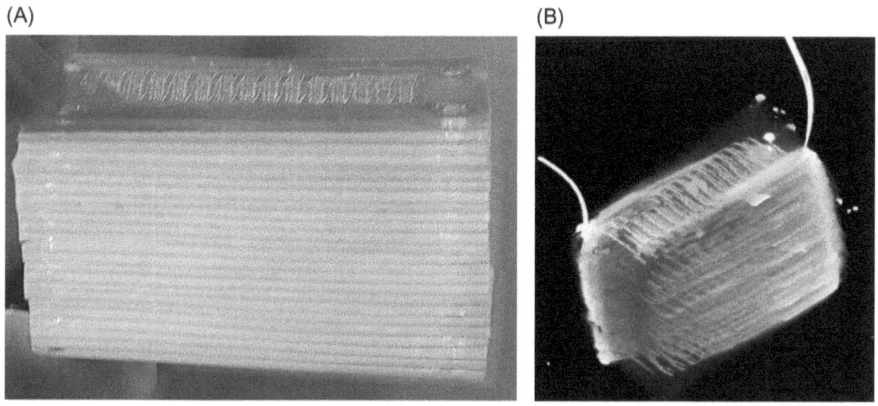

FIGURE 9.2

A 3D microfluidic device created by stacking 35 patterned slabs of PDMS: (A) an optical image of the device and (B) a 3D reconstruction of micro-CT data indicating the internal channel structure.

Source: *Images courtesy of Jeffrey Borenstein.*

larger number of interfaces. Moreover, while the 2D channel density can be determined by the fabricated pattern and easily increased, the density of channels in the third dimension depends on the number and thickness of patterned slabs used. To achieve a high channel density in 3D, a larger number of very thin slabs of material must be carefully aligned and sealed—a tedious process.

9.2.2 SFF techniques

SFF (also known as rapid prototyping) techniques are methods of manufacturing that build desired material in the 3D space utilizing additive processes. There are a wide range of SFF processes, including 3D printing (3DP), stereolithography, fused deposition modeling, and robocasting, each particularly well suited to a certain range of materials, patterns, and feature sizes.

3DP uses inkjet printer technology to selectively deposit a binder (often a pure solvent) onto a packed bed of powder to define a layer of a 3D structure [29]. After one layer is created from bound powder, further layers can be built on top by spreading a new thin layer of unbound powder and repeating the inkjet printing process. Kim et al. [30] used 3D printing techniques to create a biodegradable polymer scaffold, made from polylactide–co-glycolide (PLGA). In this work, 3DP was combined with salt leeching (incorporation and subsequent dissolution of salt crystals to form pores) to produce a microporous (45–150 μm) polymeric scaffold that also contained a series of 800 μm fabricated channels. Hepatocytes were then seeded onto the scaffold, which was then placed in a bioreactor. Increased levels of albumin were produced when the device was cultured under flow, suggesting that flow conditions improved the viability and functionality of the embedded cells.

Stereolithography is a rapid prototyping technique that employs a UV laser to cure layers of photosensitive material, such as a hydrogel precursor solution. Each individual layer can be polymerized with a slightly different pattern to build up a 3D structure. Stereolithography has been used to form a wide range of cell-culture scaffolds [31,32], and some studies have investigated the use of this technique for forming fluidic channels. Dhariwala et al. [33] used stereolithography to form simple shapes, such as a 5 mm outer diameter ring, in polyethylene glycol-based hydrogels (a commonly used hydrogel that is easily modified to allow photoinduced cross-linking using a photoinitiator). The authors demonstrated encapsulation of live cells in the hydrogel, suggesting that the curing process does not significantly affect cell viability. Improved resolution was achieved by Arcaute et al. [34], who used stereolithography to pattern more complex structures in polyethylene glycol hydrogels, including fluidic channels. However, even in this work, the smallest channel size demonstrated was several hundred microns in diameter. A similar technique based on photoablation instead of photocuring has also been used to fabricate microchannels in a hydrogel for nerve guidance [35].

Robocasting and fused deposition modeling are direct write techniques based on the controlled placement of material into a desired 3D configuration using an extruder or syringe. Lewis and coworkers [36–39] have used these techniques in conjunction with a range of fugitive inks in order to generate 3D vascular systems. In this process, the ink is first used to generate a 3D sacrificial structure using an extruder mounted on a three-axis stage. Then a low-viscosity material is allowed to infiltrate this structure and subsequently solidify. Finally, the fugitive ink is removed by heating the composite structure, leaving a channel system in its place (Figure 9.3). Recent developments have allowed the production of channels of varying diameter from a single nozzle [39], thereby allowing facile construction of branching microfluidic networks that obey Murray's law [40,41] (which states that, to optimize the energy required to both pump blood and maintain the living fluid, the sum of the cubes of the radii of daughter channels should equal the cube of the radius of the parent channel). This technique has also been adapted to produce both hydrogel-based periodic structures [42] and microfluidic structures in a hydrogel matrix (Figure 9.4) [43]. To produce channels within a hydrogel matrix, the complex phase behavior of solutions of Pluronic-127 (a triblock copolymer having both hydrophobic and hydrophilic segments) was exploited. In this work, a syringe mounted on a three-axis stage deposited a branched network of fugitive ink (Pluronic-127 above the critical micelle concentration, exhibiting shear-thinning behavior) within a reservoir (Pluronic-127 diacrylate) that could be photopolymerized. The fugitive ink was removed by lowering the temperature below its critical micelle temperature, causing liquification.

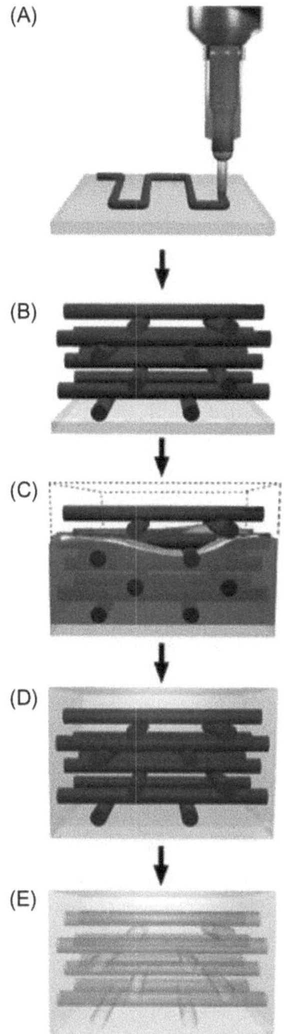

FIGURE 9.3

A schematic describing robocasting of a sacrificial structure to form a fluidic system: (A) a syringe on a three-axis stage deposits filaments of a sacrificial material in a controlled fashion, (B) multiple levels of structure are built up by the robotic deposition system, (C) the sacrificial structure is infiltrated with another material (D) which is allowed to solidify, (E) upon removal of the sacrificial structure, an embedded channel network is left behind.

Source: *Reprinted from Ref. [36] by permission of John Wiley & Sons, Inc. Copyright © 2005 by John Wiley & Sons, Inc.*

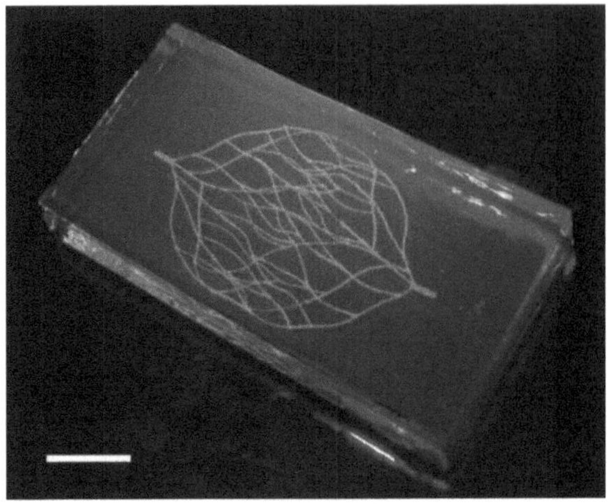

FIGURE 9.4

A fluidic network (indicated by fluorescent dye) inside a Pluronic hydrogel fabricated by robotic deposition of a sacrificial Pluronic structure.

Source: *Reprinted from Ref. [43] by permission of John Wiley & Sons, Inc. Copyright © 2011 by John Wiley & Sons, Inc.*

9.2.3 Nontraditional techniques

Other unique techniques have been developed to create vasculature within a 3D matrix. These techniques sacrifice the accurate placement of features available using the above processes in order to produce inherently 3D vascular-like structures extremely quickly. The justification for this is that the natural vascular system relies less on exact placement of vessels and more on vessel density and branching patterns. If similar density and branching/interconnectivity can be achieved in a more scalable fashion, there is reason to believe this could be beneficial if the fluidic architectures are "good enough" for the biological purpose of interest. One example of such a process involves the use of sacrificial melt-spun sugar (commonly called "cotton candy") to produce a complex interconnected microfluidic channel network with significant extent in all 3D [44]. In this process, the sugar microfiber network (connected with large sticks of sugar to form connected macrochannels for interfacing) is embedded within a polymer matrix such as PDMS or epoxy. When the polymer has solidified, the sugar can be removed by placing the entire device in a bath of warm water. Imaging of fluorescent dye and particles in this device indicates a 3D channel system with parameters, such as channel size and density, similar to those of a natural capillary bed (Figure 9.5A).

Another nontraditional technique for forming a 3D microfluidic channel system exploits the branching of electric discharge [45]. This discharge produces a

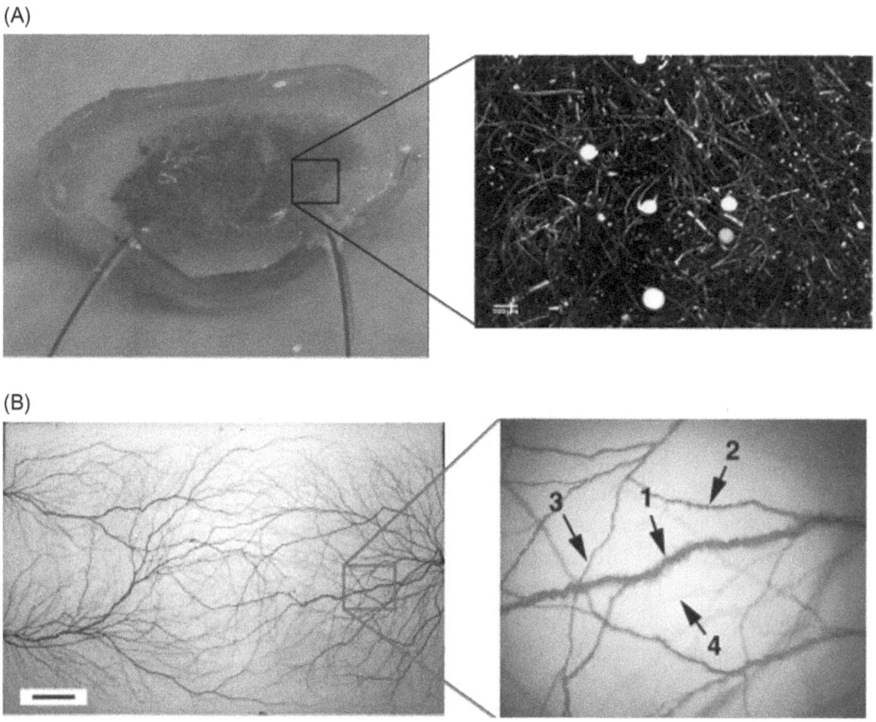

FIGURE 9.5

Images of vascular networks formed using nontraditional techniques. (A) An optical image of a PDMS device containing fluidic channels formed by sacrificial melt-spun sugar fibers, and a multiphoton microscopy image of dye filled channels at higher magnification (scale bar is 200 μm). (B) An optical image of a channel network formed in poly(lactic) acid by electronic discharge (scale bar is 2 cm), and a higher magnification optical image indicating channels of multiple dimensions (point 1, 180 μm; point 2, 70 μm; point 3, 40 μm; point 4, 20 μm diameter).

Source: *(A) From Ref. [44]. Reproduced by permission of The Royal Society of Chemistry (online version http://dx.doi.org/10.1039/B819905A). (B) Reprinted from Ref. [45] by permission of John Wiley & Sons, Inc. Copyright © 2009 by John Wiley & Sons, Inc.*

branching pattern (Lichtenberg figure) that mimics in many ways the branching of a natural vascular system (Figure 9.5B). When a material such as acrylic is charged with an electron beam and subsequently discharged (upon application of a grounded electrode or by using a charge nucleation site in the material), the electrostatic energy release generates a branching microchannel system within the material. Multiple iterations of this technique can be used to produce two or more connected branching networks. The branching of these microchannel systems is surprisingly close to the theoretical description of branching in natural vascular systems provided by Murray's law.

9.2.4 3D fabrication issues

Several issues still hinder the widespread use of 3D fabrication for any application, including the production of artificial 3D microvascular systems. An important consideration is the transition from the lab to industrial-scale manufacturing, as many of these techniques are difficult to use on a production scale level and are thus relegated to being used for prototyping purposes almost exclusively. It is necessary to find a fabrication method that can bridge the gap between discovery and commercialization.

The minimum feature size of these fabrication techniques can also be a limiting factor to the success of tissue creation. Capillary channels are on the order of 10 μm in diameter [46], a feature size easily achieved (in 2D) using photolithography. A branching artificial capillary network may also require small arteries (1–2 mm) and an arteriolar network (100–1000 μm) in addition to capillaries. Traditional microfabrication techniques offer limited capabilities in 3D, while SFF techniques are generally unable to produce feature sizes below 100 μm (with some notable exceptions). Other potential problems with fabricated structures include leakage and channel stability.

9.3 Materials for microfluidic vasculature

The functionality of a microfluidic vascular system is dependent not only on the fluidic pattern but also on the material in which the system is created. Over the past decade, there has been a marked shift from using rigid, easily manipulated, materials as a substrate for vascular networks to fabricating networks in elastomers or hydrogels, which provide material advantages such as biocompatibility, degradability, and improved diffusion between the fluidic channels.

9.3.1 Rigid materials

A wide range of rigid materials have been used in conjunction with various microfluidic fabrication techniques to create vascular networks. Rigid materials, especially silicon, can be easily and precisely etched or machined to create a wide variety of geometric patterns. Silicon became a popular material to use for this purpose because it is a familiar material with well-known properties. This familiarity stems from the material's prior widespread use in other areas, especially the field of microelectronics, for microfabrication.

To form structural patterns in/on silicon, several techniques can be used, including milling using a focused ion beam, laser micromachining, patterned deposition of material using a shadow mask, photolithographic patterning of resist layers, or the liftoff process. However, perhaps one of the most popular techniques to form fluidic designs in silicon is to etch the pattern using either a wet (liquid) etch or a dry (gas or plasma) etch. In general, when vertical sidewalls are desired (producing a roughly rectangular cross section), a dry anisotropic deep reactive ion etch (DRIE) process, such as the Bosch etch, is used. This process is commonly used in the microfabrication community, where right angles are generally desirable and common. However, it may be that cells behave differently in channels with circular cross sections and rounded edges, a geometry that better mimics natural fluidic systems. Shear stresses are more uniform in circular cross-section channels, allowing cells housed in such channels to elongate and differentiate in a more consistent manner and eliminating the dead volume spaces found at the corners of rectangular cross-sectioned channels (which may create nonuniform nutrient distribution). Circular channels allow for a continuous layer of endothelial cells to line the channels instead of the single monolayer of cells that tends to form on the bottom of a rectangular channel [47]. Therefore, making the move from rectangular cross sections to smooth, circular cross sections is advantageous for making microfluidic networks in which cells will be seeded for extended periods of time. Channels with circular cross sections can be made in a variety of ways, including wet etching (e.g., using a hydrofluoric-nitric-acetic (HNA) acid solution to isotropically etch silicon [48]), dry etching with an isotropic process (e.g. xenon difluoride etching [47]), reflow of patterned resist layers [49], and electroplating of a patterned "seed" [50].

Studies have demonstrated that cells can be grown on hard materials such as silicon, Pyrex, polycarbonate, and polystyrene [23,50,51]. Researchers also examined whether cells prefer to grow on one kind of hard material to another [23]. Silicon and Pyrex were tested with different coatings, including an uncoated trial, in order to observe the ability of cells to be removed in single sheets and to be grown successfully in flasks after removal. The authors determined that the viability of cells after removal was not significantly altered when changing materials or coatings, but that certain coatings allowed the cells to be removed in a monolayer, which is useful for growing large sheets of tissue for manipulation into various 3D structures and vascular systems.

Etched silicon templates can be used to form patterns in other materials by replica molding. This process was used to form a vascular fluidic network in the rigid biodegradable polymer poly(lactic-co-glycolic acid) (PLGA) [52], a material that has been used for resorbable implants. To pattern slabs of PLGA, pellets of the material were melted under compression onto a PDMS template formed from a patterned silicon master. The resulting thin, patterned sheets of PLGA were laminated to other sheets of PLGA by a thermal bonding process. With this technique, the authors were able to produce a multilayer fluidic system embedded within the biodegradable polymer (Figure 9.6). Templates have also been used to emboss patterns in polystyrene [50] and polycarbonate [51] substrates which were then thermally bonded to form closed fluidic systems.

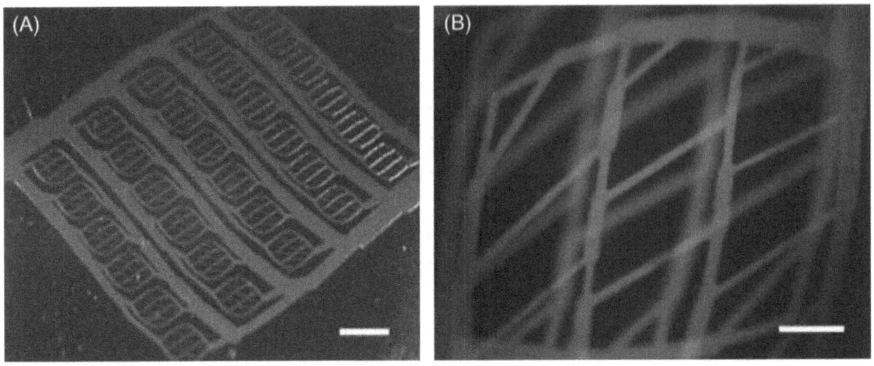

FIGURE 9.6

A microfluidic network fabricated in the biodegradable polymer PLGA. The channels are indicated by fluorescein dye. The scale bar in (A) indicates 2.5 mm and that in (B) indicates 300 μm.

A significant disadvantage in using rigid materials is the fact that they do not share many properties with soft tissue. Though complicated networks can be made in materials like silicon, glass, or acrylic, cells can only be introduced onto the surfaces (including interior surfaces such as channel walls) of these materials and not within them (without introducing microscale interconnected pores). Thus, cells grown on these materials experience attachment to a 2D environment as opposed to a 3D environment. It is well known that cell behavior can depend on the material properties, including mechanical properties, of the substrate on/in which they are growing [53–58]. Thus, culturing cells on materials that are more rigid than natural tissue could lead to nonbiomimetic behavior. Moreover, the fact that rigid materials are not flexible inhibits their use as implantable soft tissue replacements. The use of hard materials in creating microfluidic vascular networks has been a valuable model that has led researchers to investigate possibilities of creating these networks in more flexible and biocompatible materials. The replica molding technique, in which a hard material is used as a template to pattern another material (which is often flexible), has taken advantage of the precision of silicon etching and other micromachining techniques and has made the knowledge gained through research concerning vascular networks in hard materials invaluable to future research in this field.

9.3.2 Elastomers

Elastomers are natural or synthetic polymers with elastic properties (and typically low Young's moduli) and were the next materials to be used in microfabrication of vascular systems. Elastomers in general have more properties in common with living soft tissues than hard materials, including flexibility and permeability to gasses. Elastomers are also attractive materials for fabrication of microfluidic devices because they are both inexpensive and durable.

The most commonly used solid elastomeric material for microfluidic device fabrication is PDMS. Through the process of replica molding, PDMS can be patterned from a template (typically made from a hard material such as silicon) which has been patterned with the desired fluidic architecture. Using the replica molding technique, it is possible to fabricate elastomeric devices with almost the same level of precision and range of pattern geometries as can be obtained using hard materials. The formation of fluidic devices using replica molding to pattern PDMS is a straightforward process that has now become commonplace (Figure 9.1). Once the patterned elastomer is removed from the silicon template, it can be bonded to another elastomer slab (or silicon or silica) by several means, the most common of which is through the use of an oxygen plasma to activate the PDMS surface (converting $Si-CH_3$ methyl groups to $Si-OH$ silanol groups) and placing the activated surface in contact with another reactive surface (for silicon, silica, and PDMS, this causes the formation of $Si-O-Si$ bonds) [19]. Attaching the patterned PDMS portion to a flat slab (or another patterned slab) produces an enclosed channel in which liquid can flow and cells can be seeded. Two patterned elastomers can be attached together to form channels with cylindrical cross-section (if the templates are patterned with rounded trenches) or 3D fluidic systems, but aligning the two slabs perfectly can be quite difficult.

Several publications have detailed assembly techniques and cell-culture procedures associated with microfluidic PDMS devices that can house cells within the fluidic network for extended periods of time [20,22,59,60]. Typically replica molding is employed and the PDMS devices are assembled using an oxygen plasma; cells are introduced into the device along with media and are allowed to attach to the walls of the channels (Figure 9.7). Endothelial cells have been able to live inside the channels for up to 4 weeks [22], demonstrating that appropriately treated PDMS is suitably nontoxic and does not interfere with adhesion or proliferation of cells.

While PDMS is a popular material for fabricating microfluidic devices, it is not biodegradable and thus may not be suitable for implantation. To overcome this hurdle, a new biodegradable elastomer, polyglycerol sebacate (PGS), was recently developed [61]. PGS is biocompatible, biodegradable [62], and does not require any harsh chemicals during synthesis, curing, or device adhesion processes. It can also be patterned in a manner similar to PDMS and used to form microfluidic devices [21]. Endothelial cells readily adhere (without the need for a layer of adhesion proteins) and proliferate on the channel walls inside a PGS microfluidic device.

Though cells can be grown inside the channels of elastomeric devices, they cannot be seeded within the bulk material due to limited diffusion of soluble materials and water, limiting the applicability of these materials for tissue engineering. In order to move toward even more biomimetic materials, there is a

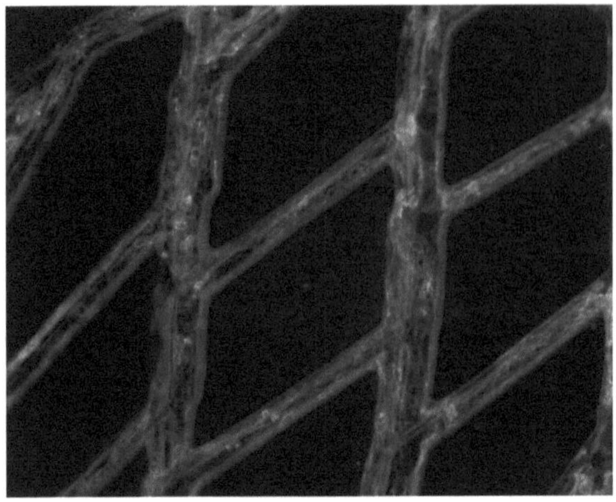

FIGURE 9.7

A microfluidic network in PDMS lined with human microvascular endothelial cells (HMEC-1) stained for CD31.

Source: *With kind permission from Ref. [20], Springer Science + Business Media Figure 9.5A.*

significant push to develop techniques to form microfluidic networks inside of hydrogels.

9.3.3 Hydrogels

Hydrogels are attractive materials to work with because they share many properties with natural soft tissue, allowing them to form better artificial tissue constructs and support cell growth [63]. Hydrogels are materials consisting of a network of cross-linked polymer chains swollen with water. They typically contain a significantly higher mass fraction of water than that of the polymer material and thus exhibit similar diffusion characteristics as soft tissue. Many different types of hydrogels, both natural and synthetic, have been used to produce vascular networks, each with different mechanical and chemical properties.

Unlike the category of elastomers, which is almost exclusively dominated by the use of PDMS, the field of hydrogels has not settled upon a single material in which to form fluidic structures. A wide range of materials, as well as corresponding patterning and assembly methods, has been used in the past few years as the field has progressed. A popular technique to use to pattern slabs of hydrogel is soft lithography, often employing PDMS templates produced by replica molding. The hydrogel (in an ungelled state) is poured into the PDMS mold, allowed to gel, and subsequently removed. The difficulty arises when attempting to adhere a patterned hydrogel slab to another slab, as the oxygen plasma treatment used to activate the surface of PDMS cannot be used with these materials.

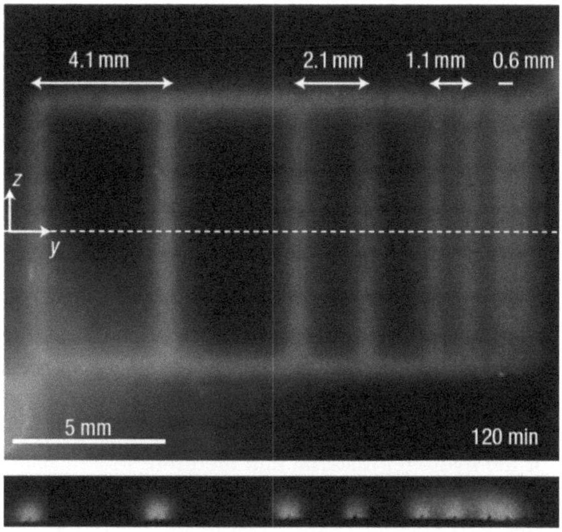

FIGURE 9.8

A microfluidic network in alginate, containing primary bovine chondrocytes embedded within the hydrogel. The live cell stain calcein AM was introduced into the fluidic network, and upon diffusing into the hydrogel was taken up by the cells and metabolized, rendering it fluorescent.

Source: *Reprinted by permission from Ref. [65], Macmillan Publishers Ltd: Nature Materials. Copyright 2007.*

Currently, there is no general treatment available to cause adhesion between two slabs of hydrogel, and thus processes have been developed that are specific to each material. To form microchannels in alginate hydrogels, Cabodi and coworkers [64] first used soft lithography to pattern slabs of alginate (cross-linked with calcium). To adhere the slabs to produce a microfluidic network, the authors then applied a solution of sodium citrate (citrate is a calcium chelator) to remove the calcium cross-links at the surfaces of the slabs, placed the slabs in contact, and then applied a solution of calcium chloride to re-gel the slabs together, forming a single gelled piece of hydrogel. A similar architecture (in which the alginate slabs were held together with a jig instead of chemically bonded) was then used to support the growth of several cell types embedded within the hydrogel using flow through the microfluidic network (Figure 9.8) [65].

Another material that has been used to form microfluidic networks in hydrogels via soft lithography is agarose [66]. Agarose is a meltable hydrogel (with a tunable melting point) that, like alginate, has been used to encapsulate cells in an attempt to protect them from host immune response [67–69]. To form slabs of agarose gel, an aqueous solution of agarose is first heated to dissolve the agarose powder. The agarose gel is combined with a suspension of cells at a high temperature (around

70°C) so that the agarose is molten and can be poured over a SU-8 patterned silicon master template. Once flat and patterned portions have been poured, they are set to gel at room temperature. To attach the flat and patterned portions together, each surface is briefly melted on a hotplate, and then the two pieces are pressed together and left to re-gel, forming a single piece of hydrogel containing a channel system. Cells embedded within the hydrogel initially showed viability of ~85% directly after lamination of the slab, but the viability dropped to 55% after 3 days of culture with media flowing through the microfluidic network.

The two techniques described above for forming microfluidic channels in hydrogels by sealing patterned slabs to flat surfaces are attractive in that they are quite similar to the standard method for producing microfluidic devices from PDMS (form two slabs, activate the surfaces, place in contact). However, unlike the oxygen plasma treatment used to activate PDMS, both sodium citrate treatment (for calcium cross-linked alginate) and surface melting (for agarose) can result in reflow of the hydrogel material and therefore loss of pattern fidelity. The processing parameters therefore must be finely tuned to allow the patterned surface to uncross-link well enough to interface to the bottom slab but prevent any significant loss of pattern features.

Another popular method to form microfluidic devices in hydrogels using patterned slabs is to hold two slabs against each other using a fixture, as opposed to chemically or physically laminating the slabs. This technique is often simpler than permanently laminating slabs and still provides a perfusable fluidic system. This technique has been used to form microfluidic systems in several materials, including polyethylene glycol diacrylate (a single fluidic channel) [70], alginate (see above) [65], and collagen [71]. While this process eliminates the need for complex surface treatments, it is not necessarily appropriate for *in vivo* applications.

An additional technique used to form channel networks in hydrogels is the use of a sacrificial structure that is removed from the hydrogel after it has solidified, forming channels where the sacrificial structure was. This technique avoids the need to laminate patterned slabs of material together. Golden and Tien [72] used gelatin as a sacrificial material in a collagen hydrogel (Figure 9.9). Gelatin mesh networks were patterned on PDMS templates by pouring a warm gelatin solution over the template and then cooling the liquid, causing physical cross-linking. The cross-linked gelatin structures were then removed from the template and then embedded within collagen which was then cross-linked. The device was then heated, causing the gelatin to melt and allowing for its easy removal via ports cut in the collagen. The resulting device is a network of patterned microfluidic channels (in the location of the gelatin structures) in a collagen hydrogel. These channels were seeded with endothelial cells that were maintained for several days. This technique is attractive because it is not specific to the embedding hydrogel, requires no high-temperature processes that could harm cells, and uses only cell-friendly materials (i.e. residual sacrificial gelatin will not harm the

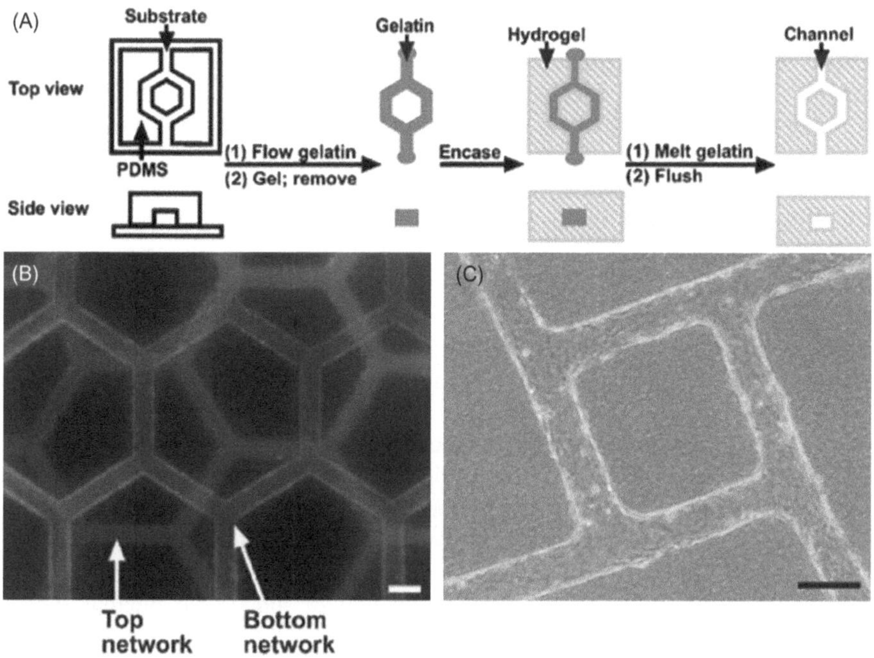

FIGURE 9.9

(A) Schematic illustrating fabrication of sacrificial gelatin structures. (B) An overlay of two confocal microscopy images of two nonconnected fluidic networks in collagen. Scale bar is 200 μm. (C) Endothelial cells lining channel walls of a fluidic network patterned in collagen. Scale bar is 200 μm.

Source: *From Ref. [72]. Reproduced by permission of the Royal Society of Chemistry (online version http://dx.doi.org/10.1039/B618409J).*

cells). Sacrificial structures can also be formed in hydrogels using SFF techniques such as robocasting (see Section 9.2.2).

The field of microfluidics has evolved in the past decade, specifically with regard to the materials that are used in fabrication of channel networks. Hydrogels appear to be the future of this field, and as handling and fabrication techniques with other gels, besides those mentioned above, are discovered and perfected, the applicability of the field of microfluidics to tissue engineering increases will increase dramatically. As biocompatibility is the most important aspect of the material one uses in this field, the utilization of materials that can mimic the human body's natural tissues in microfluidic devices brings us one step closer to being able to use these devices to grow new tissues, such as liver and kidney tissue, for medical use in patients across the world.

9.3.4 Materials issues

There are several considerations that must be taken into account when choosing a material in which to produce an artificial vascular network. If the network is to serve simply as a model to study fluid dynamics (and is not seeded with cells), rigid or elastomeric materials may suffice. However, if the material is to serve as a 3D scaffold for cell culture, it is necessary to utilize hydrogels (cells grown on the surface of channels in elastomers and rigid materials still experience a 2D environment). Even within the hydrogel category, there is a wide range of parameters that may affect how cells will grow in and on the material. Certain hydrogels can be remodeled by cells [71], allowing nature to optimize the architecture engineered by researchers and demonstrating the likelihood of significant host cell invasion into these materials. The diffusion of necessary biomolecules may depend on the concentration and porosity of the hydrogel system used, and thus it is important to determine the optimum concentration to ensure both appropriate diffusion and mechanical properties. Finally, there may be a varied level of immune response that depends on the specific hydrogel material used. Hydrogels exhibit a wide range of mechanical properties, from rigid, brittle materials to flexible, elastic gels to viscous, but flowing, gels. Thus, there are some hydrogels that are simply not mechanically robust enough to support an internal fluidic architecture.

9.4 Conclusion

There are several compelling reasons for fabricating artificial vascular systems. Scientific studies on flow in model systems will reveal more detailed information about the influence of fluidic channel network architecture on natural processes (e.g., cell elongation, adhesion, and growth; platelet behavior, etc.). Engineering vascular systems into artificial tissue constructs may enable the production of thick artificial tissue not only for clinical use but also for a wide range of *in vitro* work, including cancer studies, drug screening, and toxicity detection. While the potential applications for artificial vascular systems are attractive, the technologies for producing such system are still in their infancy. Simple single channel systems may be easily formed by casting materials around a cylinder, such as a syringe needle [73], which is subsequently removed; however, to form complex fluidic networks such a strategy is not possible and other techniques are used. The vast majority of these processes, and the materials used, have been borrowed from the semiconductor or industrial prototyping industries. Microfabrication techniques, while able to easily achieve the feature sizes of interest, are not easily scaled into the third dimension. On the other hand, SFF techniques are often unable to produce the small feature sizes needed to form capillary-like channels and cannot easily be incorporated into an industrial-scale manufacturing process due to their expense and lack of speed. Other nontraditional techniques aim to exploit naturally occurring patterns similar to vasculature to form 3D fluidic systems but sacrifice the ability to place structures in exact locations. At this point, no single technique offers the combination of 3D scalability, viability for industrial-scale manufacturing, patterning ability, and appropriate minimum feature size to commercially produce useful artificial microvascular systems. Moreover, each technique is only able to work within a restricted range of materials. While rigid materials such as silicon and Pyrex are easily fabricated, they are inappropriate in many ways (mechanical properties, biodegradability, permeability, etc.) for the construction of tissue scaffolds. Elastomeric materials such as PDMS and PGS offer improved mechanical flexibility and (for PGS) degradability but are still limited by lack of appropriate diffusion of nutrients and an inability to provide a 3D hydrated environment for cell growth. For soft tissue engineering, hydrogels are the most biomimetic class of materials but are also extremely difficult pattern with complex, small features. While a fairly large toolbox of processes exists for patterning rigid and elastomeric materials, there is a general lack of robust processes for doing so with hydrogels.

While the technologies discussed above may be considered mature for more standard applications, their use for producing vascular systems for tissue engineering applications is still primitive. Only a handful of studies have demonstrated the production of either cell-lined fabricated vascular systems or engineered vascular systems that sustain cells in a hydrogel, and the field has not settled on either a standard fabrication technique or material for these purposes. Moreover, there are several fundamental questions that have yet to be answered, likely due to the inability to fabricate the necessary constructs. For example, it is unclear how closely the geometry of an engineered microvascular system must replicate that of a natural system in order to have appropriate (i.e. sufficient) functionality. It is also not clear to what extent cells embedded within a scaffold would remodel an engineered fluidic system to provide improved functionality. To answer these questions and others, researchers will require improved techniques for microvascular network fabrication. However, though the long-term goal of this effort is primarily the production of vascularized tissue constructs for clinical use, it appears that fabricating such structures even for research purposes (when manufacturability is less of an issue) is difficult. Given the strong demand for engineered vascular networks in both clinical and research settings, however, it is likely that a wide range of additional processes will be explored and refined until fabrication of the desired structures becomes effortless.

References

[1] J.P. Vacanti, R. Langer, Tissue engineering: the design and fabrication of living replacement devices for surgical reconstruction and transplantation, Lancet 354 (1999) S32–S34.

[2] R. Langer, J. Vacanti, Tissue engineering, Science 260 (5110) (1993) 920–926.

[3] R.Y. Kannan, H.J. Salacinski, K. Sales, P. Butler, A.M. Seifalian, The roles of tissue engineering and vascularisation in the development of micro-vascular networks: a review, Biomaterials 26 (14) (2005) 1857–1875.
[4] J. Rouwkema, N.C. Rivron, C.A. van Blitterswijk, Vascularization in tissue engineering, Trends Biotech. 26 (8) (2008) 434–441.
[5] T. Kaully, K. Kaufman-Francis, A. Lesman, S. Levenberg, Vascularization—the conduit to viable engineered tissues, Tissue Eng. B Rev. 15 (2) (2009) 159–169.
[6] P. Carmeliet, Mechanisms of angiogenesis and arteriogenesis, Nat. Med. 6 (4) (2000) 389–395.
[7] M.S. Pepper, Manipulating angiogenesis: from basic science to the bedside, Arterioscl. Thromb. Vasc. Biol. 17 (4) (1997) 605.
[8] E. Cukierman, R. Pankov, K.M. Yamada, Cell interactions with three-dimensional matrices, Curr. Opin. Cell Biol. 14 (5) (2002) 633–640.
[9] L.G. Griffith, M.A. Swartz, Capturing complex 3D tissue physiology *in vitro*, Nat. Rev. Mol. Cell Biol. 7 (3) (2006) 211–224.
[10] H. Gruber, E. Hanley, Human disc cells in monolayer vs 3D culture: cell shape, division and matrix formation, BMC Musculoskeletal Disord. 1 (1) (2000) 1.
[11] A. Pitaval, Q. Tseng, M. Bornens, M. Théry, Cell shape and contractility regulate ciliogenesis in cell cycle–arrested cells, J. Cell Biol. 191 (2) (2010) 303.
[12] C.S. Chen, Geometric control of cell life and death, Science 276 (5317) (1997) 1425–1428.
[13] D.M. Supp, A.P. Supp, S.M. Bell, S.T. Boyce, Enhanced vascularization of cultured skin substitutes genetically modified to overexpress vascular endothelial growth factor1, J. Investig. Dermatol. 114 (1) (2000) 5–13.
[14] F.A. Auger, F. Berthod, V. Moulin, R. Pouliot, L. Germain, Tissue-engineered skin substitutes: from *in vitro* constructs to *in vivo* applications, Biotechnol. Appl. Biochem. 39 (3) (2004) 263–275.
[15] D. Barbone, T.-M. Yang, J.R. Morgan, G. Gaudino, V.C. Broaddus, Mammalian target of rapamycin contributes to the acquired apoptotic resistance of human mesothelioma multicellular spheroids, J. Biol. Chem. 283 (19) (2008) 13021–13030.
[16] S.S. Verbridge, E.M. Chandler, C. Fischbach, Tissue-engineered three-dimensional tumor models to study tumor angiogenesis, Tissue Eng. A 16 (7) (2010) 2147–2152.
[17] D.C. Duffy, J.C. McDonald, O.J.A. Schueller, G.M. Whitesides, Rapid prototyping of microfluidic systems in poly(dimethylsiloxane), Anal. Chem. 70 (23) (1998) 4974–4984.
[18] J.C. McDonald, G.M. Whitesides, Poly(dimethylsiloxane) as a material for fabricating microfluidic devices, Acc. Chem. Res. 35 (7) (2002) 491–499.
[19] J.C. McDonald, D.C. Duffy, J.R. Anderson, D.T. Chiu, H. Wu, O.J.A. Schueller, et al., Fabrication of microfluidic systems in poly (dimethylsiloxane), Electrophoresis 21 (2000) 27–40.
[20] M. Shin, K. Matsuda, O. Ishii, H. Terai, M. Kaazempur-Mofrad, J. Borenstein, et al., Endothelialized networks with a vascular geometry in microfabricated poly (dimethyl siloxane), Biomed. Microdevices 6 (4) (2004) 269–278.
[21] C. Fidkowski, M.R. Kaazempur-Mofrad, J. Borenstein, J.P. Vacanti, R. Langer, Y. Wang, Endothelialized microvasculature based on a biodegradable elastomer, Tissue Eng. 11 (1–2) (2005) 302–309.

[22] J.T. Borenstein, H. Terai, K.R. King, E.J. Weinberg, M.R. Kaazempur-Mofrad, J.P. Vacanti, Microfabrication technology for vascularized tissue engineering, Biomed. Microdevices 4 (3) (2002) 167–175.

[23] S. Kaihara, J. Borenstein, R. Koka, S. Lalan, E.R. Ochoa, M. Ravens, et al., Silicon micromachining to tissue engineer branched vascular channels for liver fabrication, Tissue Eng. 6 (2) (2000) 105–117.

[24] J.R. Anderson, D.T. Chiu, R.J. Jackman, O. Cherniavskaya, J.C. McDonald, H. Wu, et al., Fabrication of topologically complex three-dimensional microfluidic systems in PDMS by rapid prototyping, Anal. Chem. 72 (14) (2000) 3158–3164.

[25] D.T. Chiu, E. Pezzoli, H. Wu, A.D. Stroock, G.M. Whitesides, Using three-dimensional microfluidic networks for solving computationally hard problems, Proc. Natl. Acad. Sci. U.S.A. 98 (6) (2001) 2961.

[26] E. Leclerc, Y. Sakai, T. Fujii, In: The Sixteenth Annual International Conference on Micro Electro Mechanical Systems, 2003. MEMS-03 Kyoto. IEEE, Kyoto, Japan, pp. 415–418, doi:10.1109/MEMSYS.2003.1189774

[27] E. Leclerc, Y. Sakai, T. Fujii, P.D.M.S. Microfluidic, Polydimethylsiloxane) bioreactor for large-scale culture of hepatocytes, Biotechnol. Prog. 20 (3) (2004) 750–755.

[28] J.T. Borenstein, E.J. Weinberg, B.K. Orrick, C. Sundback, M.R. Kaazempur-Mofrad, J.P. Vacanti, Microfabrication of three-dimensional engineered scaffolds, Tissue Eng. 13 (8) (2007) 1837–1844.

[29] E.M. Sachs, J.S. Haggerty, M.J. Cima, P.A. Williams, Three-dimensional printing techniques, Google Patents, 1994.

[30] S.S. Kim, H. Utsunomiya, J.A. Koski, B.M. Wu, M.J. Cima, J. Sohn, et al., Survival and function of hepatocytes on a novel three-dimensional synthetic biodegradable polymer scaffold with an intrinsic network of channels, Ann. Surg. 228 (1) (1998) 8.

[31] V. Chan, P. Zorlutuna, J.H. Jeong, H. Kong, R. Bashir, Three-dimensional photopatterning of hydrogels using stereolithography for long-term cell encapsulation, Lab Chip 10 (16) (2010) 2062.

[32] S. Khetan, J.A. Burdick, Patterning hydrogels in three dimensions towards controlling cellular interactions, Soft Matter 7 (3) (2011) 830.

[33] B. Dhariwala, E. Hunt, T. Boland, Rapid prototyping of tissue-engineering constructs, using photopolymerizable hydrogels and stereolithography, Tissue Eng. 10 (9–10) (2004) 1316–1322.

[34] K. Arcaute, B.K. Mann, R.B. Wicker, Stereolithography of three-dimensional bioactive poly(ethylene glycol) constructs with encapsulated cells, Ann. Biomed. Eng. 34 (9) (2006) 1429–1441.

[35] O. Sarig-Nadir, N. Livnat, R. Zajdman, S. Shoham, D. Seliktar, Laser photoablation of guidance microchannels into hydrogels directs cell growth in three dimensions, Biophys. J. 96 (11) (2009) 4743–4752.

[36] D. Therriault, R.F. Shepherd, S.R. White, J.A. Lewis, Fugitive inks for direct-write assembly of three-dimensional microvascular networks, Adv. Mater. 17 (4) (2005) 395–399.

[37] C.J. Hansen, W. Wu, K.S. Toohey, N.R. Sottos, S.R. White, J.A. Lewis, Self-healing materials with interpenetrating microvascular networks, Adv. Mater. 21 (41) (2009) 4143–4147.

[38] D. Therriault, S.R. White, J.A. Lewis, Chaotic mixing in three-dimensional microvascular networks fabricated by direct-write assembly, Nat. Mater. 2 (4) (2003) 265–271.

[39] W. Wu, C.J. Hansen, A.M. Aragón, P.H. Geubelle, S.R. White, J.A. Lewis, Direct-write assembly of biomimetic microvascular networks for efficient fluid transport, Soft Matter 6 (4) (2010) 739.

[40] C.D. Murray, The physiological principle of minimum work: I. The vascular system and the cost of blood volume, Proc. Natl. Acad. Sci. U.S.A. 12 (3) (1926) 207.

[41] C.D. Murray, The physiological principle of minimum work: II. Oxygen exchange in capillaries, Proc. Natl. Acad. Sci. U.S.A. 12 (5) (1926) 299.

[42] R.A. Barry, R.F. Shepherd, J.N. Hanson, R.G. Nuzzo, P. Wiltzius, J.A. Lewis, Direct-write assembly of 3D hydrogel scaffolds for guided cell growth, Adv. Mater. 21 (23) (2009) 2407–2410.

[43] W. Wu, A. DeConinck, J.A. Lewis, Omnidirectional printing of 3D microvascular networks, Adv. Mater. 23 (24) (2011) H178–H183.

[44] L.M. Bellan, S.P. Singh, P.W. Henderson, T.J. Porri, H.G. Craighead, J.A. Spector, Fabrication of an artificial 3-dimensional vascular network using sacrificial sugar structures, Soft Matter 5 (7) (2009) 1354–1357.

[45] J.-H. Huang, J. Kim, N. Agrawal, A.P. Sudarsan, J.E. Maxim, A. Jayaraman, et al., Rapid fabrication of bio-inspired 3D microfluidic vascular networks, Adv. Mater. 21 (35) (2009) 3567–3571.

[46] R.A. Freitas Jr., Nanomedicine, Vol. I: Basic Capabilities, first ed., Landes Bioscience. Georgetown, TX, 1999; http://www.nanomedicine.com/NMI.htm.

[47] J.P. Camp, T. Stokol, M.L. Shuler, Fabrication of a multiple-diameter branched network of microvascular channels with semi-circular cross-sections using xenon difluoride etching, Biomed. Microdevices 10 (2) (2007) 179–186.

[48] C.-T. Seo, C.-H. Bae, D.-S. Eun, J.-K. Shin, J.-H. Lee, Fabrication of circular-type microchannel using photoresist reflow and isotropic etching for microfluidic devices, Jpn. J. Appl. Phys. 43 (11A) (2004) 7773–7776.

[49] G.-J. Wang, K.-H. Ho, S.-H. Hsu, K.-P. Wang, Microvessel scaffold with circular microchannels by photoresist melting, Biomed. Microdevices 9 (5) (2007) 657–663.

[50] J.T. Borenstein, M.M. Tupper, P.J. Mack, E.J. Weinberg, A.S. Khalil, J. Hsiao, et al., Functional endothelialized microvascular networks with circular cross-sections in a tissue culture substrate, Biomed. Microdevices 12 (1) (2009) 71–79.

[51] G.J. Wang, C.L. Chen, S.H. Hsu, Y.L. Chiang, Bio-MEMS fabricated artificial capillaries for tissue engineering, Microsyst. Technol. 12 (1–2) (2005) 120–127.

[52] K.R. King, C.C.J. Wang, M.R. Kaazempur-Mofrad, J.P. Vacanti, J.T. Borenstein, Biodegradable microfluidics, Adv. Mater. 16 (22) (2004) 2007–2012.

[53] R.J. Pelham, Y. Wang, Cell locomotion and focal adhesions are regulated by substrate flexibility, Proc. Natl. Acad. Sci. U.S.A 94 (25) (1997) 13661.

[54] J.Y. Wong, J.B. Leach, X.Q. Brown, Balance of chemistry, topography, and mechanics at the cell-biomaterial interface: issues and challenges for assessing the role of substrate mechanics on cell response, Surf. Sci. 570 (1–2) (2004) 119–133.

[55] C.M. Lo, H.B. Wang, M. Dembo, Y. Wang, Cell movement is guided by the rigidity of the substrate, Biophys. J. 79 (1) (2000) 144–152.

[56] X.Q. Brown, K. Ookawa, J.Y. Wong, Evaluation of polydimethylsiloxane scaffolds with physiologically-relevant elastic moduli: interplay of substrate mechanics and

surface chemistry effects on vascular smooth muscle cell response, Biomaterials 26 (16) (2005) 3123–3129.
[57] N.D. Leipzig, M.S. Shoichet, The effect of substrate stiffness on adult neural stem cell behavior, Biomaterials 30 (36) (2009) 6867–6878.
[58] D.E. Discher, P. Janmey, Y.-L. Wang, Tissue cells feel and respond to the stiffness of their substrate, Science 310 (5751) (2005) 1139–1143.
[59] E. Leclerc, Y. Sakai, T. Fujii, Cell culture in 3-dimensional microfluidic structure of PDMS (polydimethylsiloxane), Biomed. Microdevices 5 (2) (2003) 109–114.
[60] J.M. Rosano, N. Tousi, R.C. Scott, B. Krynska, V. Rizzo, B. Prabhakarpandian, et al., A physiologically realistic *in vitro* model of microvascular networks, Biomed. Microdevices 11 (5) (2009) 1051–1057.
[61] Y. Wang, G.A. Ameer, B.J. Sheppard, R. Langer, A tough biodegradable elastomer, Nat. Biotechnol. 20 (6) (2002) 602–606.
[62] Y. Wang, Y.M. Kim, R. Langer, *In vivo* degradation characteristics of poly (glycerol sebacate), J. Biomed. Mater. Res. A 66 (1) (2003) 192–197.
[63] K.Y. Lee, D.J. Mooney, Hydrogels for tissue engineering, Chem. Rev. 101 (7) (2001) 1869–1880.
[64] M. Cabodi, N.W. Choi, J.P. Gleghorn, C.S.D. Lee, L.J. Bonassar, A.D. Stroock, A microfluidic biomaterial, J. Am. Chem. Soc. 127 (40) (2005) 13788–13789.
[65] N.W. Choi, M. Cabodi, B. Held, J.P. Gleghorn, L.J. Bonassar, A.D. Stroock, Microfluidic scaffolds for tissue engineering, Nat. Mater. 6 (11) (2007) 908–915.
[66] Y. Ling, J. Rubin, Y. Deng, C. Huang, U. Demirci, J.M. Karp, et al., A cell-laden microfluidic hydrogel, Lab Chip 7 (6) (2007) 756.
[67] H. Gin, B. Dupuy, C. Baquey, D. Ducassou, J. Aurertin, Agarose encapsulation of islets of Langerhans: reduced toxicity *in vitro*, J. Microencapsulation 4 (3) (1987) 239–242.
[68] S. Sakai, K. Kawabata, T. Ono, H. Ijima, K. Kawakami, Development of mammalian cell-enclosing subsieve-size agarose capsules (<100 μm) for cell therapy, Biomaterials 26 (23) (2005) 4786–4792.
[69] H. Iwata, K. Kobayashi, T. Takagi, T. Oka, H. Yang, H. Amemiya, et al., Feasibility of agarose microbeads with xenogeneic islets as a bioartificial pancreas, J. Biomed. Mater. Res. 28 (9) (1994) 1003–1011.
[70] M.P. Cuchiara, A.C.B. Allen, T.M. Chen, J.S. Miller, J.L. West, Multilayer microfluidic PEGDA hydrogels, Biomaterials 31 (21) (2010) 5491–5497.
[71] V.L. Cross, Y. Zheng, N. Won Choi, S.S. Verbridge, B.A. Sutermaster, L.J. Bonassar, et al., Dense type I collagen matrices that support cellular remodeling and microfabrication for studies of tumor angiogenesis and vasculogenesis *in vitro*, Biomaterials (2010).
[72] A.P. Golden, J. Tien, Fabrication of microfluidic hydrogels using molded gelatin as a sacrificial element, Lab Chip 7 (6) (2007) 720.
[73] J.W. Nichol, S.T. Koshy, H. Bae, C.M. Hwang, S. Yamanlar, A. Khademhosseini, Cell-laden microengineered gelatin methacrylate hydrogels, Biomaterials (2010).

CHAPTER

10 Microfluidic Approaches Toward Pulmonary Tissue Constructs

Joshua B. White*, Nicholas J. Douville*, Christopher Moraes, James B. Grotberg and Shuichi Takayama

Department of Biomedical Engineering, University of Michigan, Ann Arbor, MI, USA

10.1 Introduction

Over 400,000 Americans die of a variety of lung diseases annually, and an additional 35 million live with a chronic lung condition. Lung diseases including obstructive conditions, infections, cancers, and fibrosis cost the US economy $154 billion annually in direct and indirect health-care expenditures and significantly detract from the quality of life of patients with these conditions [1]. Understanding and eventually addressing potential therapeutic avenues for these diseases will require an improved understanding of the cellular basis for these diseases. However, the pulmonary system remains one of the most challenging to study *in vivo*. First, the lung is a highly complex structure, both in geometry and in composition. This complex structure prevents direct manipulation and observation of cells in various regions of the lungs. Second, attempts to artificially replicate the conditions that lead to respiratory disease by introducing exogenous chemicals often introduces confounding changes in pulmonary tissue mechanics and the immune system, making it challenging to identify causative effects. Third, functional differences between humans and animals make it difficult to extrapolate results from animal models into treating human pathologies; therefore, the expenses associated with animal trials and limited translational results suggest they may be ineffective approaches to the problem.

Improved *in vitro* models could be a suitable platform to study such systems. However, the static nature of cell culture in petri dishes does not faithfully capture the complexity of the cellular microenvironment. In particular, cells associated with the pulmonary system exist in a highly complex chemical and mechanical environment, which plays a critical role in defining cellular response. Microfluidic approaches to pulmonary tissue engineering may provide the complexity required to adequately simulate features of the cellular microenvironment,

*These authors contributed equally to this chapter.

Microfluidic Cell Culture Systems. ISBN: 978-1-4377-3459-1

while addressing the issues associated with *in vivo* models: precisely modulated microenvironmental conditions that will enable direct study of the effects of various environmental factors on lung pathophysiology. In this chapter, we aim to review past and current approaches that use microengineered systems to understand and simulate various aspects of pulmonary physiology. In order to establish "design rules" for such microfluidic systems, physical and compositional aspects of lung physiology will first be discussed, followed by a discussion of the relevant pathophysiologies associated with various diseases, and a description of how microengineered platforms can be used to provide physiologically relevant models to study pulmonary systems.

10.2 Lung design

The lung is composed of a network of bifurcating airways surrounded by a dense network of connective tissue formed from walls of the alveoli. It performs various functions including gas transport and exchange with the blood, defense against inhaled pathogens, blood clot dissolution, maintenance of blood pH, air-conditioning (temperature and humidity control), phonation, and olfaction [2].

Ventilation of the lung is achieved by negative-pressure breathing: the diaphragm and external intercostal muscles contract and pull the chest wall downward and outward; the expansion of the chest wall is transmitted to the lung via the pleural fluid which occupies a thin layer (40 μm) between the lung and the chest wall. This causes the lung to expand and the pressure in the airways and alveoli to become subatmospheric; subsequently, this creates a pressure gradient that drives airflow from the mouth into the lower alveolar sacs until pressure is equalized at the end of inspiration. Unlike inspiration, expiration is a passive process during quiet breathing: the inspiratory muscles relax and the elastic recoil of the alveoli and chest wall cause a decrease in lung volume (and therefore an increased pressure); this motion establishes a new pressure gradient that favors driving air out of the lungs; expiration stops when the inward alveolar recoil is balanced by the outward recoil of the chest wall and the alveolar pressure and ambient air pressure are equalized. The airways can be divided into two distinct zones: the conducting zone that facilitates convective gas transport (roughly generations 0–16) and the respiratory zone (generations 17–23) that allows efficient mixing and diffusion of gases into and out of the pulmonary vasculature to maintain proper levels of blood oxygen and carbon dioxide (Figure 10.1).

Although the entire respiratory tract is lined by epithelial cells, the types and proportions of cells change from the upper conducting airways down to the alveolar sacs. The upper/large airways (conducting zone) contain pseudostratified columnar ciliated cells, mucus-produced goblet cells and other secretory cells, undifferentiated progenitor columnar cells, and an underlying basal layer of cells

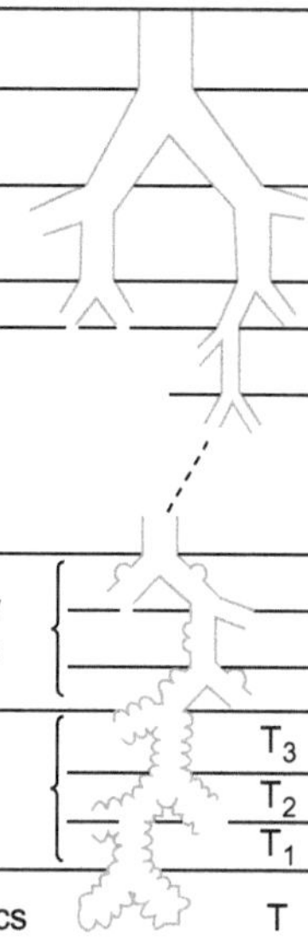

Zone	Generation			Diameter (cm)	Length (cm)	Number	Total cross-sectional area (cm²)
Conducting zone	Trachea		0	1.80	12.0	1	2.54
	Bronchi		1	1.22	4.8	2	2.33
	↓		2	0.83	1.9	4	2.13
			3	0.56	0.8	8	2.00
	Bronchioles		4	0.45	1.3	16	2.48
	↓		5	0.35	1.07	32	3.11
			↓	↓	↓	↓	↓
	Terminal bronchioles		16	0.06	0.17	6×10^4	180.0
Transitional and respiratory zones	Respiratory bronchioles		17	↓	↓	↓	↓
			18				
			19	0.05	0.10	5×10^5	10^3
	Alveolar ducts	T_3	20	↓	↓	↓	↓
		T_2	21				
		T_1	22				
	Alveolar sacs	T	23	0.04	0.05	8×10^6	10^4

FIGURE 10.1

Branching model of the airways. Larger airways are primarily responsbile for rapid gas transport. Smaller airways in the transitory and respiratory zone become alveolated and play a large role in gas mixing and exchange.

also capable of proliferation and differentiation (Figure 10.2) [3,4]. As the large airways branch into smaller airways of the bronchioles, the pseudostratified epithelium gives way to columnar and cuboidal ciliated cells; also, instead of mucus-producing goblet cells, the secretory cells are primarily Clara cells. Finally, the alveoli are lined by a simple squamous epithelium of type-I and -II alveolar cells which are optimized for gas exchange.

In addition to changes in the composition of the airway epithelium down the airway tree, distinctive features exist in the surrounding environment. In the lumen of the upper airways, the airway surface liquid (ASL) is approximately 30 μm thick; the ASL is composed of a thin layer (~7 μm) of serous periciliary liquid and a thicker layer (~20 μm) of mucus on top of the periciliary layer [5]. In healthy individuals, the ASL helps aid in host defense by trapping airborne particulates and microscopic organisms; these pathogens are then either swept out of the lungs by the ciliary escalator or forcefully expired by coughing [6]. However, in diseases such as cystic fibrosis, altered ASL properties impair mucus clearance and can lead to bacterial biofilm formation and excessive inflammatory responses [7,8]. Surrounding the epithelial cells in the conducting airways are cartilaginous

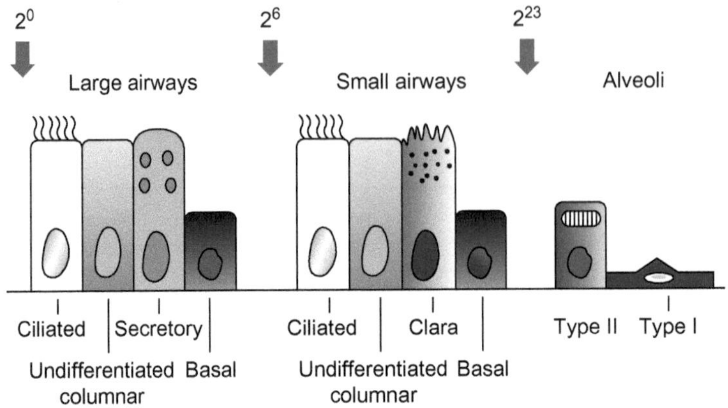

FIGURE 10.2

Cells of the airway epithelium. Large airways contain a pseudostratified epithelium with ciliated, columnar, mucus-producing secretory, and basal cells. Small airways contain similar cells but the cells are columnar in nature. Alveoli are covered by a simple squamous layer of types I and II alveolar epithelial cells and are specialized for gas exchange [3].

rings that help maintain the airways in an open state [9]. In the more distal (small) airways, the multilayer ASL gives way to a single layer of serous fluid that is enriched with numerous products secreted by Clara cells which likely aid in host defense. In the interstitium, instead of cartilage supporting these airways, a matrix of collagen I/III, elastin, fibroblasts, and smooth muscle cells surrounds the epithelium and regulates airway diameters and therefore airflow. Finally, the alveolar sacs contain a thin, surfactant-enriched fluid lining and alveolar macrophages in the lumen to minimize surface tension (and prevent alveolar collapse) and phagocytose foreign matter, respectively. The interstitium of the alveoli is highly specialized for gas exchange: the alveolar cells reside on an extremely thin basement membrane (<1 μm thick) composed primarily of laminin and collagen IV; this barrier is immediately adjacent to the pulmonary vasculature thereby allowing efficient diffusion of gases across the barrier.

Due to the distinct architectural, physical, and functional differences between the airway and alveolar regions of the lungs, microfluidic models which mimic lung function at each of these generational levels are substantially different in design and purpose. Hence, this chapter is divided by generation. Section 10.3 focuses on small airways while Section 10.4 addresses alveolar regions. Conducting airways are not addressed because the dimensions and physical parameters are better modeled using macroscale techniques. In each of the sections, we aim to provide an understanding of the important characteristics of the lung and a description of how these design parameters are incorporated in designing microfluidic *in vitro* analogs for the pulmonary system.

10.3 Engineering small airways

10.3.1 Overview

The small airways, or those with diameters less than 2 mm, refer to the distal terminal bronchioles of the conducting zone and airways of the respiratory zone of the lung [10,11]. The proximal small airways (approximately generations 10–16) function more in gas convective transport while the distal small airways (generations >17) become alveolated and also provide gas exchange with the blood. Small airways are lined by a simple epithelial monolayer of multiple cell types including ciliated, undifferentiated columnar, Clara, and basal cells (Figure 10.2). Surrounding the epithelial cells is an interstitial environment composed of extracellular matrix proteins (primarily collagens I/III and elastic fibers) and resident smooth muscle cells, fibroblasts, and monocytes. Within the lumen of the small airways, the epithelial cells are bathed in the multilayer ASL in the more proximal small airways that gives way to a single layer of serous fluid that becomes increasingly enriched in pulmonary surfactant at higher airway generations [12].

Due to the branching of the lung, small airways have a large total cross-sectional area and contribute minimally (~10%) to the total airway resistance to flow [10,11] in a healthy lung. Because the small airways lack the cartilaginous structural support found in the upper airways, and because they are surrounded by smooth muscles cells and fibroblasts that actively constrict when exposed to a variety of exogenous chemical and/or mechanical stimuli, the small airway microenvironment is predisposed to mechanical compression that leads to airway obstruction. This blockage can occur in one of two forms: (i) ASL instability forms a liquid bridge across the lumen due to the Rayleigh instability and (ii) the surface tension of the liquid lining is strong enough to pull the airway almost completely shut and, again, liquid fills the remaining cross section (Figure 10.3). However, even in diseased states, small airway collapse or blockage does not cause a drastic increase in total airway resistance but instead causes a heterogeneous distribution of air throughout the lungs [13]. This minor change in airway structure can drastically alter the mechanical and chemical milieu of the airway microenvironment and cause the development of small airway disorders or exacerbate preexisting disorders such as asthma, chronic obstructive pulmonary disease (COPD), pulmonary fibrosis, sarcoidosis, and obliterative bronchiolitis.

10.3.2 Design parameters for small airways

The physiological complexity of the small airways make it currently impossible to completely recapitulate *in vitro*; however, there are several physical and biological parameters that can guide the design of microfluidics/microengineered lung devices to answer questions about specific cell–cell or cell–environment interactions. When considering physical characteristics in designing these devices,

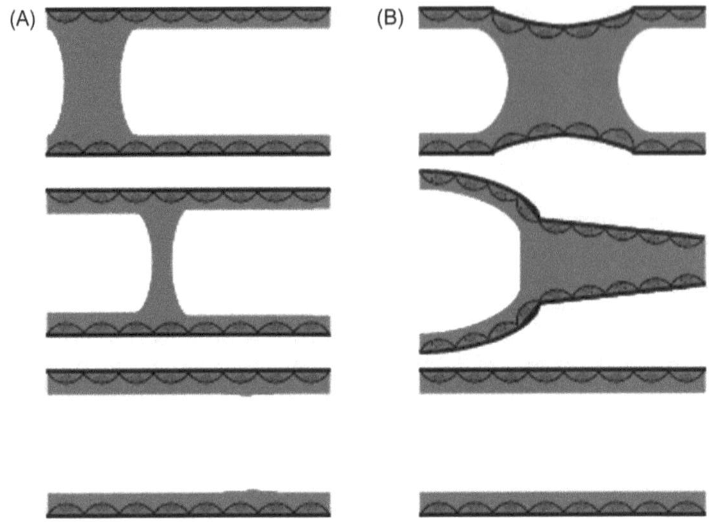

FIGURE 10.3

Schematic of models of small airway blockage. (A) Liquid plug (bridge) formation occurs when instabilities in the air–liquid interface cause a meniscus of ASL to form across the lumen of the airways. Mechanical ventilation or respiration can propagate the plug through the airway vessels which deposit liquid as it moves, leading to eventual rupture and airway reopening. (B) Compliant collapse is due to accumulation of edematous liquid in the airways; insufficient airway tethering allows the epithelium to collapse inward leading to complete flooding of the airways. Mechanical ventilation in these settings reexpands the collapsed compliant airways until the lumen is reopened.

it is important to achieve geometric, kinematic, and dynamic similarity between the model and lung microenvironment. For example, recently developed microfluidic devices have demonstrated the importance of surface tension and fluid mechanical forces on small airway pathophysiology by showing airway cell injury, death, and/or delamination of cells exposed to an air–liquid meniscus [14–16]. Here, we hope to highlight some of these important design considerations (including fluid mechanics) when engineering the microenvironments of the small airways.

10.3.2.1 Scale and geometry

Although constructed with data from the first 10 branches of the respiratory tree, the Weibel model, which is consistent with Murray's calculations for minimizing volumetric work in biological systems, can be extended to provide a first-order approximation for the diameter at each airway generation (n) [17,18]. In this model (depicted schematically in Figure 10.1), the diameter of the airways at generation n (d_n) can be estimated as a function of its generation number and the diameter of the trachea ($n = 0$; $d_o = 1.8$ cm for adult, $d_0 = 0.3$ cm for infant [19] according to Eq. (10.1). Small airway conduits undergo sequential branching from terminal bronchioles that are primarily conducting ($n = 14$) into transitional bronchioles ($n = 15-16$) and respiratory bronchioles ($n = 17-19$) where both conduction and diffusion occur. These small conduits eventually give way to highly alveolated alveolar ducts ($n = 20-22$) and the dead-end alveolar sacs ($n = 23$) which are responsible for the majority of gas exchange [18,20]. The infant respiratory tree can also be estimated from the adult model by assuming airway generation 7 of the adult lung corresponds to an infant's trachea and progresses to airway generation 18 [19]:

$$d_n = 2^{-n/3} d_0 \tag{10.1}$$

Murray also applied his principle of minimum work to determine the optimal branching angle in a symmetrically bifurcating vascular system to be 37.5° between a daughter vessel and the longitudinal axis of the parent vessel [21]. The optimal total branch angle, then, is 75°. Other measurements in dogs and humans, however, have shown that branching angle is highly variable depending on species and airway generation number [22–24].

10.3.2.2 Airflow and fluid mechanics

Based on approximate tidal volumes (V_T) and breathing rates for at-rest (500 ml, 12 breathes per minute) and intensely exercising adults (2000 ml, 60 breathes per minute) [25], and assuming an inspiration:expiration ratio of 1:1, ventilation rates for these states are estimated to be 200 and 4000 ml/s, respectively. These two states can give approximate upper and lower bounds for engineered airways. A similar approach was used to estimate infant ventilation rates as 8.5 ml/s [19]. At flow rates above approximately 500 ml/s, fluid flow in the trachea is mainly turbulent (irregular streamlines, formation of vortices, etc.); however, as air propagates to the lower airways flow becomes laminar in nature. Using the aforementioned approximated ventilation rates, airflow velocities and dimensionless fluid flow parameters can be calculated for each airway generation based on Weibel's branching model to determine flow characteristics at a specific generation.

In the equations that follow, the subscript "n" refers to the property at a specific generation and the subscript "0" refers to the flow parameters in the trachea. The principle of conservation of mass (Eq. (10.2)) is used to calculate flow velocity in systems with incompressible, Newtonian fluids and no mass accumulation over time. Similar to calculating airway diameter, the total cross-sectional area of a generation can be estimated by assuming cylindrical vessel shape ($A_n = \frac{1}{4}\pi d_n^2$) and a symmetrically bifurcating system with 2^n vessels per generation ($A_{n,\text{total}} = \frac{1}{4}\pi d_n^2 2^n$). Volumetric flow rate (Q) is the product of average axial velocity (v) and total cross-sectional area (A); therefore, Eq. (10.2) can be rearranged to calculate the average flow velocity in a single vessel (Eq. (10.3)). From Eq. (10.3) and Figure 10.1, we can quickly see why small airways typically have such low resistance and flow velocities: although individual vessel size is small, the sum of all airways results in an enormous total area and therefore minimal flow resistance and flow velocities. A similar approach can be used to calculate Reynolds (Eqs. (10.4) and (10.6)) and capillary (Eqs. (10.5) and (10.7)) numbers for each airway generation, where ρ is fluid density, μ is fluid viscosity, v is fluid velocity, D is the characteristic length of the vessel (diameter in a cylindrical airway and hydraulic diameter ($4A/P$, where A is the cross-sectional area and P is the perimeter of a rectangular conduit) in a rectangular microfluidic channel), and γ is the surface tension at the interface between the ASL and air of the lungs. These two dimensionless numbers relate the inertial to viscous forces (Re) and viscous to surface tension forces (Ca) and are important for determining the relative contributions of these stresses to the cell microenvironment. Similarly, the product of these two dimensionless numbers relates the inertial to surface tension forces:

$$\rho v_n A_n = \rho v_0 A_0 \tag{10.2}$$

$$v_n = \frac{4Q_0}{\pi d_0^2 2^{\frac{n}{3}}} \tag{10.3}$$

$$Re = \frac{\rho v D}{\mu} \tag{10.4}$$

$$Ca = \frac{\mu v}{\gamma} \tag{10.5}$$

$$Re_n = \frac{4\rho Q_0}{\pi \mu d_0 2^{\frac{2n}{3}}} \tag{10.6}$$

$$Ca_n = \frac{4\mu Q_0}{\pi \gamma d_0^2 2^{\frac{n}{3}}} \tag{10.7}$$

By using these estimations and assuming an air density ρ of 0.001 g/cm^3 and viscosity of 1.81×10^{-4} poise and a single liquid layer with properties similar to water (fluid viscosity $\mu = 0.01$ poise, surface tension $\sigma = 72.7$ dyn/cm, fluid density $\rho = 1$ g/cm^3 [26], resting (200 ml/s) and exercising (4000 ml/s)), values for each of these dimensionless parameters can be calculated for air-filled and liquid-filled airways (Figure 10.4). As airways become progressively smaller, the effects of viscosity become dominant over inertia ($Re < 1$) and surface tension becomes more important than viscosity ($Ca < 1$); therefore, these airways are inherently unstable and vulnerable to collapse and injurious surface tension forces in airway disorders.

Re and *Ca* are also dependent on fluid parameters and ventilation rates; higher flow rates result in larger *Re* and *Ca* numbers at each generation while viscosity has opposite effects on *Re* and *Ca*. Whereas increasingly viscous fluids decrease *Re* they increase *Ca* for a given generation. In diseased lungs, airways can become filled with ASL or mucus which can have viscosities thousands of times

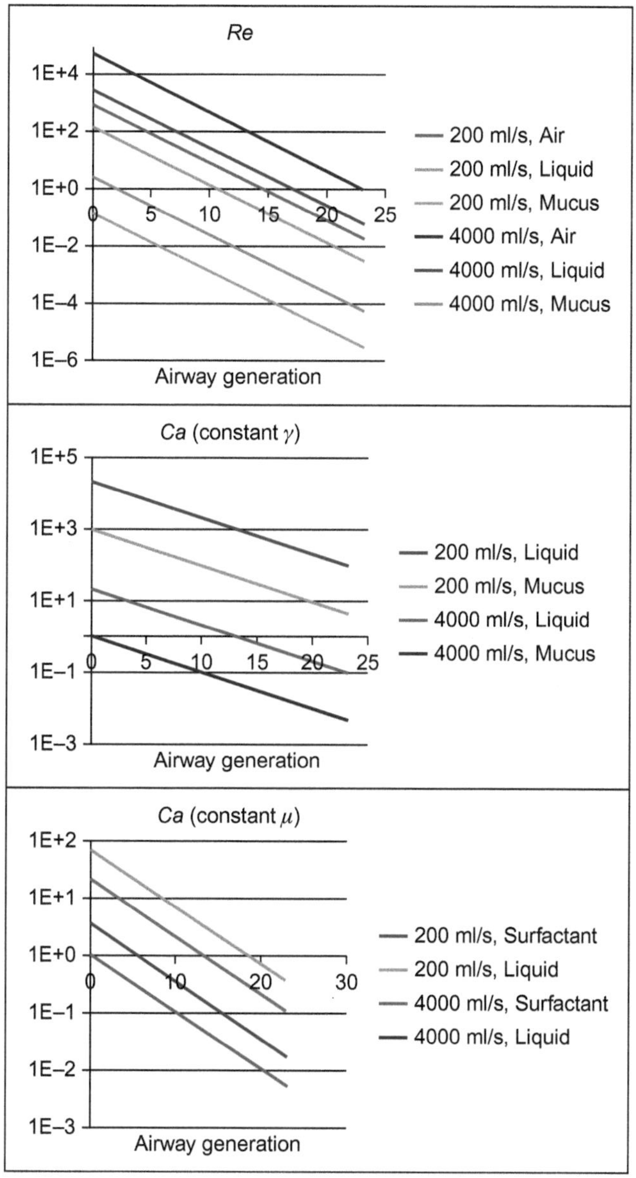

FIGURE 10.4

Fluid dynamics dimensionless numbers. *Re* and *Ca* are displayed for various airway generations based on Eqs. (10.6) and (10.7)) for an adult with tracheal diameter of 1.8 cm at resting (200 ml/s) and intensely exercising (4000 ml/s) states. (A) *Re* is tabulated and displayed for air ($\mu = 1.81 \times 10^{-4}$ poise), water ($\mu = 0.01$ poise), and mucus ($\mu = 10$ poise). (B) *Ca* is tabulated and displayed for water ($\mu = 0.01$ poise) and with surfactant-enriched liquid ($\gamma = 22$ dyn/cm) and surfactant-null liquid ($\gamma = 72.7$ dyn/cm) to show the dependence on changes in surface tension on *Ca*. (C) *Ca* is tabulated and displayed for high surface tension liquid ($\gamma = 72.7$ dyn/cm) for water ($\mu = 0.01$ poise) and mucus ($\mu = 10$ poise) to highlight the effects of increasing viscosity on *Ca*.

greater than normal [27]; examples of *Re* and *Ca* for air-filled, liquid-filled, and mucus-filled ($\mu = 10$ poise) airways are also shown in Figure 10.4. Similarly, liquid in the small airways normally contains pulmonary surfactant, a phospholipoprotein that decreases surface tension and helps maintain airways in their open state. Whereas the average surface tension of airway liquid (with active surfactant) is approximately 22 dyn/cm [28], pathophysiologies of the lung can deactivate surfactant and cause much higher surface tensions similar to water. A dimensionless plot of *Ca* (Figure 10.4) with constant viscosity (0.01 poise) shows that fluids with higher surface tension have smaller capillary numbers at all generations. Under these conditions, surface tension and viscosity become increasingly important and can further expose small airways to collapse and injury.

In cylindrical vessels, the transition from turbulent to laminar flow occurs around $Re = 2300$. Approximations of *Re* using the Weibel model of branching predict *Re* in the small airways have a maximum value in the 10's (~78.4 in air-filled airways at high flow rates) and therefore fluid flow in small airways is completely laminar and well suited for microfluidics devices which are laminar in nature.

Although beyond the scope of our discussion, the continuity equation and conservation of momentum equations can be evaluated for steady Newtonian flow in a Weibel model to describe the velocity profile (parabolic in a cylindrical tube) and wall shear stresses (Eqs. (10.8) and (10.9)). Wall shear is an important design parameter because it is the force sensed by epithelial cells that is transduced into a cell response. A similar approach can be used to obtain the velocity profile and shear in rectangular microchannels which are common to microfluidics and important for developing *in vitro* models (Eqs. (10.10) and (10.11)). Here, n is generation number, v_n and $\tau_{n,\mathrm{wall}}$ is the average velocity in the axial direction and wall shear for a generation, a_n and b_n are the radius of a cylindrical vessel or channel half-height at generation n, w is the channel width of a rectangular channel, Q_n is the generational volumetric flow rate, μ is the fluid viscosity, and r and y are the radial and y-coordinates in a cylindrical and rectangular system, respectively:

$$v_n = \frac{2Q_n}{\pi}\frac{(a_n^2 - r^2)}{a_n^4} \tag{10.8}$$

$$\tau_{n,\mathrm{wall}} = \mu\frac{dv}{dr}\bigg|_{r=a} = 4\frac{\mu v_n}{a_n} \tag{10.9}$$

$$v_n = \frac{3}{4}\frac{Q_n}{w}\frac{(b_n^2 - y^2)}{b_n^3} \tag{10.10}$$

$$\tau_{n,\mathrm{wall}} = \mu\frac{dv}{dy}\bigg|_{y=b} = \frac{3\mu v_n}{b_n} \tag{10.11}$$

10.3.2.3 Solid mechanics

Airways change length and diameter during respiration with the inflation and deflation of the lung. Using bronchoscopy, Marshall and Holden [29] found airway diameters can increase up to 28% and decrease by 13% at total lung capacity and lung residual volume during respiration, respectively. These expansions and contractions can stretch or compress the airway epithelium and surrounding interstitium and stimulate a variety of responses from airway epithelial cells, interstitial fibroblasts, smooth muscle cells, and immune cells. In fact, high transmural pressures that compress airway epithelial cells have been shown to cause an increased expression of early growth response-1 (Egr-1), endothelin-1 (ET-1), and TGF-β in epithelial cells which in turn elicit increased airway remodeling signals from fibroblasts [30,31]. These pathological signaling states can be further exacerbated in diseases such as COPD and asthma where smooth muscle cells actively constrict upward of 40% of their initial length which cause compressive strains that help lead to airway buckling [32–34]. This mechanical compression, in combination with upregulated extracellular matrix deposition due to airway remodeling, can cause a significant increase in matrix stiffness that further facilitates inflammatory factor secretion, smooth muscle and fibroblasts proliferation and migration, and increased extracellular matrix remodeling [35–38].

Because it is extremely difficult to engineer cylindrical microenvironments that would experience the three-dimensional strains associated with expansion and contraction, a linear strain model is typically used in its place. For circumferential strains, the strain magnitude, ε, is defined as the ratio of the change in circumference ($\Delta l = 2\pi a_1 - 2\pi a_0$) to the initial circumference ($2\pi a_0$) when the radius changes from a_0 to a_1 (Eq. (10.12)). Assuming the lung inflates homogeneously and isotropically, the airway radius would scale with the average lung radius. Under those conditions, the strain can be related to the tidal volume (V_T) and functional residual capacity (V_{FRC}) according to (Eq. (10.13)):

$$\varepsilon = \frac{\Delta l}{l_0} = \frac{l_1 - l_0}{l_0} \tag{10.12}$$

$$\varepsilon = \left(1 + \frac{V_T}{V_{FRC}}\right)^{\frac{1}{3}} - 1 \tag{10.13}$$

Typical values of V_{FRC} and V_T range from approximately 2–3 l and 0.5–2 l, respectively [2,25]. Using these values as an initial guide, linear strains should reach values of approximately 0.05–0.26 during inspiration. However, during mechanical ventilation, higher volume ventilation and heterogeneous distribution of airflow can cause diameter increases of 40% [39]. During forced expiration or diseases such as asthma, smooth muscle shortening of 30–40% ($\varepsilon = -0.4$–-0.3) can greatly increase airway resistance or even cause airway collapse [40,41].

10.3.2.4 Cell types

The small airways consist of cells of the airway epithelium, fibroblasts, smooth muscle cells, vascular endothelium, and immune cells. The gold standard for achieving a physiologically differentiated and functional epithelium is to culture small airway epithelial cells to confluence in submersion, then culture at an air–liquid interface for 2–3 weeks. As cells are cultured for longer times at the interface, they become more differentiated into the basal, Clara cell, and ciliated phenotypes and achieve similar genetic expression profiles to *in vivo* airways and develop strong barrier function [42,43]. However, submersion models of the epithelium can also be used to provide useful insights into response to injurious mechanical stimuli [14,44,45]. While microengineered models of the small airways have yet to utilize cocultures, larger models investigating mechanotransduction have shown that airway epithelial cells are capable of facilitating immune response and airway remodeling through the secretion of cytokines, growth factors, and other soluble factors [31,35,36,46,47].

10.3.3 Microengineered airway models

Studying small airways *in vivo* can be challenging. Current noninvasive human lung assessments rely on indirect measurements of whole lung function by spirometry, noninvasive high-resolution computed tomography (HRCT) to provide structure–function relationships of the lung parenchyma, or magnetic resonance imaging (MRI) of inhaled ^{3}He to monitor abnormalities in inhaled gas distribution [11]. However, these techniques are ineffective in capturing small airway physiology. Spirometry can only provide endpoint readouts and cannot aid in identifying the underlying causes of disorders. Similarly, HRCT lacks the spatial resolution to effectively image the small airways. Finally, MRI is limited to imaging the airway lumen, and while this is certainly a powerful tool, many small airway disorders affect the interstitial environment and further information into the lung parenchyma is needed to better understand small airway disorders.

Alternatively, using small animal models of ventilator-induced lung injury (VILI) have shown that low-volume ventilation, which is hypothesized to cause the collapse and reopening of small airways, leads to altered lung mechanics and peripheral airway injury and inflammation [48,49]. However, extrapolating these results to humans is difficult because human patients often respond differently to treatments than their small animal counterparts [50,51]. Invasive airway biopsies or bronchiolar lavage techniques have also been used to analyze small airway characteristics in diseased states [52–56]. Patients with asthma and COPD had varying levels of airway inflammation and airway remodeling as demonstrated by increased immune cell infiltration, increased staining for smooth muscle, increased collagen deposition, and goblet cell hyperplasia, among other profibrotic markers. While these *in vivo* sampling techniques shed light into the "complete" lung response to injury, they only offer a brief snapshot in time and provide no information about the underlying mechanisms that cause these disorders.

Effective *in vitro* models of the small airways can greatly contribute to our understanding of small airway injury. Recently developed platforms mimic the common phenomena of small airway closure and reopening by creating an air–liquid interface over epithelial cells in a microfluidic channel or parallel plate setup [14–16,44,45,57–59]. Using this reductionist approach, researchers were able to expose airway epithelial cells to controlled fluid stresses and model the resulting damage using numerical methods.

10.3.3.1 Small airway reopening

At low lung volumes, the injured and/or diseased lungs can become filled with edematous liquid, resulting in a variety of collapsed conditions (Figure 10.3). Patients with these collapsed airways often require the assistance of mechanical ventilators to facilitate respiration. However, it has been hypothesized that the application of low tidal volumes and lack of a positive-end expiratory pressure (PEEP) can actually lead to the cyclic closure and reopening of the fluid-filled airways (atelectrauma) [60–63]. These studies in small animals have shown that low-PEEP ventilation strategies can actually increase inflammatory signaling, increase hyaline membrane formation, and further increase airway flooding by disrupting the epithelial barrier and allowing plasma and/or interstitial liquid to enter the airway lumen.

Models of this airway reopening phenomena have recently gained particular interest to help elucidate potential mechanisms of these VILIs. To this end, a parallel plate flow chamber composed of opposing glass slides sandwiching a silicone gasket to seal the chamber was submerged in a warmed saline bath to mimic the flooded airway lumen [14,44]. The resulting chamber had a height of 1.7 mm (to approximately match the terminal bronchioles), width of 25 mm, and length of 70 mm, and fetal rat pulmonary epithelial or A549 alveolar epithelial cells were cultured to confluence on one side of the parallel plates. Using a syringe pump, a single air-finger or multiple air-fingers were passed through the flow apparatus (Figure 10.5) to achieve "reopening" velocities of approximately 0.3 or 3 cm/s to correspond to airway reopening velocities of 60–600 ml/s at the terminal bronchioles.

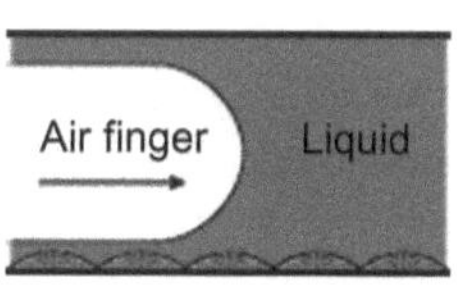

FIGURE 10.5

Model of airway reopening in a parallel plate flow system. An initially liquid-filled chamber is "reopened" by injecting airflow to model the reopening events of liquid-filled lungs.

Using calcein-AM and ethidium homodimer as markers of cell viability and cell death respectively, it was determined that slow airway reopening speeds cause a significant increase in cell death compared to faster reopening velocities. Furthermore, repeated airway reopening events resulted in a large desquamation of epithelial cells from the glass slide much like the histology seen *in vivo*. The addition of Infasurf, a biocompatible surfactant solution that decreases the surface tension of the air–liquid interface, largely protected the cells from the harmful effects of airway reopening and resulted in greater cell viability.

To explain the injurious effects caused by air-finger propagation over epithelial cells, the authors modeled the motion of a semiinfinite bubble through a Hele-Shaw cell (parallel plate flow chamber in which the channel height is much smaller than the width). Under the assumptions of steady Newtonian flow and constant surface tension, and neglecting the topological contributions of the cell monolayer, Stokes flow equations were solved using the boundary element method to obtain dimensionless parameters for shear stress, shear stress gradient, pressure, and pressure gradient in terms of Ca. These solutions were used to compute the magnitudes of each of these stresses along the epithelial surface and it was determined that an adverse pressure gradient at the front tip of the air-finger was likely most responsible for causing the most cell death.

In a modified version of the parallel plate flow chamber used by Bilek et al. and Kay et al., Yalcin et al. [45,57] investigated the effects of cell interactions and cell mechanics on injurious airway reopening events. Instead of fixing the channel height to 1.7 mm, the authors used silicone gaskets with heights of 0.5, 0.8, and 1.7 mm to monitor the effects of channel geometry on cell viability. Single or multiple airway reopening events were created by passing air-fingers with velocities of 3 mm/s (to model reopening velocity in the terminal bronchioles), 0.3 mm/s, or 30 mm/s (chosen as an order of magnitude decrease or increase of expecting reopening velocities) over an epithelial monolayer cultured on one side of the flow chamber.

Like the studies conducted by Bilek et al and Kay et al., Yalcin et al. [45] also found cells had increased staining for cell death at slow reopening velocities; furthermore, they showed that increasing the height of the channel resulted in increased cell viability for a given air-finger velocity. Also similar to previous studies, the authors showed that repeated airway reopening events resulted in increased epithelial desquamation and further increased cell death. Additionally, it was shown that cells cultured at subconfluence actually experienced a higher percentage of cell death than those cultured at high confluency [57]. Using the ratio of the cell height:cell width or cell height:cell length as an indicator of cell shape, it was determined there was no statistical significance between the subconfluent and confluent monolayer cell shapes. A computational model of a hemisphere protruding into a two-dimensional channel showed that hydrodynamic stresses applied to an "elliptical" cell are smaller than those applied to a more spherically shaped cell [64]. A second model of fluid stresses on a "sinusoidal" wall (used to mimic the airway epithelium) shows that fluid stresses applied to the cells should be essentially constant for a fixed ratio of cell height to either its length or width [65]. Therefore, the hydrodynamic stresses applied to the subconfluent and confluent cells in this study should be nearly identical and other biophysical factors such as cytoskeletal arrangement, cell–cell interactions, and cell–substrate interactions likely contribute to the ability of epithelial cells to resist fluid stress-induced cell death.

To determine the effects of cytoskeletal structure on epithelial robustness, Yalcin et al. [57] treated A549 cells with either jasplakinolide (stabilize actin) or latrunculin A (which depolymerize actin filaments) or altered experimental temperature prior to airway reopening events in their parallel plate setup. The authors found that increasing cell fluidity, which was characterized by its elastic and shear moduli, either by chemical treatment or by temperature resulted in increased cell adhesiveness and cell viability when exposed to airway reopening events. Stabilizing actin filaments by jasplakinolide treatment, however, only improved cell adhesion but had no significant improvement on cell viability.

Parallel plate models of airway reopening have led to significant insights into the potential mechanisms of epithelial damage and desquamation in small airway disorders. These models offer highly controlled and reproducible environments as well as the ability to easily image cells to monitor cell viability and adherence. Furthermore, the simple geometry allows the system to be modeled as a Hele-Shaw flow chamber which allows the computational analysis of shear stress, shear stress gradient, pressure, and pressure gradient on the epithelium. The setup, however, is limited by the rigidity and size of the glass plates. Although the 0.5–1.7 mm height falls within the range of the terminal bronchiole diameter, it would be beneficial to model even more distal airways where airway collapse is even more likely to occur. Using an elastic substrate would also be more relevant to compliant airway collapse in which the airway walls actually buckle inward. Computational modeling of reopening of compliantly collapsed airways show surface tension and elastic forces dominate low Ca flows while fluid viscous and longitudinal wall tension forces predominate at higher Ca flows; furthermore, complex stress wall pressures and shear stresses arise at the end of the bubble tip where reopening occurs that are different from the rigid parallel plate model [66]. The two-dimensional nature of the system is different from the cylindrical geometry of the airway lumen and can lead to different distribution of normal and shear stresses on the cells. Regardless, the parallel plate setup has provided useful information into the mechanisms of airway injury and provided potential therapeutic approaches to mitigate lung injury (addition of surfactant to lower surface tension, modulating epithelial properties) during mechanical ventilation.

10.3.3.2 Liquid plug rupture in small airways

During physical examinations of patients with airway disorders, a "crackling" sound is commonly heard when listening to a patients' breathing through a stethoscope [67–69]. A recently developed microfluidic device has shown that the rupture of a liquid plug (Figure 10.6) in a microchannel produces a similar sound to

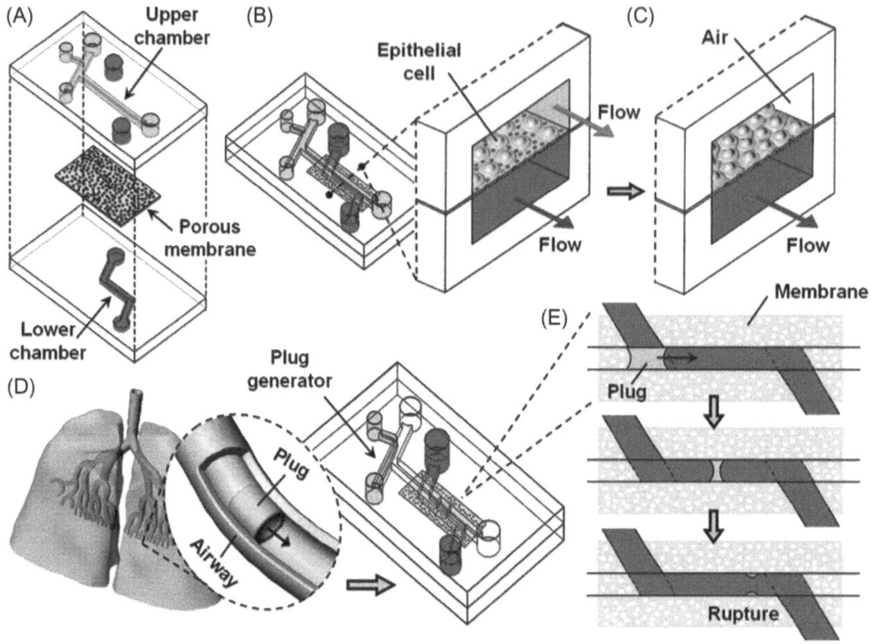

FIGURE 10.6

Microfluidic device used to model liquid plug rupture in small airways.

that heard in the clinic [15]. Therefore, this crackling sound could actually be the result of the airway reopening of a meniscus-occluded airway.

Airway occlusion by liquid plugs and their subsequent rupture (Figure 10.3A) could be a byproduct of diseased small airways, or it could further contribute to these disorders by causing injurious fluid mechanical stresses over airway epithelial cells. To investigate the contribution of liquid plug propagation and rupture to airway epithelial damage, Takayama and coworkers [15,16,58,59] have developed a microdevice to reproducibly create liquids plugs in a microchannel over airway epithelial cells (Figure 10.6). This system is composed of a wall-mounted air source regulated to deliver 10 psi airflow, a syringe pump to deliver liquid flow, a computer-controlled pinch valve, and a multilayer microfluidic device in which to culture and stimulate cells. The multilayer device is composed of 2 polydimethylsiloxane (PDMS) microfluidic channels that sandwich a 400 nm porous polyester membrane (Figure 10.6A). Airway epithelial cells are cultured to confluence in submersion culture on the porous membrane (Figure 10.6B); after reaching confluency, they are subjected to an air–liquid interface to promote differentiation and the formation of a more functional monolayer (Figure 10.6C). Once a functional monolayer is achieved, the microfluidic device is connected to both the air source and syringe pump to achieve two-layer stratified flow (Figure 10.6D). The user controls a computer that temporarily closes the pinch valve (30–100 ms), thereby blocking airflow to the device and allowing liquid to completely occlude the K-shaped channel. Upon resumption of airflow, a small bolus of liquid is propagated down the length of the cell culture channel, depositing a thin trailing liquid film along the walls until the plug eventually ruptures over the airway cells (Figure 10.6E).

Small airway monolayer functionality in the device was validated by collecting cell perfusate and analyzing by ELISA for CC10, a marker of Clara cell differentiation [15]. Similarly, barrier properties were verified by transepithelial electrical resistance (TEER) [16]. CC10 secretion into the perfusate and TEER values both increased with longer culture at an air–liquid interface, demonstrating that air–liquid interfacial culture is important to achieve a physiologically accurate model. Liquid plug propagation velocities ranging from 3 to 5 mm/s were achieved by modulating the driving air pressure and liquid volumetric flows to match the expected velocities in the terminal and respiratory bronchioles.

This microfluidic model of the small airways has demonstrated that liquid plug propagation over airway epithelial cells results in cell death and death is increased where airway reopening occurs at the point of liquid plug rupture [15]. Furthermore, cells can be protected from the harmful effects of plug propagation by culturing small airway epithelial cells at an air–liquid interface to establish a differentiated and more functional monolayer (instead of A549 or other less robust cell types) or by adding surfactant to the liquid plugs to decrease surface tension of the traveling plug [15,16]. Importantly, the system can be characterized based on several input parameters: air pressure, liquid volumetric flow rate, "pinch valve time," and channel hydrophilicity [59]. Setting these parameters in the system allows the researcher to tailor liquid plug size and propagation characteristics to achieve the desired "airway reopening" characteristics. Furthermore, the geometry of the microfluidic channel allows the system to be modeled computationally in order to evaluate the fluid stress components that cause cell injury [16,70,71]. The shear, pressure, and gradients of each have their highest magnitudes concentrated at the thin film region at the front of the liquid plug. However, the magnitude of the pressure gradient is much greater than any of the other stresses and is a likely candidate for the cause of cell death by membrane rupture.

In addition to the rigid channel geometry created by the multilayer microfluidic device, a thin membrane of PDMS can be used as one side of the microfluidic channel to allow elastic deformation of the channel [58] (Figure 10.7). As a liquid plug propagates through the channel, the membrane is able to deflect inward much like the inward deflection that occurs in diseased airways. By characterizing the elasticity of the thin PDMS membrane, the amount of inward deflection and resulting fluidic stresses can be quantified using computational modeling similar to the rigid channel geometry. Much like the airway reopening of compliantly collapsed airways, the elastic nature of the walls can actually act to amplify or

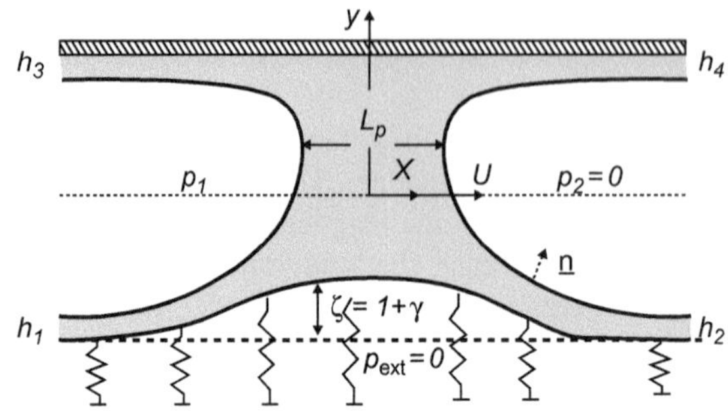

FIGURE 10.7

Model of liquid plug propagation in a microfluidic channel with an elastic PDMS substrate. The compliant nature of PDMS allows inward deflection of the membrane much like the collapse experienced in airways *in vivo*. Depending on liquid plug propagation characteristics, the inward deflection can amplify or reduce wall pressures and shear stresses.

mitigate wall shear and pressures depending on the propagation characteristics of the plug.

Microfluidic models of small airways and their accompanying computational analyses have shown that airway epithelial cells experience cell death as a result of liquid plug propagation and the mechanism is likely the pressure gradient that ruptures the cell's membrane. Cell death occurs more at the location of airway reopening where the liquid plug ruptures, and cell viability can be increased by surfactant enrichment to decrease the surface tension at the air–liquid interface or differentiating airway cells into a functional monolayer. Using PDMS as a channel material allows devices to be created that have elastic boundaries which can deflect and can be used to better model the compliant collapse and reopening of the airways.

10.3.4 Summary

Significant advances have been made in engineering the small airway microenvironment; both parallel plate flow chambers and PDMS microfluidic devices have shown airway cells die when exposed to multiple airway reopening events. Computational modeling of air-fingers and liquid plugs have both been achieved to describe the fluid mechanical stresses experienced by the airway cells and provided useful insights into managing diseased airways by decreasing surface tension with surfactants. However, the currently existing platforms are limited in several facets: (i) they do not address airway bifurcations, (ii) they are largely two-dimensional in nature, (iii) they only provide binary readouts of epithelial cell viability, and (iv) they do not address any biochemical responses of the surrounding lung environment.

Future lung microenvironments can address some or all of these shortcomings. Recent advances in microfluidics have enabled the study of liquid plugs in a microfluidic airway tree [98]. In addition to studying the nature of fluid transport at the bifurcation, it would also be interesting to analyze cell responses at the bifurcation where wall stresses are high. Similarly, analyzing the responses of airway epithelial cells to liquid bridge formation and not just propagation and rupture would offer insights into other mechanisms of damage. Modeling and experimentation (with cell incompatible oils) of liquid plug formation shows highly concentrated shear, pressure, and their respective gradients along the walls where closure first occurs [72,73]; however, experimentally this process is difficult to recreate due to the inherently stochastic nature of liquid bridge formation [74]. Current technology also makes it difficult to fabricate cylindrical microfluidic channels that are more representative of the airways; studying cells in this three-dimensional environment that is more prone to buckling and cylindrical hoop stress would be interesting. Finally, developing microfluidic systems that allow robust monitoring of biochemical signaling of the airway epithelium and other interstitial cells (fibroblasts, smooth muscle cells, immune cells) would help elucidate potential mechanisms of small airway disorders. It has been well established that airway cells can contribute to diseases through soluble factor signaling at the macroscale [31,35,36,46,47]; however, achieving a microenvironment with multiple cell types, a three-dimensional matrix environment, and the fluid stresses associated with the small airways will be essential to model airway remodeling phenomena common to asthma, COPD, IPF, and other interstitial small airway disorders.

10.4 Engineering alveolar structures

10.4.1 Overview

The exchange of oxygen and carbon dioxide occurs in the terminal divisions of the respiratory tree across the air–blood barrier (or alveolar–capillary barrier). This barrier consists of: (i) type-I alveolar epithelial cells that line air-filled alveolar sacs, (ii) capillary endothelial cells that line blood-filled pulmonary capillaries, and (iii) an interstitial region comprised of fused basement membranes for cell adherence (Figure 10.8). In addition to the type-I alveolar epithelial cells and capillary endothelial cells constituting the majority of the air–blood barrier, numerous other cell types (and accompanying cell–cell interactions) exist in alveoli and the supporting vasculature, including: (i) type-II alveolar epithelial cells, responsible for production of surfactant lining the alveolus, (ii) alveolar macrophage cells that patrol the alveolar surface to identify foreign material then recruit additional inflammatory cells, and (iii) neutrophils flowing through the pulmonary

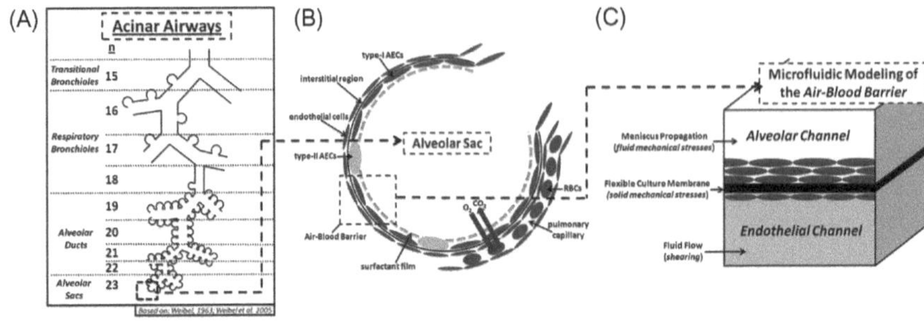

FIGURE 10.8

Schematic of alveolar structures which can be modeled using microfluidic systems. The schematic of acinar airway branching is based on the Weibel model of the airway [18] and modified to account for additional information regarding appearance of alveoli [20].

vasculature, which traverse the air–blood barrier when signaled. The air–blood barrier must be sufficiently thin to allow the passive diffusion of oxygen inward, yet sufficiently strong to prevent leakage of plasma and interstitial proteins into the air-filled alveolus [75]. When solid and fluid mechanical stresses damage the integrity of the air–blood barrier, edema-fluid fills this normally air-filled alveolar environment and pathology (such as acute respiratory distress syndrome, ARDS) results. The specific mechanisms by which these stresses impact the cells of the air–blood barrier in both physiologic and pathologic processes remain poorly understood. The role of solid mechanical stress (cyclic stretch) has been explored through traditional, well-based alveolar models; but only recently advances in microfluidic design have allowed the role of fluid mechanical stresses (and combined stress profiles) to be explored systematically in alveolar models.

10.4.2 Design parameters for the air–blood barrier

The structure and geometry of alveoli are dictated by functionality: the respiratory tree branches into hundreds of millions of alveolar sacs (estimated between 300 million [18] and 480 million [76]) to effectively maximize the surface area (143 ± 12 m^2, in a 74 kg man; greater than the size of a tennis court) available for gas diffusion [77]. Existing models have successfully captured some key cell–cell and cell–environment interactions while neglecting others based on experimental limitations. For example, when studying the impact of fluid mechanical stresses on alveolar cells, it is important to match the Reynolds and capillary numbers between the *in vivo* condition and *in vitro* model—but not inherently necessary to preserve alveolar diameter and membrane thickness to enable easier fabrication and actuation [78]. The design process is further complicated in alveolar microfluidic systems because solid and fluid mechanical stresses occur simultaneously at an extremely small scale (~250 μm diameter of alveolar sac; <10 μm diameter of pulmonary capillaries). Approximate dimensions and stresses can be easily calculated by combining elegant models of the lung's branching structure with well-established spirometry data (a commonly administered pulmonary function test). Although these dimensional models are calculated using data from larger conducting airways and spirometry data that assumes uniform distribution of stresses through the entire lung, they provide an important first-order approximation of key design parameters and illustrate key concepts about *in vivo* alveolar processes.

10.4.2.1 Scale and geometry

Dimensions of alveolar sacs

Using the Weibel model previously described in Section 10.3.2.1 and Eq. (3.1) provides an approximation for the size of the terminal alveolar sacs ($n = 23$) [18,20]. Direct measurements of lung reconstructions [79] and serial histological sectioning [80], however, demonstrate that applying the Weibel model (Eq. (3.1)) to distal airway generations leads to an underestimate of the diameter of the pulmonary acinus (~100 μm) compared to direct measurements (~250 μm).

Properties of the basement membrane

The interstitial layer is comprised of the fused basement membranes of the epithelial and endothelial layers. The thickness of the interstitial layer has been measured in humans and a variety of mammals with thicknesses normally ranging from approximately 0.2 to 0.4 μm and reaching a minimum of 0.1 μm at its narrowest point [77,81,97]. The primary proteins of the basement membrane are collagen IV and laminin with interspersed proteoglycans. The basement membrane holding the cells of the air–blood barrier is much smaller than can be reproduced through cell culture and tissue engineering platforms. In microfluidic platforms, epithelial and/or endothelial cells are routinely cultured on polyester [15] or PDMS membranes [78,82]. It is important to note that these cell culture membranes are often 1–2 orders of magnitude thicker than the basement membranes they are replicating.

Dimensions of the pulmonary capillaries

The diameter of pulmonary capillaries has been measured in humans (7.48 + 2.31 μm) and other mammals (rabbit: 6.02 + 1.95 μm; dog: 6.03 + 1.38 μm) [83]. These values, however, depend on transmural pressure and can change during exercise or with changes in gravity. In fact, Huang and coworkers [84] approximated the diameter of a capillary cross section to be approximately 5 and 8 μm at transmural pressures of 2 and 18 cmH_2O, respectively.

10.4.2.2 Mechanical stresses

Fluid stresses of the alveoli can be approximated using similar techniques outlined for stresses in the small airways. As can be seen from the calculated fluid stresses, *Re* and *Ca* for the alveolar ducts are very small (Eqs. (10.4)–(10.7), Figure 10.4). Similarly, linear strains for alveolar sacs can be calculated using Eq. (10.3). Pathologies such as ARDS, where a fraction of the alveoli may be collapsed or fluid filled, cause nonuniform ventilation and preferentially affect the alveoli which are especially prone to collapse. In these states, heterogeneous ventilation can lead to a significantly higher linear strain and the potential for overinflation [63].

10.4.3 Conventional alveolar models

Solid mechanical stretch has been implicated as a central component in nearly all alveolar pathological processes, and the role of stretch has been extensively studied in models of ARDS [85], NRDS (neonatal respiratory distress syndrome) [86,87], and VILI [88,89]. Solid mechanical stresses have been traditionally recreated by culturing alveolar cells (alveolar epithelial cells, macrophages, etc.) on the surface of a flexible membrane whose deformation can be controlled through pressure-based or mechanical actuation. The most widely used technique is the computer-controlled Flexercell (FlexCell International Corp., Hillsborough, NC) culture system that uses vacuum pressure to strain or compress (either cyclic or static) cells cultured on a flexible membrane (FlexCell International Corp.). They are also designed to interface with standard dimensions of existing cell culture systems (e.g., the 25 mm, 6-well cell culture system) which makes for easier handling and operation. Flexercell and other purely strain-based systems are the first step toward alveolar modeling within microdevices.

10.4.4 Microfluidic alveolar models

10.4.4.1 Combined stresses

Only recently advances in microfluidic design and fabrication have allowed the systematic study of the effect of fluid mechanical stresses unique to the lung. The importance of such stresses was first demonstrated in small *airway* models [14,15,45] as outlined in the previous section. Although the role of fluid stresses in alveolar models could be extrapolated from such airway models, the relative impact of fluid as compared to solid mechanical stresses could not be resolved. This is because microfluidic airway models all involve rigid cell culture membranes that do not allow dynamic manipulation of the microenvironment. Furthermore, initial alveolar models utilize an open cell culture well and therefore do not expose alveolar cells to fluid stresses generated at air–liquid interfaces.

Just recently, microfluidic alveolar models that incorporate both solid and fluid mechanical stresses, as would be experienced by alveolar cells at the air–blood barrier, have been developed. An "Alveoli-on-a-Chip" used deflection of a nonporous, flexible PDMS membrane to generate combined stresses expected in normal and pathophysiologic alveolar conditions [24]. In this system, a "dead-end" alveolar culture chamber is separated from a secondary "endothelial

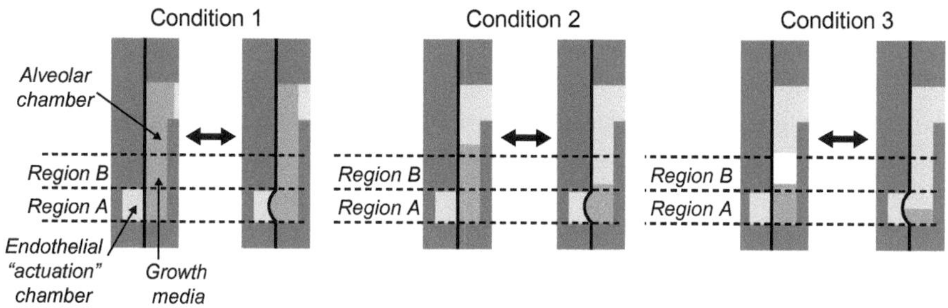

FIGURE 10.9

Schematic of stress profile in "Alveoli-on-a-Chip" as a function of growth media filling condition.

actuation channel" by a thin PDMS membrane. By varying the degree of fluid filling within the "alveolar" channel, differential strain conditions could be applied to alveolar epithelial cells cultured on different regions of the membrane. In Condition 1, (Figure 10.9) cells in region A experience solid mechanical stress from the membrane stretch and cells in region B experience minimal fluid shear stress. In Condition 2, cells in region A experience solid mechanical stretch and cells in region B experience high degrees of fluid mechanical stresses due to shear and surface tension of the air–liquid meniscus. In Condition 3, cells in region A experience combined solid and fluid mechanical stresses, while cells in region B experience neither solid nor fluid mechanical stresses.

Importantly, computational analyses of the fluid stresses along the air–liquid meniscus can help explain experimental results [78]. It is important to note that inhalation (receding stroke of the meniscus) and exhalation (advancing stroke of the meniscus) have different stress profiles. Models of a semiinfinite air-finger predict maximal values for all fluid stresses (shear stress, shear stress gradient, and pressure gradient) beneath the transient region of the meniscus in the receding direction, which supports the observed cell detachment that occurred primarily in the direction of the receding meniscus.

10.4.4.2 Gas diffusion

To effectively recapitulate the alveolar–capillary microenvironment, a barrier with high gas diffusion is desirable. Although the large surface area ($143 \pm 2\ m^2$) and high diffusion capacity (125 ± 18 to 263 ± 34 ml O_2/min mmHg) of the physiologic lung is difficult to replicate *in vitro*, a variety of microfluidic strategies have emerged to efficiently "oxygenate" blood through a thin synthetic membrane [77]. In fact, by stacking alternating "oxygen" and "vascular" layers of microchannels in parallel to maximize surface area:volume ratio, high rates of oxygen exchange through a thin PDMS membrane can be successfully achieved [90].

Burgess et al. [91] used this technique to achieve an O_2 gas permeance of 9.16×10^{-6} ml/s/cm^2/cmHg and a CO_2 value of 3.55×10^{-5} ml/s/cm^2/cmHg; similarly, Fritsche et al. [92] demonstrated the potential to utilize microfluidics for tissue engineering by culturing immortalized mouse lung epithelial cells in a microsystem mimicking the air–blood interface.

10.4.4.3 Monitoring of barrier function

Because nearly all alveolar pathophysiologies alter the permeability of the air–blood barrier, the ability to detect and quantify changes to the integrity of this barrier is an important feature of next-generation alveolar models. The presence of tight junctions can be qualitatively detected by observing characteristic proteins (occludin, ZO-1, ZO-2) or transmembrane fibrils and quantitatively measured by paracellular permeability markers or TEER [93].

Although TEER is less representative of the long-term character of the tissue than permeability markers or tight junction protein analysis, it provides an instantaneous and live-cell quantification of the integrity of a cell layer that are important for next-generation devices [94]. Using TEER as a readout, Birukov et al. [95] found pulmonary endothelial cells pretreated with pathologic levels of strain (18% elongation) and then treated with thrombin exhibit decreasing TEER while pretreatment with physiologic strain (5% elongation) maintains barrier integrity (higher TEER). TEER has also been used to show better barrier functionality of pulmonary epithelial cells cultured at an air–liquid interface compared to submersion culture (Figure 10.10) [82]. Importantly, by embedding electrodes into a multilayer microfluidic device, real-time TEER measurements can be recorded across a spectrum of frequency values to resolve the trans- and paracellular contributions to total TEER [96].

10.4.4.4 Coculture

To better understand whole-organ function, some of the latest alveolar models recreated heterocellular interactions of the air–blood barrier. For example, human alveolar epithelial cells and pulmonary microvascular endothelial cells were cocultured on opposing sides of a porous PDMS membrane within a microfluidic device (Figure 10.10B) [82]. In this coculture system, fluorescently labeled human neutrophils were incorporated into media flowing through the endothelial channel, allowing: (i) adhesion, (ii) diapedesis, and (iii) transmigration across the air–blood barrier into the alveolar channel to be monitored. The functionality of this setup was further demonstrated by incorporating bacteria into the alveolar compartment and monitoring neutrophil activation and phagocytosis of the bacterial pathogen.

10.4.5 Summary

When alveolar cells are unable to withstand solid and fluid mechanical stresses, the integrity of the air–blood barrier is compromised and pathology results. The

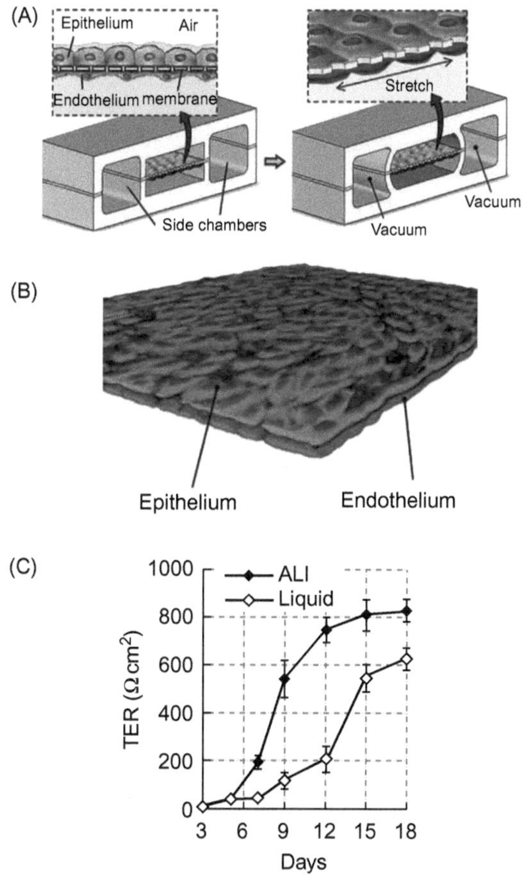

FIGURE 10.10

Next-generation "Alveoli-on-a-Chip" technologies. (A) Physiologic stresses, (B) coculture of epithelial and endothelial cells, and (C) quantification of barrier function.

mechanisms by which these stresses impact the cells of the air–blood barrier (in both physiologic and pathologic processes) remain poorly understood. Early alveolar models have revealed important insight how a single cell culture responded to a single stress type. These studies showed significant changes in the expression of surfactant proteins, inflammatory cytokines, and signaling markers in response to stretch. More recent alveolar models have taken advantage of advances in microfluidic design and fabrication to expand upon these earlier models. The microfluidic systems allow the effects of combined solid and fluid mechanical stresses to be studied in two-layered cocultures modeling the interaction of epithelial cells from the alveolar side of the air–blood barrier and endothelial cells from the vascular side. Additionally, the incorporation of sensors within the microfluidic channels allows real-time quantification of barrier function at the air–blood barrier. As microfluidic control systems continue to decrease in size and increase in complexity, a thorough understanding of the scale, stresses, and interactions occurring at the air–blood barrier becomes particularly important. We have not only highlighted the most up-to-date measurements and computational modeling available for ease of reference when designing microfluidic alveolar models, but also provided the computational framework necessary to logically estimate unforeseen variants where no direct measurements exist. As the physiologic relevance of such systems increases and the level of quantitative changes that can be detected increases in resolution, the scope of problems that can be systematically studied and disease mechanisms that can be resolved will expand significantly.

10.5 Conclusions

Understanding the cellular basis for pulmonary disease progression will require the development of novel *in vitro* models that effectively simulate the dynamic *in vivo* milieu including chemical, mechanical, and coculture conditions while avoiding the complexities inherent in studying *in vivo* model systems. Microfabrication technologies allow dynamic manipulation and monitoring of these complex microenvironments to aid in our understanding of physiology and pathophysiology of the lung. In this chapter, we have highlighted recent attempts to simulate the small airways and terminal alveolar regions of the respiratory system. Though promising, much work remains to be done in extending these systems for long-term culture, prior to establishing them as models for chronic and long-term lung conditions.

References

[1] American Lung Association, Online resource at <http://www.lungusa.org>, 2011 (accessed 03.03.11).

[2] M.G. Levitzky, Pulmonary Physiology, seventh ed., McGraw-Hill Companies, Inc, Columbus, OH, 2007.

[3] R.G. Crystal, S.H. Randell, J.F. Engelhardt, J. Voynow, M.E. Sunday, Airway epithelial cells, Proc. Am. Thorac. Soc. 5 (2008) 772–777.

[4] R.R. Mercer, M.L. Russell, V.L. Roggli, J.D. Crappo, Cell number and distribution in human and rat airways, Am. J. Respir. Cell Mol. Biol. 10 (1994) 613–624.

[5] R.C. Boucher, New concepts in the pathogenesis of cystic fibrosis lung disease, Eur. Respir. J. 23 (2004) 146–158.

[6] H. Matsui, S.H. Randell, S.W. Peretti, C.W. Davis, R.C. Boucher, Coordinated clearance of periciliary liquid and mucus from airway surfaces, J. Clin. Invest. 102 (1998) 1125–1131.

[7] H.A.W.M. Tiddens, S.H. Donaldson, M. Rosenfeld, P.D. Paré, Cystic fibrosis lung disease starts in the small airways: can we treat it more effectively? Pediatric. Pulmonol. 45 (2010) 107–117.

[8] J.C. Davies, E.W.F.W. Alton, A. Bush, Clinical review: cystic fibrosis, Br. Med. J. 335 (7632) (2007) 1255.

[9] A.T. Johnson, C.G. Lausted, J.D. Bronzino, Respiratory system, The Biomedical Engineering Handbook, second ed., CRC Press LLC, Boca Raton, FL, 2000.

[10] P.T. Macklem, The physiology of small airways, Am. J. Respir. Crit. Care Med. 157 (5) (1998) S181.

[11] R.J. Shaw, R. Djukanovic, D.P. Tashkin, A.B. Millar, R.M. Du Bois, P.A. Corris, The role of small airways in lung disease, Respir. Med. 96 (2) (2002) 67–80.

[12] J.B. Grotberg, Respiratory fluid mechanics, Phys. Fluids 23 (2) (2011) 021301, doi:10.1063/1.3517737.

[13] W. Hogg, J. Brunton, M. Kryger, R. Brown, P. Macklem, Gas diffusion across collateral channels, J. Appl. Physiol. 33 (5) (1972) 568.

[14] A.M. Bilek, K.C. Dee, D.P. Gaver III, Mechanisms of surface-tension-induced epithelial cell damage in a model of pulmonary airway reopening, J. Appl. Physiol. 94 (2003) 770.

[15] D. Huh, H. Fujioka, Y.C. Tung, N. Futai, R. Paine, J.B. Grotberg, et al., Acoustically detectable cellular-level lung injury induced by fluid mechanical stresses in microfluidic airway systems, Proc. Natl. Acad. Sci. U.S.A. 104 (48) (2007) 18886.

[16] H. Tavana, P. Zamankhan, P.J. Christensen, J.B. Grotberg, S. Takayama, Epithelium damage and protection during reopening of occluded airways in a physiologic microfluidic pulmonary airway model, Biomed. Microdevices 13 (4) (2011) 731–742, doi:10.1007/s10544-011-9543-5.

[17] C.D. Murray, The physiological principle of minimum work: I. The vascular system and the cost of blood volume, Proc. Natl. Acad. Sci. U.S.A. 12 (3) (1926) 207.

[18] E.R. Weibel, Morphometry of the human lung, J. Anat. 98 (1963) 457.

[19] K.J. Cassidy, N. Gavriely, J.B. Grotberg, Liquid plug flow in straight and bifurcating tubes, J. Biomech. Eng. 123 (6) (2001) 580, doi:10.1115/1.1406949.

[20] E.R. Weibel, B. Sapoval, M. Filoche, Design of peripheral airways for efficient gas exchange, Respir. Physiol. Neurobiol. 148 (1–2) (2005) 3–21.

[21] C.D. Murray, The physiological principle of minimum work applied to the angle of branching of arteries, J. Gen. Physiol. 9 (6) (1926) 835.

[22] K. Horsfield, G. Cumming, Angles of branching and diameters of branches in the human bronchial tree, Bull. Math. Biol. 29 (2) (1967) 245–259.

[23] K. Horsfield, G. Cumming, P. Hicken, A morphologic study of airways disease using bronchial casts, Am. Rev. Respir. Dis. 93 (6) (1967) 900–906.

[24] K. Horsfield, G. Dart, D.E. Olson, G.F. Filley, G. Cumming, Models of the human bronchial tree, J. Appl. Physiol. 31 (2) (1971) 207.

[25] W.D. McArdle, F.I. Katch, V.L. Katch, Essentials of Exercise Physiology, vol. 1, Lippincott Williams & Wilkins, Philadelphia, PA, 2006.

[26] R.W. Fox, A.T. McDonald, P.J. Pritchard, Introduction to Fluid Mechanics, sixth ed., John Wiley & Sons, Inc., 2003.

[27] S.K. Lai, Y.Y. Wang, D. Wirtz, J. Hanes, Micro- and macrorheology of mucus, Adv. Drug Deliv. Rev. 61 (2) (2009) 86–100.

[28] B.G. Hamer, R.D. Kamm, Clearance of airway liquid from the lung due to surface tension gradients Master's Thesis, Massachusetts Institute of Technology, 1995.

[29] R. Marshall, W.S. Holden, Changes in calibre of the smaller airways in man, Thorax 18 (1963) 54–58.

[30] B. Ressler, R.T. Lee, S.H. Randell, J.M. Drazen, R.D. Kamm, Molecular responses of rat tracheal cells to transmembrane pressure, Am. J. Physiol. Lung Cell. Mol. Physiol. 278 (6) (2000) L1264–L1272.

[31] M.A. Swartz, D.J. Tschumperlin, R.D. Kamm, J.M. Drazen, Mechanical stress is communicated between different cell types to elicit matrix remodeling, Proc. Natl. Acad. Sci. U.S.A. 98 (11) (2001) 6180.

[32] H.L. Hahn, P.D. Graf, J.A. Nadel, Effect of vagal tone on airway diameters and on lung volume in anesthetized dogs, J. Appl. Physiol. 41 (4) (1976) 581.

[33] W.C. Hulbert, T. McLean, B. Wiggs, P.D. Pare, J.C. Hogg, Histamine dose–response curves in guinea pigs, J. Appl. Physiol. 58 (2) (1985) 625.

[34] A.L. James, J.C. Hogg, L.A. Dunn, P.D. Paré, The use of the internal perimeter to compare airway size and to calculate smooth muscle shortening, Am. Rev. Respir. 138 (1) (1988) 136–139.

[35] M.M. Choe, P.H.S. Sporn, M.A. Swartz, An *in vitro* airway wall model of remodeling, Am. J. Physiol. Lung Cell. Mol. Physiol. 285 (2) (2003) L427–L433.

[36] M.M. Choe, Extracellular matrix remodeling by dynamic strain in a three-dimensional tissue-engineered human airway wall model, Am. J. Respir. Cell Mol. Biol. 35 (3) (2006) 306–313, doi:10.1165/rcmb.2005-0443OC.

[37] N.A. Hasaneen, S. Zucker, J. Cao, C. Chiarelli, R.A. Panettieri, H.D. Foda, Cyclic mechanical strain-induced proliferation and migration of human airway smooth muscle cells: role of EMMPRIN and MMPs, FASEB J. 19 (11) (2005) 1507.

[38] F. Le Bellego, S. Plante, J. Chakir, Q. Hamid, M.S. Ludwig, Differences in MAP kinase phosphorylation in response to mechanical strain in asthmatic fibroblasts, Respir. Res. 7 (2006) 68.

[39] L. Gattinoni, E. Carlesso, P. Cadringher, F. Valenza, F. Vagginelli, D. Chiumello, Physical and biological triggers of ventilator-induced lung injury and its prevention, Eur. Respir. J. 22 (November) (2003) 15s–25s, doi:10.1183/09031936.03.00021303.

[40] C.Y. Seow, L. Wang, P.D. Paré, Airway narrowing and internal structural constraints, J. Appl. Physiol. 88 (2) (2000) 527–533.

[41] P.D. Paré, B.R. Wiggs, A. James, J.C. Hogg, C. Bosken, The comparative mechanics and morphology of airways in asthma and chronic obstructive pulmonary disease, Am. Rev. Respir. Dis. 143 (5) (1991) 1189–1193.

[42] A.A. Pezzulo, T.D. Starner, T.E. Scheetz, G.L. Traver, A.E. Tilley, B.G. Harvey, et al., The air–liquid interface and use of primary cell cultures are important to recapitulate the transcriptional profile of *in vivo* airway epithelia, Am. J. Physiol. Lung Cell. Mol. Physiol. 300 (1) (2011) L25.

[43] A. Dvorak, A.E. Tilley, R. Shaykhiev, R. Wang, R.G. Crystal, Do airway epithelium air–liquid cultures represent the *in vivo* airway epithelium transcriptome? Am. J. Respir. Cell Mol. Biol. 44 (4) (2010) 465–473, doi:10.1165/rcmb.2009-0453OC.

[44] S.S. Kay, A.M. Bilek, K.C. Dee, D.P. Gaver III, Pressure gradient, not exposure duration, determines the extent of epithelial cell damage in a model of pulmonary airway reopening, J. Appl. Physiol. 97 (1) (2004) 269–276.

[45] H.C. Yalcin, S.F. Perry, S.N. Ghadiali, Influence of airway diameter and cell confluence on epithelial cell injury in an *in vitro* model of airway reopening, J. Appl. Physiol. 103 (5) (2007) 1796.

[46] D.J. Tschumperlin, Mechanical stress triggers selective release of fibrotic mediators from bronchial epithelium, Am. J. Respir. Cell Mol. Biol. 28 (2) (2003) 142–149, doi:10.1165/rcmb.2002-0121OC.

[47] D.J. Tschumperlin, G. Dai, I.V. Maly, T. Kikuchi, L.H. Laiho, A.K. McVittie, et al., Mechanotransduction through growth-factor shedding into the extracellular space, Nature 429 (6987) (2004) 83–86.

[48] E. D'Angelo, M. Pecchiari, P. Baraggia, M. Saetta, E. Balestro, J. Milic-Emili, Low-volume ventilation causes peripheral airway injury and increased airway resistance in normal rabbits, J. Appl. Physiol. 92 (3) (2002) 949.

[49] E. D'Angelo, A. Koutsoukou, P.D. Valle, G. Gentile, M. Pecchiari, Cytokine release, small airway injury, and parenchymal damage during mechanical ventilation in normal open-chest rats, J. Appl. Physiol. 104 (1) (2008) 41–49, doi:10.1152/japplphysiol.00805.2007.

[50] N. Krug, K.F. Rabe, Animal models for human asthma: the perspective of a clinician, Curr. Drug Targets 9 (6) (2008) 438–442.

[51] G.R. Zosky, P.D. Sly, Animal models of asthma, Clin. Exp. Allergy 37 (7) (2007) 973–988.

[52] K. Kuwano, C.H. Bosken, P.D. Paré, T.R. Bai, B.R. Wiggs, J.C. Hogg, Small airways dimensions in asthma and in chronic obstructive pulmonary disease, Am. Rev. Respir. Dis. 148 (5) (1993) 1220–1225.

[53] N. Carroll, J. Elliot, A. Morton, A. James, The structure of large and small airways in nonfatal and fatal asthma, Am. Rev. Respir. Dis. 147 (2) (1993) 405–410.

[54] N. Carroll, S. Carello, C. Cooke, A. James, Airway structure and inflammatory cells in fatal attacks of asthma, Eur. Respir. J. 9 (4) (1996) 709–715, doi:10.1183/09031936.96.09040709.

[55] N. Carroll, C. Cooke, A. James, The distribution of eosinophils and lymphocytes in the large and small airways of asthmatics, Eur. Respir. J. 10 (2) (1997) 292–300, doi:10.1183/09031936.97.10020292.

[56] Q. Hamid, Y. Song, T.C. Kotsimbos, E. Minshall, T.R. Bai, R.G. Hegele, et al., Inflammation of small airways in asthma, J. Allergy Clin. Immunol. 100 (1) (1997) 44–51.

[57] H.C. Yalcin, K.M. Hallow, J. Wang, M.T. Wei, H.D. Ou-Yang, S.N. Ghadiali, Influence of cytoskeletal structure and mechanics on epithelial cell injury during cyclic airway reopening, Am. J. Physiol. Lung Cell. Mol. Physiol. 297 (5) (2009) L881.

[58] Y. Zheng, H. Fujioka, S. Bian, Y. Torisawa, D. Huh, S. Takayama, et al., Liquid plug propagation in flexible microchannels: a small airway model, Phys. Fluids 21 (7) (2009) 071903, doi:10.1063/1.3183777.

[59] H. Tavana, Q.Yi. Chuan-Hsien Kuo, B. Lee, D. Mosadegh, P.J. Huh, J.B. Christensen, et al., Dynamics of liquid plugs of buffer and surfactant solutions in a micro-engineered pulmonary airway model, Langmuir 26 (5) (2010) 3744–3752, doi:10.1021/la903038a.

[60] L.N. Tremblay, D. Miatto, Q. Hamid, A. Govindarajan, A.S. Slutsky, Injurious ventilation induces widespread pulmonary epithelial expression of tumor necrosis factor-α and interleukin-6 messenger RNA, Crit. Care Med. 30 (8) (2002) 1693–1700.

[61] E.P. Argiras, C.R. Blakeley, M.S. Dunnill, S. Otremski, M.K. Sykes, High PEEP decreases hyaline membrane formation in surfactant deficient lungs, Br. J. Anaesth. 59 (10) (1987) 1278.

[62] B.K. Sandhar, D.J. Niblett, E.P. Argiras, M.S. Dunnill, M.K. Sykes, Effects of positive end-expiratory pressure on hyaline membrane formation in a rabbit model of the neonatal respiratory distress syndrome, Intensive Care Med. 14 (5) (1988) 538–546.

[63] R.D. Hubmayr, Perspective on lung injury and recruitment: a skeptical look at the opening and collapse story, Am. J. Respir. Crit. Care Med. 165 (12) (2002) 1647–1653, doi:10.1164/rccm.2001080-01CP.

[64] D.P. Gaver III, S.M. Kute, A theoretical model study of the influence of fluid stresses on a cell adhering to a microchannel wall, Biophys. J. 75 (2) (1998) 721–733.

[65] A.M. Jacob, D.P. Gaver, An investigation of the influence of cell topography on epithelial mechanical stresses during pulmonary airway reopening, Phys. Fluids 17 (3) (2005) 031502, doi:10.1063/1.1862642.

[66] D.P. Gaver III, D. Halpern, O.E. Jensen, J.B. Grotberg, The steady motion of a semi-infinite bubble through a flexible-walled channel, J. Fluid Mech. 319 (1996) 25–65.

[67] J.C. Hogg, F. Chu, S. Utokaparch, R. Woods, W.M. Elliott, L. Buzatu, et al., The nature of small-airway obstruction in chronic obstructive pulmonary disease, N. Engl. J. Med. 350 (26) (2004) 2645–2653.

[68] D.P. Gaver III, O.E. Jensen, D. Halpern, Surfactant and airway liquid flows, Lung Biol. Health Dis. 201 (2005) 191.

[69] N. Yasuda, K. Gotoh, Y. Yagi, K. Nagashima, T. Sawa, M. Nomura, et al., Mechanism of posturally induced crackles as predictor of latent congestive heart failure, Respiration 64 (5) (1997) 336–341.

[70] H. Fujioka, J.B. Grotberg, Steady propagation of a liquid plug in a two-dimensional channel, J. Biomech. Eng. 126 (5) (2004) 567, doi:10.1115/1.1798051.

[71] H. Fujioka, J.B. Grotberg, The steady propagation of a surfactant-laden liquid plug in a two-dimensional channel, Phys. Fluids 17 (8) (2005) 082102, doi:10.1063/1.1948907.

[72] C.-F. Tai, S. Bian, D. Halpern, Y. Zheng, M. Filoche, J.B. Grotberg, Numerical study of flow fields in an airway closure model, J. Fluid Mech. 677 (2011) 483–502.

[73] S. Bian, C.-F. Tai, D. Halpern, Y. Zheng, J.B. Grotberg, Experimental study of flow fields in an airway closure model, J. Fluid Mech. 647 (2010) 391–402.

[74] D. Halpern, J.B. Grotberg, Fluid-elastic instabilities of liquid-lined flexible tubes, J. Fluid Mech. 244 (1992) 615.

[75] J.N. Maina, J.B. West, Thin and strong! the bioengineering dilemma in the structural and functional design of the blood-gas barrier, Physiol. Rev. 85 (2005) 811–844.

[76] M. Ochs, J.R. Nyengaard, A. Jung, L. Knudsen, M. Voigt, T. Wahlers, et al., The number of alveoli in the human lung, Am. J. Respir. Crit. Care Med. 169 (2004) 120.

[77] P. Gehr, M. Bachofen, E.R. Weibel, The normal human lung: ultrastructure and morphometric estimation of diffusion capacity, Respir. Physiol. 32 (1978) 121.

[78] N.J. Douville, P. Zamankhan, Y.C. Tung, R. Li, B.L. Vaughan, C.F. Tai, et al., Combination of fluid and solid mechanical stresses contribute to cell death and detachment in a microfluidic alveolar model, Lab Chip 11 (2011) 609.

[79] B. Haefeli-Bleuer, E.R. Weibel, Morphometry of the human pulmonary acinus, Anat. Rec. 220 (1988) 401.

[80] J.E. Hansen, E.P. Ampaya, Human air space shapes, sizes, areas, and volumes, J. Appl. Physiol. 38 (6) (1975) 990.

[81] E.K. Birks, O. Mathieu-Costello, Z. Fu, W.S. Tyler, J.B. West, Comparative aspects of the strength of pulmonary capillaries in rabbit, dog, and horse, Respir. Physiol. 97 (1994) 235.

[82] D. Huh, B.D. Matthews, A. Mammoto, M. Montoya-Zavala, H.Y. Hsin, D.E. Ingber, Reconstituting organ-level lung functions on a chip, Science 328 (2010) 1662.

[83] C.M. Doerschuk, N. Beyers, H.O. Coxson, B. Wiggs, J.C. Hogg, Comparison of neutrophil and capillary diameters and their relation to neutrophil sequestration in the lung, J. Appl. Physiol. 74 (6) (1993) 3040.

[84] Y. Huang, C.M. Doershuk, R.D. Kamm, Computational modeling of RBC and neutrophil transit through pulmonary capillaries, J. Appl. Physiol. 90 (2001) 545.

[85] Q.M. Ning, X.R. Wang, Response of alveolar type II epithelial cells to mechanical stretch and lipopolysaccharide, Respiration 74 (2007) 579.

[86] I.B. Copland, D. Reynaud, C. Pace-Asciak, M. Post, Mechanotransduction of stretch-induced prostanoid release by fetal lung epithelial cells, Am. J. Physiol. Lung Cell Mol. 291 (2006) L487.

[87] I.B. Copland, M. Post, Stretch-activated signaling pathways responsible for early response gene expression in fetal lung epithelial cells, J. Cell Physiol. 210 (2007) 133.

[88] D.J. Tschumperlin, S.S. Margulies, Equibiaxial deformation-induced injury of alveolar epithelial cells in vitro, Am. J. Physiol. 275 (1998) L1173.

[89] N.E. Vlahakis, M.A. Schroeder, A.H. Limper, R.D. Hubmayr, Stretch induces cytokine release by alveolar epithelial cells in vitro, Am. J. Physiol. Lung Cell. Mol. Physiol. 277 (1999) 167.

[90] T. Kniazeva, J.C. Hsiao, J.L. Charest, J.T. Borenstein, A microfluidic respiratory assist device with high gas permanence for artificial lung applications, Biomed. Microdevices 13 (2011) 315–323.

[91] K.A. Burgess, H.H. Hu, W.R. Wagner, W.J. Federspiel, Towards microfabricated biohybrid artificial lung modules for chronic respiratory support, Biomed. Microdevices 11 (2009) 117.

[92] C.S. Fritsche, O. Simsch, E.J. Weinberg, B. Orrick, C. Stamm, M.R. Kaazempur-Mofrad, et al., Pulmonary tissue engineering using dual-compartment polymer scaffolds with integrated vascular tree, Int. J. Artif. Org. 32 (2009) 701.

[93] B.M. Gumbiner, Breaking through the tight junction barrier, J. Cell Biol. 123 (6) (1993) 1631–1633.

[94] M.S. Balda, J.A. Whitney, C. Flores, S. Gonzalez, M. Cereijido, K. Matter, Functional dissociation of paracellular permeability and transepithelial electrical resistance and disruption of the apical-basolateral intramembrane diffusion barrier by expression of a mutant tight junction membrane protein, J. Cell Biol. 134 (1996) 1032.

[95] K.G. Birukov, J.R. Jacobson, A.A. Flores, S.Q. Ye, A.A. Birukova, A.D. Verin, et al., Magnitude-dependent regulation of pulmonary endothelial cell barrier function by cyclic stretch, Am. J. Physiol. Lung Cell. Mol. Physiol. 285 (2003) L785.
[96] N.J. Douville, Y.C. Tung, R. Li, J.D. Wang, M.E.H. El-Sayed, S. Takayama, Fabrication of two-layered channel system with embedded electrodes to measure resistance across epithelial and endothelial barriers, Anal. Chem. 82 (2010) 2505.
[97] E.R. Weibel, B.W. Knight, A morphometric study on the thickness of the pulmonary air–blood barrier, J. Cell Biol. 21 (3) (1964) 367.
[98] Y. Song, M. Baudoin, P. Manneville, C.N. Baroud, The air-liquid flow in a microfluidic airway tree. Med. Eng. Phys. 33 (7) (2011) 849.

CHAPTER

Microfabricated Kidney Tissue Models

11

Else M. Frohlich and Joseph L. Charest

Department of Biomedical Microsystems, Draper Laboratory, Cambridge, MA, USA

11.1 Introduction

Microfabricated *in vitro* kidney tissue models replicate essential components of *in vivo* kidney physiology *in vitro*, providing a platform for direct observation of controlled yet physiologically representative kidney tissue. In general, *in vitro* kidney tissue models consist of four main components. Cells, such as cell lines or cells derived or harvested from kidneys, comprise the first main component. Materials, such as synthetic polymers or biological macromolecules or proteins, comprise the second component. Chemistry, such as the composition of media flowing through the device or present on the surface of the materials, comprise the third component. Mechanical parameters, such as surface topographic structures, environmental architecture, and fluid properties, comprise the fourth component. Microfabricated kidney tissue models improve upon other *in vitro* kidney tissue models by precisely controlling the geometry of device components via high-resolution fabrication and forming processes. Control over device component geometry consequently dictates control over mechanical parameters which influence and guide kidney cell and tissue structure and function. In addition, microfabrication methods create platforms compatible with the various cells, materials, and chemistries which also provide cues leading to replication of critical kidney function *in vitro*.

This chapter provides a background of kidney structure and function and how they translate to signaling elements which are crucial to replicate in an *in vitro* model of kidney tissue. The significance of kidney tissue models as platforms to study disease is emphasized as well as their importance in applications such as toxicity tests and drug screenings. A brief examination of current *in vitro* kidney cell culture methods is given followed by an in-depth review of current microfabricated kidney tissue models. This chapter aims to provide an understanding of the complex relationship between kidney structure and function, the crucial elements of *in vitro* tissue models, how those models have evolved, and what to expect from the more advanced current microfabricated models of kidney tissue.

Microfluidic Cell Culture Systems. ISBN: 978-1-4377-3459-1

11.2 Significance of microfabricated kidney tissue models

The kidneys filter over 1700 l of blood per day [1,2], removing toxins and waste products from the body and maintaining concentrations of water, salt, and other electrolytes [1,3] through complex mechanisms of filtration, absorption, and secretion. In order to carry out these functions, kidney physiology consists of a high degree of structural complexity, which when disrupted often leads to kidney malfunction and disease. Chronic kidney disease is a prevalent, life-threatening condition that currently affects millions of people worldwide [4]. There are over 50,000 patients on the kidney transplant waiting list in the United States alone, and as a result, the majority of patients suffering from end-stage renal disease (ESRD) depend on kidney replacement therapies or hemodialysis treatments. These treatments are costly and expected to rise, with ESRD treatment costs accounting for 6.3% of all Medicare expenditures and new ESRD cases per year projected to increase by 460,000 in 2030 [5].

Kidney tissue modeling provides a means to prevent kidney disease through a better understanding of basic kidney biology and the screening of toxic and harmful effects from potential insults, such as pharmaceuticals. Kidney tissue modeling also provides a means to enhance and accelerate kidney therapy development by providing a convenient platform for therapy discovery and development research. As kidney tissue models increase in sophistication, they replicate a larger fraction of native kidney function, providing a path to regenerative kidney therapies culminating in a synthetic source for organ repair or replacement. Microfabricated kidney tissue models currently represent a sophisticated version of kidney tissue modeling, since they allow a high degree of control and consistency in modeling kidney tissue. The resulting microfabricated platforms offer physiologic realism in cell and tissue function, well-controlled and repeatable test conditions, and ease of access for direct observation and data collection.

11.3 Kidney structure and function relationship

Kidney tissue modeling requires replication of the complex kidney tissue structure *in vitro* as tissue structure is both influenced by and is a driver for kidney function. The human kidney consists of an outer cortex and an inner medulla and contains 1–4 million nephrons, which are the basic, functional unit of the kidney [2]. The nephron comprises a glomerulus that filters incoming blood and a tubule that modifies the filtrate before excreting it as urine. The tubule contains four distinct sections: the proximal convoluted tubule, the loop of Henle, the distal convoluted tubule (DCT), and the collecting duct, each with distinct and specialized transport functions. The tubule is lined by polarized epithelial cells which control bidirectional transport of solutes and water between the filtrate and the interstitium. The structure of renal tubular epithelial cells, shown in Figure 11.1, varies considerably

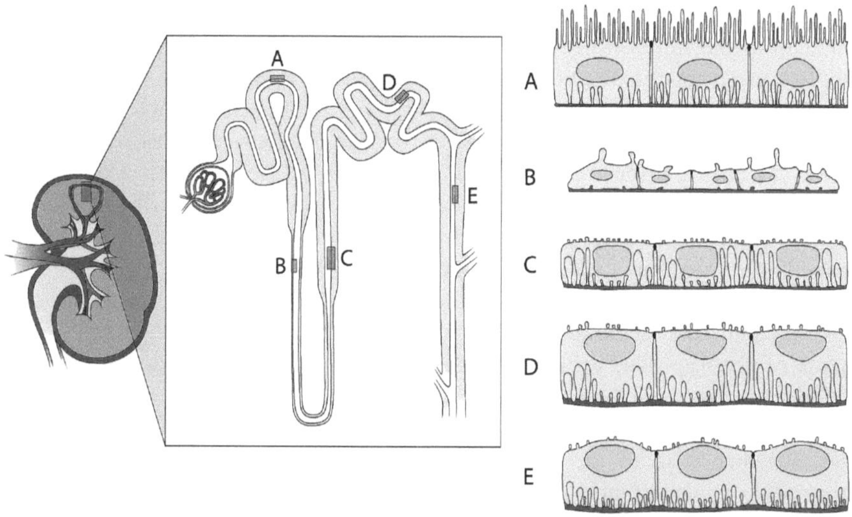

FIGURE 11.1

The functional heterogeneity of the nephron correlates with the varying renal epithelial cell type and structure along the tubule. All tubular epithelial cells are in direct contact with a BM and are interconnected by TJs that separate the luminal compartment from the lateral intercellular space. (A) PTCs display a characteristic brush border of numerous microvilli on their apical face and contain interdigitating basolateral folds resulting in a moderate surface area. These structures allow PTC epithelium to reabsorb a major fraction of water and solutes from the tubular fluid and secrete various organic compounds into the tubular lumen. (B) The descending thin limb of the loop of Henle contains relatively flat cells in a simple organization with sparse microvilli on their apical surface and few basal infoldings. (C) Cells of the thick ascending limb (TAL) of the loop of Henle are typically interdigitated, contain conspicuous lateral cell processes and infoldings, and have short, stubby microvilli. The resulting water impermeable TAL actively reabsorbs salt to concentrate urine. (D) Cells of the DCT are taller than those of the TAL, have high basolateral surface area, and large lateral interdigitating processes. The large basolateral surface area contains active Na/K pumps that are responsible for sodium and calcium reabsorption into the blood. (E) Cortical collecting duct (CCD) cells are typically noninterdigitated, contain basal infoldings that are restricted to the basal cell portion, and have only a few short microvilli on their apical surface. CCD epithelium is responsible for the remaining water and solute reabsorption before excreting the filtrate as urine.

at different sections of the nephron and correlates with function [1,6,7]. For example, healthy primary rat proximal tubular cells (PTCs) exhibit cuboidal or columnar shapes, well-defined nuclei, and uniform cytoplasm with close contact between neighboring cells [8,9]. PTCs have a highly developed structure that includes microvilli, numerous mitochondria, and extensive interlocking between cells [1,7], which permits the active reabsorption of up to 80% of solutes and water from the filtrate [2,10]. Furthermore, epithelial lumen size along the renal tubule corresponds to variations in glomerular filtration rate and is reflected in the morphology of the nephron segments [7,11]. The relationship between the rate of filtrate flow and lumen diameter allows tubular sections, such as the proximal tubule, to alter transport activity in proportion to filtered volume. Conceivably, *in vitro* models of kidney tissue must replicate the complex and varying *in vivo* kidney environment to elicit natural cell response and tissue formation in order to properly represent renal function, injury, and diseases.

Renal injury and pathology often depend on and alter kidney tissue structure and function. Due to the functional and morphological heterogeneity among segments of the nephron, cellular response to structural damage and pathological conditions may vary between renal cell populations. However, the interdependence between the components of the kidney implies that damage to one structure, whether it is to the blood vessels, glomeruli, tubules, specific nephron segments, or interstitium, will almost always secondarily affect other structures and segments [12]. Structural changes in kidney tissue can be due to direct injury to cells, tissue inflammation, or obstruction of renal excretion [13–16], all of which can cause buildup of renal toxicity and lead to acute or chronic renal failure. Obstruction of renal filtration mechanisms stems from localized structural changes, such as tubular cyst formation, found in polycystic kidney diseases [1,17,18], and abnormal accumulation of extracellular matrix (ECM) proteins, known as fibrosis [12,15,19]. Autosomal-dominant polycystic kidney disease results in fluid-filled cysts in the kidneys [17] that arise from preexisting nephrons as defects in the structure of intact tubules [18,20]. Cyst growth and formation ultimately result in chronic renal failure in most patients [17,21]. Fibrosis can lead to complete destruction of kidney tissues and renal failure and commonly stems from changes in the basement membrane (BM) [22]. Regardless of the origin of the disease, structural changes due to kidney cell and tissue damage disrupt renal function, impacting all kidney components and ultimately leading to renal failure.

Toxicity of chemical compounds to the kidney depends on the structure and function of kidney tissues. Drug-induced kidney toxicity accounts for 20% of acquired kidney failures in the general population and up to 66% in older adults [15]. Antibiotics, anticancer drugs, certain chemicals, and metals [6] are often associated with toxic injury to kidneys with the incidence of nephrotoxic injury as high as 36% due to antibiotics [14]. Kidney functional requirements necessitate high blood-flow rates and a large number of cellular transport systems, resulting in high exposure levels of toxins in proximal tubule epithelial, glomerular, and interstitial cells [15]. Therefore, these cells are particularly vulnerable to drug-induced nephrotoxicity, chemical injury, and changes in renal structure and function [6,14,23]. Generally, drug-induced tubulointerstitial disease results from direct toxic injury to the tubules or inflammation resulting from an immunogenic response [15,16]. These toxic responses are enabled by the structure and function of the kidney tubule. In a similar fashion, the toxic responses can drive changes in structure and function of the kidney. Drugs attach to components of tubular cells, resulting in damage to tubular and BM structure [1,16,24]. Covalent binding to critical cellular proteins, oxidative stress, and disruption of lysosomal structures are additional mechanisms through which nephrotoxicants can damage renal tubular cells. Toxins may also impair the normal function of mitochondria and disrupt cell membranes, which leads to alterations in epithelia permeability and shifts in ion gradients. Furthermore, direct tubular injury can disrupt cell polarity, a crucial element in maintaining tubular reabsorption, a key function of the kidney tubule [14,25]. *In vitro* models of kidney tissue provide an opportunity for direct observation of kidney tissue structure and function during dosing of drugs, thereby allowing novel insight into levels of drug toxicity and generating unique opportunities for drug safety evaluation. Current kidney toxicity testing methods, both *in vivo* and *in vitro*, do not effectively combine physiological accuracy or well-controlled parameters in one platform. Improved microfabricated kidney tissue models, however, would provide more physiological structure and function, thereby improving the accuracy of toxicity evaluations.

11.4 Traditional kidney tissue models

Traditional kidney tissue models can be separated into two types of systems: conventional *in vitro* models using static cell culture surfaces and *in vivo* animal models. Both models allow basic biological studies of cell function and tissue structure formation, and tests for therapeutic and drug-screening applications. Traditional *in vitro* models commonly use petri dishes, glass slides, or tissue culture flasks and utilize either primary cultures of renal cells or transformed cell lines originally derived from tubular epithelium [23]. Traditional *in vitro* models provide convenient, cultured kidney epithelial cell layers, yet limit accuracy due to a lack of a realistic microenvironment, physiologically relevant parameters, and complexity as seen *in vivo*. Animal models, on the other hand, provide realistic physiology and complexity, but have limited control of parameters, innate variations, and several technical limitations. For example, in renal toxicity screening, animal models require large quantities of drug, do not reveal toxicity until severe kidney damage occurs, and may demonstrate varying tissue response when compared to human-derived tubular epithelium [26]. *In vitro* cell culture models and animal models provide valuable insights, though they limit direct observations of tubular response to kidney diseases and lack accuracy and control of cell–microenvironment conditions.

Advances in cell culturing techniques enabled by microfabricated devices can drive *in vitro* models toward the physiologic complexity of an *in vivo* animal model system. Microfabrication can facilitate *in vitro* models that incorporate physical and chemical parameters including flow-induced shear stress (FSS), surface topography, stiffness, or chemical patterning to mimic physiological events such as flow in vessels, small structures in the ECM, tension or compression of tissue, and chemical signaling. Additionally, cell types with adequate function needed for *in vitro* devices are becoming more accessible, as are growth factors and ECM proteins [27]. Improved *in vitro* models will enhance and lengthen the timespan of expression of kidney-specific function [28], remain easily converted to high-throughput analysis applications, and provide easy access for direct observation of cell and tissue response.

11.5 Crucial signaling elements for kidney tissue models

The kidney is a complex organ with unique structures and conditions that influence function of multiple cell types. The *in vivo* microenvironment, which signals kidney cells, contains both mechanical and biochemical elements. The crucial signaling elements found in the native kidney microenvironment largely depend on two main components of the kidney: the BM and fluid flows. BMs contact all endothelial and epithelial monolayers [29], while fluid flows, specifically intratubular filtrate flow, which results in FSS, contacts epithelial cells. Microfabricated kidney tissue models can deliver crucial signaling elements by tailoring cell substrates to mimic BM and controlling perfusate flow to mimic intratubular filtrate flow. Historically [30,31], cells cultured in flat, static recepticles resulted in many discoveries in molecular and cell biology, although limitations of these methods are now being recognized. Microfabricated kidney tissue models can leverage precise control of the microenvironment to suit one or more kidney cell types. The crucial signaling elements delivered via cell substrate and fluid flow, in particular, can elicit improved kidney-specific cell function from *in vitro* cultures and better mimic native kidney tissue.

Microfabricated *in vitro* models of kidney tissue can deliver crucial signaling elements to a wide range of cell types. The selection of cell type for *in vitro* kidney models is critical, and special attention should be given to the identification of a cell type originating from relevant areas of the kidney. This is especially true in nephrotoxicity studies as not all the different cell populations will display the same response or degree of damage after exposure to toxins [6]. Both primary renal cells and immortalized renal cell lines from human or animal subjects can serve as cell types for *in vitro* models of kidney tissue. Primary renal cells typically possess more kidney-specific function than immortalized cell lines and are direct samples of viable human tissue. Commonly used primary renal cell types include glomerular mesangial cells [32] and tubular epithelial cells originating from the proximal tubule [23,33,34], loop of Henle [35], distal tubule, and collecting ducts [36]. Alternatively, immortalized renal epithelial cell lines offer large populations of cells with relatively uniform function, consistent long-term properties, and an almost unlimited potential for population expansion. Common kidney cell lines include Madin–Darby canine kidney (MDCK) cells derived from collecting duct epithelium [37,38], rat inner medullary collecting duct (IMCD) cells, human embryonic kidney cells (HEK-293) [39], and human kidney epithelial cells (HK-2) [40] from the proximal tubule. However, cells from immortalized cell lines nearly always differ in important ways from their normal progenitors in the tissue from which they were derived [41]. Cell lines and primary cells cultured *in vitro* reproduce some, but not all, of the properties of native kindey cells, as they do not receive significant physiologic stimuli as they do *in vivo*. The role of microfabricated *in vitro* models is to provide significant physiological signaling elements to influence all cell types and guide them closer to physiolgically representative function.

11.5.1 Signaling elements delivered via cell substrate

Cell substrates can provide cues similar to those contributed by the kidney BM and ECM in order to influence cell function and facilitate maintenance of kidney tissue structure *in vitro*. The ECM of the kidney comprises structural proteins and proteoglycans which provide both mechanical and biochemical signals. These signals, properties of the ECM and BM, determine cell polarity, influence cell metabolism, organize proteins in adjacent cells, and promote cell survival [41]. The signaling elements possessed by kidney ECM and BM *in vivo* are categorized into surface chemistry, substrate stiffness, and substrate topography. Signaling elements of the kidney microenvironment have a strong influence over cell and tissue behavior both *in vivo* from ECM and BM and *in vitro* from cell substrates.

11.5.1.1 Surface chemistry

Kidney cells *in vivo* receive biochemical signals from the surrounding ECM, mainly from proteins and growth factors that compose the BM. As the BM directly contacts the basolateral side of all epithelial and endothelial cells in the kidney [29], its surface chemistry and composition significantly affect kidney tissue structure and function. The kidney BM mainly contains not only type-IV collagen [29] but also laminin, fibronectin, osteonectin, fibulin, and agrin [29,42]. Additionally, the BM can sequester growth factors and bind to cell surface receptors to influence cell behavior, resulting in interdependence between the effects of growth factors and ECM proteins [43]. The specific and complex composition of ECM in the kidney enables its specialized function. For example, the predominantly collagen IV- and laminin-containing BM of the glomerulus contains a negatively charged proteoglycan, heparin sulfate, which restricts the passage of cationic molecules, allowing the BM to act as a selective filter. Furthermore, collagen IV and bone morphogenic protein (BMP-7) promote differentiation in PTCs [44], while collagen I and the growth factor transforming growth factor TGF-β promote dedifferentiation of PTCs to a myofibroblastic state [45], which often leads to loss of cell–cell contact and epithelial dysfunction. The complex chemical composition of the kidney ECM provides crucial signaling elements *in vivo*, that when replicated *in vitro*, can provide direct control over cell behavior while promoting and sustaining the formation of kidney tissue.

Various ECM coatings and additives significantly influence kidney epithelial cell monolayer formation and epithelial cell differentiation *in vitro*. Human PTCs from the cell line HK-2 remain in a confluent monolayer for up to 3 weeks when cultured on synthetic substrates coated with collagen IV and laminin. Primary PTCs, however, respond optimally to substrates coated with collagen IV, laminin, nidogen, and nephronectin, with confluent monolayers lasting 2 weeks and well-formed tight junctions (TJ) lasting 3 weeks. Combination of a collagen IV and laminin surface with BMP-7 results in primary cell monolayers sustaining up to 4 weeks, with TJs disintegrating after only 1 week [43]. It is important to note, however, that ECM coatings alone are not sufficient to improve PTC performance on non-PTC-compatible materials [46]. Substrate chemistry not only directly signals cells but also serves as a cell–substrate binding mechanism to enable cell mechanotransduction of substrate mechanical properties.

11.5.1.2 Substrate stiffness

Cell response to matrix stiffness is a feedback loop that involves cell-exerted forces and the degree of substrate elasticity. *In vivo*, adherent cells reside in a solid microenvironment that can range from stiff and rigid to soft and elastic. Healthy tissue cells push and pull at their surroundings, thereby transmitting forces to substrates via adhesion molecules, such as integrins and cadherins. In turn, the degree of resistance of the substrate dictates the extent to which a cell can contract and migrate, causing the cell to respond through adhesion adjustment and cytoskeletal reorganization [47–50]. The mechanisms that allow cells to communicate and probe their microenvironment are complex and allow cells to be sensitive to changes in substrate stiffness. In the kidney, a change in glomerular BM rigidity may influence ECM accumulation [51], which leads to kidney structure deformation, obstruction, and disease [22]. Similarly, over-crosslinking of ECM protein molecules increases mechanical stiffness and impairs kidney function [52]. The mechanical properties of the matrix facilitate cell–matrix interaction and directly results in endothelial and epithelial heterogeneity in the kidney [29]. To control cell function and tissue structure formation in a synthetic environment, *in vitro* models of kidney tissue can employ an appropriate, controlled level of substrate stiffness as an influential signaling element.

As with *in vivo* tissue matrices, cells on synthetic substrates use tactile sensing of substrate stiffness as feedback for adhesion and cytoskeletal organization [48,53]. Various cell types cultured on synthetic substrates demonstrate a range of behavior in response to substrate stiffness. For example, aortic smooth muscle cells and endothelial cells respond to rigid substrates with increased spreading and highly organized cytoskeleton and focal adhesions [50,54], while neurons grown on softer substrates form more than three times as many branches than those on stiffer gels [55]. Substrate stiffness also has a profound effect on kidney cells in an *in vitro* environment; renal epithelial cells from the rat proximal tubule show reduced spreading, and irregularly shaped and highly dynamic focal adhesions when cultured on flexible substrates, as compared to rigid ones [56,57]. Furthermore, distention of kidney tubular epithelial cells cultured on an elastic membrane results in secretion of transforming growth factor-β and nitric oxide [58]. In both natural matrix and synthetic cell substrates, kidney and other cell types have a pronounced ability to survey the stiffness of the surrounding microenvironment in order to receive critical cues.

11.5.1.3 Substrate topography

Naturally occurring topographic structures and mechanical features in the ECM provide cells with mechanotransductive cues to influence cell behavior. The supporting meshwork of specialized protein found in the kidney ECM exhibits topographical features on the nano- and microscale that are directly involved in cell–matrix communication [42,59] and serve in structural and filtering roles. The kidney ECM consists mainly of the BM, a flexible thin mat 50–100 nm thick, containing ridge structures in the submicron regime with protein fiber dimensions ranging from 6–7 nm to several microns wide and pore sizes averaging 14.1 nm in diameter [60]. These topographical and mechanical features interact directly with cells adherent to the kidney BM, indicating potential signaling of cells *in vivo* and their importance as critical elements of an *in vitro* model of kidney tissue.

Topographic patterns influence morphological and functional cell response *in vitro* [61], but their likely impact on kidney cell function in an *in vitro* environment has yet to be thoroughly explored. Well-defined topographic patterns influence cell response, most notably through contact guidance, and result in controlled, directed cell migration, alignment, elongation, and structural reorganization [62,63]. Cell response to topography depends on many factors, including cell type, feature size, and feature geometry [64]. Three basic nanotopographic geometries, gratings, posts, and pits, have effects on cell morphology, migration, attachment, adhesion, and proliferation [59]. In addition, cell–topography interactions have the ability to control higher order cell function such as stem cell differentiation and cellular superstructure formation [61]. In general, epithelial cells show increased alignment and elongation along topographic features [65]. Cells from a highly modified HEK-293 demonstrate enhanced adhesion, proliferation, and alignment to substrate topography [66], though the impact of topography on more physiologically relevant kidney cells is not well documented. As the BM of the kidney tubule contains micro- and nanoscale topographic structures [42,67,68], topographic patterns provide a promising avenue to influence kidney epithelial cell function *in vitro*.

11.5.2 Signaling elements delivered via fluid flow

Cells, in addition to receiving biochemical and mechanical signals from synthetic and natural substrates, also receive signals from fluid flow. This is particularly true for renal epithelial cells and vascular endothelial cells, since they receive fluid flow stimuli in their native environment. Vascular flow modifies ion channel properties and growth factor synthesis in endothelial cells [69–71], while flow influences proliferation and differentiation in mesenchymal stem cells [68] and human osteoblasts [72]. In the kidney, as blood enters the glomerulus and filtrate is funneled into the renal tubule to be modified into urine, the surrounding cells receive chemical signals from hormones and fluctuations in solute concentration as well as mechanical signals from glomerular filtrate rate and pressure [73]. Signals present due to filtrate flow are responsible for maintaining kidney function by regulating electrolyte balance, body fluid concentration, and excretion of waste products. Appropriate chemical and mechanical signals naturally delivered via fluid flow can be replicated in a synthetic microenvironment to encourage and control specific tissue function *in vitro*.

11.5.2.1 Chemical composition of fluid

Since kidney epithelial tissue serves as a selectively permeable barrier between the lumen space and the interstitial space, it receives unique chemical cues from the fluids occupying those spaces. Signaling elements delivered via fluid flow are often chemical in the form of peptides, solutes, and small molecules carried by the renal filtrate, or hormones delivered via blood flow. Adverse signals are also delivered via renal filtrate, including toxic effects from therapeutic agents, which can cause drug-induced kidney disease [15]. The function of kidney cells depends on chemical signaling elements, but the signaling elements also depend on specific kidney function. In the loop of Henle, for example, a hypertonic gradient occurs due to the filtering properties of the tubular epithelium. The gradient, in turn, signals the intracellular pumps to pump NaCl from the lumen into the interstitium, creating conditions required to form concentrated urine [2,73]. *In vitro* kidney tissue models incorporate fluid flow of growth media solutions, dependent on cell type and application, to mimic renal filtrate. Media can contain a designed combination of small molecules, such as salts, glucose, amino acids, and vitamins. More specialized solutions can contain growth factors, which stimulate cell proliferation, transferrin, which carries iron into cells [41], or pharmaceutical compounds which allow for nephrotoxicity testing. Selecting appropriate media composition provides a realistic chemical environment *in vitro*, with controls over the media flow providing the mechanical cues to surrounding tubular epithelial cells.

11.5.2.2 Flow-induced shear stress

Renal filtrate flow and resulting FSS act as stimuli to epithelial cells that line the tubular lumen. The filtrate flow rate changes according to endocrine mechanisms and exerts a shear stress between 0.2 and 20 dyn/cm^2 [74] on tubular epithelial cells. The mechanical signal generated by FSS is transmitted into cellular function via mechanosensory mechanisms by mechanosensors, such as microvilli [75–79] and cilia [80,81], located on the surface of kidney cells. FSS affects overall tubular function and may impact tubular quality and cell-regulated ion transport by enhancing cell–cell junctions and cell layer integrity. *In vitro* kidney cell studies show FSS induces reorganization of the cytoskeleton [82] by reinforcing the apical and lateral domains of actin filaments [67,83], spurs the formation of TJs and adherens junctions, and dramatically reinforces expression of vinculin [67]. Sections of mouse proximal tubule *in vitro* change diameter and alter sodium ion absorption with changes in FSS [78], and the diameter of primary rabbit collecting duct tubules increases by approximately 14% with an acute increase in FSS [74]. Nitric oxide production *in vitro* is also modulated by FSS in IMCD cells and may play a role in the regulation of sodium and water excretion within the renal tubular system [84]. FSS influences cells in predictable ways both *in vivo* and *in vitro*, thereby providing a user-controlled means to control kidney cell response *in vitro*.

11.6 Review of current microfabricated kidney tissue models

Microfabrication techniques provide a means to control microscale architecture and FSS of *in vitro* tissue models. Specifically, micro- and nanoscale polymer molding creates small-scale architectures to influence cells and subcellular structures [85] and permits subsequent chemical functionalization for increased control of cell function [86,87]. Microfabrication also creates user-defined geometries for microfluidic channels, which serve a variety of applications [88]. For example, molding techniques can create defined, microfluidic channels mimicking vasculature [89] and complex channel architectures found in nephron segments [90]. Since small-scale architecture of a device can influence cell function and physiological structure formation [91], control of device architecture can lead to a consistent platform with significant control of cell function and tissue structure.

11.6.1 Microfabricated kidney models with controlled FSS

Since fluid flow and resulting FSS have profound effects on kidney cells, microfabricated *in vitro* kidney tissue models leverage microfabrication and micromolding to precisely control fluid flow. Microfabricated *in vitro* kidney models incorporate a microfluidic channel mounted on a flat, cell culture substrate [67,79] to control flow contacting the cells. In these *in vitro* tissue models, kidney epithelial cells are cultured on a glass slide and grown to confluency prior to device assembly. A prefabricated microfluidic chamber is subsequently placed over the cell population, and fluid flow controlled by inlet flow rate is established across the cell monolayer. Mouse PTCs exposed to laminar flow for 24 h demonstrated cytoskeletal reorganization, indicated by reinforcement of a lateral actin network [67]. Similarly, actin distribution reorganized after 5 h of flow at

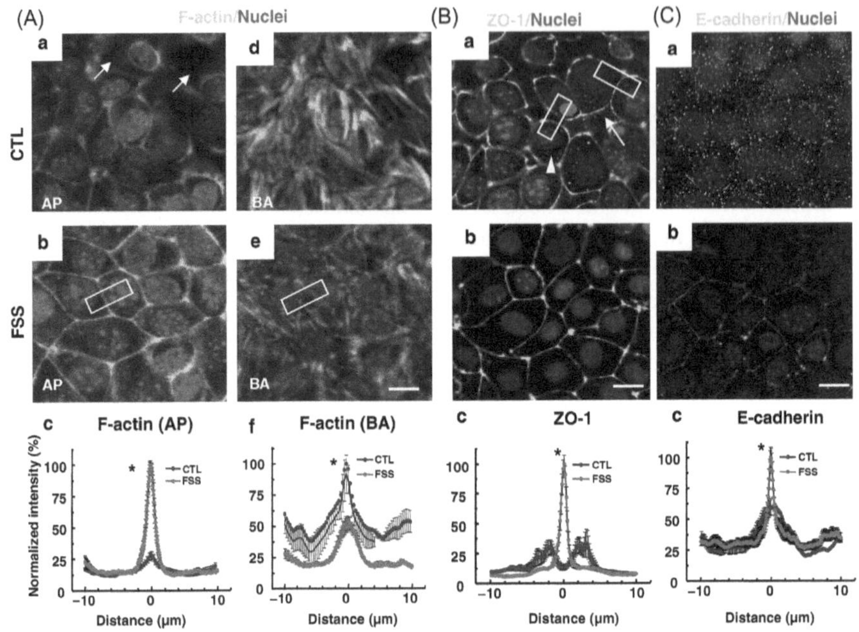

FIGURE 11.2

Microfluidic models with controlled shear stress can be used to study cytoskeletal reorganization and intracellular junction distribution. (A) Mouse PTCs respond to FSS with F-actin redistribution in the apical (Aa, Ab, Ac) and basal (Ad, Ae, Af) regions of the cell. (B) In the absence of FSS, the TJ-associated protein, ZO-1, did not form apical cell–cell contacts (Ba), which is demonstrated graphically in the intensity profile (Bc). FSS exposure for 5 h induced significant reassembly of intercellular junctions (Bb). (C) Similarly, the punctuate expression of E-cadherin under static conditions (Ca) became continuous and localized at cell–cell junctions under the presence of shear (Cb). Scale bar 10 μm.

Source: *Reprinted from Ref. [79] with permission from the National Academy of Sciences of the United States of America, Copyright 2008.*

1 dyn/cm^2 [79], with effects differing between apical and basal sides of the cells. Compared to cells under static conditions, 5 h of FSS induced a continuous distribution of intercellular junctions, demonstrated by a dramatically reinforced pattern of ZO-1 TJs, as shown in Figure 11.2. These models deliver uniform FSS across a simple, flat substrate, with FSS levels controlled via inlet flow rate, demonstrating the ability of FSS to enhance kidney cell function *in vitro*.

In addition to delivering uniform flow fields across a substrate and controlling FSS by inlet flow rates, microfluidic channels can control ranges of FSS within a channel by employing various channel geometries [92,93]. A kidney tissue model with a microfluidic channel containing an oblique barrier in the path of fluid flow

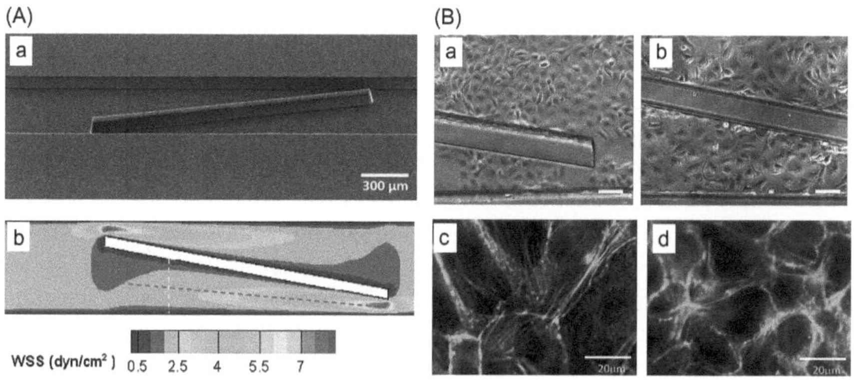

FIGURE 11.3

Microfluidic devices can control a range of FSS within a single channel to study the effect of shear on cells. (Aa) An oblique barrier in the path of fluid flow generates a wall shear stress distribution within the microfluidic channel. (Ab) Simulation shows a uniform shear stress outside the barrier region and a shear gradient between the barrier and the walls. (B) MDCK cells grow in confluent monolayers within the channel and around the barrier under static (Ba) and perfusion (Bb) conditions. Scale bars 50 μm for Ba and Bb. Shear stress has an effect on actin cytoskeleton reorganization. (Bc) Cells located in the broader region between the barrier and the wall experience a shear stress of 0.5 dyn/cm^2. (Bd) Cells located in the narrow region between the barrier and the wall experience a shear stress of 3.0 dyn/cm^2 and show a more lateral accumulation of actin filaments.

Source: *Reprinted from Ref. [94] with permission from the Royal Society of Chemistry.*

generates a wide range of FSS within the channel [94], enabling precise control over local FSS levels. As illustrated in Figure 11.3, the oblique barrier creates a uniform shear stress outside the obstructed region, a stable and linear shear gradient above and below the barrier, and a nonuniform flow field at the tip of the barrier. MDCK cells cultured within the device and exposed to a constant media flow rate of 2.7 μl/min receive an FSS of ~0.5 dyn/cm^2 in the wide region below the barrier and ~3.0 dyn/cm^2 in the narrow region above the barrier. After 2.5 h of continuous flow, cells located in the lower shear stress region exhibit partial cytoskeletal reorganization, while cells under high shear demonstrate more lateral accumulation of actin filaments. In general, intracellular Ca^{2+} uptake rapidly increases in response to an increase of FSS. After prolonged exposure to FSS, however, cells in higher FSS regions show a diminished increase of Ca^{2+} concentration, indicating that Ca^{2+} uptake rate in cells with higher degrees of cytoskeletal reorganization may be less responsive to flow change. By focusing on geometrical control of FSS, this device generates and controls complex FSS patterns within a single chamber. The result enables high-throughput examination of the effect of shear stress on cell growth and function while keeping culture conditions constant.

11.6.2 Microfabricated kidney models with porous membranes and scaffolds

To capture FSS cues of the *in vivo* microenvironment, while allowing observation of cell-mediated transport and gradients, microfabricated *in vitro* kidney models can incorporate synthetic porous BMs. A multilayer microfluidic device integrates a layer of microchannels, a porous polyester membrane, and a reservoir to serve as a collecting-duct-on-a-chip [82,95]. The device allows the recreation of natural transepithelial osmotic gradients across the membrane, differential FSS, and hormonal stimulation via the microchannels, as seen in Figure 11.4A. Primary rat IMCD cells form a confluent monolayer on the porous membrane with enhanced cell polarization and cytoskeletal reorganization due to controlled fluidic conditions. Results show that F-actin depolymerization occurs after 3 h of exposure to a FSS of 1 dyn/cm^2 but does not become fully depolymerized until 5 h of FSS exposure. F-actin depolymerization and aquaporin-2 (AQP2) translocation increase under increased FSS, and the combination of FSS with arginine–vasopressin stimulation from the basolateral side further heightens this effect, as seen in Figure 11.4B. In addition, a transepithelial osmotic gradient of 300 mOsm/kg H_2O increases AQP2 trafficking to the plasma membrane. Using a simple 2D synthetic membrane as a cell scaffold, this device controls FSS, hormonal stimulation, and osmotic gradients across the membrane, replicating both flow and transport functions of the kidney.

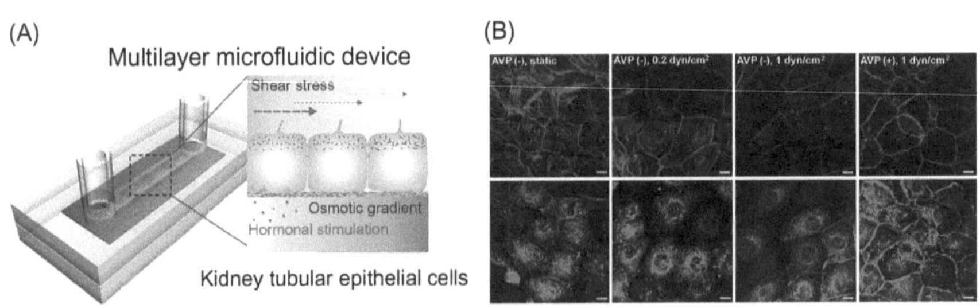

FIGURE 11.4

The multilayer microfluidic device (A) mimics natural transepithelial osmotic gradients, shear stress, and hormonal stimulation by integrating microchannels, a porous polyester membrane, and a reservoir. (B) F-actin depolymerization and AQP2 translocation increase with greater FSS. Combining FSS with hormonal stimulation further heightens the effect. Scale bars 10 μm.

Source: *Reprinted from Ref. [95] with permission from the Royal Society of Chemistry.*

The heterogeneity and complexity of the native renal epithelial BMs suggest that its function allows not only for generation of gradients but also for further signaling functionality and structural guidance properties. To capture part of the signaling functionality and structural guidance, *in vitro* kidney models can incorporate membranes with engineered surface cues [42,96] or porous scaffolds of structural biological materials [97,98]. An electro-spun membrane, composed of nano- and microfibers synthesized with a urea-modified ureido-pyrimidinone (UPy), low molecular weight LMW polycaprolactone base polymer, PCLdi(U-UPy), serves as a cell scaffold in a perfusion chamber [42]. ECM-derived bioactive peptides, which were derived from the renal BM-specific proteins of laminin, collagen I, collagen IV, and fibronectin, are combined with the fibrous mesh to create a bioactive membrane, shown in Figure 11.5, for the culture of human renal primary tubular epithelial cells (PTEC). This membrane, with the inclusion of bioactive molecules, is able to sustain PTECs in tight monolayers for 19 days while maintaining renal epithelial phenotype and function. Under perfusion culture, and in contrast to cells cultured on polymer membranes without bioactive molecules, cells on the bioactive membranes express genes which code markers of renal epithelial function, including the expression of the epithelial cell marker, E-cadherin, and various transporter proteins, such as peptide transporters, organic anion/cation transporters, and adenosine triphosphate ATP-driven transporters. Fibrous membranes with engineered surface peptides mimic physiological properties of the renal epithelial BM, providing appropriate morphology, structure, and chemistry to maintain a level of renal epithelial phenotype and function over extended culture periods.

Porous scaffolds of structural biological materials provide structural guidance and possess inherent biochemical cues. Typically, biomaterial scaffolds are composed of collagen or matrigel, which provide some structural support and matrix-mediated cell signaling. A collagen I gel matrix has been shown to support angiogenic sprouting in embryonic kidneys [98], and kidney progenitor cells show markers for tubular formation when cultured in layers of polyester fleece and perfused with hormone-containing media [99]. An aqueous silk scaffold [97] with infused collagen−matrigel provides some of the structural and chemical properties of the renal ECM in addition to porous features for transport and robust mechanical properties associated with silk. Mouse embryonic kidney epithelial cells derived from different tubular segments and seeded in the infused silk scaffolds exhibit branching and morphogenesis, resulting in cystic structures. The formation of structures are sustained for at least 6 weeks in collagen−matrigel-infused silk scaffolds under perfusion conditions, demonstrated by the presence of the marker proteins E-cadherin, N-cadherin, and Na^+ K^+ ATPase pumps. These protein-infused scaffolds offer limited confinement for growth of kidney-like structures, allowing direct observation of normal or diseased tissue formations. Due to their fabrication methods, fibrous and porous scaffolds possess inherent nano- and microstructures, which allow tissue structure formation and provide a level of structural guidance.

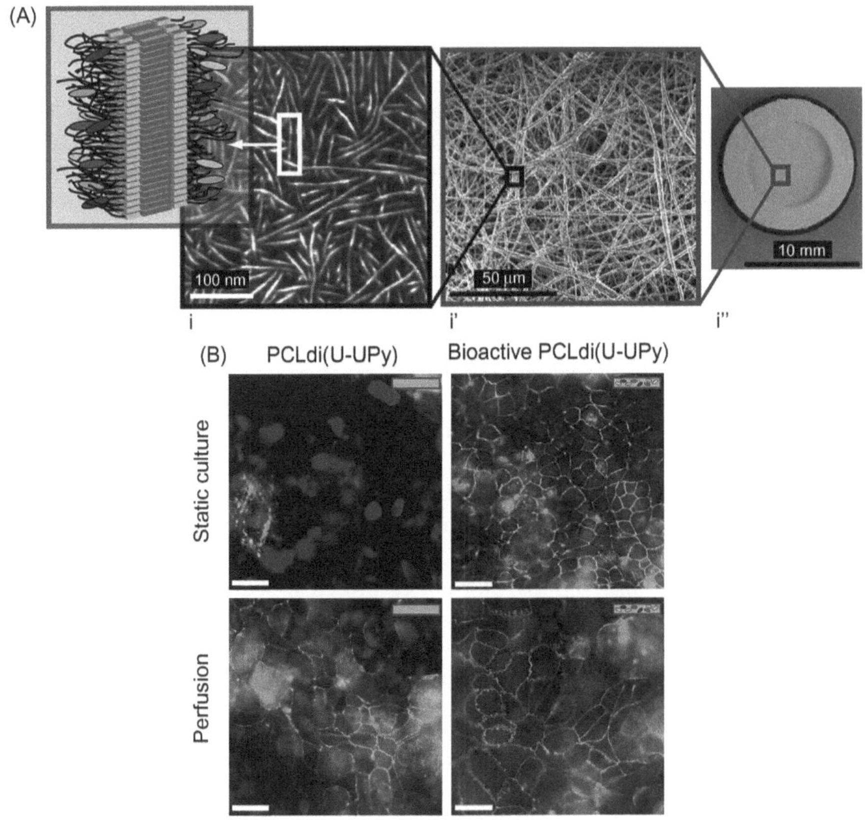

FIGURE 11.5

The bioactive membrane (A) contains nano- and microscale fibers with tethered ECM-derived bioactive peptides. (B) The formation of TJs is scarce in cell populations on nonbioactive substrates and under perfusion conditions, whereas PTECs cultured on bioactive membranes show high ZO-1 expression under both static and perfusion conditions. Scale bars are 10 μm unless otherwise labeled.

Source: *Reprinted from Ref. [42] with permission from Elsevier, Copyright 2011.*

11.6.3 Microfabricated kidney models with user-defined substrate topography

To create direct control of structural guidance, while maintaining control over FSS, deliberately formed topographical patterns can be incorporated into a cell scaffold within a microfabricated kidney tissue model. A molded cell substrate with submicrometer scale topographical features serves as a kidney cell scaffold within a microfluidic system (Frohlich, et al., 2011). A microfluidic flow channel system, shown in Figure 11.6, attaches to the molded cell substrate to form flow chambers which contain the user-defined topography of the substrate. The

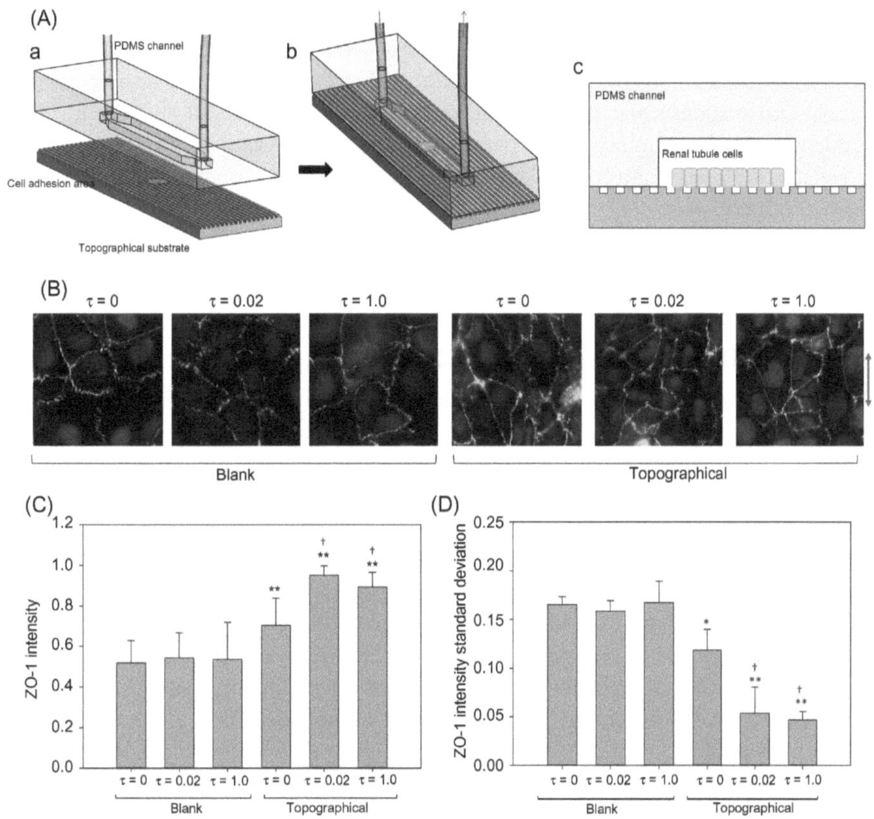

FIGURE 11.6

The combination and independent control over surface topography and FSS enhance kidney epithelial cell response to signaling elements. The disassembled (Aa) microscale tissue modeling device assembles (Ab) to form a flow chamber which contains user-defined topography of the substrate (cross section: Ac). FSS and topography synergistically influence expression and distribution of ZO-1. (B) Representative images of ZO-1 expression for cells cultured on blank or topographical substrates and exposed to either 0, 0.02, or 1.0 dyn/cm^2 FSS. ZO-1 intensity (C) and continuity, measured by the standard deviation of intensity (D), increase significantly in cell populations cultured on topographical substrates, with higher levels of shear enhancing this effect.

modular nature of the system combines and independently controls cell substrate surface topography and FSS signals to cells. Collagen IV is deliberately patterned on the cell substrates to control selective placement of HK-2 cells cultured within the microfluidic channels. The mechanical parameters of surface topography and FSS work in concert to elicit nuclear alignment and TJ formation, with topography enhancing cell response to FSS. FSS alone does not signal nuclear alignment or TJ formation of cells; however, the combination with topography enhances nuclear alignment as well as ZO-1 expression and continuity, indicating mature TJ formation. Results indicate that topographical substrates may accelerate the rate of cell response and increase cell sensitivity to FSS. By administering independently controlled mechanical parameters to cell populations, this device provides user-defined control over signaling elements based on both substrate and flow, driving toward deliberate enhancement of *in vitro* human renal tissue.

11.7 Summary and future direction

In vitro kidney tissue models provide platforms to directly observe kidney tissue function, tissue structure formation, and tissue response to toxicity exposure in a consistent, controlled environment. Microfabricated kidney tissue models elucidate and replicate the complex relationship between kidney structure and kidney function by mimicking native chemistry and mechanical parameters within an *in vitro* tissue model. As opposed to traditional *in vitro* kidney tissue models, microfabricated models precisely control crucial signaling elements such as surface chemistry, stiffness and topography, chemical signals delivered via fluid flow, and FSS while providing compatibility with primary cells and multiple cell lines. Microfabricated microfluidic systems with controlled FSS, porous membranes or scaffolds, and user-defined scaffolds provide some relevant complexity of the *in vivo* kidney microenvironment. Certain biophysical parameters are present in all microfabricated *in vitro* kidney tissue models indicating their importance in models of kidney tissue. Specifically, FSS and the delivery and exchange of chemical signals via fluid flow are critical for the sustainment of functional kidney tissue, whether the cell population resides on a flat surface, porous membrane, or in a user-defined scaffold. Microfabricated kidney tissue models provide extreme architectural precision, leading to precise and user-defined control of crucial cell-signaling elements.

Careful combination of some or all of the crucial signaling elements found *in vivo* leads to the maintenance of physiologically accurate kidney tissue in microfabricated kidney tissue models. Microfluidic models with controlled FSS cause cytoskeletal and intracellular junction reorganization of kidney cell monolayers. Gradients of FSS generated within a single chamber provide a range of FSS to spatially vary cell response within one model. Microfabricated tissue models with porous membranes simulate natural transepithelial osmotic gradients and hormonal stimulation and allow kidney cell-mediated transport across the membrane. Microfabricated fibrous membranes with tethered ECM peptides sustain cell monolayer growth for up to 19 days due to their tailored biochemical signaling. Protein-infused, highly structural scaffolds encourage branching, morphogenesis, and structure formation of embryonic kidney cells. Finally, deliberate enhancement of human renal tissue results from administering user-defined control of cell substrate surface topography and FSS to cells. Deliberately controlled elements within microfabricated kidney tissue models cue and control significant kidney cell and tissue response

Ultimately, the high degree of control over the synthetic microenvironment provided by microfabricated kidney tissue models, including mechanical and chemical signaling delivered via substrates and fluid flow, will provide *in vitro* platforms in which kidney tissue structure and function can be studied. These platforms enhance many applications including drug screening, therapy efficacy and toxicity tests, disease mechanism research, and opportunities for understanding kidney regeneration mechanisms. As microfabrication provides increasingly sophisticated kidney tissue models, a larger fraction of native kidney function will be replicated, providing deeper insights into kidney function and eventually culminating in a synthetic source for organ repair or replacement.

References

[1] V. Kumar, A.K. Abbas, N. Fausto, Robbins and Cotran Pathologic Basis of Disease, seventh ed., Elsevier Saunders, Philadelphia, PA, 2005.

[2] L. Junqueira, J. Carneiro (Eds.), Basic Histology, McGraw-Hill, New York, NY, 2005.

[3] A. Saito, R. Kaseda, M. Hosojima, H. Sato, Proximal tubule cell hypothesis for cardiorenal syndrome in diabetes, Int. J. Nephrol. 2011 (2010) 1–9.

[4] W.L. St Peter, A.C. Schoolwerth, T. McGowan, W.M. McClellan, Chronic kidney disease: issues and establishing programs and clinics for improved patient outcomes, Am. J. Kidney Dis. 41 (5) (2003) 903–924.

[5] A.J. Collins, R.N. Foley, C. Herzog, B.M. Chavers, D. Gilbertson, A. Ishani, et al., Excerpts from the US Renal Data System 2009 Annual Data Report, Am. J. Kidney Dis. 55 (1 Suppl. 1) (2010) S1–S420, A426–A427.

[6] W. Pfaller, G. Gstraunthaler, Nephrotoxicity testing *in vitro*—what we know and what we need to know, Environ. Health Perspect. 106 (Suppl. 2) (1998) 559–569.

[7] A. Vasilyev, Y. Liu, S. Mudumana, S. Mangos, P.Y. Lam, A. Majumdar, et al., Collective cell migration drives morphogenesis of the kidney nephron, PLoS Biol. 7 (1) (2009) 101–114.

[8] W.C. Prozialeck, J.R. Edwards, P.C. Lamar, J. Liu, V.S. Vaidya, J.V. Bonventre, Expression of kidney injury molecule-1 (Kim-1) in relation to necrosis and apoptosis during the early stages of Cd-induced proximal tubule injury, Toxicol. Appl. Pharmacol. 238 (3) (2009) 306–314.

[9] W.C. Prozialeck, P.C. Lamar, S.M. Lynch, Cadmium alters the localization of N-cadherin, E-cadherin, and beta-catenin in the proximal tubule epithelium, Toxicol. Appl. Pharmacol. 189 (3) (2003) 180–195.

[10] H. Rennke, B. Denker, Renal Pathophysiology, second ed., Lippincott Williams and Wilkins, Philadelphia, PA, 2007.

[11] I. Drummond, Making a zebrafish kidney: a tale of two tubes, Trends Cell Biol. 13 (7) (2003) 357–365.

[12] H.T. Cook, The origin of renal fibroblasts and progression of kidney disease, Am. J. Pathol. 176 (1) (2010) 22–24.

[13] K.J. Ullrich, G. Rumrich, S. Kloss, Contraluminal organic anion and cation transport in the proximal renal tubule: V. Interaction with sulfamoyl- and phenoxy diuretics, and with beta-lactam antibiotics, Kidney Int. 36 (1) (1989) 78–88.

[14] D. Choudhury, Z. Ahmed, Drug-associated renal dysfunction and injury, Nat. Clin. Pract. Nephrol. 2 (2) (2006) 80–91.

[15] C.A. Naughton, Drug-induced nephrotoxicity, Am. Fam. Physician 78 (6) (2008) 743–750.

[16] G.S. Markowitz, M.A. Perazella, Drug-induced renal failure: a focus on tubulointerstitial disease, Clin. Chim. Acta 351 (1–2) (2005) 31–47.

[17] M.A. Arnaout, Molecular genetics and pathogenesis of autosomal dominant polycystic kidney disease, Annu. Rev. Med. 52 (2001) 93–123.

[18] I.A. Drummond, Polycystins, focal adhesions and extracellular matrix interactions, Biochim. Biophys. Acta 1812 (10) (2011) 1322–1326.

[19] N. Shweke, N. Boulos, C. Jouanneau, S. Vandermeersch, G. Melino, J.C. Dussaule, et al., Tissue transglutaminase contributes to interstitial renal fibrosis by favoring accumulation of fibrillar collagen through TGF-beta activation and cell infiltration, Am. J. Pathol. 173 (3) (2008) 631–642.

[20] V. Osathanondh, E.L. Potter, Pathogenesis of polycystic kidneys, Arch. Pathol. 77 (1964) 459–512.

[21] P.D. Wilson, Polycystic kidney disease, N. Engl. J. Med. 350 (2) (2004) 151–164.

[22] M. Zeisberg, G. Bonner, Y. Maeshima, P. Colorado, G.A. Muller, F. Strutz, et al., Renal fibrosis: collagen composition and assembly regulates epithelial–mesenchymal transdifferentiation, Am. J. Pathol. 159 (4) (2001) 1313–1321.

[23] W.C. Prozialeck, J.R. Edwards, P.C. Lamar, C.S. Smith, Epithelial barrier characteristics and expression of cell adhesion molecules in proximal tubule-derived cell lines commonly used for *in vitro* toxicity studies, Toxicol. In Vitro 20 (6) (2006) 942–953.

[24] M.A. Perazella, Renal vulnerability to drug toxicity, Clin. J. Am. Soc. Nephrol. 4 (7) (2009) 1275–1283.

[25] S.L. Karp, B.A. Molitoris, Establishment of polarity in epithelial cells of the developing nephron, in: P.D. Vize, A.S. Woolf, J.B.L. Bard (Eds.), The Kidney, Academic Press, Inc., 2003, pp. 211–220.

[26] A. Stroope, B. Radtke, B. Huang, T. Masyuk, V. Torres, E. Ritman, et al., Hepato-renal pathology in pkd2ws25/-mice, an animal model of autosomal dominant polycystic kidney disease, Am. J. Pathol. 176 (3) (2010) 1282–1291.

[27] M. Baker, Tissue models: a living system on a chip, Nature 471 (7340) (2011) 661–665.

[28] R. Baudoin, L. Griscom, M. Monge, C. Legallais, E. Leclerc, Development of a renal microchip for *in vitro* distal tubule models, Biotechnol. Prog. 23 (5) (2007) 1245–1253.

[29] V.S. LeBleu, B. Macdonald, R. Kalluri, Structure and function of basement membranes, Exp. Biol. Med. (Maywood) 232 (9) (2007) 1121–1129.

[30] F. Hesse, R. Wagner, Developments and improvements in the manufacturing of human therapeutics with mammalian cell cultures, Trends Biotechnol. 18 (4) (2000) 173–180.

[31] M. Raab, J.W. Shin, D.E. Discher, Matrix elasticity *in vitro* controls muscle stem cell fate *in vivo*, Stem. Cell Res. Ther. 1 (5) (2010) 38–39.

[32] T. Nishikawa, S. Suematsu, J. Saito, A. Soyama, H. Ito, T. Kino, et al., Human renal mesangial cells produce aldosterone in response to low-density lipoprotein (LDL), J. Steroid Biochem. Mol. Biol. 96 (3–4) (2005) 309–316.

[33] S.E. Kays, C.D. Berdanier, A.R. Swagler, E.A. Lock, R.G. Schnellmann, An *in vitro* model of renal proximal tubule cell regeneration, J. Pharmacol. Toxicol. Methods 29 (4) (1993) 211–215.

[34] P.J. Boogaard, J.P. Zoeteweij, T.J. van Berkel, J.M. van't Noordende, G.J. Mulder, J.F. Nagelkerke, Primary culture of proximal tubular cells from normal rat kidney as an *in vitro* model to study mechanisms of nephrotoxicity. Toxicity of nephrotoxicants at low concentrations during prolonged exposure, Biochem. Pharmacol. 39 (8) (1990) 1335–1345.

[35] U.M. Rose, A. Hartog, J.W. Jansen, C.H. Van Os, R.J. Bindels, Anoxia-induced increases in intracellular calcium concentration in primary cultures of rabbit thick ascending limb of Henle's loop, Biochim. Biophys. Acta 1226 (3) (1994) 291–299.

[36] L. Khandrika, F.J. Kim, A. Campagna, S. Koul, R.B. Meacham, H.K. Koul, Primary culture and characterization of human renal inner medullary collecting duct epithelial cells, J. Urol. 179 (5) (2008) 2057–2063.

[37] C.R. Gaush, W.L. Hard, T.F. Smith, Characterization of an established line of canine kidney cells (MDCK), Proc. Soc. Exp. Biol. Med. 122 (3) (1966) 931–935.

[38] J.M. Arthur, The MDCK cell line is made up of populations of cells with diverse resistive and transport properties, Tissue Cell 32 (5) (2000) 446–450.

[39] F.L. Graham, J. Smiley, W.C. Russell, R. Nairn, Characteristics of a human cell line transformed by DNA from human adenovirus type 5, J. Gen. Virol. 36 (1) (1977) 59–74.

[40] M.J. Ryan, G. Johnson, J. Kirk, S.M. Fuerstenberg, R.A. Zager, B. Torok-Storb, HK-2: an immortalized proximal tubule epithelial cell line from normal adult human kidney, Kidney Int. 45 (1) (1994) 48–57.

[41] B. Alberts, Molecular Biology of the Cell, Garland, New York, NY, 2000.

[42] P.Y. Dankers, J.M. Boomker, A. Huizinga-van der Vlag, E. Wisse, W.P. Appel, F.M. Smedts, et al., Bioengineering of living renal membranes consisting of hierarchical, bioactive supramolecular meshes and human tubular cells, Biomaterials 32 (3) (2011) 723–733.

[43] H. Zhang, F. Tasnim, J.Y. Ying, D. Zink, The impact of extracellular matrix coatings on the performance of human renal cells applied in bioartificial kidneys, Biomaterials 30 (15) (2009) 2899–2911.

[44] M. Zeisberg, J. Hanai, H. Sugimoto, T. Mammoto, D. Charytan, F. Strutz, et al., BMP-7 counteracts TGF-beta1-induced epithelial-to-mesenchymal transition and reverses chronic renal injury, Nat. Med. 9 (7) (2003) 964–968.

[45] J. Yang, Y. Liu, Dissection of key events in tubular epithelial to myofibroblast transition and its implications in renal interstitial fibrosis, Am. J. Pathol. 159 (4) (2001) 1465–1475.

[46] M. Ni, J.C.M. Teo, M. Shahrudin bin Ibrahim, K. Zhang, F. Tasnim, P.-Y. Chow, et al., Characterization of membrane materials and membrane coatings for bioreactor units of bioartificial kidneys, Biomaterials 32 (6) (2011) 1465–1476.

[47] L.G. Griffith, M.A. Swartz, Capturing complex 3D tissue physiology *in vitro*, Nat. Rev. Mol. Cell Biol. 7 (3) (2006) 211–224.

[48] D.E. Discher, P. Janmey, Y.L. Wang, Tissue cells feel and respond to the stiffness of their substrate, Science 310 (5751) (2005) 1139–1143.

[49] A. Buxboim, I.L. Ivanovska, D.E. Discher, Matrix elasticity, cytoskeletal forces and physics of the nucleus: how deeply do cells "feel" outside and in? J. Cell Sci. 123 (Pt 3) (2010) 297–308.
[50] A. Engler, L. Bacakova, C. Newman, A. Hategan, M. Griffin, D. Discher, Substrate compliance versus ligand density in cell on gel responses, Biophys. J. 86 (1 Pt 1) (2004) 617–628.
[51] P. Cortes, X. Zhao, B.L. Riser, R.G. Narins, Role of glomerular mechanical strain in the pathogenesis of diabetic nephropathy, Kidney Int. 51 (1) (1997) 57–68.
[52] J.D. Furber, Extracellular glycation crosslinks: prospects for removal, Rejuvenation Res. 9 (2) (2006) 274–278.
[53] A. Saez, A. Buguin, P. Silberzan, B. Ladoux, Is the mechanical activity of epithelial cells controlled by deformations or forces? Biophys. J. 89 (6) (2005) L52–L54.
[54] C.F. Deroanne, C.M. Lapiere, B.V. Nusgens, *In vitro* tubulogenesis of endothelial cells by relaxation of the coupling extracellular matrix-cytoskeleton, Cardiovasc. Res. 49 (3) (2001) 647–658.
[55] L.A. Flanagan, Y.E. Ju, B. Marg, M. Osterfield, P.A. Janmey, Neurite branching on deformable substrates, Neuroreport 13 (18) (2002) 2411–2415.
[56] R.J. Pelham Jr., Y. Wang, Cell locomotion and focal adhesions are regulated by substrate flexibility, Proc. Natl. Acad. Sci. U.S.A. 94 (25) (1997) 13661–13665.
[57] Y.K. Wang, Y.H. Wang, C.Z. Wang, J.M. Sung, W.T. Chiu, S.H. Lin, et al., Rigidity of collagen fibrils controls collagen gel-induced down-regulation of focal adhesion complex proteins mediated by alpha2beta1 integrin, J. Biol. Chem. 278 (24) (2003) 21886–21892.
[58] A. Miyajima, J. Chen, I. Kirman, D.P. Poppas, E.J. Darracott Vaughan, D. Felsen, Interaction of nitric oxide and transforming growth factor-beta1 induced by angiotensin II and mechanical stretch in rat renal tubular epithelial cells, J. Urol. 164 (5) (2000) 1729–1734.
[59] C.J. Bettinger, R. Langer, J.T. Borenstein, Engineering substrate topography at the micro- and nanoscale to control cell function, Angew. Chem. Int. Ed. Engl. 48 (30) (2009) 5406–5415.
[60] K. Hironaka, H. Makino, Y. Yamasaki, Z. Ota, Renal basement membranes by ultrahigh resolution scanning electron microscope, Kidney Int. 43 (2) (1993) 334–345.
[61] J.L. Charest, W.P. King, Engineering biomaterial interfaces through micro and nano-patterning, in: P. Hesketh (Ed.), BioNanoFluidic MEMS, Springer Science, New York, NY, 2008.
[62] P. Clark, P. Connolly, A.S. Curtis, J.A. Dow, C.D. Wilkinson, Topographical control of cell behaviour: II. Multiple grooved substrata, Development 108 (4) (1990) 635–644.
[63] P. Weiss, Cell contact, Int. Rev. Cytol. 7 (1958) 391–423.
[64] R.G. Flemming, C.J. Murphy, G.A. Abrams, S.L. Goodman, P.F. Nealey, Effects of synthetic micro- and nano-structured surfaces on cell behavior, Biomaterials 20 (6) (1999) 573–588.
[65] E.K. Yim, K.W. Leong, Significance of synthetic nanostructures in dictating cellular response, Nanomedicine 1 (1) (2005) 10–21.
[66] E. Rebollar, I. Frischauf, M. Olbrich, T. Peterbauer, S. Hering, J. Preiner, et al., Proliferation of aligned mammalian cells on laser-nanostructured polystyrene, Biomaterials 29 (12) (2008) 1796–1806.

[67] M. Essig, F. Terzi, M. Burtin, G. Friedlander, Mechanical strains induced by tubular flow affect the phenotype of proximal tubular cells, Am. J. Physiol. Renal Physiol. 281 (4) (2001) F751–F762.

[68] J.R. Glossop, S.H. Cartmell, Effect of fluid flow-induced shear stress on human mesenchymal stem cells: differential gene expression of IL1B and MAP3K8 in MAPK signaling, Gene Expr. Patterns 9 (5) (2009) 381–388.

[69] P.F. Davies, Flow-mediated endothelial mechanotransduction, Physiol. Rev. 75 (3) (1995) 519–560.

[70] S.L. Diamond, S.G. Eskin, L.V. McIntire, Fluid flow stimulates tissue plasminogen activator secretion by cultured human endothelial cells, Science 243 (4897) (1989) 1483–1485.

[71] Y. Kawai, Y. Matsumoto, K. Watanabe, H. Yamamoto, K. Satoh, M. Murata, et al., Hemodynamic forces modulate the effects of cytokines on fibrinolytic activity of endothelial cells, Blood 87 (6) (1996) 2314–2321.

[72] S. Kapur, D.J. Baylink, K.H. Lau, Fluid flow shear stress stimulates human osteoblast proliferation and differentiation through multiple interacting and competing signal transduction pathways, Bone 32 (3) (2003) 241–251.

[73] S. Kardasz, The function of the nephron and the formation of urine, Anaesth. Intensive Care Med. 10 (6) (2009) 265–270.

[74] W. Liu, S. Xu, C. Woda, P. Kim, S. Weinbaum, L.M. Satlin, Effect of flow and stretch on the [Ca2 +]i response of principal and intercalated cells in cortical collecting duct, Am. J. Physiol. Renal Physiol. 285 (5) (2003) F998–F1012.

[75] P. Guo, A.M. Weinstein, S. Weinbaum, A hydrodynamic mechanosensory hypothesis for brush border microvilli, Am. J. Physiol. Renal Physiol. 279 (4) (2000) F698–F712.

[76] S.M. Nauli, F.J. Alenghat, Y. Luo, E. Williams, P. Vassilev, X. Li, et al., Polycystins 1 and 2 mediate mechanosensation in the primary cilium of kidney cells, Nat. Genet. 33 (2) (2003) 129–137.

[77] S. Weinbaum, Y. Duan, L.M. Satlin, T. Wang, A.M. Weinstein, Mechanotransduction in the renal tubule, Am. J. Physiol. Renal Physiol. 299 (6) (2010) F1220–F1236.

[78] Z. Du, Y. Duan, Q. Yan, A.M. Weinstein, S. Weinbaum, T. Wang, Mechanosensory function of microvilli of the kidney proximal tubule, Proc. Natl. Acad. Sci. U.S.A. 101 (35) (2004) 13068–13073.

[79] Y. Duan, N. Gotoh, Q. Yan, Z. Du, A.M. Weinstein, T. Wang, et al., Shear-induced reorganization of renal proximal tubule cell actin cytoskeleton and apical junctional complexes, Proc. Natl. Acad. Sci. U.S.A. 105 (32) (2008) 11418–11423.

[80] B.W. Bisgrove, H.J. Yost, The roles of cilia in developmental disorders and disease, Development 133 (21) (2006) 4131–4143.

[81] W. Kriz, B. Kaissling, Structural and functional organization of the kidney, in: Geibisch Sa (Ed.), The Kidney, Elsevier, Inc., 2008, pp. 479–563.

[82] K.J. Jang, K.Y. Suh, A multi-layer microfluidic device for efficient culture and analysis of renal tubular cells, Lab Chip 10 (1) (2009) 36–42.

[83] M. Essig, G. Friedlander, Tubular shear stress and phenotype of renal proximal tubular cells, J. Am. Soc. Nephrol. 14 (Suppl. 1) (2003) S33–S35.

[84] Z. Cai, J. Xin, D.M. Pollock, J.S. Pollock, Shear stress-mediated NO production in inner medullary collecting duct cells, Am. J. Physiol. Renal Physiol. 279 (2) (2000) F270–F274.

[85] J.L. Charest, L.E. Bryant, A.J. Garcia, W.P. King, Hot embossing for micropatterned cell substrates, Biomaterials 25 (19) (2004) 4767–4775.
[86] J.L. Charest, M.T. Eliason, A.J. Garcia, W.P. King, Combined microscale mechanical topography and chemical patterns on polymer cell culture substrates, Biomaterials 27 (11) (2006) 2487–2494.
[87] J.L. Charest, A.J. Garcia, W.P. King, Myoblast alignment and differentiation on cell culture substrates with microscale topography and model chemistries, Biomaterials 28 (13) (2007) 2202–2210.
[88] J.T. Borenstein, E.J. Weinberg, B.K. Orrick, C. Sundback, M.R. Kaazempur-Mofrad, J.P. Vacanti, Microfabrication of three-dimensional engineered scaffolds, Tissue Eng. 13 (8) (2007) 1837–1844.
[89] J.T. Borenstein, H. Terai, K.R. King, E.J. Weinberg, M.R. Kaazempur-Mofrad, J.P. Vacanti, Microfabrication technology for vascularized tissue engineering, Biomed. Microdevices 4 (3) (2002) 167–175.
[90] J.L. Charest, E.J. Weinberg, J.C. Hsiao, J.T. Borenstein, Numerical modeling, microfabrication, and cell adhesion testing of a tissue engineered loop of Henle device, Society for Biomaterials Translational Biomaterial Research Symposium, Atlanta, GA, 2008.
[91] C.J. Bettinger, Z.T. Zhang, S. Gerecht, J.T. Borenstein, R. Langer, Enhancement of *in vitro* capillary tube formation by substrate nanotopography, Adv. Mater. 20 (1) (2008) 99–103.
[92] M. Rossi, R. Lindken, B.P. Hierck, J. Westerweel, Tapered microfluidic chip for the study of biochemical and mechanical response at subcellular level of endothelial cells to shear flow, Lab Chip 9 (10) (2009) 1403–1411.
[93] J.Y. Park, S.J. Yoo, C.M. Hwang, S.H. Lee, Simultaneous generation of chemical concentration and mechanical shear stress gradients using microfluidic osmotic flow comparable to interstitial flow, Lab Chip 9 (15) (2009) 2194–2202.
[94] J. Wang, J. Heo, S.Z. Jua, Spatially resolved shear distribution in microfluidic chip for studying force transduction mechanisms in cells, Lab Chip 10 (2009) 235–239.
[95] K.J. Jang, H.S. Cho, H. Kang do, W.G. Bae, T.H. Kwon, K.Y. Suh, Fluid-shear-stress-induced translocation of aquaporin-2 and reorganization of actin cytoskeleton in renal tubular epithelial cells, Integr. Biol. (Camb.) 3 (2) (2010) 134–141.
[96] P.Y. Dankers, J.M. Boomker, A.H. der Vlag, F.M. Smedts, M.C. Harmsen, M.J. van Luyn, The use of fibrous, supramolecular membranes and human tubular cells for renal epithelial tissue engineering: towards a suitable membrane for a bioartificial kidney, Macromol. Biosci. 10 (11) (2010) 1345–1354.
[97] B. Subramanian, D. Rudym, C. Cannizzaro, R. Perrone, J. Zhou, D.L. Kaplan, Tissue-engineered three-dimensional *in vitro* models for normal and diseased kidney, Tissue Eng. A 16 (9) (2010) 2821–2831.
[98] I. Barkefors, S. Thorslund, F. Nikolajeff, J. Kreuger, A fluidic device to study directional angiogenesis in complex tissue and organ culture models, Lab Chip 9 (4) (2009) 529–535.
[99] A. Roessger, L. Denk, W.W. Minuth, Potential of stem/progenitor cell cultures within polyester fleeces to regenerate renal tubules, Biomaterials 30 (22) (2009) 3723–3732.
[100] E.M. Frohlich, X. Zhang, J.L. Charest, The use of controlled surface topography and flow-induced shear stress to influence renal epithelial cell function, Integr. Biol. 4 (1) (2012) 75–83.

CHAPTER

Microfluidic Cell Culture Techniques

12

Yun Xiao[1], Boyang Zhang[1], Anne Hsieh[2], Nimalan Thavandiran[3], Cristina Martin[4] and Milica Radisic[5]

[1]*Department of Chemical Engineering and Applied Chemistry, Institute of Biomaterials and Biomedical Engineering, University of Toronto, Ontario, Canada*

[2]*Department of Chemical Engineering and Applied Chemistry, University of Toronto, Ontario, Canada*

[3]*Institute of Biomaterials and Biomedical Engineering, University of Toronto, Ontario, Canada*

[4]*Department of Chemical Engineering, University of Salamanca, Spain*

[5]*Department of Chemical Engineering and Applied Chemistry, Institute of Biomaterials and Biomedical Engineering, Heart & Stroke/Richard Lewar Centre of Excellence, University of Toronto, Ontario, Canada*

12.1 Fundamentals of microscale cell culture

12.1.1 Cell culture

Ever since Harrison [1] deposited fragments of frog embryo in hanging drops of coagulated frog lymph and observed the formation of nerve fibers from protoplasmic extensions in 1907, the use of cell culture has increased exponentially and is an essential tool in almost every biological and medical research laboratories. As one of the cornerstones of modern biology, cell culture has now emerged as a technique providing better understanding of human biology and disease for molecular genetics, immunology, surgery, bioengineering, and pharmaceutics.

12.1.2 Difference between macro- and microscale cell culture

Although conventional cell culture systems have served as very important tools for biological and medical studies, the methods surprisingly did not change a lot during the first 100 years. Most of the studies are still conducted in traditional containers such as dishes, bottles, and flasks with macroscale control over culturing conditions such as nutrient supply, pH, and temperature. These macroscale cell culture systems, however, cannot meet the need for current biological and medical studies, which aim at a more detailed understanding of cell behavior under well controlled microenvironment.

Microfluidic Cell Culture Systems. ISBN: 978-1-4377-3459-1

Polydimethylsiloxane (PDMS) based microfluidic cell culture models have been intensively studied recently as they provide better temporal and spatial control over the local cellular microenvironment in micrometer scale (0.1–100 μm) [2]. A wide variety of microfluidic cell culture models have been developed to mimic and control the cellular microenvironment. The strengths of microfluidic cell culture mainly come from changes of some physical properties when the scale of culture systems is reduced.

Most apparently, microfluidic cell culture provides the possibility to culture small cell clusters, or even single cells, which requires significantly smaller amount of reagents. This enables complex studies on cell behavior against different stimuli with higher throughput, which may be useful in pharmaceutical researches [3]. Meanwhile, the ratio between cell volume and medium volume in microfluidic cell culture system is usually greater than one, which provides more cell–cell interactions. This reflects the conditions of *in vivo* cellular microenvironment, where the cell/medium volume ratio is usually also greater than one.

Another significant difference between traditional macroscale cell culture and microfluidic cell culture models is the surface area to volume (SAV) ratio, which results in higher sensitivity to surface interactions on cell behavior. The improved interactions between cells and surfaces provide better platform to study different stimuli, both biochemical and physical, on cell behavior. Microelectromechanical systems (MEMS) also enable integration of other stimuli such as electrical or mechanical as well as sensing capabilities within the microfluidic cell culture platforms.

12.1.3 Microscale cell culture considerations

Due to the substantial differences between macroscale cell culture and microfluidic systems, understanding of fundamental physical principles of microfluidics is required to develop efficient microfluidic cell culture systems. In microfluidic cell culture systems, a small microchannel is usually designed when continuous nutrient supply is required. Therefore, for a given maximum fluid velocity, the shear stress present in the microchannel is significantly greater than in macroscale cell cultures [4]. The increased shear stress should be considered carefully since certain cell types respond to shear stress differently and there is a maximum amount of shear that can be tolerated by cells.

Microfluidic system often operates in low Reynolds number domain ($10 > Re > 0.001$) at which the fluid flow is highly laminar and follows well-defined streamline. Thus mass transfer relies mainly on diffusion [5,6], while convection is the dominant transport mechanism in traditional macroscale cell culture. This phenomenon results in better maintenance of secreted microenvironment around a cell but makes nutrient supply and waste removal more difficult at the same time.

To characterize cell culture microenvironment, Walker et al. [4] introduced the concept of an "effective culture volume" (ECV) as an indicator of the cell's ability to control its microenvironment during culture. It is a function of the magnitude of mass transport, the effects of diffusion and convection, and the degree of protein adsorption. For example, microfluidic systems are characterized by one-dimensional mass transport, higher levels of protein adsorption, and mainly diffusional mass transport. Thus cells in microfluidic systems have better control over their local environment, which results in smaller ECV.

12.2 Microfluidic cell culture systems

One of the greatest strengths of microfluidic cell culture systems is that while it provides better microenvironmental control, it still enables similar operations as traditional cell culture such as medium change and cell passaging. Moreover, with improved monitoring techniques and better control on handling, microfluidic cell culture systems can accomplish those procedures more precisely [7].

12.2.1 Cell seeding

The first crucial step to culture cells in microfluidic devices is cell seeding. With numerous studies on microfluidic cell culture techniques, the seeding density in microfluidic cell culture systems varied over a large range from extremely high cell density (1×10^8 cells/ml) to single cells. In general, a more homogenous distribution compared to traditional macroscale cell cultures has been achieved

Morel et al. [8] proposed a "microfluidic sticker" method to load cells into microfluidic devices. They took advantage of UV curable resin, which can tightly adhere on wet surfaces without any substrate modification, to fabricate a "sticker" on cells cultured using standard methods (Figure 12.1). After exposing the "sticker" to UV with inverted mask, the observation channel for cell culture was maintained. The simplicity and efficiency of this method has been demonstrated with several cell lines and tissues and hold the potential to be adapted for general microfluidic cell culture systems.

To improve the biological relevance of microfluidic cell culture systems, 3D cell culture has been studied intensively. Gel encapsulation is a common strategy to support the cells by surrounding hydrogels, either natural or synthetic, as a mimic of extracellular matrix (ECM) surrounding cells *in vivo* [9,10]. Cells can be embedded in gels and then seeded into microfluidic culturing systems more easily. However, gel encapsulation method has several limitations, such as insufficient mass transport of oxygen and nutrients. Particularly, this method is not ideal to build cell-dense and ECM-poor *in vitro* models, e.g., solid tumor models.

Ong et al. [11] developed a gel-free microfluidic 3D cell culture system by synthesized intercellular polymeric linker, polyethyleneimine-hydrazide (PEI-hy), and microfabricated pillar arrays. Briefly, the aldehyde groups on the sodium periodate treated cells reacted with the hydrazides on the intercellular linker to form cell aggregations and these aggregations accumulated by the micropillar arrays.

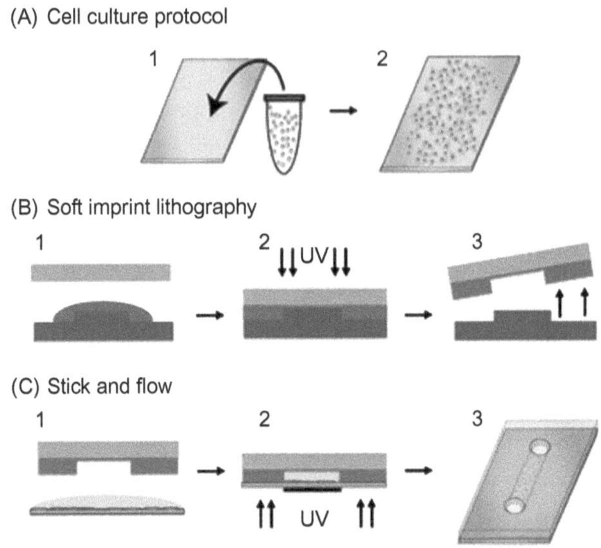

FIGURE 12.1

A novel method to seed cells into microfluidic devices with "microfluidic sticker." Cells are seeded on glass surface using standard protocols (A). The "microfluidic sticker" is fabricated right before experiments (B). A drop of resin (dark blue) is molded between a glass slide (light blue) and the invert replica of the device (gray) (B1), exposed to UV for cross-linking (B2) and eventually detached from the mold (B3). The prepared sticker is aligned on the cell-seeded glass surface (C1) and then sealed with gentle mechanical pressure and short UV exposure (C2) with a mask (black) to block direct UV exposure on the cells [8]. (For interpretation of the references to color in this figure legend, the reader is referred to the web version of this book.)

Source: *Reproduced by permission of the Royal Society of Chemistry.*

Optical tweezers is a powerful technique that can select and position individual cell on surface of microfluidic cell culture devices. Many recent studies combined microfluidic systems with optical tweezers, including image-based sorting of mammalian cells [12], developing biosensors [13], and measuring pico Newton force of biomolecules [14].

12.2.2 Cell culture maintenance

12.2.2.1 Cell density monitoring

Cell density in culture has been shown to influence cell responses from primitive cell differentiation [15,16] to primary cell interaction [17] and migration [18]. Meanwhile, the exponential growth of cells would result in quick buildup of cellular waste.

Scaling-up and controlling cell density is critically important in conventional bioprocesses focusing on the bacterial and yeast cell culture [19]. The integration of microfluidic cell culture systems with optical transducers is the most popular method to monitor the cell density. Briefly excitation fibers provide optical signal through cell culture and the optical density is measured by a transmission configuration incorporating low numerical aperture optics that provide a linear correlation between optical density and cell density [20].

12.2.2.2 Nutrient delivery

Culturing cells on chip requires the delivery of nutrient/culture medium to the cells. Such delivery can be done with channel networks. However, complications arise when culturing different types of cells. For instance, when culturing shear-sensitive cell types, to avoid the effects of shear stress on the cells, the cell culture chamber must be shielded from the flow of the medium while allowing cells to uptake sufficient nutrient from the medium at the same time. In such situations, multiple parallel microchannels in the range of few microns (1–2 μm in width) are typically used to bridge the flow channel to the cell culture chamber to provide sufficient nutrient delivery through diffusion while restraining convective flow to the culturing chamber [21–23]. Delivery of medium can be done in a continuous fashion [21–23] or in a periodic fashion [7] with programmed pneumatic pump that can deliver precisely controlled amounts of culture medium at a given time.

When culturing cells on a large scale with cell chamber arrays, multiple issues must be considered, such as cross-contamination between each compartment and the ability to individually address each culture chamber. Therefore, the complexity of the channel network must be increased as the cell culture array increases in size. These issues can be solved with bridge-and-underpass of the channel architecture, where flow can travel vertically to pass over (or under) the intersected channels (Figure 12.2). This way channels can cross each other and thus more complex channel networks can be incorporated to address each individual culture chamber without being restricted to the planar layout [22]. In addition, more complex fluid manipulation can be carried out with the incorporation of on-chip pneumatic valves.

12.2.2.3 Oxygen delivery

One of the requirements for cell culture is to be able to control and maintain the culture system at the desired oxygen and carbon dioxide environment. Delivery of oxygen and carbon dioxide at correct concentration is as important as delivering sufficient nutrients. Different types of cells demand different oxygen content to grow optimally. For instance, human embryonic stem cells (ESCs) are often cultured at a low oxygen (<5%) environment in order to initiate differentiation, while other mammalian cells are typically cultured with normal 21% oxygen concentration [24]. Furthermore, even within the acceptable range of oxygen concentration for cell culture, transient changes in oxygen concentration can have profound effects on cell behaviors such as grow factor production, gene expression, and differentiation.

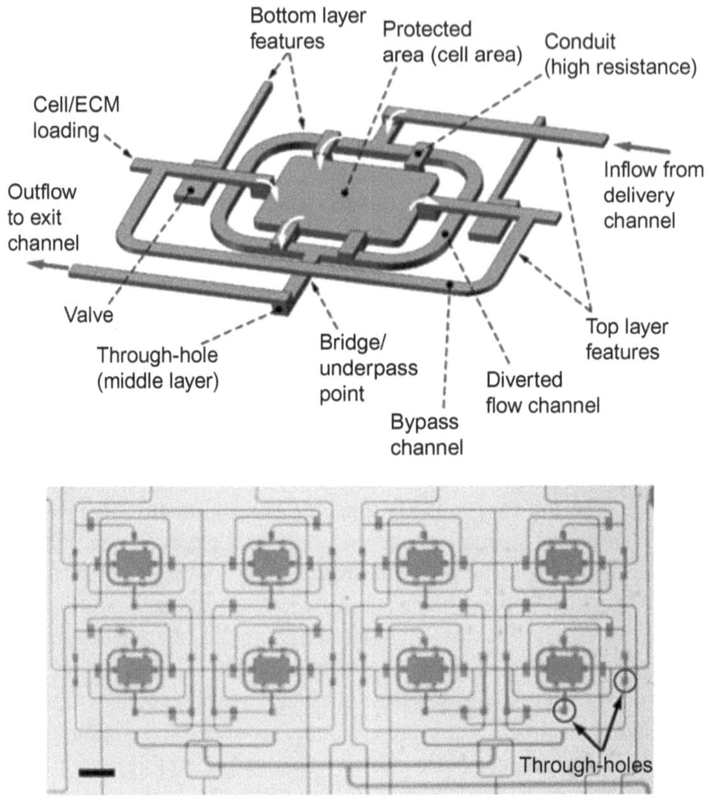

FIGURE 12.2

Schematics of a microfluidic cell culture system with intersecting fluid delivery channels. (A) Culture chamber schematics that show intersected fluid delivery channels crossing each other through bridges/underpass. White arrows indicate where fluid flows vertically. (B) Overall layout of the culture chamber array where blue dye marks the fluid delivery channels and red dye marks the pneumatic valve control channels. Black arrows indicate the locations of through-holes where fluid travels vertically in a different plane [22]. (For interpretation of the references to color in this figure legend, the reader is referred to the web version of this book.)

Source: *Reproduced by permission of the Royal Society of Chemistry.*

The advantage of PDMS-based microfluidic cell culture system is its high diffusivity ($D = 4.1 \times 10^{-5}$ cm^2/s) of oxygen in PDMS [25]. Due to this high permeability of PDMS, passive diffusion of oxygen from the incubating environment through the PDMS to the cells is often assumed to be sufficient. However, complications arise if the surface of the PDMS is modified. For instance, with up to 5 min of plasma oxidation, which is a common step for device bonding, the mass transfer coefficient of oxygen through PDMS can be reduced 1000 folds [26].

Moreover, coating PDMS with bovine serum albumin which is often performed to reduce nonspecific cell adhesion can reduce the mass transfer coefficient up to five times [26]. As the permeability of oxygen through PDMS decreases, oxygen gradient can form along the cell culture channel, especially at typical flow perfusion rates [27,28]. Therefore, proper monitoring and control of oxygen content becomes necessary.

Numerous integrated oxygen sensing method have been developed. Optical oxygen sensors based on immobilizing oxygen-sensitive fluorescent dye, such as tris(2,2A-bipyridyl) dichlororuthenium(II) hexahydrate [29,30] and platinum octaethylporphyrin ketone [31], can be used to quantify both the dissolved and gaseous oxygen content through either the degree of the luminescent intensity or the luminescent lifetime change [30]. The advantages of such method are easy integration, quick responses, and high sensitivity.

Furthermore, there have been studies on controlling the oxygen content in microfluidic culturing channel in the microfluidic devices with the purpose of conducting fundamental biological studies. For instance, Allen and Bhatia [32] constructed a parallel plate reactor with a gas exchanger built in the flow circuit. The gas content of the inlet to the reactor was controlled by the gas exchanger. The gas content of the outlet was monitored with an oxygen sensor. Rat hepatocytes were culture in the microreactor and the consumption of oxygen facilitates the formation of a linear oxygen gradient along the microreactor from the inlet to the outlet. This model was used to simulate the oxygen gradient in the liver which has been implicated in the maintenance of regional metabolic and detoxification functions in the liver, termed zonation [32].

Oxygen gradients have also been generated directly by on-chip gas mixing. Mark et al. have used branched channels to mix oxygen with nitrogen and form nine different oxygen concentration conditions for *Escherichia coli* culturing simultaneously. Gas was delivered to the culture chamber from the top by permeation through a thin layer of PDMS. *In situ* oxygen concentration was measured using optical oxygen sensor with oxygen-sensitive fluorescence dyes. *E. coli* was then cultured in semipermeable microchamber with nutrient delivery from the side channels. Within this system, the growth rate of bacteria had shown to increase with oxygen concentration as expected.

12.2.2.4 Temperature control

Temperature control for cell culture in microfluidic system is typically done off-chip by placing the microfluidic system in a cell culture incubator. However, integration of a heating unit on chip may sometimes be necessary, especially when a temperature gradient is required. Numerous heating methods have been introduced. For instance, a temperature gradient across a flow channel can be formed by simply incorporating a hot fluid channel on one side as heat source and a cold fluid channel on the other as heat sink [33]. Furthermore, a conductive wire can also be integrated next to a microchannel to provide heating. Vigolo et al. [34] have used a silver-filled epoxy that can be injected and solidified in a microchannel as a conducting wire. Using this method, they can maintain the temperature of the flow channel as well as build a temperature gradient across the channel. Lastly, more sophisticated microelectrode array can also be integrated on chip to provide highly controllable local heating on chip [35].

12.2.3 Cell passaging

Cell passaging, essential for long-term proliferative adherent cell culture, is composed of repeated processes including cell seeding, growth, detachment, and reseeding on a fresh surface. In translating the process to microfluidic system, a device needs to integrate the processes of reagent introduction and repeated cell growth and passage cycles onto a single system. Hung et al. [21] developed a self-contained microfluidic cell culture array using soft-lithography and replicate molding, capable of conducting periodic trypsin stripping of cells from a culture chamber and replating. Later, with the advance in the development in microfabrication technology, which allows for integrated on-chip design such as dielectrophoresis, Barbulovic-Nad et al. [36] created a digital microfluidic (DMF) platform that can grow adherent cells on patterned islands, exchange media rapidly and efficiently, detaching and collecting cells to be replated. DMF offers advantages over channel microfluidic devices by its ability to manipulate discrete samples on the surface of the actuation electrodes, providing discontinuous fresh sterile surface for replating. Liu et al. [37] integrated PDMS diaphragms with microfluidic cell culture system to detach adherent cells purely through hydrodynamic shearing forces. The passaging diaphragms were programmed to apply alternating forces, detaching a customizable number of the cells from the surface. We have developed microfluidic devices coated with RGD modified alginate for controlled cell capture and release. The cells adhering to the RGD-alginate in a microfluidic channel can easily be detached without the need for trypsinization simply by removing Ca^{2+} from the media in the microfluidic channel. This is easily accomplished by application of a dilute solution of ethylene diamine tetraacetic acid (EDTA) or a similar chelator of divalent ions [38].

12.2.4 Cell concentration and dilution

In macroscale, cell concentration and dilution are typically conducted by centrifugation and resuspension. However, it is often challenging when working with minute number of rare cells to form proper pellet for supernatant aspiration. In lieu of this, an on-chip cell concentrator will allow the precise control of the cell culture density inside the microenvironment and reduction of the lost number of rare cells required for cell loading. Technologies that have been developed for cell concentration in microfluidic chips include applying physical size-based impediments [39,40], electric charge [41], or surface functionalized with recognizable biomolecules [42]. Furthermore, microfluidic systems offer additional methods for cell enrichment by changing parameters in the channel dimension

[43] and applying principles of inertial focusing [44]. Balasubramanian et al. [45] utilized bacterial cells' surface charge and polarizability to transport cells toward the anode, and van der Waals and electrostatic interactions to concentrate the cells. Without any manipulation on the cells, Warrick et al. [46] developed a gravity-based cell centrifugation on chip which minimized the amount of shear stress and manipulation that may affect cell physiology.

Some cell culture approaches require minute cell numbers or single-cell cultivation such as cloning. The conventional cloning by limiting dilution that relies on the crude statistical sampling and cell counting in the original cell suspension [47] is labor intensive, yields low-throughput, brings low single-cell loading efficiency, and lacks reproducibility. However, utilizing the versatile design and fabrication techniques, microfluidic devices with various differential hydrodynamic guiding structures can provide high-throughput manipulation with accurate control at single-cell level [48]. The microfluidic design can further incorporate active or passive single-cell capture schemes. Active single-cell capture schemes include electrokinetic methods such as dielectrophoresis [49] and optical image-driven dielectrophoresis [50]. These methods offer accurate cell manipulation and cell capturing; however, they require cell suspension media of high conductivity to generate dielectrophoresis, limiting their potential applications. These methods also require the connection to external equipment with precise control.

Alternatively, passive methods using hydrodynamic weirs [51] are attractive because of their simplicity. In this approach, individual particles are sequestered to designated on-chip locations using a mechanical barrier (Figure 12.3). Subsequent particle loading to the same site is prevented by the altered local

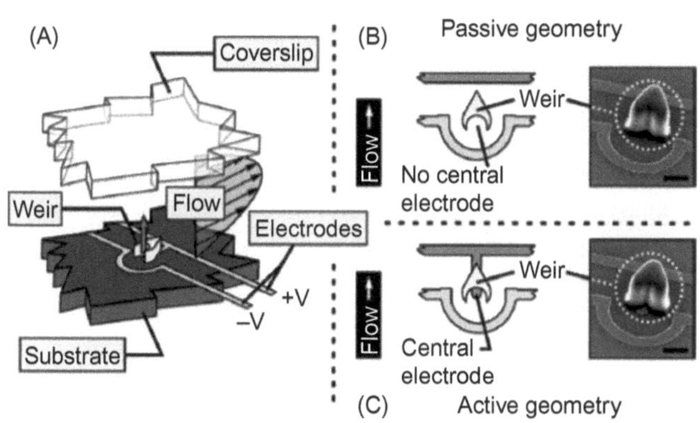

FIGURE 12.3

Single particle capture and *n*-DEP electroactive sorting and patterning. (A) Microfabrication of electrodes to exert *n*-DEP forces and fluid flow regime. (B and C) Schematic and SEM images of passive and active designs of the electric weirs in presence and absence of central electrodes, respectively, to direct particle mobility (scale bar = 20 μm).

Source: *Reprinted with permission from Ref. [51]. Copyright 2009, American Institute of Physics.*

hydrodynamic flow profiles derived from the presence of the trapped particle. In addition, passive methods do not require any external equipment. Most weir systems, however, have relatively poor capturing efficiencies, with most of the injected cells going to the outlet, making it difficult to apply to rare cell populations or requiring a recycling mechanism.

12.3 Microenvironmental stimuli

The cellular microenvironment comprises of complex of biochemical and physical factors that work in concert to influence cell behaviors. With its unique micro-scale control and integration with other techniques, microfluidic cell culture systems serve as ideal platforms to study the effects of different stimuli on cell behavior.

12.3.1 Biochemical stimuli

In the scope of a microfluidic device, mass transfer can be performed in either chemostatic perfusion for fast media exchange [20] or diffusive replacement [52] with minimum shear stress applied to cell. Nevertheless, perfusion cell cultures may hamper cell-to-cell communication through intrinsic and extrinsic growth factors owing to the continuous washing away of these biomolecules [53]. A recirculatory system can be applied to better imitate the *in vivo* conditions [54].

The ability to manipulate solubility factors temporally and spatially has been utilized to generate various biomolecular gradients involved in biological signaling phenomena [55]. The flexible and controllable microfluidic gradient generators with various stable profiles [56] and proper concentration resolution [57] offer many advantages over other macroscale devices. Utilizing the laminar flow and a microfluidic mixer, a gradient of a biological molecule such as vascular endothelial growth factor (VEGF) can be generated on a micrometer scale. This is a powerful approach that enables screening of the chemotactic ability of a particular growth factor and a selected cell population [58].

The flow channel can be driven by conventional external fluidic pumps such as syringe and peristaltic pumps, or on-chip piezoelectric [59], pneumatic [60], and electromagnetic [61] micropumps. However, the fabrication processes for integrating the micropumps on chip are highly complicated with bulky power sources to operate the micropumps. Pumpless perfusion cell culture chips use horizontally patterned source [62] vertically aligned with drain reservoirs at the lower level accompanies by a hydraulic pressure difference–induced gravity-driven flow [63]. Often, a resistance barrier structural element is incorporated between the flow chamber and the cell culture chamber to ensure only perfusion mass transfer is taking place, free from flow regime disturbances. Furthermore, transient and spatial biochemical distribution and gradient can be created applying the fluid dynamic in a multiconditioning cell cultural array [21,64] where many different experimental conditions can be tested at the same time in parallel.

12.3.2 Physical stimuli

12.3.2.1 Topographical stimuli

It is well known that the architecture of the extracellular environment influences cell behavior with respect to morphology, cytoskeletal structure, and functionality [65]. Thus, substrates with different architectures can serve as platforms to study the effects of cell–substrate interactions on cell adhesion, orientation, migration, and differentiation. It is also necessary to replicate, as closely as possible, the *in vivo* environment in terms of dimensional, configurational, and compositional properties when designing tissue replacements [66]. With excellent control of substrate micro- and nanoarchitecture, microfluidic cell culture systems could replicate with high fidelity *in vivo* microstructure and serve as a customizable platform to study cell–substrate interactions [67,68].

A large number of studies have been focused on substrate topography, especially in muscle and cardiac tissue engineering since the alignment of myocytes is important for the contraction and impulse propagation along the long axis of the cells [69]. It has been demonstrated that microfabricated substrates with microscale grooves can significantly improve cell alignment [70,71] and result in greater expression of cardiac genes and cardiomyocyte markers [72,73]. Meanwhile, the elongation in fibroblasts caused by micropatterned substrates inhibited fibroblasts proliferation [74], which would be beneficial for cardiac cell culture since fibroblasts are considered as contaminating nonmyocytes.

12.3.2.2 Mechanical stimuli

Mechanical stimulation and perturbation of live cells via the manipulation of flow regimes in microfluidic devices can be induced through normal and shear stresses [75]. To introduce various mechanical stimuli onto these trapped cells, either fluid flow or substrate deformations can be utilized.

Shear-induced alignment of multiple cell types has been reported along with proposed associated mechanotransduction-mediated mechanisms with which the cell response is elicited. Typically, endothelial cells align either perpendicular or parallel to the direction of experienced shear in fluid flow [76]. Blood flow in arteries and capillaries exhibits the same sort of shear stress *in vivo* and affects endothelial cell phenotype and alignment. Additionally, endothelial cells need to be able to interact with other smooth muscle cells which either dilate or compress blood vessels in response to receptor-mediated signaling through endothelial cells. Flow rates can be adjusted in just about any microfluidic device to apply various shear stresses to cells, and since microfluidic devices are easily characterized with their simple geometries, complex models can be generated to predict effects of simulated shear stresses on different cell types. Moreover, shear stress has been reported to influence the differentiation of ESCs. Toh el al. [77] utilized a multiplex microfluidic array to study the shear effects on self-renewing mouse embryonic stem cells (mESCs) and demonstrated that mESCs possess molecular machinery to sense shear stress.

12.3.2.3 Electrical stimuli

Electrical stimulation of cells in microfabricated devices has been done through macroscale wire electrode techniques and surface patterned electrodes. Many cells, especially electromechanically coupled cells like myocytes, respond to point and field stimuli. Maturation of cardiomyocytes through field stimulation has been repeatedly shown through changes in morphology and protein expression [78]. Moreover, it has been demonstrated that electrical stimulation has an interactive effect with topographic cues on orientation and elongation of fibroblasts and cardiomyocytes [79,80]. In a recent study, a commonly used metal, indium tin oxide (ITO), was used to pattern alternating electrodes on a glass surface [81]. ITO is convenient for microfabricated cell culture substrates due to its superior conductivity, optical transparency, and ability to be micropatterned. In this study, electrical stimulation of cardiomyocytes plated onto the electrodes was shown to enhance proliferation, elongation, and alignment. This study also investigated adipose derived stem cells which exhibited higher numbers of Connexin-43-composed gap junctions under stimulated conditions.

12.3.3 Compartmentalization

Compartmentalization in microfluidic devices allows the ability to separately treat specific parts of cells or tissues with various biochemicals or other localized stimuli. In the neuroscience field, much interest has been shown in separately treating the cell body and the neurite outgrowths [82,83]. These studies utilize separated compartments which allow the seeding of dissociated cells in suspension. Once the cells adhere and begin to grow, grooved extensions topographically guide neurite and axon growth from the initial cell body compartment to another independently treatable compartment for neuritis and axons. Figure 12.4 shows neurons in a neuritic chamber with cell bodies on the left and neurite extensions through the narrow channels to the other side where the ends can be treated with specific biochemicals and the elicited response can be studied in isolation. Using conventional macroscale culture methods, these studies would be impractical or could be performed only with low-throughput and with low reproducibility. One can also couple other technologies with compartmentalized devices including the use of multiple laminar flow streams to treat subcellular areas with small molecules [84]. Additionally, one can also use electrodes in compartmentalized devices to either record or stimulate at a subcellular level [83]. In general, these various technologies can be mixed and matched to produce a system whereby high content screening can be performed fairly robustly and at a low cost.

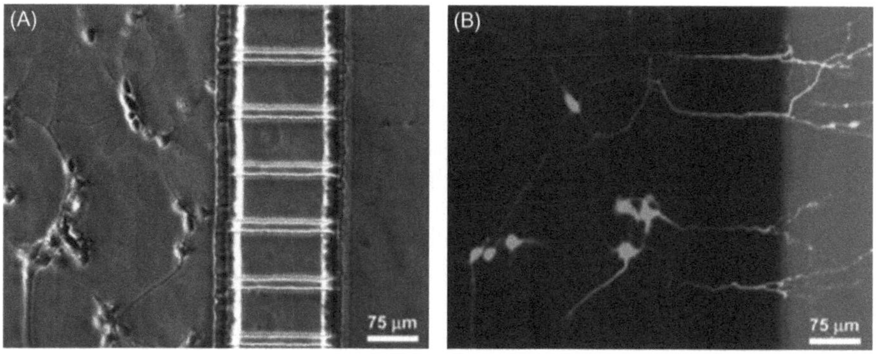

FIGURE 12.4

Demonstration and fluorescent micrograph of neuronal culture inside the microfabricated device. (A) Phase micrograph of neurons extended to the neuritic chamber (right) after 4 days. (B) Fluorescent micrograph of the same region with cell stained with calcein AM (green) and isolation of Texas Red dextran (red) in the neuritic chamber (Texas Red dextran is not detectable in the somal chamber). (For interpretation of the references to color in this figure legend, the reader is referred to the web version of this book.)

Source: *Reprinted with permission from Ref. [82]. Copyright 2003, American Chemical Society.*

12.4 Microfluidic cell and tissue culture systems for drug discovery and studies in physiology

There is a clear drive to minimize the use of animal testing in drug research for ethical considerations. In addition, animal testing is often not a reliable way to predict human responses. During the past several years, significant effort has been made on combining cell culture techniques with microscale engineering technologies, biomaterials science, and tissue engineering for the creation of microchips that mimic biological systems, specifically human tissues and organs. These microfluidic cell/tissue culture systems not only provide a more realistic 3D-microarchitecture than conventional cell culture techniques but also recreate tissue-specific physical environment. The goal is to create adequate experimental model systems to efficiently screen and validate potential drug candidates that can replace, or at least reduce, the costly and time-consuming animal studies.

The lung-on-a-chip [85] is one of the most sophisticated microfluidic cell/tissue culture systems so far. Other organs that have been mimicked using microfluidic culture models include liver [86], kidney [87], gastrointestinal tract [88–90], and vasculature [91]. Sophisticated microchip for testing of small caliber blood vessels, i.e., artery-on-a-chip, has also been developed [92]. Researchers at the Wyss Institute for Biological Inspired Engineering at Harvard University are also working on a more predictive human kidney-on-a-chip and an intestine-on-a-chip that exhibits peristalsis, i.e., the series of muscle contractions that occur as food moves through the digestive tract [93].

Another advantage of microfluidic cell culture systems for drug research is that they can integrate different organs, all linked by microfluidic channels [88,94–96]. Shuler and coworkers [94] have used these "microscale cell culture analogs" to investigate the pharmacokinetics–pharmacodynamics of a given drug, i.e., to predict how the drug moves through and accumulates in various organs.

These microfluidic cell culture systems are being seen as a great platform for drug research and physiological studies and there is a clear movement from proof-of-concept research to commercialization of the chips. Although there are not products in the market yet, the potential of this technology is currently leading to collaborations between researchers and pharmaceutical companies in order to rapidly demonstrate and validate these microfluidic cell culture models for drug-development applications [97]. For example, Wyss Institute is pursuing a heart–lung micromachine project funded by FDA and NIH. The participation of FDA in the early stages of the development process could enhance the acceptance of these new organ-on-a-chip technologies [93]. The ultimate goal of these efforts is the "human-on-a-chip," a chip that will contain linked mimics of the most relevant human organs that might be possibly used to carry out preclinical trials on chips. This approach would help advance the personalized medicine by growing the specific tissue of a patient in a microchip to study the effects of different drugs.

It is unlikely that microfluidic cell/tissue culture technology will ever completely replace animal and human trials, but it is very likely that it will have a significant contribution in shortening the time and cost of drug development as well as increasing the success in human trials, which will translate in higher efficiency in finding new treatments.

12.5 Conclusions

Different microfluidic cell culture systems have been successfully utilized to maintain cell microenvironment and provide various types of stimuli to the cells. The integration of microfluidic devices with other microscale techniques (e.g., microtopography or electrical stimulation) enables both basic cell biology research and development of engineered tissues. Although integrations with other microscale techniques are very helpful and sometimes required, the microfluidic cell culture systems should be simple and convenient to handle for biologists to use without the help from engineers. Recently, several high impact publications describe development of sophisticated systems (e.g., lung-on-a-chip or artery-on-a-chip) that mimic the *in vivo* biological environment with high fidelity and act as a new platform for drug testing and studies of cellular/tissue responses to biochemical stimuli. In the future, such systems may accelerate the discovery of new drugs and minimize the need for preclinical animal testing.

References

[1] R.G. Harrison, et al., Observations of the living developing nerve fiber, Anat. Rec. 1 (5) (1907) 116–128.

[2] I. Meyvantsson, D.J. Beebe, Cell culture models in microfluidic systems, Ann. Rev. Anal. Chem. 1 (2008) 423–449.

[3] M.-H. Wu, S.-B. Huang, G.-B. Lee, Microfluidic cell culture systems for drug research, Lab Chip 10 (8) (2010) 939–956.

[4] G.M. Walker, H.C. Zeringue, D.J. Beebe, Microenvironment design considerations for cellular scale studies, Lab Chip 4 (2) (2004) 91–97.

[5] T. Gervais, K.F. Jensen, Mass transport and surface reactions in microfluidic systems, Chem. Eng. Sci. 61 (2006) 1102–1121.

[6] A.E. Kamholz, P. Yager, Theoretical analysis of molecular diffusion in pressure-driven laminar flow in microfluidic channels, Biophys. J. 80 (1) (2001) 155–160.

[7] R. Gómez-Sjöberg, et al., Versatile, fully automated, microfluidic cell culture system, Anal. Chem. 79 (22) (2007) 8557–8563.

[8] M. Morel, et al., Microfluidic stickers for cell- and tissue-based assays in microchannels, Lab Chip 9 (7) (2009) 1011–1013.

[9] A.P. Wong, et al., Partitioning microfluidic channels with hydrogel to construct tunable 3-D cellular microenvironments, Biomaterials 29 (12) (2008) 1853–1861.

[10] C.-H. Choi, et al., Generation of monodisperse alginate microbeads and in situ encapsulation of cell in microfluidic device, Biomed. Microdevices 9 (6) (2007) 855–862.

[11] S.-M. Ong, et al., A gel-free 3D microfluidic cell culture system, Biomaterials 29 (22) (2008) 3237–3244.

[12] J.R. Kovac, J. Voldman, Intuitive, image-based cell sorting using optofluidic cell sorting, Anal. Chem. 79 (24) (2007) 9321–9330.

[13] L. Tong, et al., Optical aggregation of metal nanoparticles in a microfluidic channel for surface-enhanced Raman scattering analysis, Lab Chip 9 (2) (2009) 193–195.

[14] K. Uhrig, et al., Optical force sensor array in a microfluidic device based on holographic optical tweezers, Lab Chip 9 (5) (2009) 661–668.

[15] W. Song, et al., Gradient patterning and differentiation of mesenchymal stem cells on micropatterned polymer surface, J. Bioact. Compat. Polym. 26 (3) (2011) 242–256.

[16] M. Bitar, et al., Effect of cell density on osteoblastic differentiation and matrix degradation of biomimetic dense collagen scaffolds, Biomacromolecules 9 (1) (2008) 129–135.

[17] R.L. Mauck, et al., The role of cell seeding density and nutrient supply for articular cartilage tissue engineering with deformational loading, Osteoarthr. Cartilage 11 (12) (2003) 879–890.

[18] A.B. Khoshfetrat, et al., Seeding density modulates migration and morphology of rabbit chondrocytes cultured in collagen gels, Biotechnol. Bioeng. 102 (1) (2009) 294–302.

[19] F.K. Balagaddé, et al., Long-term monitoring of bacteria undergoing programmed population control in a microchemostat, Science 309 (5731) (2005) 137–140.

[20] K.S. Lee, et al., Microfluidic chemostat and turbidostat with flow rate, oxygen, and temperature control for dynamic continuous culture, Lab Chip 11 (10) (2011) 1730–1739.

[21] P.J. Hung, et al., Continuous perfusion microfluidic cell culture array for high-throughput cell-based assays, Biotechnol. Bioeng. 89 (1) (2005) 1–8.

[22] E.S. Park, et al., Continuously perfused, non-cross-contaminating microfluidic chamber array for studying cellular responses to orthogonal combinations of matrix and soluble signals, Lab Chip 10 (5) (2010) 571–580.

[23] P.J. Hung, et al., A novel high aspect ratio microfluidic design to provide a stable and uniform microenvironment for cell growth in a high throughput mammalian cell culture array, Lab Chip 5 (1) (2005) 44–48.

[24] T. Ezashi, P. Das, R.M. Roberts, Low O_2 tensions and the prevention of differentiation of hES cells, Proc. Natl. Acad. Sci. U.S.A. 102 (13) (2005) 4783–4788.

[25] T.C. Merkel, et al., Gas sorption, diffusion, and permeation in poly(dimethylsiloxane), J. Polym. Sci. B Polym. Phys. 38 (3) (2000) 415–434.

[26] H. Shiku, et al., Oxygen permeability of surface-modified poly(dimethylsiloxane) characterized by scanning electrochemical microscopy, Chem. Lett. 35 (2) (2006) 234–235.

[27] J.W. Allen, S.R. Khetani, S.N. Bhatia, In vitro zonation and toxicity in a hepatocyte bioreactor, Toxicol. Sci. 84 (1) (2005) 110–119.

[28] J. Park, et al., A microsystem for sensing and patterning oxidative microgradients during cell culture, Lab Chip 6 (5) (2006) 611–622.

[29] D.A. Chang-Yen, B.K. Gale, An integrated optical oxygen sensor fabricated using rapid-prototyping techniques, Lab Chip 3 (4) (2003) 297–301.

[30] Q.L. Chen, et al., Dissolved oxygen sensing using organometallic dyes deposited within a microfluidic environment – art. no. 68860O, in: W. Wang, C. Vauchier (Eds.), Microfluidics, Biomems, and Medical Microsystems Vi, Spie-Int Soc Optical Engineering, Bellingham, 2008. p. O8860

[31] A.P. Vollmer, et al., Development of an integrated microfluidic platform for dynamic oxygen sensing and delivery in a flowing medium, Lab Chip 5 (10) (2005) 1059–1066.

[32] J.W. Allen, S.N. Bhatia, Formation of steady-state oxygen gradients in vitro: application to liver zonation, Biotechnol. Bioeng. 82 (3) (2003) 253–262.

[33] H. Mao, T. Yang, P.S. Cremer, A microfluidic device with a linear temperature gradient for parallel and combinatorial measurements, J. Am. Chem. Soc. 124 (16) (2002) 4432–4435.

[34] D. Vigolo, et al., A portable device for temperature control along microchannels, Lab Chip 10 (6) (2010) 795–798.

[35] H. Yang, et al., An independent, temperature-controllable microelectrode array, Anal. Chem. 76 (5) (2004) 1537–1543.

[36] I. Barbulovic-Nad, S.H. Au, A.R. Wheeler, A microfluidic platform for complete mammalian cell culture, Lab Chip 10 (12) (2010) 1536–1542.

[37] L. Liu, et al., A microfluidic device for continuous cancer cell culture and passage with hydrodynamic forces, Lab Chip 10 (14) (2010) 1807–1813.

[38] B.D. Plouffe, et al., Controlled capture and release of cardiac fibroblasts using peptide-functionalized alginate gels in microfluidic channels, Lab Chip 9 (11) (2009) 1507–1510.

[39] S. Zheng, et al., Membrane microfilter device for selective capture, electrolysis and genomic analysis of human circulating tumor cells, J. Chromatogr. A 1162 (2) (2007) 154–161.

[40] J.S. Kuo, et al., Deformability considerations in filtration of biological cells, Lab Chip 10 (7) (2010) 837–842.

[41] P.R. Gascoyne, et al., Isolation of rare cells from cell mixtures by dielectrophoresis, Electrophoresis 30 (8) (2009) 1388–1398.

[42] D.J. Sherman, et al., A differential cell capture assay for evaluating antibody interactions with cell surface targets, Anal. Biochem. 401 (2) (2010) 173–181.

[43] H. Mohamed, et al., Isolation of tumor cells using size and deformation, J. Chromatogr. A 1216 (2009) 8289–8295.

[44] S.S. Kuntaegowdanahalli, et al., Inertial microfluidics for continuous particle separation in spiral microchannels, Lab Chip 9 (20) (2009) 2973–2980.

[45] A.K. Balasubramanian, et al., A microfluidic device for continuous capture and concentration of microorganisms from potable water, Lab Chip 7 (10) (2007) 1315–1321.

[46] J. Warrick, et al., A microfluidic cell concentrator, Anal. Chem. 82 (19) (2010) 8320–8326.

[47] R. Staszewski, Murphy's law of limiting dilution cloning, Stat. Med. 9 (4) (1990) 457–461.

[48] J. Chung, Y.J. Kim, E. Yoon, Highly-efficient single-cell capture in microfluidic array chips using differential hydrodynamic guiding structures, Appl. Phys. Lett. 98 (12) (2011) 123701.

[49] M.S. Jaeger, et al., Contact-free single-cell cultivation by negative dielectrophoresis, J. Phys. D Appl. Phys. 41 (2008) 1–8.

[50] P.Y. Chiou, A.T. Ohta, M.C. Wu, Massively parallel manipulation of single cells and microparticles using optical images, Nature 436 (7049) (2005) 370–372.

[51] B.M. Taff, S.P. Desai, J. Voldman, Electroactive hydrodynamic weirs for microparticle manipulation and patterning, Appl. Phys. Lett. 94 (8) (2009) 84102.

[52] Y. Zeng, et al., Numerical simulation of mass transport in a microchannel bioreactor with cell micropatterning, J. Biomech. Eng. 130 (3) (2008) 031018.

[53] D. Ellison, A. Munden, A. Levchenko, Computational model and microfluidic platform for the investigation of paracrine and autocrine signaling in mouse embryonic stem cells, Mol. Biosyst. 5 (9) (2009) 1004–1012.

[54] J. Atencia, D.J. Beebe, Capillary inserts in microcirculatory systems, Lab Chip 6 (4) (2006) 575–577.

[55] A.L. Paguirigan, D.J. Beebe, Microfluidics meet cell biology: bridging the gap by validation and application of microscale techniques for cell biological assays, Bioessays 30 (9) (2008) 811–821.

[56] A. Shamloo, et al., Endothelial cell polarization and chemotaxis in a microfluidic device, Lab Chip 8 (8) (2008) 1292–1299.

[57] T.M. Keenan, A. Folch, Biomolecular gradients in cell culture systems, Lab Chip 8 (1) (2008) 34–57.

[58] D. Odedra, et al., Endothelial cells guided by immobilized gradients of vascular endothelial growth factor on porous collagen scaffolds, Acta Biomater. 7 (8) (2011) 3027–3035.

[59] W. Gu, et al., Computerized microfluidic cell culture using elastomeric channels and Braille displays, Proc. Natl. Acad. Sci. U.S.A. 101 (45) (2004) 15861–15866.

[60] M.H. Wu, et al., A high throughput perfusion-based microbioreactor platform integrated with pneumatic micropumps for three-dimensional cell culture, Biomed. Microdevices 10 (2) (2008) 309–319.

[61] H. Kimura, et al., An integrated microfluidic system for long-term perfusion culture and on-line monitoring of intestinal tissue models, Lab Chip 8 (5) (2008) 741–746.
[62] T. Kim, Y.H. Cho, A pumpless cell culture chip with the constant medium perfusion-rate maintained by balanced droplet dispensing, Lab Chip 11 (10) (2011) 1825–1830.
[63] M.W. Ashraf, S. Tayyaba, N. Afzulpurkar, Micro electromechanical systems (MEMS) based microfluidic devices for biomedical applications, Int. J. Mol. Sci. 12 (2011) 3648–3704.
[64] B. Harink, et al., Microfluidic platform for the simultaneous generation of four independent gradients: towards the high throughput screening of trace elements for bone tissue engineering, in: 14th International Conference on Miniaturized Systems for Chemistry and Life Sciences, 2010, pp. 638–640.
[65] R. Singhvi, G. Stephanopoulos, D.I. Wang, Effects of substratum morphology on cell physiology, Biotechnol. Bioeng. 43 (8) (1994) 764–771.
[66] T.A. Desai, Micro- and nanoscale structures for tissue engineering constructs, Med. Eng. Phys. 22 (9) (2000) 595–606.
[67] G.M. Whitesides, et al., Soft lithography in biology and biochemistry, Annu. Rev. Biomed. Eng. 3 (2001) 335–373.
[68] H. Andersson, A.V.D. Berg, Microfabrication and microfluidics for tissue engineering: state of the art and future opportunities, Lab Chip 4 (2) (2004) 98–103.
[69] G. Vunjak-Novakovic, et al., Challenges in cardiac tissue engineering, Tissue Eng. B Rev. 16 (2) (2010) 169–187.
[70] J. Deutsch, et al., Fabrication of microtextured membranes for cardiac myocyte attachment and orientation, J. Biomed. Mater. Res. 53 (3) (2000) 267–275.
[71] E. Entcheva, H. Bien, Tension development and nuclear eccentricity in topographically controlled cardiac syncytium, Biomed. Microdevices 5 (2) (2003) 163–168.
[72] D. Motlagh, et al., Microfabricated grooves recapitulate neonatal myocyte connexin43 and N-cadherin expression and localization, J. Biomed. Mater. Res. A 67A (1) (2003) 148–157.
[73] D. Motlagh, et al., Microtextured substrata alter gene expression, protein localization and the shape of cardiac myocytes, Biomaterials 24 (14) (2003) 2463–2476.
[74] S.Y. Boateng, et al., Inhibition of fibroblast proliferation in cardiac myocyte cultures by surface microtopography, Am. J. Physiol. Cell Physiol. 285 (1) (2003) C171–C182.
[75] S.S. Kohles, Y. Liang, A.K. Saha, Volumetric stress–strain analysis of optohydrodynamically suspended biological cells, J. Biomech. Eng. 133 (1) (2011) 011004.
[76] E.W.K. Young, C.A. Simmons, Macro- and microscale fluid flow systems for endothelial cell biology, Lab Chip 10 (2) (2010) 143–160.
[77] Y.-C. Toh, J. Voldman, Fluid shear stress primes mouse embryonic stem cells for differentiation in a self-renewing environment via heparan sulfate proteoglycans transduction, FASEB J. 25 (4) (2011) 1208–1217.
[78] M. Radisic, et al., Functional assembly of engineered myocardium by electrical stimulation of cardiac myocytes cultured on scaffolds, Proc. Natl. Acad. Sci. U.S.A. 101 (52) (2004) 18129–18134.
[79] H.T.H. Au, et al., Interactive effects of surface topography and pulsatile electrical field stimulation on orientation and elongation of fibroblasts and cardiomyocytes, Biomaterials 28 (29) (2007) 4277–4293.

[80] Au, H.T. Heidi, et al., Cell culture chips for simultaneous application of topographical and electrical cues enhance phenotype of cardiomyocytes, Lab Chip 9 (4) (2009) 564–575.

[81] N. Tandon, et al., Surface-patterned electrode bioreactor for electrical stimulation, Lab Chip 10 (6) (2010) 692–700.

[82] A.M. Taylor, et al., Microfluidic multicompartment device for neuroscience research, Langmuir 19 (5) (2003) 1551–1556.

[83] S.K. Ravula, et al., Spatiotemporal localization of injury potentials in DRG neurons during vincristine-induced axonal degeneration, Neurosci. Lett. 415 (1) (2007) 34–39.

[84] S. Takayama, et al., Selective chemical treatment of cellular microdomains using multiple laminar streams, Chem. Biol. 10 (2) (2003) 123–130.

[85] D. Huh, et al., Reconstituting organ-level lung functions on a chip, Science 328 (5986) (2010) 1662–1668.

[86] P.M. van Midwoud, E. Verpoorte, G.M.M. Groothuis, Microfluidic devices for in vitro studies on liver drug metabolism and toxicity, Integr. Biol. 3 (5) (2011) 509–521.

[87] K.-J. Jang, K.-Y. Suh, A multi-layer microfluidic device for efficient culture and analysis of renal tubular cells, Lab Chip 10 (1) (2010) 36–42.

[88] Y. Imura, K. Sato, E. Yoshimura, Micro total bioassay system for ingested substances: assessment of intestinal absorption, hepatic metabolism, and bioactivity, Anal. Chem. 82 (24) (2010) 9983–9988.

[89] J.H. Sung, et al., Microscale 3-D hydrogel scaffold for biomimetic gastrointestinal (GI) tract model, Lab Chip 11 (3) (2011) 389–392.

[90] G.J. Mahler, et al., Characterization of a gastrointestinal tract microscale cell culture analog used to predict drug toxicity, Biotechnol. Bioeng. 104 (1) (2009) 193–205.

[91] J. Borenstein, et al., Functional endothelialized microvascular networks with circular cross-sections in a tissue culture substrate, Biomed. Microdevices 12 (1) (2010) 71–79.

[92] A. Gunther, et al., A microfluidic platform for probing small artery structure and function, Lab Chip 10 (18) (2010) 2341–2349.

[93] D. Ingber, "Organ-on-a-chip" technology: on trial, Chem. Ind. 13 (2011) 18–20.

[94] F. Sonntag, et al., Design and prototyping of a chip-based multi-micro-organoid culture system for substance testing, predictive to human (substance) exposure, J. Biotechnol. 148 (1) (2010) 70–75.

[95] J. Sung, M. Shuler, In vitro microscale systems for systematic drug toxicity study, Bioproc. Biosyst. Eng. 33 (1) (2010) 5–19.

[96] J.H. Sung, C. Kam, M.L. Shuler, A microfluidic device for a pharmacokinetic–pharmacodynamic (PK–PD) model on a chip, Lab Chip 10 (4) (2010) 446–455.

[97] M. Baker, Tissue models: a living system on a chip, Nature 471 (7340) (2011) 661–665.

PART

In Vitro Models

3

CHAPTER

Functionalized Microfluidic Devices for Separation of Cell Phenotypes 13

Beili Zhu, Brian D. Plouffe and Shashi K. Murthy
Department of Chemical Engineering, Northeastern University, Boston, MA, USA

13.1 Introduction

Over the past 10 years, microfluidic devices have emerged as effective tools for cell separation. The systematic nature of microfluidic cell separation systems allows for precise manipulation of various parameters which may ultimately influence the separation process in terms of purity, efficiency, and/or throughput [1]. Over the last 5 years, cell isolation limits have been pushed lower and lower, now achieving separation of single cells from large volumes of unprocessed whole blood—a cell threshold that is difficult with conventional separation platforms.

Similar to traditional isolation techniques like fluorescence-activated cell sorting (FACS) and magnet-activated cell sorting (MACS), many microfluidic separation devices isolate cells based on differences in cell phenotype. By harnessing the unique surface receptor expression of an individual cell population in a heterogeneous cell suspension, one can specifically target one or more cell types. This chapter describes the ability of functionalized microfluidic platforms to isolate and enrich a target cell population via an affinity-based methodology. Adhesion-based cell separation systems behave in a similar fashion as chromatographic separation instrumentation, where a mixture is passed through a column packed with beads or other materials capable of binding to the selected constituents of the feed. In the case of cell separation, biomolecules (e.g., peptides, lectins, and antibodies) immobilized on surfaces are used for binding. In contrast to separation techniques that rely on differences in physical attributes, such as size or density, adhesion-based separation relies on specific receptor–ligand interactions and thus acts independently of physical characteristics. Furthermore, this approach requires no preprocessing incubation of the starting cell mixture with fluorescent or magnetic antibody tags, unlike FACS and MACS. This fact is especially important in separations where it is essential to minimize cell activation and disruption of protein homeostasis [2,3]. Furthermore, by

Microfluidic Cell Culture Systems. ISBN: 978-1-4377-3459-1

eliminating the preprocessing step, the overall separation duration is significantly reduced. Adhesion-based cell separation within microfluidic devices can be categorized into either positive selection or negative selection. Negative selection is an attractive approach when the target cells are not fully characterized. Of course, since this approach requires 100% removal of the nontarget cells with minimal target cell capture, extensive characterization of the adhesion profiles of all cell types in the heterogeneous suspension is needed. On the other hand, positive selection involves the direct capture of the target cell type and thus is generally easier to implement. However, a major limitation of this approach is the lack of techniques to recover cells with minimal perturbation from the microfluidic device following isolation. Both negative and positive selections, as well as recovery techniques, are discussed in this chapter.

13.2 Negative selection for enrichment of target cells

In negative selection separation, a heterogeneous cell suspension flows through the device and unwanted cell types adhere to the device surface while target cells are allowed to pass directly through the device. The target cells collected at the outlet of the devices have an increased purity due to the depletion of non-target cells inside the devices. Adhesion-based negative selection is extremely useful when the target cell population is not fully characterized. One example of where negative selection has an application is the isolation of cardiac progenitor cells from digested myocardium. To date, little is known of the marker expression of cardiac progenitors. Digested cardiac muscle tissue usually includes smooth muscle cells (SMCs), endothelial cells (ECs), fibroblasts (FBs), cardiomyocytes, and possible cardiac progenitor cells [4]. Therefore, to isolate valuable progenitor cells from the heterogenous cell suspension, a negative selection approach may be useful. First, it has been well described that size-based techniques can be used for isolation of cardiomyocytes [5], due to their large size, but an adhesion-based approach needs to be investigated in order to separate the remaining cells (i.e., SMCs, ECs, FBs, and progenitor cells). One methodology to effectively isolate these cells is based on their unique receptor chemistries. SMCs and ECs, for example, are known to preferentially adhere to surfaces coated with peptides val-ala-pro-gly (VAPG) and arg-glu-asp-val (REDV), respectively [6,7]. Surfaces coated with the peptide arg-gly-asp-ser (RGDS) are known to bind to a number of cell types, including FBs [8,9]. In addition to ligand choice, for an adhesion-based separation strategy, the effect of shear stress on cell capture must also be considered. For adhesion-based separation in a flow system, the effect of shear stress on cell adhesion was first characterized by Murthy et al. [10]. Here, cell adhesion as a function of shear stress was investigated using surface modified microfluidic device design. A variable shear stress device [11] was fabricated and the tetrapeptides were covalently attached using traditional silane chemistry [12]. This work demonstrated the ability to achieve specific adhesion of one cell type from a mixed suspension within antibody-coated microfluidic channels.

As first illustrated by Plouffe et al. [13,14] and later confirmed by Green and Murthy [15], microfluidic devices coated with REDV, VAPG, and RDGS can be used for the adhesion-based capture of ECs, SMCs, and FBs from heterogeneous suspensions (Figure 13.1). As shown in Figure 13.1, four different cells were investigated over a range of shear stresses on three different peptide coatings. In addition to SMCs, ECs, and FBs, a model cell type, adipose-derived stem cells (ADSCs), was also investigated as the nontarget cell population in the variable shear stress device. At high shear stresses, there is no distinguishable difference in cell adhesion, but the adhesion of the target cells on cell-specific biomolecule-coated devices becomes significantly greater than that of the nontarget cells with

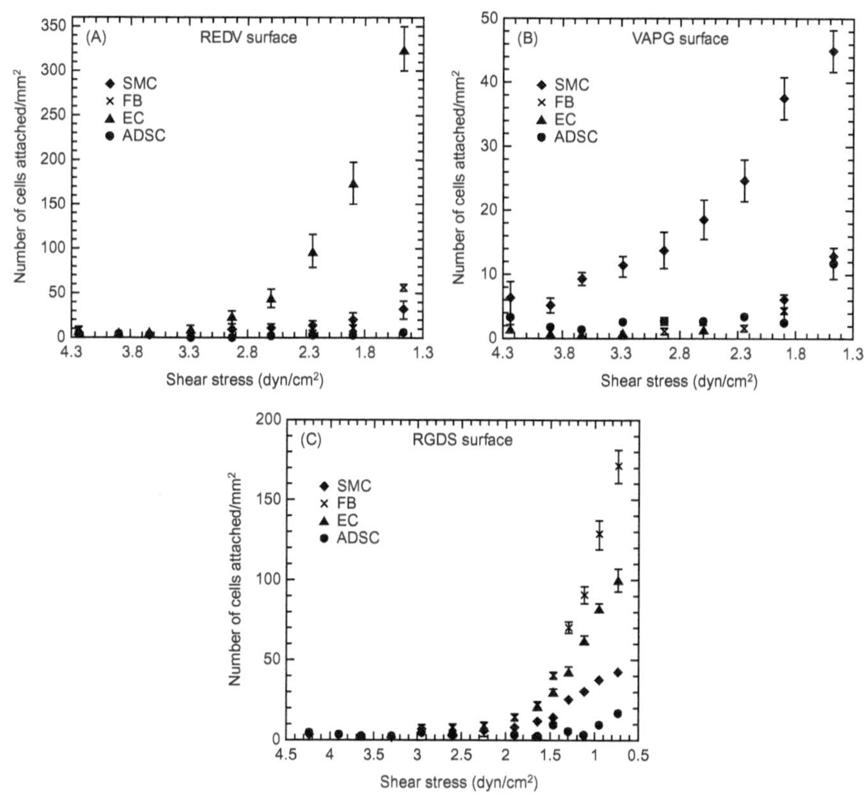

FIGURE 13.1

Cell adhesion as a function of shear stress observed with ECs, SMCS, FBs, and ADSCs on microfluidic devices coated with (A) REDV, (B) VAPG, and (C) RGDS. Each data set represents a distinct experiment performed with only one cell type. Error bars denote standard errors for each point based on five repetitions.

Source: *Ref. [15]. Reproduced by permission of the Royal Society of Chemistry.*

decreasing shear stress. Importantly, ADSCs show much lower adhesion on all three coatings even at low shear stresses.

In order to effectively remove large numbers of nontarget cells, a microfluidic device with a large surface area and at single shear stress was needed. To determine the optimum device geometry, computation fluid dynamic (CFD) modeling was conducted as to determine the shear stress profiles as a function of curve geometry [16]. The CFD results illustrate that curved turns are preferable to obtain uniform velocity profiles (Figure 13.2). Additionally, the gap spacing of the curvature was

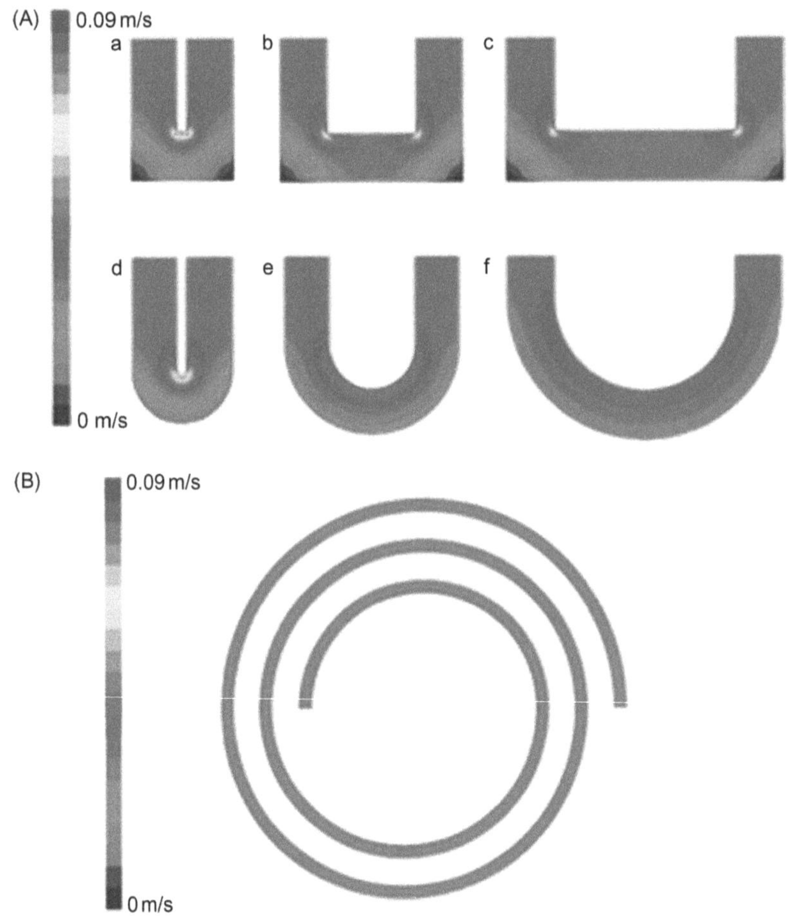

FIGURE 13.2

(A) Velocity profiles of sharp-turn and curved-turn microchannels generated using CFD software. (B) Velocity profile for a microchannel with spiral geometry obtained by CFD modeling. The profile shown is uniform throughout the length and cross section of the channel at all locations.

Source: *Ref. [5]. Reproduced by permission of the Royal Society of Chemistry.*

shown to play a role in the flow profile and larger gap spacings result in more predictable adhesion versus narrow, tight turns. The CFD results were validated with cell adhesion studies and compared well with traditional straight channel microfluidic chambers. All these results show that the high surface area device required for efficient negative selective cell isolation should be a spiral-like design as shown in Figure 13.2B. The spiral design has a constant, predictable velocity profile through the entire device and thus the prior cell adhesion characterizations (Figure 13.1) can be directly paralleled over to a constant shear stress device.

Based on the above design criteria, a new three-stage spiral device (Figure 13.3A) was designed to deplete ECs, SMCs, and FBs from a model

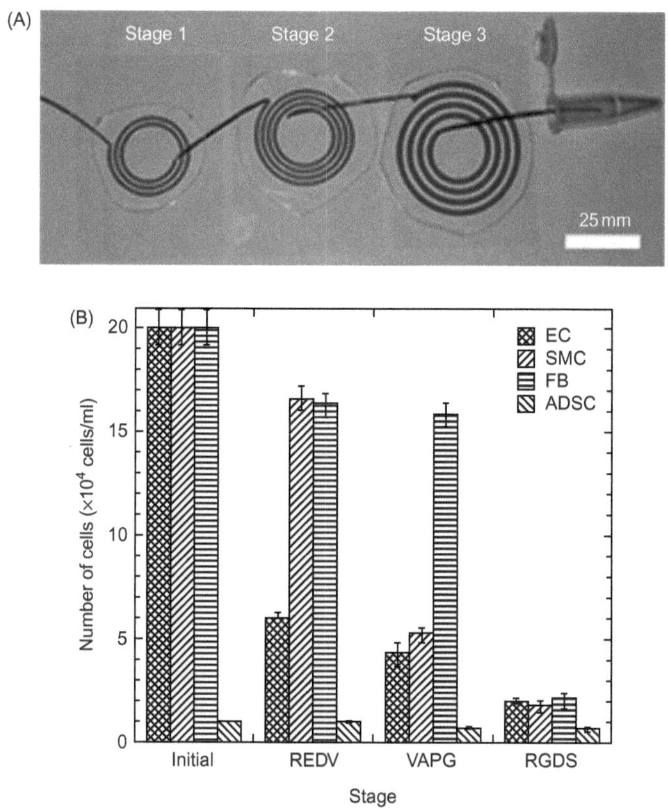

FIGURE 13.3

(A) Experimental setup of the ADSC recovery experiments. From left to right: REDV stage (266 mm^2), VAPG stage (491 mm^2), and the RGDS stage (793 mm^2). (B) Depletion of endothelial cells, smooth muscle cells, and fibroblasts in a three-stage separation process. The "initial" stage represents the heterogeneous suspension input into the first stage (REDV-coated device). *Y*-axis values for each subsequent stage represent the number of cells in the output collected from each stage.

Source: *Ref. [15]. Reproduced by permission of the Royal Society of Chemistry.*

cardiac muscle digestate cell suspension. The heterogeneous cell suspension contains 10,000 ADSCs which composed of 1.6% of cell population in suspension. Surface of each stage of device was coated with REDV to remove ECs, VAPG for SMCs, and RGDS for FBs. The channel widths were 1 mm for REDV and VAPG stages, and 1.7 mm for the RGDS stage. In order to maximize negative cell depletion and minimize ADSC adhesion, the optimal surface areas were 266 mm^2, 491 mm^2, and 793 mm^2 for REDV, VAPG, and RGDS stages, respectively. After the third stage depletion, the initial composition of ADSCs (1.6%) had increased to 8.9%. This indicated significant target cell enrichment. The final collected stream from stage 3 (as illustrated in Figure 13.3B) showed that over 90% of the nontarget cells were depleted from the initial solution while maintaining over 50% of the ADSCs. The work illustrates that a negative selection approach can be developed to enrich a fourth cell type whose markers may not be fully characterized. Effective enrichment can be accomplished by just ensuring that little to no adhesion of the target cell type occurs with the ligands for nontarget cell capture. This process requires no labeling of the cells in the sample and utilizes a very simple and user-friendly design scheme.

13.3 Positive selection of target cells for diagnostic purposes

Using a similar concept to negative selection, antibodies specific to surface markers on target cells can be immobilized onto microchannels. Target cells can thus captured inside devices for diagnostic purposes. Positive selection is useful when surface markers of interested cells are known. Little characterization needs to be done on the rest of the cell populations in suspension.

This concept was recently utilized to the capture of endothelial progenitor cells (EPCs) from whole blood. EPCs are a subtype of bone-marrow-derived cells and are present in peripheral blood [17,18]. EPC number inversely correlates with endothelial dysfunction and has been proposed as a biomarker for risk assessment, disease progression, and response to therapy in cardiovascular diseases [19–21].

The first step is to choose the antibody that has the strongest affinity to EPCs for effective cell capture. For this purpose, a microfluidic device generating variable shear stress was used to determine the adhesion affinity of ovine EPCs to some candidate antibodies specific to the cell surface profile [22]. Surface of devices was functionalized using traditional silane chemistry.

Variable shear stress microfluidic chambers were utilized to compare the adhesion levels of different antibodies at a fixed surface density. Results showed that as the shear decreased along the channel, the density of cells adhering to the surface increased. An optimum shear stress of 1.47 dyn/cm^2 captured the largest number of EPCs within the shear stress range investigated. Antibodies against CD34, CD31, and kinase domain receptor (KDR) showed an increased affinity to ovine EPCs versus the nontarget cells investigated, mesenchymal stem cells, mature vascular ECs, and mature SMCs. Anti-CD34 showed the highest EPC adhesion versus the nontarget cells in the devices and thus indicates its potential as a capture molecule for EPCs.

The above work was quickly advanced to study human subjects. A micropost array device precoated with anti-CD34 antibody was used to capture and enumerate EPCs directly from human whole blood in healthy subjects and patients with pulmonary arterial hypertension (PAH) [23]. A single pass of a 200 μl volume of whole blood flowed through functionalized posts for EPC capture experiments. Such small testing volume is important especially in pediatric clinical care and small animal research.

Briefly, peripheral blood was collected in ethylene diamine tetraacetic acid (EDTA) tubes and directly injected into the polymeric microfluidic chips at a flow rate of 0.6 ml/h (Figure 13.4A and B). Following capture, cells were identified and enumerated via immunofluorescent staining (Figure 13.4C and F) for expression of CD31, KDR, and CD45 antigens. Data was tabulated for cell numbers which stained for KDR, CD31, while negative for CD45—a frequently described EPC phenotypes [17,24,25]. A comprehensive diagnostic readout was attained in approximately 1 h, significantly faster than traditional techniques such as flow cytometry [20,26], magnetic-bead-based approaches [18], or colony forming cell assays [20,27].

Captured cells with surface markers of CD34+/KDR+ were recognized as EPCs, CD34+/KDR+/CD31+/CD45− were identified as late EPCs. Figure 13.4G shows that the cells captured inside postarrays were positive for EPC surface markers. The results from EPC chips were compared to traditional three-color flow cytometry measurements. A 1:2.7 ratio in EPC number (EPC chip versus flow cytometry) revealed that the chip had an efficiency of about 37% capture rate. The two techniques had an excellent correlation with Pearson's correlation of 0.83. Considering 20 min time duration for 200 μl whole blood to make a single pass through the device without incubation time, the diagnostic potential of this positive selection approach is appealing.

After validation, EPC numbers were investigated in patients diagnosed with PAH. In this study, EPCs were found to be significantly low in PAH patients versus controls. Among patients, postmenopausal women did have decreased EPC numbers than young females, and obese/overweight PAH patients had a lower number of EPCs compared to patients with normal BMI, the resulting EPC numbers were also inversely associated with two potential disease modifiers: body mass index (BMI) and postmenopausal status. In summary, the new EPC chip applied a polymeric cell affinity microfluidic platform for the diagnosis of cardiovascular diseases.

In another study, Vickers et al. [28] used positive-selection-based devices to capture and release lymphocytes from whole blood for the diagnosis of acute and chronic lymphocytic leukemia. Both acute and chronic leukemia are marked by an increased production of abnormal white blood cells, where the acute type of

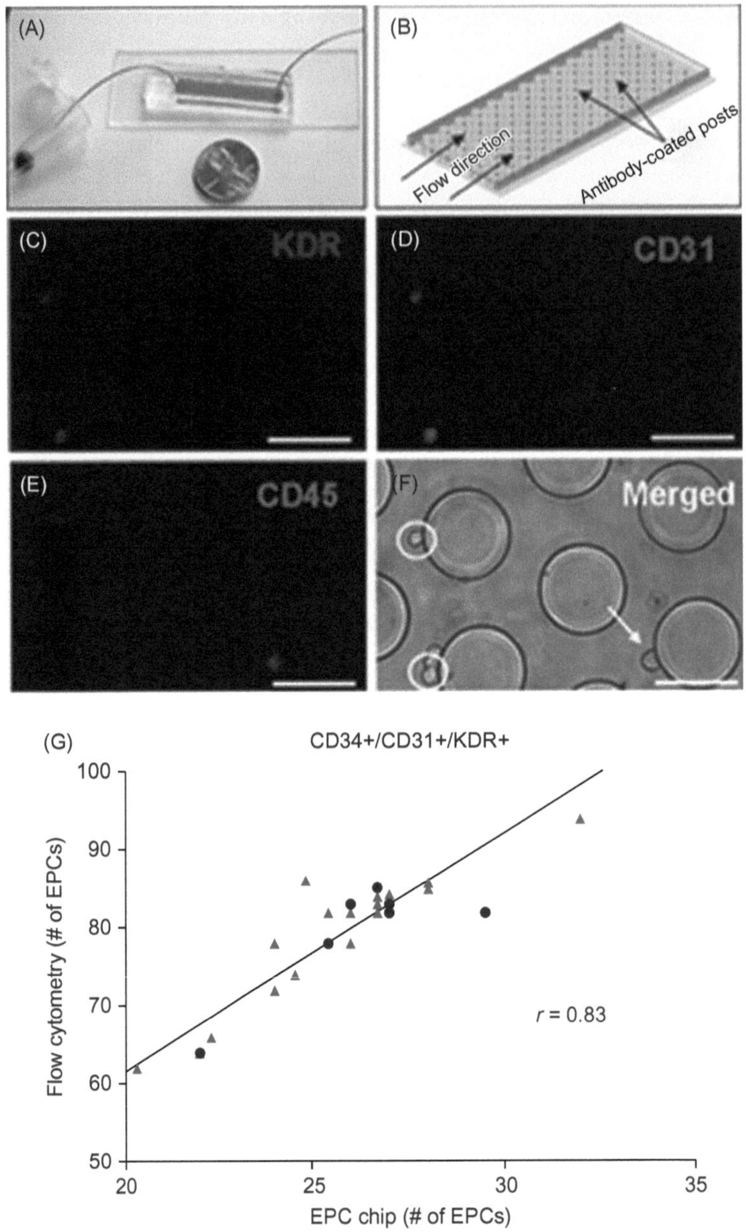

FIGURE 13.4

Isolation of EPCs from human peripheral whole blood using a microfluidic platform. (A) The EPC capture chip requires the injection of 200 μl of whole blood; shown alongside a US penny for comparison of device dimensions. (B) The microfluidic platform consists of a dense array of 100 μm posts which are coated with anti-CD34 antibody.

(Continued)

leukemia incapacitates the body's ability to fight infections. Current methods of diagnosing leukemia include bone marrow biopsy, bone marrow aspiration, and complete blood cell count. These existing techniques are quite intensive, invasive, and require a trained technician. Therefore, the strategy of this study was to utilize lectins immobilized onto device surfaces to capture lymphocytes from a mixed cell population in flow. Lectins are a group of proteins that bind specifically and reversibly to carbohydrate structures, which exist on cell surfaces. Since each different cell type displays a unique pattern of carbohydrates, certain lectins that have highest affinity to its matched carbohydrate structures can be used to selectively bind, and isolate, a specific cell type. Secondly, selected sugars can then be used to release lymphocytes from devices. These sugars compete with cell surface carbohydrates for lectin binding and thus the release process will not damage cells.

As a practical example to isolate T and B lymphocytes from blood, a variable shear stress device was first used to determine the shear stress and the optimal lectins for binding Jurkat T and Raji B lymphocytes as model cells. This pilot experiment determined that out of a wide selection of lectins, *peanut agglutinin* (PNA) provided the highest level of binding affinity for Jurkat T cells, and *Lotus tetragonolobus lectin* (LTL) showed highest affinity for Raji B lymphocytes. Therefore, postarray microfluidic devices, similar to those used in the EPC study, were coated with these two lectins individually. To test the efficiency of these devices, these two types of lymphocytes were spiked into whole blood. Since 2×10^4 model cells were spiked into every 2 ml of blood, it was estimated that a starting purity of these model cells to be approximately 0.0002%, based on the fact that there are 5×10^9 erythrocytes per ml blood for an average person. The samples were allowed to flow through these devices. For Jurkat T cells, 5% of the total T cells from the input samples were captured, at a purity of 23%. For Raji B cells, the capture efficiency was 11% and the purity was 52%. These results represent a 10–20 million fold enrichment of the T and B lymphocytes. Finally, after capture, to release cells from devices, galactose and L-fucose [29] were chosen as the competitive binding sugars for PNA-coated and LTL-coated devices, respectively. This release method was able to detach about half of the

◀ (C–E) Fluorescence micrographs of captured cells which express KDR (blue), CD31 (green), and CD45 (red). (F) Merged image with bright field image of postarray, illustrating KDR+/CD31+/CD45− cells (circled) and a KDR−/CD31−/CD45+ cell (arrow). Scale bar = 100 μm. (G) The EPC chip cell capture was compared to traditional three-color flow cytometry cell counts (CD34+/CD31+/KDR+). The red triangles represent individual experiments ($n = 21$ devices; two to four separate blood draws from seven subjects) and the black circles represent the median for each subject ($n = 7$). (For interpretation of the references to color in this figure legend, the reader is referred to the web version of this book.)

Source: *Ref. [23]. Reprinted with kind permission from Springer Science and Business Media.*

captured cells from both types of devices. Overall, this study shows that lectins and their binding sugars can be utilized to isolate lymphocytes from blood in microfluidic devices.

13.4 Capture and release of target cells from positive selection for tissue engineering purposes

As described earlier, one of the shortcomings associated with position selective cell separation is the need for a capture and release system [30]. Few studies have focused on the detachment of cells after capture in microfluidic devices. The detachment of cells adhered to any surface requires the application of a force that is greater in magnitude to that of adhesion. Fluid shear forces have been shown to be simplest method for cell detachment [31–33]. Although this is a local and simple method of cell release, excessive exposure to fluid shear may result in cell damage and reduction in viability. An alternative approach is to cleave the protein ligand that is bound to the capture surface using enzymes, such as trypsin. However, enzymatic exposure can cause morphological changes due to a disruption of the cell membrane and glycocalyx, leading to losses in cellular activity [34,35]. Other approaches have also been utilized but require external interfacing such as heating/cooling [36–38] or electrical potential [39]. These limitations illustrate the need to establish a general technique to capture and release cells in microscale devices without extensive physical or chemical perturbations to the cell environment. A positive capture and release system should not affect cells or change cellular environment.

One approach that has shown some promise is the conjugation of a capture molecule to a dissolvable hydrogel, such as alginate. It has been shown that alginate will form a gel in the presence of Ca^{2+} ion and can then be reversed back to its native state by scavenging the Ca^{2+} ions using chelators like EDTA. To test this experimental system, RGDS was covalently conjugated to an alginate backbone [22]. The alginate–RGDS complex was adsorbed to the device geometry and then exposed to Ca^{2+} to form a thin gel within the device. Cardiac FBs were injected into the microscale device and subsequently captured from the stream. Conjugation of the tetrapeptide (RGDS) to the alginate resulted in an over two-fold increase in cell capture in comparison to the native alginate. Upon rinsing with EDTA, captured cells were then released with greater than 95% efficiency. In summary, this work has demonstrated the application of a peptide-functionalized alginate hydrogel as a method of capture and release of cells in microfluidic channels. When combined with the capability of selective cell capture from a heterogeneous cell suspension within a microfluidic device, as described earlier, the cell release methodology described could provide viable, functional cells for further application.

In subsequent work, alginate hydrogels were modified by adding four-arm amine terminated polyethylene glycol (PEG) molecules within the alginate hydrogel [40]. Four-arm PEG-NH_2 is known for its biological nonadhesiveness, and the addition of the amine moieties allows for an enhancement in the biomolecule conjugation via amide binding. PEG molecules are supposed to enhance functionalization with cell-adhesive antibodies as well as to suppress nonspecific cell binding. In the new study, the RGDS peptide was replaced by antibodies against CD34 to specifically capture EPCs directly from whole blood. In contrast to the previously described work with positive selection of EPCs [23], this work focused on both selective adhesion of the EPCs from whole blood, and release and collection of the rare cell population.

Briefly, whole blood (300 μl) was collected in heparin tubes and directly injected into individual microfluidic devices. Cells were captured and subsequently released from each device, followed by pooling of 10 devices into a single suspension to allow enumeration by flow cytometry. Although several different gels were tested, the best-case scenario resulted in approximately 1000 cells per device being recovered (~30% recovery) and a purity of the final suspension of approximately 75%. Therefore, it was shown that alginate hydrogels functionalized with four-arm PEG molecules and antibodies can be utilized for effective capture and release of EPCs from whole blood in microfluidic devices. The significance of this work is that using a relatively simple synthesis protocol, microfluidics achieved high purity of capture and efficient release of target cells from a complex, heterogeneous cell suspension. This capture/release approach can potentially be extended to the isolation of other rare cells from complex solutions, such as stem or progenitor cells from digested tissue.

13.5 Effect of shear on changes of receptor expression in cells

When it comes to separating subpopulations of ECs, e.g., human umbilical vein endothelial cells (HUVECs), adult human microvascular endothelial cells (HMVECs), and endothelial colony forming cells (ECFCs), it becomes extremely difficult as they all have the same EC surface markers. In this case, using conventional method of fluorescent labeling of surface markers on ECs followed by FACS analysis is not sufficient enough for separating different types of ECs. But ECs are known to respond to shear stress, and remodeling of cytoskeleton by shear stress might profoundly affect other aspects of cell physiology [41,42]. Thus, Vickers and Murthy [43] aimed to probe the influence of shear stress on receptor expressions of different subpopulations of ECs and how that can be used as a potential method for EC separation.

REDV is a tetrapeptide that is known to selectively bind to ECs [7] and is used in this study as the adhesion ligand on microfluidic devices [41]. At first, when these three types of ECs were flowed through an REDV-coated surface in variable shear stress devices, all of them showed very similar adhesion affinity to

REDV in the shear stress range of 0.6–2 dyn/cm^2. Then, a new method was developed to test cell receptor changes under shear stress. Three types of ECs were first incubated with different concentrations of REDV (0–100 μg/ml) in solutions. Cells were removed from the solution and flowed through an REDV-coated straight channel at a shear stress level of 1.1 dyn/cm^2. Here, normal human dermal fibroblasts (NHDFs) served as a negative control as they do not selectively bind to REDV nor do they change receptor expression, when being exposed to shear stress.

Figure 13.5A shows that when cells were incubated with increasing concentrations of REDV from 0 to 50 μg/ml, the amount of cells adhered in microchannels decreased reciprocally for all three types. This is expected, as fewer cells would bind to REDV-coated devices when more of their receptors were blocked by the same REDV peptide solutions beforehand. However, when cells were incubated with higher concentration of REDV at 100 μg/ml, more cells adhered to microfluidic channels, thus number of binding sites is significantly different among the three types of ECs. Figure 13.5B showed that after incubating HUVECs with various concentrations of REDV in solution, the number of unoccupied REDV-binding sites on cells changed before and after shear stimulation.

This study clearly showed that the number of receptors will change among EC types under shear stress, and shear stimulation is an effective tool to isolate EC subpopulations on microfluidic devices.

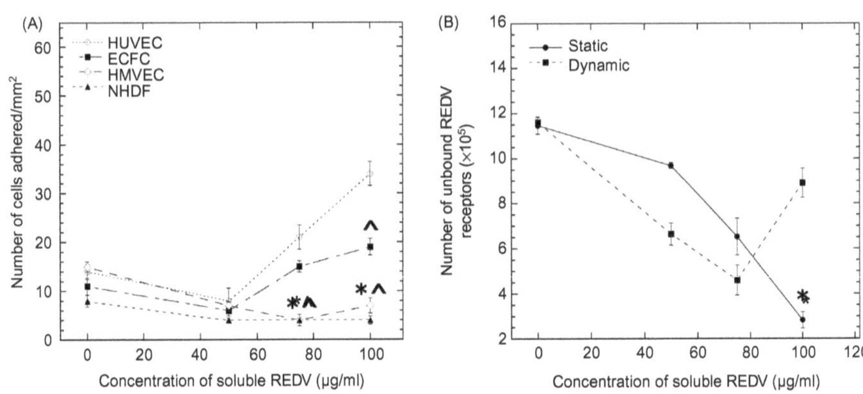

FIGURE 13.5

(A) Cell adhesion comparison between cell types on REDV surface at 1.1 dyn/cm^2. ^ denotes significant difference with $P<0.001$ compared to HUVECs at specified REDV concentrations and * denotes significant difference with $P<0.001$ compared to ECFC at specified REDV concentrations. (B) REDV-binding site quantification with HUVECs prior to (static) and after flow through microchannels at a shear stress of 1.1 dyn/cm^2 (dynamic). * denotes significant difference with $P<0.001$ compared with the dynamic data point for 100 μg/ml soluble REDV.

Source: *Ref. [41]. Reproduced by permission of the Royal Society of Chemistry.*

13.6 Conclusions

In conclusion, it has been demonstrated that microfluidics can be a valuable tool in the isolation of key cell populations for applications in diagnostics and tissue engineering. There still remain many challenges in the field of microfluidics cell separation based on cell phenotype. Although negative selection has illustrated some promise, there is limited use for a separation technique that relies on near 100% adhesion of the unwanted cell population. Furthermore, the inherent challenge of biofouling and nonspecific binding of undesired cell populations continues to limit the effectiveness of negative selection. On the same note, nonspecific binding is always a challenge in positive selection. The adhesion of nontarget cells reduces purity and, in the case of rare cells, overwhelms the isolated population. These challenges have limited the application of functionalized microfluidic channel separation in the clinic and the general biomedical market to date. We and others continue to try to improve capture via selection of unique capture moieties and manipulation of the fluid dynamics and aim to push the limits of microfluidics, both efficiency and collection purity.

References

[1] M. Radisic, R.K. Iyer, S.K. Murthy, Micro- and nanotechnology in cell separation, Int. J. Nanomed. 1 (1) (2006) 3–14.

[2] Z. Sha, H.C. Yen, H. Scheel, J. Suo, K. Hofmann, E.C. Chang, Isolation of the Schizosaccharomyces pombe proteasome subunit Rpn7 and a structure–function study of the proteasome-COP9-initiation factor domain, J. Biol. Chem. 282 (44) (2007) 32414–32423.

[3] Z. Sha, L.M. Brill, R. Cabrera, O. Kleifeld, J.S. Scheliga, M.H. Glickman, et al., The eIF3 interactome reveals the translasome, a supercomplex linking protein synthesis and degradation machineries, Mol. Cell 36 (1) (2009) 141–152.

[4] A.C. Nag, Study of non-muscle cells of the adult mammalian heart—a fine-structural analysis and distribution, Cytobios 28 (109) (1980) 41–61.

[5] J.V. Green, T. Kniazeva, M. Abedi, D.S. Sokhey, M.E. Taslim, S.K. Murthy, Effect of channel geometry on cell adhesion in microfluidic devices, Lab Chip 9 (5) (2009) 677–685. <http://dx.doi.org/10.1039/B813516A>.

[6] A.S. Gobin, J.L. West, Val-ala-pro-gly, an elastin-derived non-integrin ligand: smooth muscle cell adhesion and specificity, J. Biomed. Mater. Res. A 67A (1) (2003) 255–259.

[7] J.A. Hubbell, S.P. Massia, N.P. Desai, P.D. Drumheller, Endothelial cell-selective materials for tissue engineering in the vascular graft via a new receptor, Bio. Technol. 9 (6) (1991) 568–572.

[8] L. Kam, W. Shain, J.N. Turner, R. Bizios, Selective adhesion of astrocytes to surfaces modified with immobilized peptides, Biomaterials 23 (2) (2002) 511–515.

[9] S.E. Sakiyama-Elbert, J.A. Hubbell, Functional biomaterials: design of novel biomaterials, Ann. Rev. Mater. Res. 31 (2001) 183–201.

[10] S.K. Murthy, A. Sin, R.G. Tompkins, M. Toner, Effect of flow and surface conditions on human lymphocyte isolation using microfluidic chambers, Langmuir 20 (26) (2004) 11649–11655.
[11] S. Usami, H.H. Chen, Y.H. Zhao, S. Chien, R. Skalak, Design and construction of a linear shear-stress flow chamber, Ann. Biomed. Eng. 21 (1) (1993) 77–83.
[12] S.K. Bhatia, L.C. Shriverlake, K.J. Prior, J.H. Georger, J.M. Calvert, R. Bredehorst, et al., Use of thiol-terminal silanes and heterobifunctional crosslinkers for immobilization of antibodies on silica surfaces, Anal. Biochem. 178 (2) (1989) 408–413.
[13] B.D. Plouffe, D.N. Njoka, J. Harris, J.H. Liao, N.K. Horick, M. Radisic, et al., Peptide-mediated selective adhesion of smooth muscle and endothelial cells in microfluidic shear flow, Langmuir 23 (9) (2007) 5050–5055.
[14] B.D. Plouffe, M. Radisic, S.K. Murthy, Microfluidic depletion of endothelial cells, smooth muscle cells, and fibroblasts from heterogeneous suspensions, Lab Chip 8 (3) (2008) 462–472.
[15] J.V. Green, S.K. Murthy, Microfluidic enrichment of a target cell type from a heterogenous suspension by adhesion-based negative selection, Lab Chip 9 (15) (2009) 2245–2248. <http://dx.doi.org/10.1039/B906768J>.
[16] J.V. Green, M. Radisic, S.K. Murthy, Deterministic lateral displacement as a means to enrich large cells for tissue engineering, Anal. Chem. 81 (21) (2009) 9178–9182.
[17] T. Asahara, T. Murohara, A. Sullivan, M. Silver, R. van der Zee, T. Li, et al., Isolation of putative progenitor endothelial cells for angiogenesis, Science 275 (5302) (1997) 964–966.
[18] M. Hristov, W. Erl, P.C. Weber, Endothelial progenitor cells—mobilization, differentiation, and homing, Arterioscler. Thromb. Vasc. Biol. 23 (7) (2003) 1185–1189.
[19] J.M. Hill, G. Zalos, J.P. Halcox, W.H. Schenke, M.A. Waclawiw, A.A. Quyyumi, et al., Circulating endothelial progenitor cells, vascular function, and cardiovascular risk, New Engl. J. Med. 348 (7) (2003) 593–600.
[20] E.M. Van Craenenbroeck, V.M. Conraads, D.R. Van Bockstaele, S.E. Haine, K. Vermeulen, V.F. Van Tendeloo, et al., Quantification of circulating endothelial progenitor cells: a methodological comparison of six flow cytometric approaches, J. Immunol. Methods 332 (1–2) (2008) 31–40.
[21] N. Werner, S. Kosiol, T. Schiegl, P. Ahlers, K. Walenta, A. Link, et al., Circulating endothelial progenitor cells and cardiovascular outcomes, New Engl. J. Med. 353 (10) (2005) 999–1007.
[22] B.D. Plouffe, T. Kniazeva, J.E. Mayer Jr., S.K. Murthy, V.L. Sales, Development of microfluidics as endothelial progenitor cell capture technology for cardiovascular tissue engineering and diagnostic medicine, FASEB J. 23 (10) (2009) 3309–3314.
[23] G. Hansmann, B.D. Plouffe, A. Hatch, A. von Gise, H. Sallmon, R.T. Zamanian, et al., Design and validation of an endothelial progenitor cell capture chip and its application in patients with pulmonary arterial hypertension, J. Mol. Med. (2011) doi:10.1007/s00109-011-0779-6.
[24] V.L. Sales, G.C. Engelmayr Jr., J.A. Johnson Jr., J. Gao, Y. Wang, M.S. Sacks, et al., Protein precoating of elastomeric tissue-engineering scaffolds increased cellularity, enhanced extracellular matrix protein production, and differentially regulated the phenotypes of circulating endothelial progenitor cells, Circulation 116 (Suppl. I) (2007) I-55–63.

[25] O.M. Tepper, R.D. Galiano, J.M. Capla, C. Kalka, P.J. Gagne, G.R. Jacobowitz, et al., Human endothelial progenitor cells from type II diabetics exhibit impaired proliferation, adhesion, and incorporation into vascular structures, Circulation 106 (22) (2002) 2781–2786.

[26] D.G. Duda, K.S. Cohen, D.T. Scadden, R.K. Jain, A protocol for phenotypic detection and enumeration of circulating endothelial cells and circulating progenitor cells in human blood, Nat. Protocols 2 (4) (2007) 805–810.

[27] J. Leor, M. Marber, Endothelial progenitors: a new tower of babel? J. Am. Coll. Cardiol. 48 (8) (2006) 1588–1590.

[28] D.A. Vickers, M. Hincapie, W.S. Hancock, S.K. Murthy, Lectin-mediated microfluidic capture and release of leukemic lymphocytes from whole blood, Biomed. Microdevices 13 (3) (2011) 565–571.

[29] T. Zheng, H. Yu, C. Alexander, D. Beebe, L. Smith, Lectin-modified microchannels for mammalian cell capture and purification, Biomed. Microdevices 9 (4) (2007) 611–617.

[30] S.K. Murthy, M. Radisic, Cell adhesion and detachment, in: D. Li (Ed.), Encyclopedia of Microfluidics and Nanofluidics, Springer, New York, NY, 2008, pp. 202–208.

[31] H. Lu, L.Y. Koo, W.C.M. Wang, D.A. Lauffenburger, L.G. Griffith, K.F. Jensen, Microfluidic shear devices for quantitative analysis of cell adhesion, Anal. Chem. 76 (18) (2004) 5257–5264.

[32] S.P. Wankhede, Z.Q. Du, J.M. Berg, M.W. Vaughn, T. Dallas, K.H. Cheng, et al., Cell detachment model for an antibody-based microfluidic cancer screening system, Biotechnol. Prog. 22 (5) (2006) 1426–1433.

[33] X. Zhang, P. Jones, S.J. Haswell, Attachment and detachment of living cells on modified microchannel surfaces in a microfluidic-based lab-on-a-chip system, Chem. Eng. J. 135 (Suppl. 1) (2008) S82–S88.

[34] N. Fujioka, Y. Morimoto, K. Takeuchi, M. Yoshioka, M. Kikuchi, Difference in infrared spectra from cultured cells dependent on cell-harvesting method, Appl. Spectrosc. 57 (2) (2003) 241–243.

[35] K. Jung, G. Hampel, M. Scholz, W. Henke, Culture of human kidney proximal tubular cells—the effect of various detachment prodecures on viability and degree of cell detachment, Cell Physiol. Biochem. 5 (5) (1995) 353–360.

[36] O. Ernst, A. Lieske, M. Jager, A. Lankenau, C. Duschl, Control of cell detachment in a microfluidic device using a thermo-responsive copolymer on a gold substrate, Lab Chip 7 (10) (2007) 1322–1329.

[37] M. Yamato, M. Utsumi, A. Kushida, C. Konno, A. Kikuchi, T. Okano, Thermo-responsive culture dishes allow the intact harvest of multilayered keratinocyte sheets without dispase by reducing temperature, Tissue Eng. 7 (4) (2001) 473–480.

[38] M. Yamato, C. Konno, M. Utsumi, A. Kikuchi, T. Okano, Thermally responsive polymer-grafted surfaces facilitate patterned cell seeding and co-culture, Biomaterials 23 (2) (2002) 561–567.

[39] W.S. Yeo, C.D. Hodneland, M. Mrksich, Electroactive monolayer substrates that selectively release adherent cells, Chembiochem 2 (7–8) (2001) 590–593.

[40] A. Hatch, G. Hansmann, S.K. Murthy, Engineered alginate hydrogels for effective microfluidic capture and release of endothelial progenitor cells from whole blood, Langmuir 27 (7) (2011) 4257–4264.

[41] R. Cabrera, Z. Sha, T.J. Vadakkan, J. Otero, F. Kriegenburg, R. Hartmann-Petersen, et al., Proteasome nuclear import mediated by Arc3 can influence efficient DNA damage repair and mitosis in Schizosaccharomyces pombe, Mol. Biol. Cell 21 (18) (2010) 3125–3136.

[42] E.A. Osborn, A. Rabodzey, C.F. Dewey Jr., J.H. Hartwig, Endothelial actin cytoskeleton remodeling during mechanostimulation with fluid shear stress, Am. J. Physiol. Cell Physiol. 290 (2) (2006) C444–C452.

[43] D.A. Vickers, S.K. Murthy, Receptor expression changes as a basis for endothelial cell identification using microfluidic channels, Lab Chip 10 (18) (2010) 2380–2386. <http://dx.doi.org/10.1039/C004870D>.

CHAPTER

Microfluidic Hepatotoxicity Platform 14

Philip Lee, Molly Allen and Paul Hung
CellASIC Corp., CA, USA

14.1 Introduction

Liver toxicity is a critical issue for the development of new biopharmaceuticals, commercial chemical substances, and environmental exposure to synthetic compounds. In many cases, the major health hazard of human exposure to foreign substances is liver damage. This is due to the central role the liver plays in removing toxic compounds from the circulation [1,2]. Regulatory pressure from government institutes such as the US Food and Drug Administration (FDA) and the National Institute of Environmental Health Sciences (NIEHS) are pushing biopharmaceutical and chemical companies to develop more stringent methods to predict potential safety problems with new products [3,4].

The traditional method to predict liver toxicity is through the use of animal models. Typical species used for toxicity testing are rat, dog, and monkey. Toxicology studies generally require dosing the animals with high concentrations of test compound, sacrificing the animal, and performing pathology analysis of various organs. For the liver, pharmacokinetics and metabolism studies are also performed to determine how chemicals are broken down within the body over time. In the case of biopharmaceuticals, promising animal studies are followed by human clinical trials. In recent years, the ability of animal tests to predict human toxicity has been questioned [5,6]. It is estimated that up to one-third of drug failures in human clinical trials result from liver toxicity, leading to significant waste of time and resources and putting human patients at unnecessary risk [7]. Government regulations, notably in the United States and European Union, are calling for reduction or elimination of animal testing in the near future.

There is significant interest in developing improved *in vitro* models to predict liver toxicity [8–10]. The advent of *in vitro* methods in the high throughput screening (HTS) industry has allowed companies to test hundreds of thousands of potential compounds at a fraction of the time and cost of prior methods. However, for complex responses such as liver toxicity, existing screening protocols are limited in providing predictive results. The central problem is that in order to model

Microfluidic Cell Culture Systems. ISBN: 978-1-4377-3459-1

liver tissue activity, it is necessary to recreate physical aspects of the liver cell environment. Virtually all *in vitro* cell cultures are performed in a "petri dish" format—static batch culture in 2D monolayers on a plastic surface [11]. It is well established that individual liver cells isolated from a healthy organ will rapidly dedifferentiate when cultured in a 2D monolayer. Minor improvement can be obtained by coating the surface with biomolecules or gels, as well as to maintain a high density to promote cell–cell contact.

Microfluidic cell culture technology provides the ability to control aspects of the cell environment not possible with traditional culture techniques [12,13]. In our work, we have found that the two most fundamental improvements are: (1) providing a dynamic, continuous flow transport mechanism similar to the blood circulation and (2) defining a 3D microstructure to recapitulate tissue architecture. By designing the microfluidic liver models appropriately, we have been able to demonstrate long-term maintenance of liver-specific activity in a format that is amenable to automation and screening.

14.2 Liver tissue microenvironment

The liver is the primary site for xenobiotic metabolism in the human body. This function is carried out by hepatocytes, which constitute roughly 80% of the mass of the liver ($\sim 10^{11}$ cells in a human liver). In order for the liver to process the large amounts of blood circulating through the body, hepatocytes are arranged into millions of parallel structures enabling a high degree of mass transport. The cells are organized as cords or sheets, with the outer surface exposed to the blood and the inner surface exposed to the biliary system. Movement of xenobiotic compounds and metabolites across the hepatocyte membranes is mediated by various families of transporter proteins. Hepatocytes are rich in drug metabolizing enzymes, categorized as Phase I (oxidation, reduction, hydrolysis) and Phase II (conjugation), responsible for detoxifying and deactivating chemicals from the bloodstream. Blood flows to the liver from the hepatic artery (nutrient/oxygen rich) as well as from the portal vein (carries substances absorbed by the small intestines) and exits through the hepatic vein.

14.2.1 Liver acinus

The basic functional unit of the liver is the acinus. This structure is typically 1–2 mm in length and is defined by the entrance of blood from the microcirculation (branches of the hepatic artery and portal vein) to the exit through the central vein. Within the acinus, blood perfuses through a set of sinusoid channels in contact with sheets of hepatocytes. As blood moves through the sinusoids, there is a gradient of oxygen, nutrients, and metabolites. Besides hepatocytes, the acinus also contains endothelial cells, Kupffer cells (macrophages), and stellate (fat storing) cells. A typical human liver contains over 10^7 parallel acini.

14.2.2 Sinusoid properties

The sinusoids are the sites of mass transport between the circulating blood and hepatocytes. They consist of narrow, low-pressure vascular channels that carry blood from the hepatic artery as well as from the portal vein (originating from the small intestines). This allows absorbed xenobiotics in the body to become metabolized by hepatocytes. The sinusoids are lined with an endothelial layer that is highly fenestrated with pores allowing free flow of plasma to the hepatocytes. Additionally, there is a fluid-filled gap between the endothelial cells and hepatocytes (space of Disse) to allow hepatocytes to extend microvilli to promote chemical absorption. The flow rate through an individual sinusoid is about 1.5 nl/min, with approximately 10^9 sinusoids in the liver.

14.2.3 Cell–cell interactions

Hepatocytes form tight junctions with neighboring hepatocytes to define bile canaliculi (that feed into the bile duct) separated from the sinusoidal flow. A key function of hepatocytes is to transport certain waste metabolites out of the blood into the bile duct. In order to carry out this function, hepatocytes are polarized (have an apical and basal surface) with selective expression of transport proteins on either side [14]. Additionally, hepatocytes form gap junctions to mediate intercellular communication as well as adherens junctions to promote 3D architecture.

14.2.4 Cell–matrix interactions

The interaction between liver cells and surrounding extracellular matrix is an important factor for proper hepatocyte function. For example, integrin binding to collagen has been shown to be important for cell survival, while the presence of basement membrane proteins induce polarization and formation of bile canaliculi [15]. Hepatocytes are capable of synthesis of extracellular matrix proteins including collagen (Figure 14.1).

14.3 Microfluidic liver design

The goal of the microfluidic hepatocyte culture system is to replicate the key aspects of the natural liver cell environment. Traditional *in vitro* hepatocyte studies are performed using static monolayer culture formats. The best current approach is to coat the bottom of a plastic dish with collagen I, seed isolated hepatocytes into a dense monolayer, and overlay with a basement membrane gel. This protocol enables the cells to form cell–cell contacts and retain some liver-specific functions. The key limitation is the short functional lifetime of hepatocytes in this arrangement. Generally, the cells will dedifferentiate between 4 and 14 days post isolation.

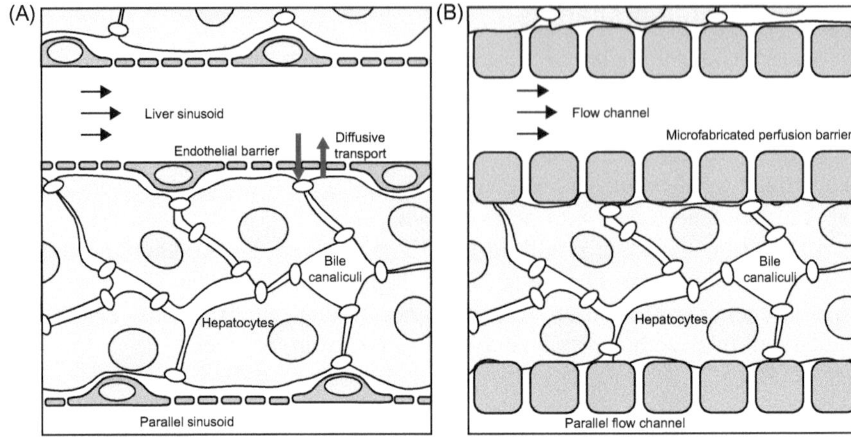

FIGURE 14.1

Cell environment.

The microfluidic hepatocyte culture design has two key aspects not possible with standard methods: (1) continuous flow of culture media and (2) microscale control of cell architecture. The central hypothesis is that replicating the cell environment *in vivo* will lead to more relevant behaviors of hepatocytes *in vitro*. To achieve this goal, we utilized microfabrication technology to create an environment that mimics the liver acinus. By placing primary hepatocytes into this environment, it was possible to maintain long-term liver-specific activity. In addition, the format of the microfluidic plate makes it amenable to standard experiment processes and automation.

14.3.1 Microfabrication

Microfluidic cell culture plates were fabricated using a modified polydimethyl siloxane (PDMS) molding method developed by CellASIC (Hayward, CA,). Computer-assisted design software was used to create the two-layer design of the biomimetic microfluidic liver architecture, with 4 μm tall barriers to decouple the downstream cell loading channels from the medium perfusion channels and a 60 μm height layer to define the hepatocyte culture regions, air diffusion channels, perfusion barriers, and flow channels. After photomasks of each layer were generated, different tones of SU-8 epoxy photoresists (SU-8 2005 and SU-8 2035, Microchem, Newton, MA) were used to create the two-layer microfluidic structures on a 6″ silicon substrate. After each layer of the epoxy was spin-coated to the specified thickness (4 and 60 μm), a mask aligner was used to photolithographically transfer the patterns from the mask onto the epoxy layer.

The basic principle behind the soft lithography replicate molding process is to cast the PDMS over a master template. After the polymer is cured, the soft

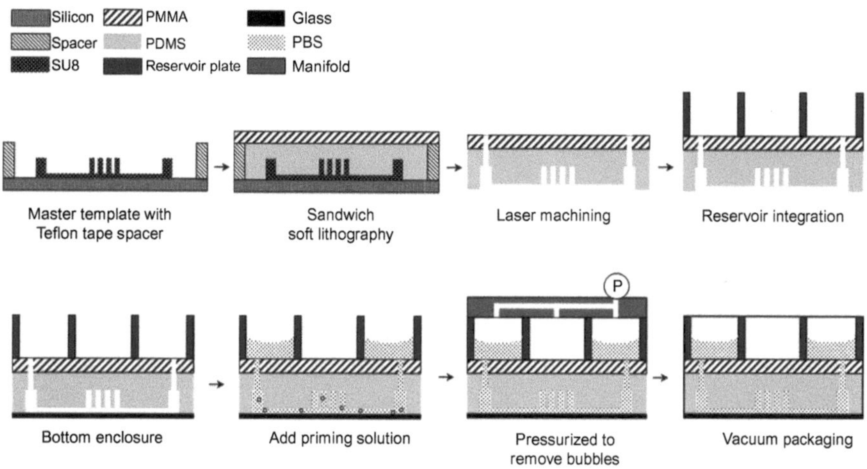

FIGURE 14.2

Microfab.

polymer will replicate the geometry of the master template with the opposite polarity. PDMS (Sylgard 184, Dow Corning, Midland, MI) was mixed 10:1 (monomer:curing agent) and placed under vacuum for 30 min to remove air bubbles. After dispensing 2 ml of the PDMS in the center of the master template, a sheet of chemically modified polymethyl methacrylate (PMMA) was pressed against the mold to sandwich the PDMS between the sheet and the mold, with the final PDMS thickness determined by Teflon spacers. The assembly was then cured in a 60°C oven for 4 h to replicate the microfluidic structures in PDMS. The cured PDMS microfluidic structures adhered preferably to the PMMA as a result of the chemical modification. After detaching the microfluidic structures from the master template, a CO_2 laser cutter was used to create through-holes at certain locations on the PMMA/PDMS composite sheet. A biocompatible UV adhesive was used to attach the microfluidic device to an open bottom 96-well plate frame.

After the reservoir integration, the surface of the PDMS was oxidized by oxygen plasma to covalently bond to a $103 \times 97\ \text{mm}^2$ #1.5 thickness (170 μm) cover glass to enclose the microfluidic structures. Sterile phosphate buffered saline (PBS) solution was dispensed into each reservoir and a custom pressure manifold was used to remove bubbles in the microfluidic channels through outgassing across the gas-permeable PDMS. The plate was sterilized in a UV-ozone chamber and vacuum packaged in a sterile medical-grade pouch (Figure 14.2).

14.3.2 Single unit layout

The basic microfluidic culture unit consists of three well positions: a flow inlet, a microfluidic cell culture region, and a flow outlet. The cells are loaded into a

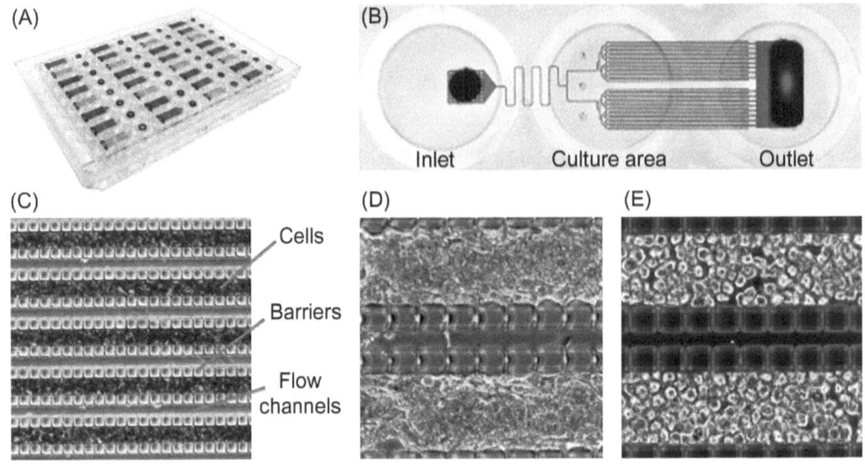

FIGURE 14.3

Design.

set of $60 \times 60 \times 3{,}000\ \mu m^3$ chambers defined by microfabricated perfusion barriers. In between the cell culture regions are 50 μm wide sinusoid channels for medium flow. Each flow unit houses between 20,000 and 50,000 hepatocytes, with a medium perfusion rate of 100 μl/day.

Hepatocytes are prepared in suspension and loaded into the microfluidic chambers. The barrier structure prevents cells from passing through, while concentrating the cells into a high density tissue-like cord. The high degree of cell–cell contact in 3 dimensions induces the cells to form an aggregate structure during culture. This morphology is known to preserve liver-specific activity [16]. The aggregates are roughly 3 cells thick, 3 cells wide, and a few hundred cells long (Figure 14.3).

14.3.3 Perfusion barriers

An important design feature is the microfabricated perfusion barrier. These serve as an artificial endothelial-like structure to separate the cells from the flow channels. The barriers were modeled after the fenestrated liver endothelium, with 3 μm openings to permit free solution exchange. The barriers are formed out of PDMS and are 55 μm cubical pillars with 3 μm gaps between posts. This dimension has been found to effectively localize cells without causing stress, channel blockage, or cell death. Smaller pore sizes prevent adequate loading of cells, cause membrane stress, and limit nutrient diffusion. Larger pores cause cells to squeeze through during loading (Figure 14.4).

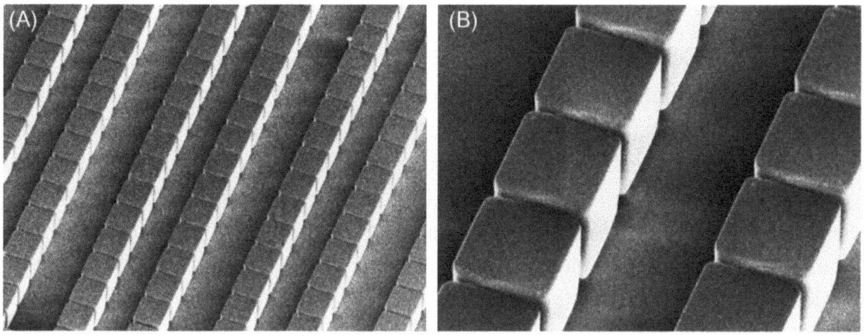

FIGURE 14.4

Barriers.

14.3.4 Air channels

Hepatocytes have a high oxygen demand, which is necessary for metabolic functions. In order to maximize oxygen transport to the hepatocytes, microfluidic air diffusion channels were patterned within 50 μm of the flow channels. These channels carry ambient air into the microfluidic layer for diffusion across the gas-permeable PDMS walls to the medium and cell chambers. The oxygen delivery is highest at the edges of the sinusoid channels, with a gradient across the unit similar to the liver acinus. In preliminary studies, it was found that cell behavior is indeed modified by the oxygen gradient.

14.3.5 Cell loading

Hepatocytes are loaded into the microfluidic chambers using pressure-driven flow. Cells can be obtained from a variety of sources, including fresh isolation, cryopreserved lots, or cell lines. The cell suspension is concentrated to approximately 5×10^6 cells/ml (varies by preparation type) and 8 μl is dispensed into the outlet well. Pressure (or vacuum) driven flow is initiated using a custom pneumatic manifold to create a pressure differential between the outlet and inlet wells of approximately 8 kPa. This is sufficient to move the cells into the microfluidic regions within about 1 min to create a solid mass of cells without cell death. The loading of cells into a high-density configuration with a high degree of cell–cell contact is critical for establishing long-term functional activity. Loading at too high pressure will lead to cell death and too low pressure results in low cell–cell contact. The density of cells in suspension is also important for adequate loading, with too high density leading to clumping and channel blockage and too low density leading to insufficient cell–cell contact.

14.3.6 Gravity perfusion

Once loaded, the cells are fed with continuous perfusion of culture medium through the flow sinusoid channels. Perfusion is driven by gravity, eliminating the need for external pumps during operation. The liquid level difference between the inlet well and the outlet well drives flow through the hepatocyte culture area. The fluidic resistance of the channels is designed such that 100 μl of solution will flow through in 24 h when 300 μl is added to the inlet well and the outlet is empty. Continuous flow is maintained by refilling the inlet with 100 μl every day and removing the waste from the outlet. The flow through the sinusoid channels is approximately 6 nl/min (60 μm/s velocity), which is similar to the flow in the human liver. The flow rate from the beginning of the perfusion period will be faster than at the end of the period (due to change of liquid levels); however, this range does not seem to affect hepatocyte behavior.

14.3.7 Multiplexed plate

The culture units are formatted to a standard 96-well plate frame, with 32 independent flow units per plate. As described above, each flow unit contains 20,000–50,000 hepatocytes with a flow through volume of 100 μl per unit per day. The inlet and outlet wells contain up to 300 μl of solution. The bottom surface of the plate is a 170 μm thick glass slide optimized for microscopy. Since the cells are in contact with the glass slide, high-quality images can be taken with a standard inverted microscope on live and fixed cells.

A key benefit of the gravity-driven flow is that once loaded, the plates can be operated just like any other culture plate. No special equipment or handling is necessary to maintain the microfluidic culture. This also allows unlimited plates to be operated in parallel.

14.3.8 Automated system

In order to facilitate long-term hepatocyte culture on multiple microfluidic plates, we developed an automated system to perform cell loading and feeding operations. The automated platform uses off-the-shelf instrumentation with a custom cell loading module. The key components are a liquid dispenser and an automated incubator. Every 24 h, the plates are taken from the incubator and refilled with medium. Since the inlet and outlet positions conform to standard 96-well plate dimensions, this operation is compatible with existing automation instruments.

In order to automate the cell loading and lysing functions, we developed a pneumatic flow system to be operated with the automated system. The module allows plates to be placed into a nest using a robotic arm and utilizes a manifold to vacuum seal to the wells of the 96-well plate, pneumatically isolating each well. The manifold can then apply positive or negative pressure to selected well positions to cause flow thorough the microfluidic channels. Once the flow operation is complete, the plate is returned to the nest for subsequent automation operations (Figure 14.5).

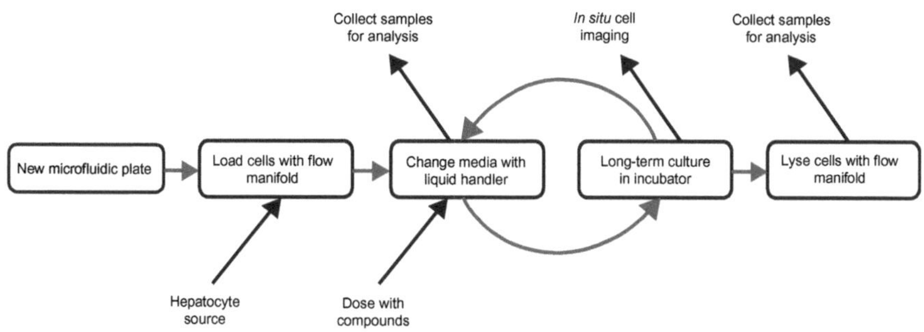

FIGURE 14.5

Operation.

14.3.9 Operation

The operation of the microfluidic hepatocyte culture plates was designed to be similar to a standard 96-well plate. The microfluidic plates are shipped primed with PBS to eliminate air bubbles in the channels. The plates are further vacuum sealed to prevent evaporation. Once opened, the PBS can be removed and replaced with the desired solution by the user.

The first step is to load hepatocytes into the microfluidic culture units. This is achieved by aspirating the PBS from the outlet wells and pipetting 8 μl of hepatocyte suspension into the port. Hepatocytes can be freshly isolated or cryopreserved. The optimal suspension density depends on the hepatocyte size; for human hepatocytes 7×10^6 cells/ml is preferred and for rat hepatocytes, 5×10^6 cells/ml. Immediately after dispensing the cells into the port, the plate should be sealed to the pneumatic manifold and loaded under pressure-driven flow. In the standard loader configuration, a negative pressure is created by vacuum on the inlet wells at 5 kPa for 30 s. As the cells flow into the culture chambers, the perfusion barriers serve to pack them into a high-density tissue-like configuration. Loading quality can be checked on a standard tissue culture microscope.

Once the cells are loaded, culture medium is added to the inlet well to initiate gravity-driven flow. Generally, any serum-free hepatocyte culture medium is suitable for long-term culture experiments. The plates are then placed into a standard cell culture incubator (37°C, 5% CO_2) for long-term maintenance. Every 24–48 h, the medium should be replenished in the inlet well and removed from the outlet.

Analysis of cells over time can be performed in a variety of formats. For metabolism analysis, the flow through medium can be collected from the outlet and measured by chromatography or mass spectrometry. Fluorescent measurements using a plate reader are also amenable with standard substrates, such as live/dead assays. The cells can be fixed in the microfluidic plate and immunostained or used in live cell imaging. The cells can also be lysed from the microfluidic channels with lysis buffer (using the opposite flow as the cell loading) and analyzed with methods such as quantitative real-time polymerase chain reaction (qPCR) and protein quantification.

14.4 Long-term hepatocyte culture

The ideal *in vitro* liver model should demonstrate stable liver-specific activity for over 1 month. This timescale enables studies of mechanisms of toxicity, chronic exposure to low doses, and changing combinations of substances over time. Static 2D cultures are generally only useful for a few days, with loss of liver activity in as little as a few hours. A relatively new approach that uses micropatterned coculture of hepatocytes with endothelial cells has been shown to maintain activity for over a month but does not incorporate aspects of flow or 3D architecture [17]. In our system, we have demonstrated continuous microfluidic perfusion culture of hepatocytes for over 1 month with maintained liver activity.

The current gold standard for *in vitro* hepatotoxicity studies is to use freshly isolated human hepatocytes. These are obtained from healthy donor tissue and used immediately after isolation. This cell source provides the most accurate representation of the native human liver activity. Key limitations of fresh human hepatocytes include limited availability, high cost, genetic variation, altered behavior based on donor history, and inability to perform repeat studies on the same tissue. An alternate hepatocyte source that is gaining in popularity is the use of cryopreserved human hepatocytes. These are isolated from donors in the same way as fresh hepatocytes and immediately frozen in liquid nitrogen for long-term storage. Because a single tissue donation can yield thousands of identical tubes of hepatocytes, cryopreserved cells enable the researcher to precharacterize the lot and perform repeat experiments over time. A limitation of cryopreserved cells is that they are more difficult to maintain in long-term culture, presumably due to the shock of the freeze/thaw cycle. Besides human cells, animal hepatocytes are commonly used in screening, including rat, mouse, dog, and monkey. There has also been interest in immortal hepatocyte-like cell lines; however, there is insufficient evidence that these can replace primary hepatocytes. Various commercial sources for hepatocytes are now easily available from reputable vendors (Figure 14.6).

14.4.1 Comparison with standard culture methods

In our studies, we compared microfluidic hepatocyte culture with 2D static culture on collagen coated plates with Matrigel (BD Biosciences, San Jose, CA) overlay. In both cases, about 40,000 cells were seeded per unit and 100 μl of medium was exchanged daily. Comparisons were performed using the same cell source over the same time period. The key physical differences between the two culture formats were continuous flow versus static incubation and 3D architecture versus 2D monolayer.

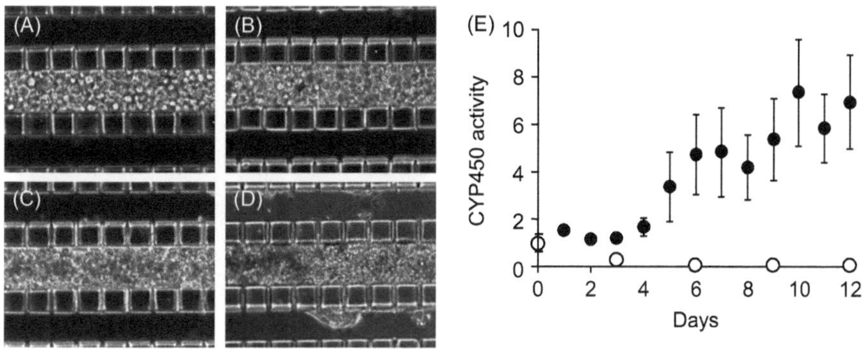

FIGURE 14.6

Culture.

14.4.2 Morphology

In the microfluidic plates, primary hepatocytes recovered within the first 4 days and formed a tissue aggregate morphology. This is evidenced by cell–cell adhesions and loss of clear cell boundaries. As primary hepatocytes do not divide in culture, the aggregated state is maintained for the duration of the culture period. From previous work with hepatocyte aggregate culture, this configuration is indicative of long-term maintenance of native liver activity. Because each culture channel is $60 \times 60\ \mu m^2$ in cross section, the cells form a connected cord structure roughly 2–3 cells thick, similar to the liver. Hepatocytes stay within the microfabricated endothelial barrier, recreating a space of Disse for medium perfusion from the sinusoid flow.

Immunostaining of the aggregated hepatocyte structure shows the presence of 3D morphological features found in liver tissue. Cell–cell attachments result from the formation of adherens junctions with staining of E-cadherin. Communication between cells is facilitated by gap junctions that allow transport of fluids from one cell to the next. The fusion of cell membranes by tight junctions creates polarized compartments between cells that serve as the bile canaliculi. The expression of transporter proteins specific to the bile canaliculi are further evidence of the morphological integrity of hepatocytes in the microfluidic format (Figure 14.7).

14.4.3 Viability and CYP450 activity

Long-term stability of cellular activity is an important aspect for an *in vitro* liver model. In the microfluidic culture, viability of hepatocytes was measured with a luminescent substrate for cellular reduction (energy consumption). In the microfluidic format, we observed a characteristic increase in activity as the cells

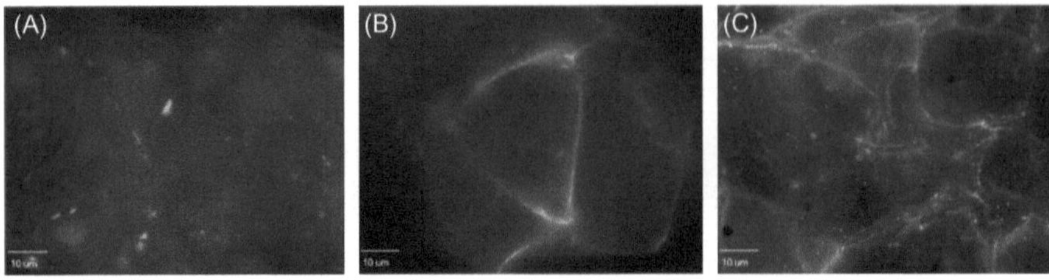

FIGURE 14.7

Immunostain.

recovered over the first few days with maintenance of stable viability for over 1 month. In the 2D overlay culture, the cells showed an immediate drop-off of viability at day 1, presumably from cells that were unable to adhere to the 2D substrate. The attached cells showed a slow decline of viability and activity over 2 weeks. Similarly, measurement of a probe substrate for cytochrome P450 (CYP450) 3A activity was maintained in microfluidic culture. CYP450s are a key family of enzymes highly expressed in hepatocytes responsible for hydroxylation reactions to remove foreign chemicals from the blood. CYP450 3A1 (in rat hepatocytes) is the key enzyme involved in drug metabolism. Monitoring of CYP450 activity is important in determining drug circulation times, toxic metabolites, and drug–drug interactions.

14.4.4 Gene expression

Maintenance of hepatocyte phenotype can also be monitored by measuring expression of liver-related genes. We used qPCR to measure mRNA levels in microfluidic culture after lysing the cells at 1-week intervals and compared with static 2D overlay culture. Expression levels were normalized to day 0 activity (freshly isolated cells) which is believed to be representative of liver tissue. In this study, we charted the relative expression of three liver-specific genes: CYP450 3A1, PXR, and UGT1A1. As described above, CYP450 3A1 is the rat analog of the human CYP450 3A4 enzyme, which is involved in the metabolism of the majority of biopharmaceuticals. PXR, the pregnane X receptor, senses the presence of toxic substances and regulates the expression of proteins (such as CYP450 3A1) to remove the toxins. UGT1A1, a UDP-glucuronosyltransferase, is involved in converting lipophilic molecules into soluble metabolites to be excreted from the bloodstream. The glucurondation pathway is a key mechanism for clearing foreign chemicals from the body and is a part of the Phase II conjugation reactions found in the liver. The results from these gene expression experiments lend further evidence that the microfluidic culture environment is suitable for preserving long-term hepatocyte functions (Figure 14.8).

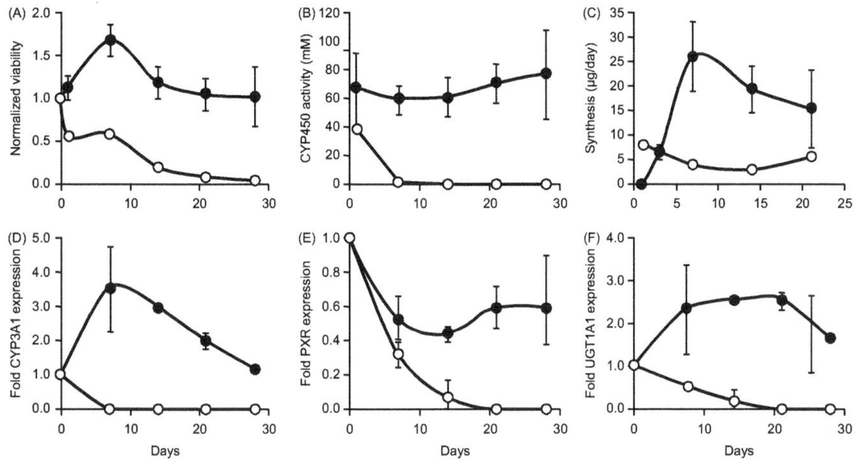

FIGURE 14.8

Plots.

14.4.5 Albumin/urea production

Hepatocytes are responsible for the synthesis and secretion of albumin and urea into the bloodstream. These are critical functions of a healthy liver. In the microfluidic culture system, we were able to collect the outlet medium at various times in culture and measure the amount of albumin and urea. This data shows that hepatocytes cultured in this system are able to maintain elevated levels of synthesis and can therefore be used as a model for albumin and urea regulation in response to external signals or drugs.

14.4.6 Hepatocyte sources

The microfluidic culture format has been demonstrated to maintain both fresh and cryopreserved hepatocytes for long-term culture. In general, fresh hepatocytes recover more quickly and have higher levels of liver-specific activity. This may be due to possible loss of functions resulting from the freeze/thaw cycle. While the exact differences between fresh and cryopreserved cells are yet to be determined, we find reasonably consistent results working with both types of cell sources.

We have also evaluated the performance of human and rat hepatocytes. In both cases, the hepatocytes maintain longevity and activity for over a month in culture. We note differences in cell size, morphology, and lot variability. As expected, human cell sources are more variable, while rat cells tend to be fairly consistent. There are also known differences in hepatocyte biology between species, including expression of different CYP450 enzymes, rates of metabolism, and sensitivity to toxins. This data has not yet been collected in the microfluidic culture format and is an avenue of ongoing investigation.

14.5 Summary

The application of microfluidic technologies for *in vitro* hepatotoxicity studies provides many advantages over current methods. Most importantly, the ability to model critical aspects of the hepatocyte microenvironment will lead to more clinically relevant and predictive data. The scalable nature of the microfluidic culture system will enable more experimentation to be performed earlier in the safety evaluation process and for a lower cost. This should also reduce the burden of animal testing while preventing toxicity failures in human clinical trials. Additional applications of the microfluidic liver model include investigation of mechanisms of toxicity of environmental chemicals as well as for research regarding liver diseases.

Acknowledgments

This work was supported in part by funding from The Hamner Institutes for Health Sciences through the Long-Range Research Initiative (LRI) of the American Chemistry Council (ACC).

References

[1] H. Jaeschke, et al., Mechanisms of hepatotoxicity, Toxicol. Sci. 65 (Feb) (2002) 166.
[2] W.M. Lee, Drug-induced hepatotoxicity, New Engl. J. Med. 349 (July) (2003) 474.
[3] F.S. Collins, G.M. Gray, J.R. Bucher, Toxicology. Transforming environmental health protection, Science 319 (2008) 906.
[4] S. Bhattacharya, Q. Zhang, P.L. Carmichael, K. Boekelheide, M.E. Andersen, Toxicity testing in the 21 century: defining new risk assessment approaches based on perturbation of intracellular toxicity pathways, PLoS One 6 (2011) e20887.
[5] T.S. Peters, Do preclinical testing strategies help predict human hepatotoxic potentials? Toxicol. Pathol. 33 (2005) 146.
[6] A. Knight, Systematic reviews of animal experiments demonstrate poor contributions toward human healthcare, Rev. Recent Clin. Trials 3 (May) (2008) 89.
[7] F.P. Guengerich, Mechanisms of drug toxicity and relevance to pharmaceutical development, Drug Metab. Pharmacokinet. 26 (2010) 3.
[8] L. Jia, X. Liu, The conduct of drug metabolism studies considered good practice (II): *in vitro* experiments, Curr. Drug Metab. 8 (December) (2007) 822.

[9] M.L. Greer, J. Barber, J. Eakins, J.G. Kenna, Cell based approaches for evaluation of drug-induced liver injury, Toxicology 268 125 (February 9).

[10] C. Guguen-Guillouzo, A. Guillouzo, General review on *in vitro* hepatocyte models and their applications, Methods Mol. Biol. 640 (2010) 1.

[11] N.J. Hewitt, et al., Primary hepatocytes: current understanding of the regulation of metabolic enzymes and transporter proteins, and pharmaceutical practice for the use of hepatocytes in metabolism, enzyme induction, transporter, clearance, and hepatotoxicity studies, Drug. Metab. Rev. 39 (2007) 159.

[12] E.W. Young, D.J. Beebe, Fundamentals of microfluidic cell culture in controlled microenvironments, Chem. Soc. Rev. 39 (2010) 1036 (March).

[13] P.J. Lee, P.J. Hung, L.P. Lee, An artificial liver sinusoid with a microfluidic endothelial-like barrier for primary hepatocyte culture, Biotechnol. Bioeng. 97 (August) (2007) 1340.

[14] J. Sahi, S. Grepper, C. Smith, Hepatocytes as a tool in drug metabolism, transport and safety evaluations in drug discovery, Curr. Drug Discov. Technol. 7, (2010) 188 (September).

[15] B. Swift, N.D. Pfeifer, K.L. Brouwer, Sandwich-cultured hepatocytes: an *in vitro* model to evaluate hepatobiliary transporter-based drug interactions and hepatotoxicity, Drug Metab. Rev. 42 (2010) 446 (August).

[16] Q. Meng, Three-dimensional culture of hepatocytes for prediction of drug-induced hepatotoxicity, Expert. Opin. Drug Metab. Toxicol. 6 (2010) 733 (June).

[17] S.R. Khetani, S.N. Bhatia, Microscale culture of human liver cells for drug development, Nat. Biotechnol. 26 (2008) 120.

CHAPTER

Live Cell Analysis Under Shear Flow

15

Carolyn G. Conant, J. Tanner Nevill and Cristian Ionescu-Zanetti

Fluxion Biosciences, CA, USA

15.1 Introduction

An important aspect of improving *in vitro* models for *in vivo* biological processes is the use of appropriate mechanical forces such as shear. The effect of shear on live cells is well documented in developmental biology, vascular biology, cancer biology, and immunology [1–6]. In order to create a physiologically relevant environment outside of the body, the dynamic nature of many of these processes must be considered. For example, recruitment of a white blood cell to the site of a tissue injury occurs as a series of events that may include cell rolling, arrest, and transmigration [7]. This cannot be studied in a static system due to the inherent influence of blood flow. Likewise development of stem cells occurs under many types of stimuli, including both chemical and mechanical. Mesenchymal stem cells can be differentiated to endothelial cells in the presence of shear flow [8]. The biochemistry of platelet behavior is another example of a biological system that is particularly affected by shear forces [9].

Systems used to study live cells under flow include laminar flow chambers, cone and plate viscometers, microtiter plate assays in which the microplate is agitated to create shear inside the well, and microfluidic devices. Laminar flow chambers and cone and plate viscometers can model *in vivo* conditions very well; both offer the ability to coat the surfaces with substrates of interest. Cone and plate viscometers consist of a rotating cone and a stationary plate. By rotating the cone on the plate, a uniform field of laminar shear stress is applied to the fluid retained between the two [10]. Following the application of shear stress, a number of analyses such as particle counting may be performed to determine the extent of aggregate formation. If the surface was coated, additional analyses can be performed such as staining and quantitation of surface coverage. Key advantages of cone and plate viscometers include low sample volumes of blood or platelet rich plasma and the ability to operate at varying shear rates. While valuable for many studies of lower throughput end point analysis, the cone and plate method lacks real-time data acquisition. This eliminates the possibility of analyzing very early events of interest. Low sample throughput and parallelization is another drawback to the cone and plate method, as only one condition can be examined at a time.

Microfluidic Cell Culture Systems. ISBN: 978-1-4377-3459-1

A parallel plate flow chamber and its derivative apparatus address this void by allowing real-time high-resolution data capture with flow present [11,12]. Common perfusion configurations include parallel plate flow chambers or glass capillary tube flow chambers. Typically, the flow is driven by a syringe or peristaltic pump placed alongside the chamber to generate wall shear stress that is similar to blood flow in the vasculature. These configurations are flexible in terms of shear rates applied as well as substrate in the flow path. One of the distinct disadvantages of perfusion chambers is that a relatively large volume of fluid must be used. These volumes typically fall in the 10–50 ml range per each experimental condition. This amount can be prohibitive in many cases. Long setup times are another drawback of perfusion chambers. Typical chambers require a lengthy sterilization process and assembly of the chamber housing to a glass coverslip. For example, in the study of platelet perfusion over a surface using whole blood, this can be a severe limitation, considering the working lifetime of one blood draw can be as little as 3 h. Only a small and limited number of conditions can be examined from the same donor using such conventional flow chambers.

Many laboratories have devised microfluidic devices that can be employed in lieu of parallel plate flow chambers for shear studies [13–16]. These devices generally provide workflow advantages over conventional technologies, which include controlled laminar flow, compatibility with real-time microscopy, and microliter-scale fluidic requirements. The latter benefit enables multiple experiments from small samples. Although the microfluidic format has provided several key advantages over conventional platelet assays, many of these devices do not address the need for a simple, reliable, and repeatable fluidic interface. As such, these devices have limited capacity for widespread use outside of the developer's laboratory.

Here, we present a novel microfluidic system for live cell analysis that enables precise control over shear stress to emulate physiological conditions encountered in the vasculature and other niches. It addresses critical limitations with cone and plate assay technologies by providing real-time, high-resolution microscopy data in a low-volume microfluidic format. The system utilizes a pneumatic pumping system instead of liquid displacement pumps, which obviates the need to sterilize components between runs and contributes to higher throughput and parallelization [17]. The microfluidic device is contained within an Society for Biomolecular Screening (SBS)-standard well plate format that remedies many of the interfacing issues often encountered with handmade microfluidic devices.

15.2 Flow control and well plate microfluidics

15.2.1 Principles of well plate microfluidics

The BioFlux system leverages the advantages of microfluidics to create a network of laminar flow cells integrated into standard well plates. Each BioFlux plate utilizes an SBS-standard well plate to ensure compatibility with common microscope stages, liquid handlers, and plate readers. In the manufacturing process, a network of microfluidic channels becomes integrated to the bottom of the plate. These channels connect the wells together into groupings of independent experiments. Reagents are added to the wells of the plate using conventional means (i.e., pipette, syringe, liquid handler). Pressures can then be applied to each individual well using the automation software, which applies the pressure from the controller through to the interface and into the well. As that pressure is applied, it drives the reagent in that well into the corresponding channel underneath the plate. After the reagent goes through the channels, it ends up in a separate well on the plate, i.e., an "output" well (Figure 15.1).

In this manner, it is possible to set up networks of microfluidic channels that run between wells of the plate. In the BioFlux system, one experimental channel typically consists of an "input" well and an "output" well. This is the case with

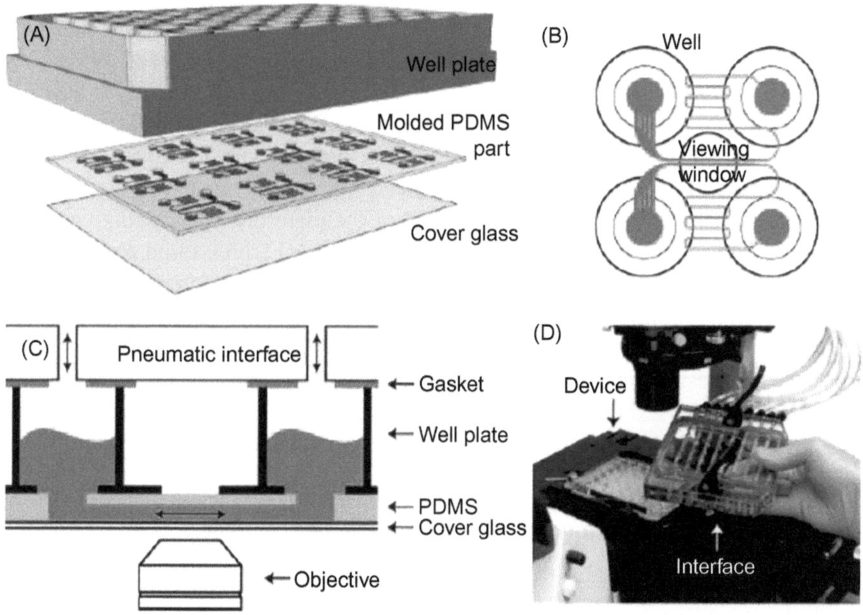

FIGURE 15.1

Overview of the well plate microfluidic (WPM) system. (A) Fabrication of the WPM device involves creating a laminate consisting of an SBS well plate, a molded PDMS layer which forms the ceiling and side walls of the channel, and a #1.5 borosilicate glass coverslip which forms the floor of the microfluidic channel. (B) Multiple experiment parallelization is enabled via repeating a pattern of channels over the full plate resulting in up to 24 independent assay channels. Two independent channels of an example device (with a range of 20–200 dyn/cm^2) are shown here. (C) A set range of pneumatic pressures is used to create fluid flow within the device; altering the geometry of the channel enables a variable shear range with a set pressure range. (D) The WPM device can be used as part of an inverted microscope.

Source: *Adapted with permission from Ref. [18].*

the 48-well BioFlux plate, which contains 24 experimental channels (i.e., each with an input and an output well). In some assays, it is desirable to have multiple input wells for compound addition or other reagents that need to be added during the course of an experiment. This is the case with the 24-well BioFlux plate, which offers two input wells and one output well per experimental channel. As such, the 24-well plate provides eight experimental channels.

15.2.2 WPM device fabrication

Channel patterns were lithographically patterned onto substrates as previously described [19]. Profilometry was used to confirm that channel depths were consistent within $\pm 1\ \mu m$ across the wafer. Microscopy plus image analysis was used to quantify the channel widths. Polydimethylsiloxane (PDMS) was dispensed through a custom dispensing tool that mixes the two components on demand with a static mixer. The material was cured in an oven and then demolded. Vias were created with a custom jig. The patterned PDMS part was permanently bonded to a glass slide (#1.5 borosilicate cover glass). The PDMS-glass sandwich was then glued to the bottom of a well plate. Glue is dispensed with a custom three-axis dispensing robot. This process resulted in a channel array under the footprint of the 48-well SBS microplates where the well of the plate provided a contained liquid reservoir to feed the microfluidic channels (Figure 15.1A). Fluid flow through the microfluidic channels was controlled using an airtight interface connected to a pneumatic continuous displacement pump (BioFlux Controller, Fluxion Biosciences, S. San Francisco, CA) (Figure 15.1C and D). The pump was designed and tested under typical working conditions; pneumatic flow control was found to be ± 0.01 psi (± 68.95 Pa). There are several different designs available; the typical dimensions are 60–75 μm high and 90–350 μm wide.

15.2.3 WPM device modeling and characterization

Movement of fluid through the device is achieved by applying a precise pressure to the inlet headspace, while the outlet headspace is vented to atmosphere. Flow rate of fluid through a channel in this situation is a function of applied pressure, channel dimensions (height, width, and length), and fluid properties (density and viscosity). For flow at low Reynolds numbers, the Navier–Stokes equation was simplified to

$$\Delta p = \eta \nabla^2 v$$

The Hagen–Poiseuille law, which is considered to be the equivalent of Ohm's law for fluid flow, introduces the proportionality factor of hydrodynamic resistance (R_{hyd}).

$$\Delta p = R_{hyd} Q$$

The hydrodynamic resistance is dependent upon channel dimensions, just like the resistivity of a wire; for a given pressure, fluid in long, narrow channels will flow slower than short, wide channels. R_{hyd} can be determined for a channel with rectangular cross section by Fourier expansion of the Navier–Stokes equation.

Assuming the channel height, h, is always less than or equal to width, w, the resulting approximation is

$$R_{hyd} = \frac{1}{C}\frac{12\eta L}{h^3 w}$$

where h is viscosity and C is a correction factor that takes into account the relative aspect ratio of each channel. Bruus points out that significant errors can be noticed unless the aspect ratio is properly taken into account [20]. The correction factor was determined to be

$$C = 1 - \frac{0.63}{w}\sum_{n,\text{odd}}^{\infty}\frac{1}{n^5}\left(\frac{1-e^{-n\pi(w/h)}}{1+e^{-n\pi(w/h)}}\right)$$

Individual R_{hyd} values were determined for each section of a channel using the above equations and then summed together to determine the total hydrodynamic resistance, just like resistors in series. In this case, the values for channel lengths were determined based upon the design of the channels. The channel heights were collected from profilometry data of channel depths during the fabrication process; channel widths were determined via image analysis of microscope images. Additionally, the shrinkage of material upon curing was taken into account.

Fluid flow testing of fabricated well plate microfluidic (WPM) devices was performed and compared to the numerical model. Distilled water was metered into inlet wells of the devices and perfused at known pressures for set time periods. Flow was stopped and the amount of liquid that passed through the channel was measured using an analytical balance (Mettler Toledo, Columbus, OH). Fluid temperatures were recorded for each test to correct for density and viscosity differences. A comparison between the experimental results and the model are shown in Figure 15.3 for both devices. To determine shear at the bottom center of the channel, the following equation was used:

$$\tau = \frac{1}{C}\frac{6\eta\Delta p}{R_{hyd}}\frac{1}{h^2 w}$$

Importantly, this equation is not valid when approaching the edges of the channel, as the shear stress drops precipitously to zero in the corners. Finite element models suggest that the shear stress is at least 80% of the maximum value across 82.5% of the channel width. Therefore, we ignore anything that is outside of this center region.

The software is designed to receive the desired shear stress, fluid viscosity, and plate design as inputs from the user. Using the above equations, the software determines and applies the appropriate pressure.

Measured flow rates for all 24 channels in triplicate were averaged for both device designs. The coefficient of variation for interplate repeatability is less than 2.5% for both devices tested. Intraplate variability was determined to be less than 5% for the low shear device and less than 3.5% for the high shear device. This was determined by measuring the flow rates of all 24 channels for five different low shear devices and nine different high shear devices on different days.

15.3 Cell biology applications

15.3.1 *In vitro* cells under shear

Live cell imaging has emerged as a preferred method for cell biology research and drug screening. While commercial systems exist for high content screening, it is difficult or impossible to run these assays under continual shear flow. Applications such as neuronal development, stem cell research, and vascular physiology can all benefit from being run under flow conditions. In these cases, monolayers of cells can be cultured in the BioFlux channels with shear flow being applied during or after monolayer formation. The presence of flow and associated mechanical stimuli drive changes in protein expression, cell morphology, and function. This is especially important for the endothelium, where cells are exposed to shear forces *in vivo*. For endothelial cell monolayers, there are clear differences in cell morphology and polarity that is dependent on the shear flow regime (Figure 15.2A).

Not only the magnitude of flow, but the time-dependent characteristics of the flow rate are important. For example, we describe a method of specialized endothelial cell culture using controlled shear flow to generate cyclic waveforms to mimic athero-prone and athero-protected sites in the vasculature (Dai, G. et al., 2004). We grew endothelial cells under different flow regimens for a period of 24 h. We then evaluated temporal changes in protein expression and morphology during application of programmed shear and found significant differences in actin localization, cell density, and connexin 37 and 43 expression in response to pulsatile flow (Figure 15.2B). This method is extensible to atherosclerosis modeling as well as investigation of other aspects of endothelial cell biology.

15.3.2 Stem cell differentiation

Stem cell research has the potential to produce novel treatments for previously incurable diseases and injuries. The application of controlled shear flow to undifferentiated embryonic stem cells promotes enhanced expansion of cell lines (Fok, E. and Zandstra, P., 2005). Shear stress is also used as a stimulus for

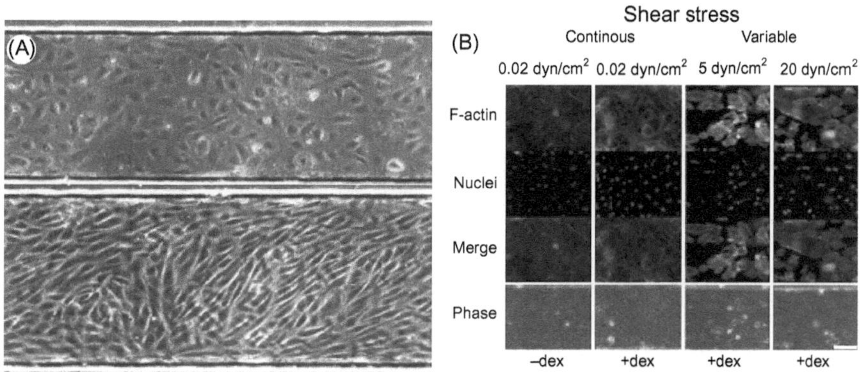

FIGURE 15.2

Cellular response to physiological flow. (A) A good example of where shear plays an important role *in vivo* is the biology of the endothelium, which is always under blood flow. When endothelial cells are grown under shear flow, there are changes in cell signaling leading to morphological and functional changes in cell behavior. Endothelial cells were exposed either to very low shear greater than 0.02 dyn/cm^2 (top) or to relatively high shear 2 dyn/cm^2 (bottom) for a duration of 16 h. A change in cellular morphology and polarity was observed. (B) HUVECs expressing actin-GFP (green) and NLS-RFP (red, nuclei) were imaged in time lapse during pulsatile flow at a maximum amplitude of 5 dyn/cm^2 for 1.2 Hz. Cells were found to begin reorientation within an hour of flow commencement. At the end point of pulsatile shear experiments (24 h), cells were stained with phalloidin–Alexa 488. There is a marked difference in actin polarity as well as monolayer density. High molecular weight dextran was added to extend flow times in channels with high flow magnitudes to extend duration of experiments. Scale bar is 100 μm microns and all flow is from left to right. (For interpretation of the references to color in this figure legend, the reader is referred to the web version of this book.)

differentiation especially for cell types that naturally respond to physiological shear, such as endothelial cells (Illi et al., 2005; [4,5]). Differentiation of cells into specific cell types and subsequent production of biomaterials is also facilitated by mechanical forces such as shear. This is the case with chondrocytes used to produce cartilage (Schumann et al., 2006).

Mesenchymal stem cells may be differentiated under different conditions into several types of specialized cells. Application of shear stress to mesenchymal stem cells or addition of vascular endothelial growth factor (VEGF) may influence their differentiation to endothelial cells, which exclusively express and secrete von Willebrand factor (vWF) [6,8]. We used application of shear stress in the BioFlux system to differentiate mesenchymal stem cells into endothelial cells under flow conditions. We found that even after 48 h of shear stress without addition of VEGF that cells began to express vWF in the cytosol. The most marked

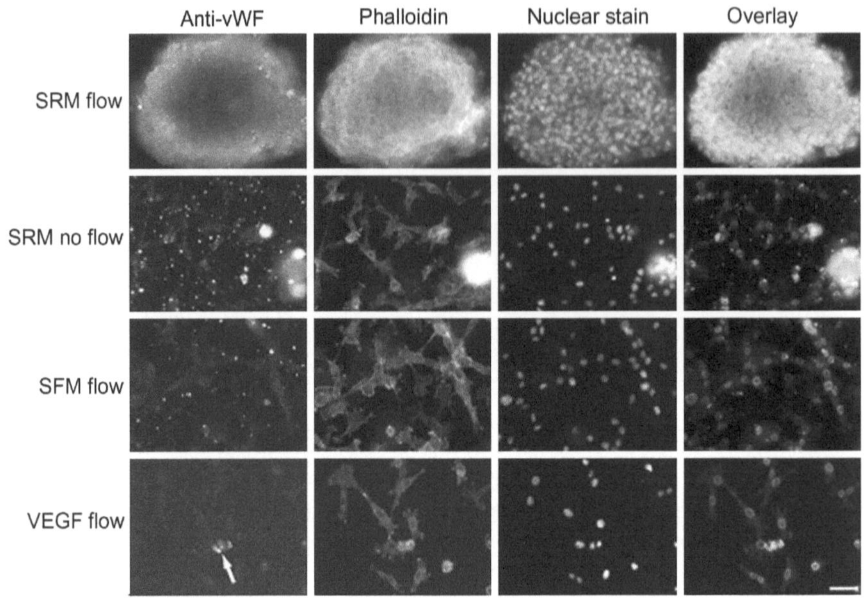

FIGURE 15.3

Stem cell differentiation under flow. Micrographs of mesenchymal stem cells before (start) and after flow. Scale bar is 50 μm. Arrow indicates vWF expressing cells in the VEGF-flow condition. All fluorescence micrographs are after 48 h of continuous flow or no flow as indicated.

differentiation or expression of cytosolic vWF occurred without the addition of VEGF in rich media under shear (Figure 15.3, SRM flow). However, differentiation also occurred in the VEGF/shear stress-treated cells as well (Figure 15.3, VEGF flow).

In conclusion, the shear flow and the WPM systems (i.e., BioFlux) may be used to drive stem cell differentiation in controlled microenvironments

15.3.3 Platelet biology: adhesion, aggregation, and thrombus formation

There are few places in the adult human anatomy that are more influenced functionally in health and disease by mechanical forces than the vasculature. In particular, the biochemistry of platelet function is vastly influenced by blood flow-induced shear stress within the blood vessels and cannot be fully appreciated out of this dynamic context. In hemostasis, physiological blood flow influences vWF's conformation and allows platelets to operate in surveillance mode with no consequent abnormal blood clotting [21–23]. Undesirable and potentially

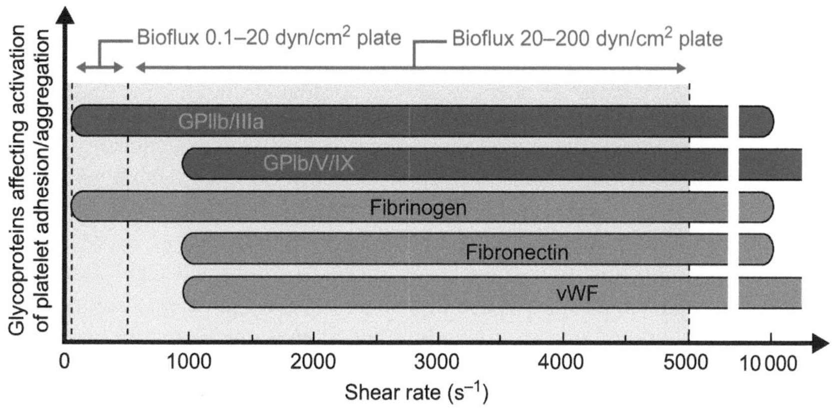

FIGURE 15.4

Shear regimes and platelet aggregation. Receptor–ligand interactions together with shear rate are important for platelet aggregation. Receptors (gray bars) and ligands (orange) are shown with respect to the shear range in which they play a dominant role during platelet aggregation. (For interpretation of the references to color in this figure legend, the reader is referred to the web version of this book.)

Source: *Adapted with permission from Ref. [18].*

life-threatening thrombus formation occurs when the following conditions are met: high shear due to an obstruction (stenosis) and/or exposure of subendothelium due to a variety of factors (e.g., desquamation from a plaque rupture [24]).

The biochemistry and mechanobiology of platelet aggregation has a complex dependence upon shear. At low shear rates, activated platelets aggregate by fibrinogen and GPIIb/IIIa. As shear increases, due *in vivo* to stenosis or injury [25–27], molecules involved in aggregation change to include a combination of vWF, fibrinogen, fibronectin (FCN), and finally at very high shear, vWF alone (Figure 15.4) [3,9]. Therefore, in order to evaluate potential antithrombotic therapeutics, a wide variety of subsurface and controlled shear conditions must be tested.

Several microfluidic design systems have been used to investigate various conditions of matrix deposition, shear flow, and online agonist addition [13–16,28–31]. Advantages of microfluidic systems for vascular studies include low reagent consumption (especially important for murine and pediatric samples, where sample volumes are limited), physiologically relevant channel dimensions, the promise of quick sample-to-result, accurate flow control, easy fluidics interfacing, and the opportunity for highly parallel studies. It is the latter three points that have been lacking in previous microfluidic vascular platforms. Most works have focused on a single channel or small (up to 6) arrays that require manual tubing-to-chip connections [13–16,28–31]. Additionally, because the working life span of a fresh blood draw is often short (minutes to hours depending on the type of anticoagulant treatment used), any optimized device would allow for a large number of samples to be queried quickly and preferably simultaneously with minimal device preparation[1] .

Here, we use a WPM system to deliver laminar flow over a wide shear range (covering physiological and pathological shear values) for evaluation of platelet behavior under different conditions as experienced at the platelet/vessel wall boundary in flowing blood (Figure 15.4). The principle distinguishing features of this system are tightly controlled shear stress, self-containment of experiments (i. e., no tubing, connectors, or syringe pumps needed), "plug and play" fast assembly, optimization for microscopy, and higher throughput than previously described designs for immobilized ligands. Two different designs are presented to cover the wide range of shear values desired. Both designs were characterized by flow testing and comparing experimental results with a computational model. The physiological range, or low shear, device consisted of an array of simple channels with one inlet and one outlet each. This device delivered a range of 0.1–20 dyn/cm^2 or 3–500 s^{-1} at 4 cP. The pathological shear range, or high shear, device was validated within the range of 20–200 dyn/cm^2 or 500–5000 s^{-1} at 4 cP. Both devices had 24 parallel channels, which could be actuated independently or together, with all fluids contained within the device. To determine the proper working design for the high shear device, the effects of microfluidic geometry on platelet biology were evaluated using vWF as a substrate with the application of various shear ranges. We found that a gradual increase in shear is required at the entrance to the channel to prevent thrombus formation in undesired locations. In the low shear range plate, a variety of substrates, including collagen I, FCN, and vWF, were tested to demonstrate the breadth of applicability of the design for platelet adhesion and aggregation. As well, the dose response to a GPIIb/IIIa inhibitor was determined. The low and high shear devices were also found to be amenable to growth of endothelial cell monolayers and subsequent cell–platelet interactions. The final WPM devices are self-contained, high-throughput tools for multiparametric whole blood assays.

15.3.3.1 High shear device evolution and characterization

A specialized set of microfluidic channel designs were used in order to deliver higher shear in the pathological range. A microfluidic channel design was proposed to deliver pathological shear up to 200 dyn/cm^2 using a single channel geometry informed by the previous low shear design. This device was fabricated and tested as described above. While the design provided the correct shear range and biological response within the main experimental area, we found that the inlet area required a different geometry as unwanted thrombi were preferentially forming in this region, presumably due to the extremely high shear and resultant large shear drop at the inlet (Nesbitt et al., 2009).

[1]J. Dong, Professor, Department of Medicine, Section of Cardiovascular Research, Division of Thrombosis, 2009.

To address this issue, the WPM channel geometry was reconfigured to meet three criteria: thrombus formation in the region of interest, no inappropriate areas of thrombus formation before the experimental area of the channel, and maximum shear stress must be $\geq 100\ \text{dyn/cm}^2$. Several inlet designs were fabricated and tested using vWF as the adhesion substrate to determine where thrombus formation would preferentially occur (Figure 15.5C). Model designs with large inlets at the well-channel border met the inlet criteria for the high shear design, whereas smaller width inlet designs showed signs of thrombus formation in undesirable locations upstream of the viewing area. The final high shear design (shown in Figure 15.5B, inset) met both criteria and was fabricated and tested as above using vWF as a substrate (Figure 15.5). The result was shear-dependent thrombus

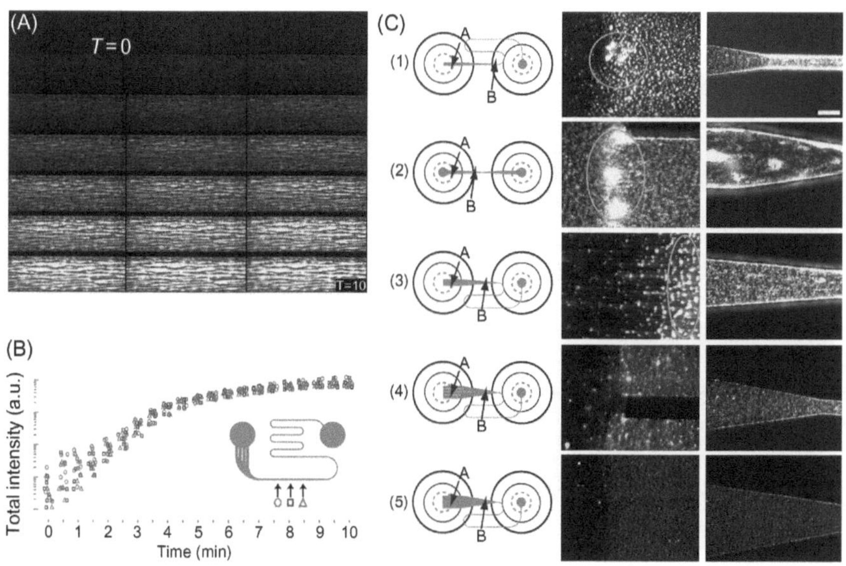

FIGURE 15.5

Thrombus formation in channels at high shear. (A) Time lapse of 10 min at 125 dyn/cm^2 at 30 s intervals. The width of the channel is 250 μm, and flow is from left to right. Channel is coated with vWF prior to flow. Shown (B) is the intensity (arbitrary units) over the time lapse for six replicate channels. Three locations per channel were imaged; inset legend depicts approximate location of interrogation. (C) WPM inlet designs for high shear flow and effect on thrombus formation. CAD designs of test channels (left panels). Channel vias are represented by dashed red circles. Position of micrograph is indicated with arrow A (microfluidic channel entrance) and arrow B (neckdown region prior to viewing window; middle panels) corresponding platelet behavior under maximum shear at well-channel border (right panels), and platelet behavior at neckdown region. Scale bar is 100 μm. (For interpretation of the references to color in this figure legend, the reader is referred to the web version of this book.)

Source: *Adapted with permission from Ref. [18].*

formation over time (Figure 15.5B) with no emboli forming prior to the main experimental region.

15.3.3.2 Evaluation of platelet function in the low shear WPM devices

Platelet adhesion, rolling, and aggregation were evaluated on three different physiologically relevant matrices using the low shear device: collagen I, FCN, and vWF [9], at either 10 or 20 dyn/cm^2. On vWF, we observed massive and uniform platelet attachment and rolling at both shear values. Rolling velocity was tabulated for 10 dyn/cm^2 at 2.95 μm/s. The attachment was reversible; cessation of shear resulted in platelets floating free of the ligand. No aggregation was observed at either shear stress with vWF as the ligand (Figure 15.6A). The interaction of platelets with FCN is thought to be important at shear rates slightly higher than are delivered within the low shear device. Thus as expected, platelets

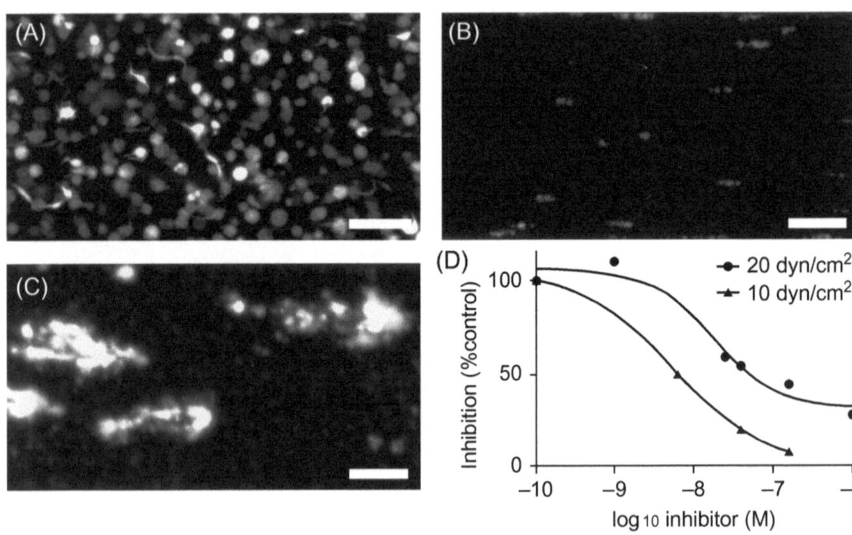

FIGURE 15.6

Shear dependence of antithrombotic compound effects. (A–C) Platelet aggregation at 20 dyn/cm^2. (A) Platelets rolling on vWF at physiological shear, scale bar is 20 μm (supplementary video). (B) FCN, moderate platelet aggregation was observed with a mixture of reversible and irreversible binding to the matrix, scale bar is 100 μm. (C) Collagen I, massive irreversible aggregation, scale bar is 100 μm. Images are taken from the center of the channel. (D) Dose–response with GPIIb/IIIa inhibitors. Multiple concentrations of antibody were tested under either 10 or 20 dyn/cm^2. Fluorescence microscopy data were collected at five positions in the channel covering 100% of the viewing area. Total pixel count was determined for each channel and compared against a no antibody control. Background fluorescence was set using a no collagen control, which did not exhibit any platelet adhesion.

Source: *Adapted with permission from Ref. [18].*

within whole blood perfused over the FCN substrate formed unstable, partially reversible small aggregates at both shear values with an average aggregate size of 95 μm^2 (Figure 15.6B). Consistent with the literature (Baumgartner, 1977; Nakamura et al., 1998), we found the most potent mediator of platelet aggregation was collagen I. Exposure to collagen I mediated rapid and large stable platelet aggregates within 2–3 min. The average size of aggregates under control conditions was 2000 μm^2 (Figure 15.6C). No platelet adhesion or aggregation was observed on bovine serum albumin BSA-coated channels as expected.

15.3.3.3 Investigation of platelet aggregation inhibition

The action of a platelet aggregation inhibitor, anti-GPIIb/IIIa, was investigated using the low shear device at two shear values. GPIIb/IIIa is potent mediator of platelet–platelet interactions and aggregation stabilization. Adhesion to collagen, in turn, activates the GPIIb/IIIa complex to elicit this response (Nakamura et al., 1999). Whole blood was incubated with anti-GPIIb/IIIa for 1 h prior to shear exposure. Treated platelets in whole blood were exposed to collagen-coated channels at either 10 or 20 dyn/cm^2 for 5 min. A dose-dependent response was observed at both shear stress values (Figure 15.6D). The IC50 values were shifted from 17 nM at 10 dyn/cm^2 to 92 nM at 20 dyn/cm^2, indicating that the increase in shear decreases the overall effectiveness of inhibition with this particular antibody. Maximum inhibition compared to the no antibody control was 11% at 10 dyn/cm^2 and 28% at 20 dyn/cm^2.

15.3.3.4 Discussion of platelet biology applications

Here, we demonstrated design and testing of several microfluidic designs for use in platelet biology experiments under shear flow. Different physiological conditions dictate the simulation of different shear conditions in devices intended for *in vitro* studies. *In vivo*, under conditions of physiological shear ranging up to 20 dyn/cm^2, the pathologic adhesion of platelets to the endothelium is precluded except under conditions of injury of the wall and exposure of substratum [11]. Testing of the 0.1–20 dyn/cm^2 (0–500 s^{-1}) device on different components of the substratum or conditions of simulated injury yielded results comparable to both *in vivo* and *in vitro* generated data (Bergmeier et al., 2006; [3,12]; Savage et al., 2002). Further, the design required no special modifications to inlet or outlet regions joining the channel to the well and as tested none of the geometries precipitated aggregation (data not shown). The microfluidic device used for physiological flow is extensible to studies under shear loading conditions for potential experiments including evaluation of murine models and clinical trial samples including small volume blood specimens.

Modeling of a device for higher shear forces required additional design considerations besides the reduction in fluidic resistance when compared to the lower shear design. One aspect of the design that needed the most optimization was the conduit region between the well and the channel; this area caused spontaneous thrombus formation with frequent release of large platelet aggregates (emboli) into the channel (data not shown). This is consistent with the findings of Nesbitt et al. (2009) who showed that drastic shear changes dependent on geometry precipitated thrombus formation as a primary process irrespective of soluble agonists. In the case of the first design with a pinch point at the well-channel border, a very favorable geometry was present for platelet adhesion and aggregation. By changing the aspect ratio at the well-channel border, we modified the flow pattern so that the biology of interest would occur at the desired location in the design. Subsequent testing of the design with vWF as a substrate yielded biologically appropriate results (Savage et al., 2002).

Recapitulation of an *ex vivo* platelet biology system using human primary cells is useful for dissecting pathways with mutant cells, human blood donors with clotting disorders, and other considerably complex studies with host or blood mutations. Most studies to understand these phenomena are based on murine models which do not completely correlate to the biochemistry of the human hemostasis and coagulation systems (Ware, 2004). For instance, the viscosity of mouse blood is typically lower than human which bears an important influence on the behavior of platelets (Windberger et al., 2003). Certain receptor–ligand interactions described in murine models are disparate to the human biochemistry, such as those involved in PAR (protease-activated receptors) and thrombin signaling (Ware, 2004). Therefore, there is value in assessing these interactions using a human–human model. The presented WPM devices have been shown to be well suited for growth of cellular monolayers as the inlet well of the plate functions as a reservoir for fresh cell culture media. Cells can be grown under active perfusion (endothelial cells) or light gravity-driven perfusion to successful outcomes. Once a confluent monolayer is established within the microfluidic channel, then whole blood can be perfused over the monolayer. Because the well plate devices are self-contained, this enables high-throughput experimental setup whereby many plates of cells can be prepared for experiments in advance and then used as needed. This ability confers a significant dimension of function over simply perfusing blood over a substrate; high content data can be captured from both the platelet and the substrate monolayer.

Overall, the benefits of microfluidics are put to good use for these assays, especially the use of high-density experimental arrays and low-volume consumption. Comparable assays which utilize traditional parallel plate flow chambers tend to use milliliters of blood per test [11]. This essentially eliminates the possibility of large clinical screens and single animal murine studies. The designs presented within this study use less than 1 ml of blood per channel for a thrombosis assay with each chip containing 24 assays run in parallel.

In conclusion, the effects of different factors on platelet adhesion, aggregation, and thrombus formation provide a higher physiological relevance if studied under shear flow. In this context, WPM have been successfully used to determine the effects of either genetic mutations or small molecules on thrombus formation (Figure 15.7A and B, respectively).

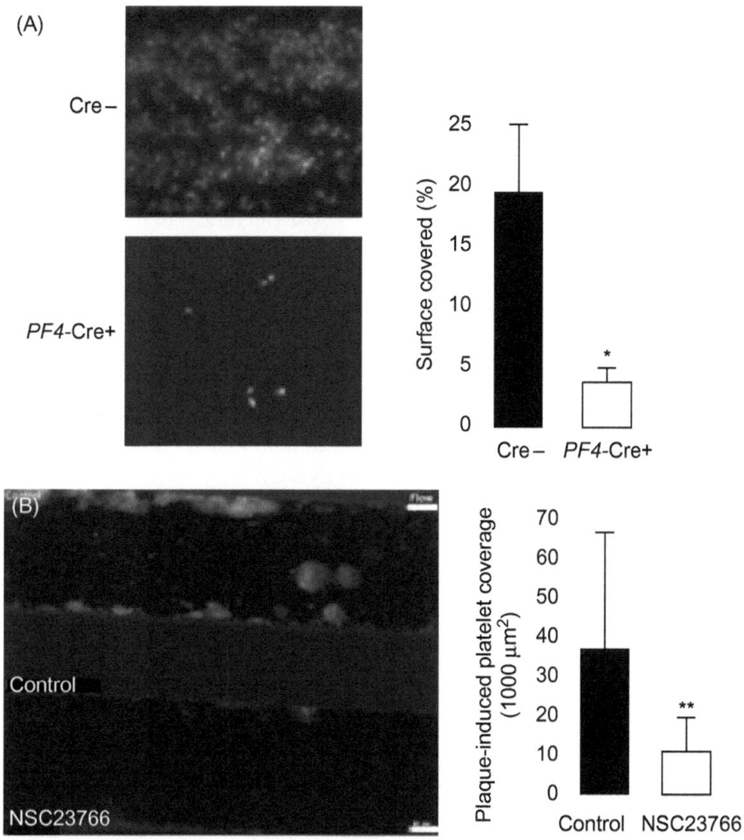

FIGURE 15.7

Platelet aggregation modulation under flow (A) DIOC6-labeled platelets in whole blood from p110flox/flox control (Cre) or *PF4*-Cre/p110flox/flox (PF4-Cre) mice were perfused through fibrinogen-coated BioFlux plates at a shear rate of 1500 s^{-1} for 5 min. Representative images at 5 min are shown. Images obtained with an epifluorescence microscope (Axiovert 200, Carl Zeiss; 40/1.3 NA objective; 37°C) were captured with a CCD camera (CoolSnap HQ; Roper Scientific) and Metamorph software Version 6.2r6 (Universal Imaging Corp). Area covered by platelets was measured. Results shown are the mean SEM of five experiments. Statistical analysis: $^{*}P < 0.005$. (B) Effect of NSC23766 on atherosclerotic plaque-induced platelet thrombus formation under arterial flow conditions. Hirudin anticoagulated blood preincubated with H_2O or with NSC23766 (300 μM) for 5 min was perfused over plaque-coated surfaces for 10 min at 37°C at a shear rate of 1500 s^{-1}. (A) Representative flow images of control (upper channel) and NSC23766 treated blood (lower channel) 10 min after flow initiation; platelets are visualized by mepacrine fluorescence; bar diagram (values are mean ± SD; $n = 5$). $^{**}P < 0.002$.

Source: *(A) Adapted with permission from Ref. [32]. (B) Adapted with permission from Ref. [33].*

15.3.4 Rolling and static cell adhesion

Cell adhesion through protein–protein or protein–carbohydrate interactions plays an important role in directing cell migration, gene expression, cell growth, morphological changes during development, and programmed cell death. It plays a particularly important role in immune response, where adhesion of lymphocytes regulates a number of processes in the inflammation cascade.

We exploited the interaction between VLA-4 on Jurkat cells and recombinant VCAM-1, a cell adhesion molecule, to model cell rolling and adhesion assays in the BioFlux system. VCAM-1 is a cell adhesion molecule expressed on the endothelial surface during inflammation. VLA-4 is a 1 integrin expressed on the surface of lymphocytes, monocytes, and eosinophils. VLA-4 has been shown to interact with VCAM-1 in the context of lymphocyte homing to inflammation. Jurkat cells rolled on VCAM-1 at a shear of 1 dyn/cm^2 (Figure 15.8) and ultimately attached to the substrate. After firm attachment, cells could be removed from the surface between shears of 7 and 30 dyn/cm^2.

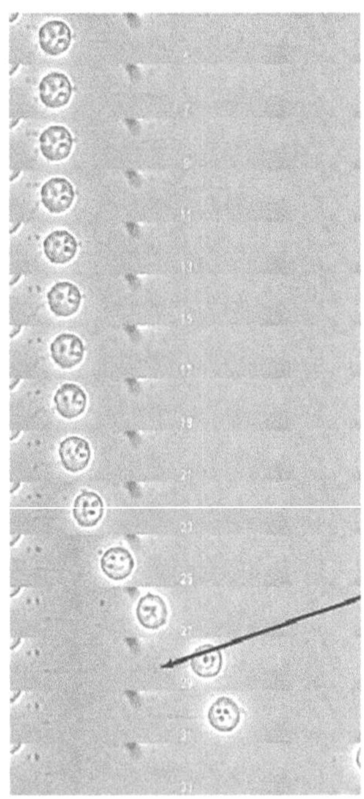

FIGURE 15.8

Cell rolling from the tethered state. Cells were introduced into channels under variable flow conditions to assess rolling behavior. Cells were relieved from shear for 5 min and then shear was ramped up from 2 to 36 dyn/cm^2. Cell tethers are indicated (black arrow). Frame number is shown in white and encompasses shear from ~7 to ~28 dyn/cm^2.

Static cell adhesion is also a research area, and the molecular interactions governing adhesion between white blood cells and cancer cells and the cells of the endothelium are the focus of targeted therapeutic drug development. Cell adhesion molecules can be co-opted by other cell types in the body, contributing to inflammation and metastatic disease. In an example study, the authors used the VLA-4–VCAM-1 cell adhesion model as the basis for a rapid, higher throughput adhesion inhibition screen of monoclonal antibodies against VLA-4. Using the WPM device, they generated IC50 dose–response curves 96 times faster than conventional flow cells (Figure 15.9). The same WPM device was also used to study transmigration of mononuclear cells through endothelial cell monolayers. Twenty-four channels of static adhesion and transmigration data were generated in a single experiment.

15.4 Microbiology applications

15.4.1 Bacterial adhesion and biofilm development

A key aspect of microbiology research is bacterial colonization of various physiological environments, which is critically dependent on bacterial adhesion to surfaces. As an example, in the oral cavity, colonization of the tooth surface involves adhesin-mediated interactions between different species of bacteria, and between bacteria and salivary components adsorbed onto the acquired enamel pellicle [35]. The dynamic adhesion of wild type *Streptococcus gordonii* DL1 to glass coated with different concentrations of fetuin or 25% saliva was determined for various shear levels. The study detected the presence of shear-enhanced binding of *S. gordonii* to adsorbed sialic acid-containing receptors raises the possibility suggested a set of possible binding mechanisms and may contribute to the persistence of streptococci in a turbulent environment (Figures 15.10 and 15.11)

Another important area of research is the development and suppression of bacterial biofilms. Bacterial biofilms comprise communities of bacteria attached to a surface and encased in a polymeric matrix. Although biofilms were identified as early as the 1930s [37], the importance of the biofilm mode of growth has only been recognized relatively recently [38]. Biofilms are responsible for biofouling of industrial water systems, biocorrosion of metals, and many chronic diseases including cystic fibrosis, otitis media [18], urinary tract infections, chronic wounds, and infections associated with medical implants. Diseases with a biofilm component are often difficult to eradicate because they resist antibiotic treatment and clearance by the host immune system. However, biofilms also have useful roles, for example, in wastewater treatment, bioremediation of heavy metals, and microbial fuel cells. Thus, understanding differences between bacterial biofilms and free-living (planktonic) bacteria is important. Historically, study of biofilms has lagged behind that of planktonic bacteria, partly due to the need for techniques suitable for biofilm study in a laboratory. Recent technological advances

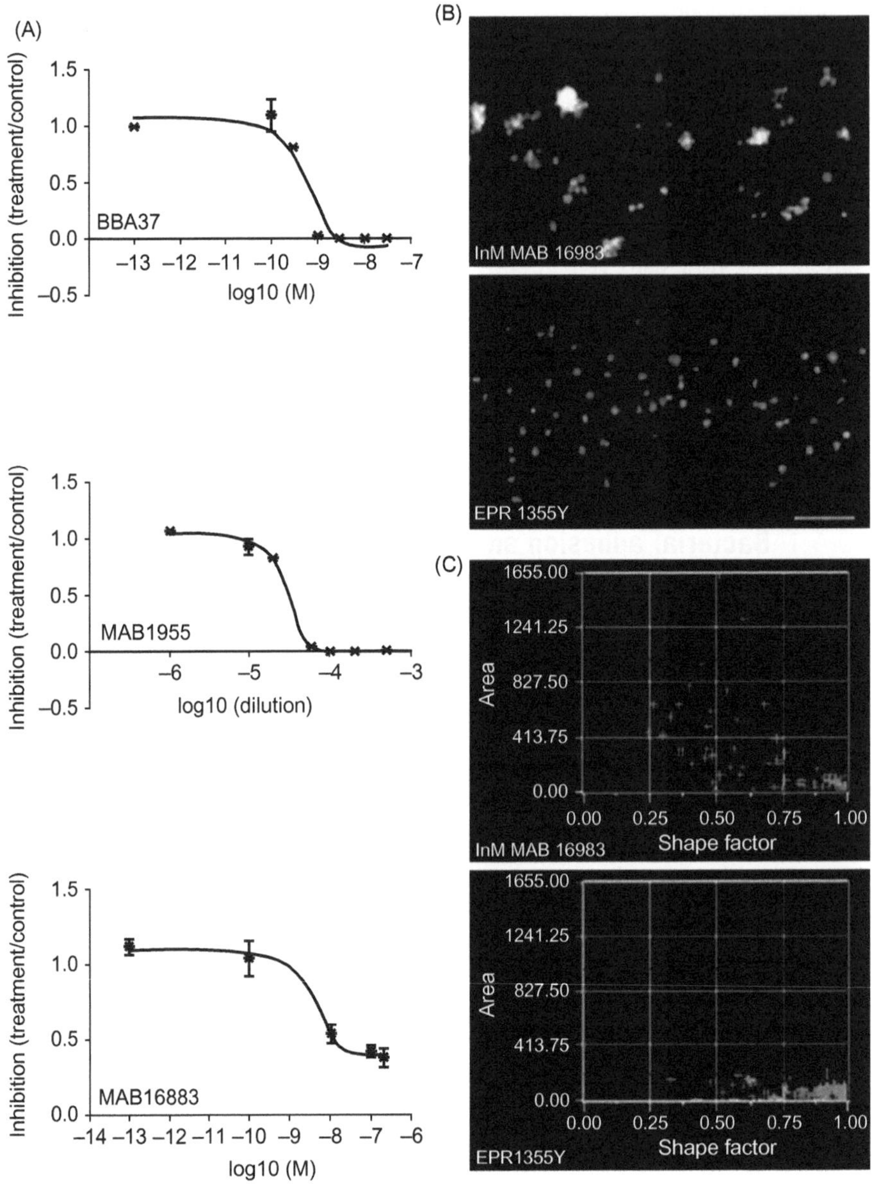

FIGURE 15.9

Static adhesion under flow. Cell adhesion to VCAM-1 following treatment with anti-VLA-4 antibodies. (A) Inhibition level was expressed as number of cells adhering after treatment over number of cells attached in control samples. Error bars represent standard error of the mean for three experiments. IC50 values are shown. (B) Homotypic aggregation was

(*Continued*)

have allowed for detailed study of biofilm growth; however, only limited high-throughput screening methods exist for biofilms.

The traditional high-throughput method for assessing biofilm formation involves crystal violet (CV) staining of biofilms grown in microtiter wells, followed by rinsing, then eluting the CV with ethanol or acetic acid [39]. The remaining CV is then read with a spectrophotometer. This method can provide information regarding the amount of biofilm material but does not give details regarding the number or viability of bacteria within the biofilm. Wheat germ agglutinin–Alexa Fluor 488 conjugate has also been used for staining biofilms, which evidently is more specific than CV staining because the conjugate selectively binds to *N*-acetylglucosamine residues in biofilms. CV staining can also be performed in parallel with a respiratory indicator, giving an indirect indication of viability. However, the method requires two screens performed in parallel and respiratory activity is not always a good indication of viability, especially for persister cells, which are known to play a role in biofilm survival. The Calgary Biofilm Device (CBD) consists of plastic pegs that are immersed within microtiter biofilms [40]. Biofilms form on the pegs, which can be removed by sonication and enumerated counting colony forming units (CFUs). Advantages of the CBD include the option for shear flow over the pegs (by rocking the plate back and forth) and information about cell viability. Disadvantages include the lack of fresh medium to accompany shear flow and CFU counting, which is slow and laborious.

We sought to take advantage of the LIVE/DEAD® BacLight™ stain from Molecular Probes, which is commonly used to assess bacterial cell viability of planktonic cultures. The stain functions by using two fluorescent dyes: green for viable cells and red for dead ones. The BacLight stain has previously been used to visualize bacterial biofilms with confocal laser scanning microscopy (CLSM) by the Matin lab and others. The red and green signal intensity from the CLSM images can then be used to quantify the percentage of viable cells. BacLight has previously been used in a fluorescence-based assay to determine viability of dental plaque biofilms. However, the assay suffered from a critical flaw by not taking into account the number of bacteria remaining in the biofilms after treatment. Bacteria are commonly removed from microtiter well biofilms during treatment and rinsing; without accounting for this removal, fluorescence values become unreliable. This was likely a contributing factor as to why the method could not be correlated with epifluorescence imaging.

◀ elicited with MAB16983 at 1 nM. Jurkat cells stained with calcein-AM treated with 1 nM MAB16983 (top) or an antibody against an intracellular epitope of VLA-4 (EPR1355Y; bottom) are shown. Scale bar is 75 μm. (C) Cell morphology metrics plotted for 1 nM MAB16983-treated cells (top) and EPR1355Y (bottom) are shown. A value of 1 for shape factor denotes a perfect circle, and units of area are pixels^2.

Source: *Adapted with permission from Ref. [34].*

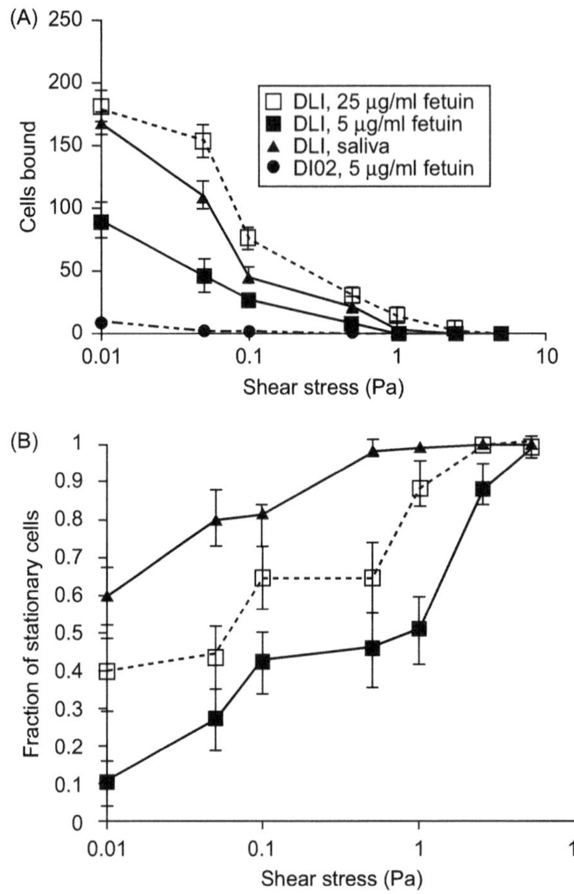

FIGURE 15.10

Microbial adhesion under flow. (A) Dynamic adhesion of wild type *S. gordonii* DL1 to glass coated with different concentrations of fetuin or 25% saliva. Cell suspensions of this strain or Hsa-deficient mutant strain D102 were pumped through channels of a BioFlux 100 plate at flow rates corresponding to shear forces ranging from 5 to 0.01 Pa. Total cells (i.e., rolling plus stationary) bound per field of view were determined after 5 min at each shear force. Mean values for cells bound per field of view and exact 95% confidence intervals for a Poisson variable were calculated from six repetitions. (B) Shear-enhanced adhesion of *S. gordonii* DL1. Cell suspensions were pumped through channels of a BioFlux 100 plate at a shear force of 0.01 Pa for 5 min to allow accumulation of bound cells as in panel A. Shear force was then set at a value ranging from 0.01 to 5 Pa, as indicated on the logarithmic *X*-axis. After 5 min at each shear force, the fraction of stationary cells was determined by counting stationary and rolling cells. Mean values for the fraction of stationary cells and exact 95% confidence intervals for a binomial distribution were calculated from 10 repetitions.

Source: *Adapted with permission from Ref. [35].*

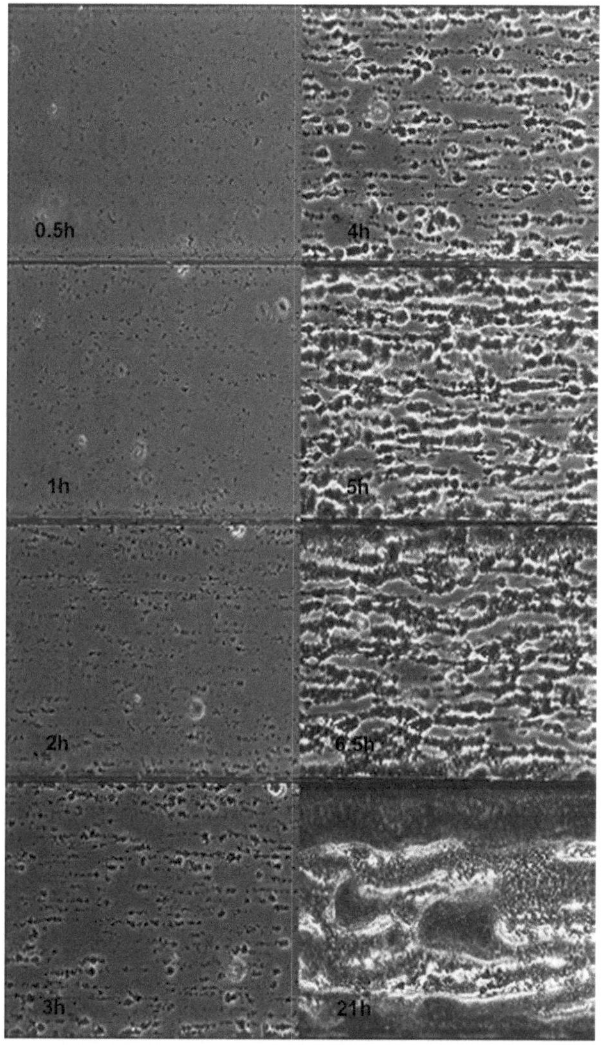

FIGURE 15.11

Biofilm development. Biofilm formation and development in a BioFlux channel. The images are phase-contrast images obtained after flow began (the flow was from left to right; each channel was 370 μm wide) and show initial attachment (0.5 and 1 h), microcolony formation (2 and 3 h), and development (4, 5, and 6.5 h) into fully formed biofilms (21 h).

Source: *Adapted with permission from Ref. [36].*

Furthermore, no other method was used to confirm the results. By taking into account the amount of bacteria remaining in microtiter biofilms after treatment, we were able to accurately determine percent survival, which closely resembled CFU counts.

In addition to the microtiter plate biofilm assay, we characterized a novel device for screening flow-through biofilms. Microtiter plate biofilms are static, i.e., they are not continually replaced with fresh medium. Thus, the results must later be confirmed with flow-through (i.e., flow-cell) experiments. In some cases, results between microtiter and flow-cell environments do not correlate [41], illustrating the importance for performing screens with flow-through biofilms. The device allows for biofilms to be grown in microfluidic channels, treated, and stained with BacLight (or any fluorescent stain). The biofilms can then be scanned with a fluorescence plate reader, and the green and red fluorescence can be quantified and used to estimate viability. To validate our results, CLSM was used to image the biofilms. The green and red fluorescence from the resulting images was quantified and showed a high correlation with the plate-reader results. The two methods described here should prove useful for testing compound libraries against biofilms and for probing genes involved in biofilm resistance mechanisms by screening mutant libraries.

15.4.2 Biofilm compound screening

Control of biofilms requires rapid methods to identify compounds effective against them and to isolate resistance-compromised mutants for identifying genes involved in enhanced biofilm resistance. While rapid screening methods for microtiter well plate ("static") biofilms are available, there are no methods for such screening of continuous flow biofilms ("flow biofilms"). Since the latter biofilms more closely approximate natural biofilms, development of a high-throughput method for screening them is desirable. We describe here a new method using a device comprised of microfluidic channels and a distributed pneumatic pump (BioFlux) that provides fluid flow to 96 individual biofilms. This device allows fine control of continuous or intermittent fluid flow over a broad range of flow rates, and the use of a standard well plate format provides compatibility with plate readers. We show that use of green fluorescent protein (GFP)-expressing bacteria, staining with propidium iodide, and measurement of fluorescence with a plate reader permit rapid and accurate determination of biofilm viability. The biofilm viability measured with the plate reader agreed with that determined using plate counts, as well as with the results of fluorescence microscope image analysis. Using BioFlux and the plate reader, we were able to rapidly screen the effects of several antimicrobials on the viability of Pseudomonas aeruginosa PAO1 flow biofilms (Figure 15.12).

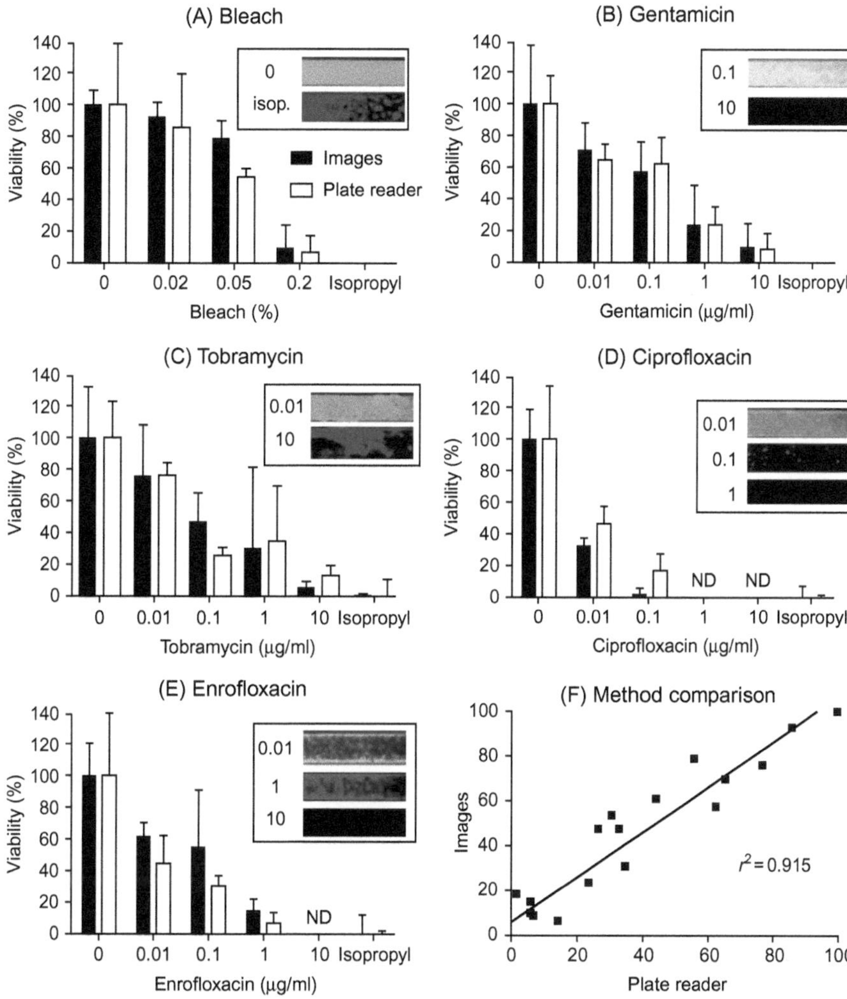

FIGURE 15.12

Biocide screening. Estimates of the viability of GFP-producing *P. aeruginosa* PAO1 biofilms cultivated in the BioFlux device following treatment with bleach (A) or various antibiotics (B–E) and then stained with PI. Levels of viability were calculated by using the green fluorescence/red fluorescence ratios determined from quantified microscope images (black bars) and by using plate-reader measurements (white bars). Isopropyl alcohol (70%) was used to ensure complete loss of viability. Insets show representative microscope images, including images of untreated (0) and isopropyl-alcohol treated (isop.) controls. Note that no biofilm remained after treatment with the highest dose of enrofloxacin (E) and the two highest doses of ciprofloxacin (D). Fluorescence signals were not detected (ND) with the plate reader, and no biofilms were observed using microscopy (insets). (F) Linear regression analysis for determinations of viability by the two methods.

Source: *Adapted with permission from Ref. [36].*

15.5 Summary

The effect of shear on live cells is well documented in developmental biology, vascular biology, cancer biology, and immunology. An important consideration is the dynamic nature of many of these processes. For example, recruitment of a white blood cell to the site of a tissue injury occurs as a series of events that may include cell rolling, arrest, and transmigration. This cannot be studied in a static system due to the inherent influence of blood flow. Likewise development of stem cells occurs under many types of stimuli, both chemical and mechanical. Mesenchymal stem cells can be differentiated to endothelial cells in the presence of shear flow.

Despite the documented importance shear, an overwhelming majority of *in vitro* cell biology experiments are conducted in static conditions. This is largely due to the laboriousness and high cost of performing cell biology experiments under flow using traditional macroscopic laminar flow chambers.

This chapter presents emerging research using well plate microfluidics technology and the BioFlux system; the approach allows an increase in throughput of the shear-based systems by over an order of magnitude and reduces cost dramatically by 10–100 fold reduction in the amount of reagent required per experiment, which is important where *ex vivo* samples such as whole blood are being used. The laborious and low-throughput nature of traditional flow assays are reduced, making shear-based assays more practical and dramatically faster.

Validation and original research data is presented for a number of different applications: In cell biology, the determination of shear regime effects on adherent cell development was followed by stem cell differentiation in the presence of shear forces. A large portion of the research presented deals with blood physiology, namely platelet activation, adhesion, and aggregation into thrombi, both static and rolling leukocyte adhesion as a part of the inflammation cascade. Finally, for microbiology, shear-mediated adhesion to surfaces is studied as important step in the bacterial colonization; development of bacterial biofilms necessitates the presence of shear and shear-based experiments and can be used to provide important insights into the antibiotic resistance of bacterial biofilm colonies.

In conclusion, the use of well plate microfluidic technology represents an important step forward in improving the relevance of *in vitro* models for *in vivo* biological processes by enabling controlled shear stress experiments to be run with a high degree of parallelism.

References

[1] C. Gaucher, C. Devaux, C. Boura, P. Lacolley, J.F. Stoltz, P. Menu, *In vitro* impact of physiological shear stress on endothelial cells gene expression profile, Clin. Hemorheol. Microcirc. 37 (2007) 99–107.

[2] N. Gomes, M. Berard, J. Vassy, N. Peyri, C. Legrand, F. Fauvel-Lafeve, Shear stress modulates tumour cell adhesion to the endothelium, Biorheology 40 (2003) 41–45.

[3] Z.M. Ruggeri, J.N. Orje, R. Habermann, A.B. Federici, A.J. Reininger, Activation-independent platelet adhesion and aggregation under elevated shear stress, Blood 108 (2006) 1903–1910.

[4] H. Wang, G.M. Riha, S. Yan, M. Li, H. Chai, H. Yang, et al., Shear stress induces endothelial differentiation from a murine embryonic mesenchymal progenitor cell line, Arterioscler. Thromb. Vasc. Biol. 25 (2005) 1817–1823.

[5] K. Yamamoto, T. Sokabe, T. Watabe, K. Miyazono, J.K. Yamashita, S. Obi, et al., Fluid shear stress induces differentiation of Flk-1-positive embryonic stem cells into vascular endothelial cells *in vitro*, Am. J. Physiol. Heart Circ. Physiol. 288 (2005) H1915–H1924.

[6] L. Zeng, Q. Xiao, A. Margariti, Z. Zhang, A. Zampetaki, S. Patel, et al., HDAC3 is crucial in shear- and VEGF-induced stem cell differentiation toward endothelial cells, J. Cell. Biol. 174 (2006) 1059–1069.

[7] P. Alcaide, S. Auerbach, F.W. Luscinskas, Neutrophil recruitment under shear flow: it's all about endothelial cell rings and gaps, Microcirculation 16 (2009) 43–57.

[8] J. Oswald, S. Boxberger, B. Jorgensen, S. Feldmann, G. Ehninger, M. Bornhauser, et al., Mesenchymal stem cells can be differentiated into endothelial cells *in vitro*, Stem Cells. 22 (2004) 377–384.

[9] S.P. Jackson, The growing complexity of platelet aggregation, Blood 109 (2007) 5087–5095.

[10] D. Varon, N. Savion, Impact cone and plate(let) analyzer, in: A.D. Michelson (Ed.), Platelets, Academic Press, Burlington, MA, 2007.

[11] P.G. De Groot, J.J. Sixma, Perfusion chambers, in: A.D. Michelson (Ed.), Platelets, Academic Press, Burlington, MA, 2007.

[12] S. Kulkarni, W.S. Nesbitt, S.M. Dopheide, S.C. Hughan, I.S. Harper, S.P. Jackson, Techniques to examine platelet adhesive interactions under flow, Methods Mol. Biol. 272 (2004) 165–186.

[13] E. Gutierrez, B.G. Petrich, S.J. Shattil, M.H. Ginsberg, A. Groisman, A. Kasirer-Friede, Microfluidic devices for studies of shear-dependent platelet adhesion, Lab Chip 8 (2008) 1486–1495.

[14] C. Minelli, A. Kikuta, N. Tsud, M.D. Ball, A. Yamamoto, A micro-fluidic study of whole blood behaviour on PMMA topographical nanostructures, J. Nanobiotechnol. 6 (2008) 3.

[15] K.B. Neeves, S.L. Diamond, A membrane-based microfluidic device for controlling the flux of platelet agonists into flowing blood, Lab Chip 8 (2008) 701–709.

[16] K.B. Neeves, S.F. Maloney, K.P. Fong, A.A. Schmaier, M.L. Kahn, L.F. Brass, et al., Microfluidic focal thrombosis model for measuring murine platelet deposition and stability: PAR4 signaling enhances shear-resistance of platelet aggregates, J. Thromb. Haemost. 6 (2008) 2193–2201.

[17] M. Schwartz, BioFlux technology white paper, <http://www.fluxionbio.com/library-c1-Product_Information.aspx>, 1 December, 2008.

[18] C.G. Conant, M.A. Schwartz, J.E. Beecher, R.C. Rudoff, C. Ionescu-Zanetti, J.T. Nevill, Well plate microfluidic system for investigation of dynamic platelet behavior under variable shear loads, Biotechnol. Bioeng. (2011).

[19] Y. Xia, E.W. Whitesides, Soft lithography, Annu. Rev. Mater. Sci. 28 (1998) 158–184.

[20] H. Bruus, Theoretical microfluidics, Oxford University Press, 17 November, 2007.
[21] J.C. Gebhardt, M. Rief, Biochemistry. Force signaling in biology, Science 324 (2009) 1278–1280.
[22] S.W. Schneider, S. Nuschele, A. Wixforth, C. Gorzelanny, A. Alexander-Katz, R.R. Netz, et al., Shear-induced unfolding triggers adhesion of von Willebrand factor fibers, Proc. Natl. Acad. Sci. U.S.A. 104 (2007) 7899–7903.
[23] X. Zhang, K. Halvorsen, C.Z. Zhang, W.P. Wong, T.A. Springer, Mechanoenzymatic cleavage of the ultralarge vascular protein von Willebrand factor, Science 324 (2009) 1330–1334.
[24] M.H. Kroll, J.D. Hellums, L.V. McIntire, A.I. Schafer, J.L. Moake, Platelets and shear stress, Blood 88 (1996) 1525–1541.
[25] D. Bluestein, L. Niu, R.T. Schoephoerster, M.K. Dewanjee, Fluid mechanics of arterial stenosis: relationship to the development of mural thrombus, Ann. Biomed. Eng. 25 (1997) 344–356.
[26] A. Mailhac, J.J. Badimon, J.T. Fallon, A. Fernandez-Ortiz, B. Meyer, J.H. Chesebro, et al., Effect of an eccentric severe stenosis on fibrin(ogen) deposition on severely damaged vessel wall in arterial thrombosis. Relative contribution of fibrin(ogen) and platelets, Circulation 90 (1994) 988–996.
[27] J. Strony, A. Beaudoin, D. Brands, B. Adelman, Analysis of shear stress and hemodynamic factors in a model of coronary artery stenosis and thrombosis, Am. J. Physiol. 265 (1993) H1787–H1796.
[28] C.J. Ku, T. D'Amico Oblak, D.M. Spence, Interactions between multiple cell types in parallel microfluidic channels: monitoring platelet adhesion to an endothelium in the presence of an anti-adhesion drug, Anal. Chem. 80 (2008) 7543–7548.
[29] S.F. Maloney, L.F. Brass, S.L. Diamond, P2Y12 or P2Y1 inhibitors reduce platelet deposition in a microfluidic model of thrombosis while apyrase lacks efficacy under flow conditions, Integr. Biol. (Camb.) 2 (2010) 183–192.
[30] U.M. Okorie, S.L. Diamond, Matrix protein microarrays for spatially and compositionally controlled microspot thrombosis under laminar flow, Biophys. J. 91 (2006) 3474–3481.
[31] D.P. Sarvepalli, D.W. Schmidtke, M.U. Nollert, Design considerations for a microfluidic device to quantify the platelet adhesion to collagen at physiological shear rates, Ann. Biomed. Eng. 37 (2009) 1331–1341.
[32] V. Martin, J. Guillermet-Guibert, G. Chicanne, C. Cabou, M. Jandrot-Perrus, M. Plantavid, et al., Deletion of the p110 isoform of phosphoinositide 3-kinase in platelets reveals its central role in Akt activation and thrombus formation *in vitro* and *in vivo*, Blood 115 (10) (2010) 2008–2013.
[33] S. Dwivedi, D. Pandey, A.L. Khandoga, R. Brandl, W. Siess, Rac1-mediated signaling plays a central role in secretion-dependent platelet aggregation in human blood stimulated by atherosclerotic plaque, J. Trans. Med. 8 (1) (2010) 128.
[34] C.G. Conant, M.A. Schwartz, C. Ionescu-Zanetti., Well plate-coupled microfluidic devices designed for facile image-based cell adhesion and transmigration assays, J. Biomol. Screen 15 (1) (2010) 102–106.
[35] A.M. Ding, R.J. Palmer, J.O. Cisar, P.E. Kolenbrander, Shear-enhanced oral microbial adhesion, Appl. Environ. Microbiol. 76 (4) (2010) 1294–1297.
[36] M.R. Benoit, C.G. Conant, C. Ionescu-Zanetti, M. Schwartz, A. Matin, New device for high-throughput viability screening of flow biofilms, Appl. Environ. Microbiol. 76 (13) (2010) 4136–4142.

[37] A.T. Henrici, Studies of freshwater bacteria: I. A direct microscopic technique, J. Bacteriol. 25 (3) (1933) 277–287.

[38] L. Hall-Stoodley, J.W. Costerton, P. Stoodley, Bacterial biofilms: from the natural environment to infectious diseases, Nat. Rev. Microbiol. 2 (2) (2004) 95–108.

[39] G.A. O'Toole, R. Kolter, Initiation of biofilm formation in *Pseudomonas fluorescens* WCS365 proceeds via multiple, convergent signalling pathways: a genetic analysis, Mol. Microbiol. 28 (3) (1998) 449–461.

[40] H. Ceri, M.E. Olson, C. Stremick, R.R. Read, D. Morck, A. Buret, The calgary biofilm device: new technology for rapid determination of antibiotic susceptibilities of bacterial biofilms, J. Clin. Microbiol. 37 (6) (1999) 1771–1776.

[41] A. Ito, T. May, K. Kawata, S. Okabe, Significance of rpoS during maturation of *Escherichia coli* biofilms, Biotechnol. Bioeng. 99 (6) (2008) 1462–1471.

[42] G. Dai, M.R. Kaazempur-Mofrad, S. Natarajan, Y. Zhang, S. Vaughn, B.R. Blackman, et al., Distinct endothelial phenotypes evoked by arterial waveforms derived from atherosclerosis-susceptible and -resistant regions of human vasculature, Proc. Natl. Acad. Sci. U.S.A. 101 (41) (2004) 14871–14876.

[43] E.Y. Fok, P.W. Zandstra, Shear-controlled single-step mouse embryonic stem cell expansion and embryoid body-based differentiation, Stem. Cells. 23 (9) (2005) 1333–1342.

[44] W.S. Nesbitt, E. Westein, F.J. Tovar-Lopez, E. Tolouei, A. Mitchell, J. Fu, et al., A shear gradient-dependent platelet aggregation mechanism drives thrombus formation, Nat. Med. 15 (6) (2009) 665–673.

[45] B. Savage, J.J. Sixma, Z.M. Ruggeri, Functional self-association of von Willebrand factor during platelet adhesion under flow, Proc. Natl. Acad. Sci. U.S.A. 99 (1) (2002) 425–430.

[46] B. Illi, A. Scopece, S. Nanni, A. Farsetti, L. Morgante, P. Biglioli, et al., Epigenetic histone modification and cardiovascular lineage programming in mouse embryonic stem cells exposed to laminar shear stress, Circ. Res. 96 (5) (2005) 501–508.

[47] U. Windberger, A. Bartholovitsch, R. Plasenzotti, K.J. Korak, G. Heinze, Whole blood viscosity, plasma viscosity and erythrocyte aggregation in nine mammalian species: reference values and comparison of data, Exp. Physiol. 88 (3) (2003) 431–440.

[48] D. Schumann, R. Kujat, M. Nerlich, P. Angele, Mechanobiological conditioning of stem cells for cartilage tissue engineering, Biomed. Mater. Eng. 16 (4 Suppl) (2006) S37–S52.

[49] J. Ware, Thromb dysfunctional platelet membrane receptors: from humans to mice., Haemost. 92 (3) (2004) 478–485.

CHAPTER

16 Microfluidic Platforms for Evaluating Angiogenesis and Vasculogenesis

Jessie S. Jeon[1], Seok Chung[2] and Roger D. Kamm[3]

[1]*Department of Mechanical Engineering, Massachusetts Institute of Technology, Cambridge, MA, USA*

[2]*School of Mechanical Engineering, Korea University, Korea*

[3]*Cecil and Ida Green Distinguished Professor of Biological and Mechanical Engineering, Massachusetts Institute of Technology, Cambridge, MA, USA*

16.1 Introduction

Angiogenesis, the development of new blood vessels from preexisting vasculature, and vasculogenesis, the development of vascular networks from endothelial precursor cells are essential events in development and in a vast range of physiologic and pathologic conditions. And despite the existence of numerous routine procedures to culture endothelial cells under various culture conditions, our ability to study the angiogenic process *in vitro* has progressed only slowly.

One of the primary motivations for the *in vitro* simulation of angiogenesis stems from the seminal work of Dr. Judah Folkman who identified suppression of tumor vascularization as a new approach to the treatment of cancer. Indeed, tumor growth and metastatic dissemination are critically dependent on the tumor's blood supply [1]. And while the promise of antiangiogenic therapies has yet to be fully realized, interest remains high in the identification of new factors that either promote or inhibit the formation of new microvascular networks. Angiogenesis also plays an important role more generally in vascular biology because it is an essential process that occurs in various normal and pathological events. Wound healing and the reperfusion of ischemic regions following myocardial infarct or stroke are cases where angiogenesis needs to be stimulated. Cancer and diabetes are cases in which antiangiogenic therapies are targeted. In addition to these, there are more than 70 disorders that are directly or indirectly linked with angiogenesis, and the list of conditions continues to grow [2]. A better understanding of the biology of angiogenesis (depicted in Figure 16.1) may reveal new targets for treating these diseases.

There are multiple steps involved in angiogenesis, often classified as vasodilation, endothelial cell proliferation and migration, survival/maturation, and

Microfluidic Cell Culture Systems. ISBN: 978-1-4377-3459-1

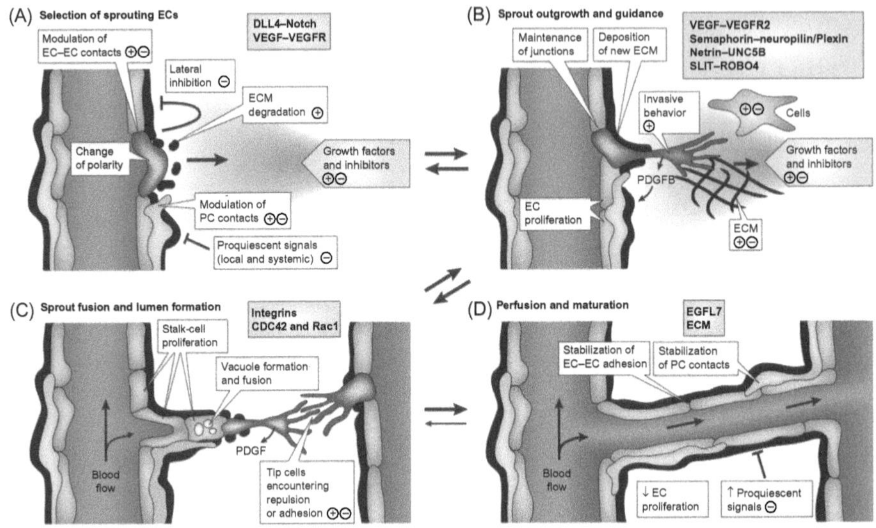

FIGURE 16.1

The progressive sequence by which a new vessel is formed. (A) Endothelial cells (ECs) in a preexisting vessel locally alter their phenotype, change their polarity, and begin to degrade matrix locally, setting the stage for invasion into the surrounding matrix. (B) These stimulated cells become tip cells and start to sprout out into the surrounding ECM. (C) As the tip cell proceeds, it recruits other cells from the original vessel that become the stalk cells, later forming a lumen as more sprouts form, fuse, and cells proliferate. (D) Finally, ECs stably adhere to each other and mature into a perfusable vessel.

Source: *Adapted by permission from Macmillan Publishers Ltd: Ref. [3], copyright 2007.*

remodeling [4] (Figure 16.1). It has long been known that hypoxic conditions stimulate the local cells to release various factors that induce the recruitment of new blood vessels. A variety of factors have been identified (Table 16.1), many of which can be traced to the upregulation of hypoxia inducible factor (HIF-1) [7,8]. Tumor growth and metastatic dissemination are critically dependent on the tumor's supply of blood vessels [1], making angiogenesis a putative target for cancer therapy. It is commonly believed that blocking vessel formation into a tumor mass will limit tumor growth as well as metastasis, although the *in vivo* situation is apparently much more complex [5]. Therefore, many inhibitors of angiogenesis are under intense clinical investigation as antiangiogenic factors. Current efforts in antiangiogenic therapies either act by inhibiting one of the various receptors of these factors or by suppressing the degradation of HIF-1, but the approaches are numerous and a need clearly exists for a method that is capable of screening many factors rapidly in an *in vitro* setting that recapitulates many of the features of the *in vivo* microenvironment.

In this chapter, we will explore how angiogenesis is replicated in a microenvironment formed within a microfluidic system, and how these systems can be used

Table 16.1 Angiogenic Activators and Inhibitors as well as their functions are listed [5,6]

Activators/Inhibitors	Functions
VEGF	Stimulate angio/vasculogenesis, permeability, leukocyte adhesion
ANG-1	Stabilize vessels, inhibit permeability
PDGF-BB	Recruit smooth muscle cells
TNF-α	Induces production of bFGF in endothelial cells and enhances its secretion, activates macrophages
TGF-α	Stimulate endothelial mitogenicity
TGF-β	Stimulate ECM production, inhibits endothelial mitogenicity
aFGF, bFGF	Stimulate angio/arteriogenesis
NOS	Stimulate angiogenesis and vasodilation
ANG2	Antagonist of ANG-1
TSP-1, -2	Inhibit endothelial migration, growth, adhesion, and survival
Endostatin	Inhibit endothelial survival and migration
Vasostatin; calreticulin	Inhibit endothelial growth
Platelet factor-4	Inhibit binding of bFGF and VEGF
IFN-α, -β, -γ, IP-10, IL-4, -12, -18	Inhibit endothelial migration, downregulate bFGF
Prothrombin kringle-2; antithrombin III	Suppress endothelial growth

NOS, nitric oxide synthase; TSP, thrombospondin; IFN, interferon; IP, inducible protein; IL, interleukin; other abbreviations are defined in the text.

to investigate the numerous stimuli that are known to induce angiogenesis. Since true angiogenesis *in vivo* involves a coordinated effort of different cell types, the coculturing of multiple cell types will be discussed. These interactions between heterotypic cells not only induce angiogenesis when particular chemokines are secreted, but also help the newly created vessels to mature and stabilize. Finally, we will describe various methods for quantifying the angiogenic response and discuss future directions.

16.2 Current methods in microfluidics

Microfluidics has the potential to overcome many of the practical limitations that have impeded progress in the study of angiogenesis. Among the major advantages of microfluidic systems in studying angiogenesis is that they have the potential to mimic the true three-dimensional (3D) nature of the *in vivo* situation while allowing for high-resolution, multimodal imaging. In addition, because of the ready access to the cell-seeded regions, media can be changed frequently with little disturbance to the cells, or lows can be generated to create the shear stress conditions that are found *in vivo*. And perhaps most importantly, multiple cell types can be seeded into these devices so that they can interact over physiologically relevant length scales. These interactions between endothelial cells and heterotypic cell types not only induce angiogenesis when particular chemokines are secreted but also play a critical role in the maturation of the newly created vessels. Finally, the conditions of culture can be closely regulated in a spatiotemporal manner through access enabled via the microfluidic channels.

16.2.1 Mimicking angiogenesis in microfluidics

16.2.1.1 Microfluidic platform design

In recent years, there has been considerable progress in the development of microfluidic designs to replicate a diverse range of physiological events. Microfabrication technology using soft lithography has enabled tremendous utilization of submicron scale designs for micropatterns, narrow channels, and chambers as well as microscale design to accommodate optimal cell–cell or cell–matrix interactions. Larger designs including arrays of hundreds of micron scale features can also be fabricated using this technique, providing the opportunity of high-throughput testing, but at the cost of added complexity [9,10]. Furthermore, not only can heterotypic cell cocultures be accommodated in certain microfluidic designs, but so can living tissue or biopsy specimens be introduced [11], raising the prospect of even greater realism and the potential to screen for patient-specific therapies.

The three dimensionality of microfluidic platforms arises as a result of the integration of various hydrogels or nanofiber constructs, which allows the matrix regions in the microfluidic system to mimic the role of extracellular matrix (ECM) comprising the microenvironment of cell or tissue samples. The gel regions in a microfluidic device can take several forms, depending on the most suitable design for a particular experiment. One approach that has been proposed incorporates microchannels surrounded on three sides by matrix, and that could contain either tumor cells or other types of cells [12]. These latter cells might serve the purpose of simulating the natural organ or they could be pericytes or smooth muscle cells, introduced for the purpose of creating a more realistic environment for a sprouting or preformed vessel. Another approach that has recently been put into practice introduces matrix on one wall of a channel imprinted in the microfluidic device [13,14] (Figure 16.2). These designs utilize multiple polydimethylsiloxane (PDMS) posts to contain the hydrogel solution within the gel region through the forces of surface tension and thus confine the matrix keeping it separate from the media channels where culture medium is introduced and endothelial cells can be seeded.

The most commonly used hydrogels are Matrigel, collagen type I, or fibrin gel all of which are commercially available [17–21]. In the case of Matrigel, it may be used directly as obtained commercially, while for the others, simple synthesis of at most a few reagents is required for gelation. The introduction of a region of

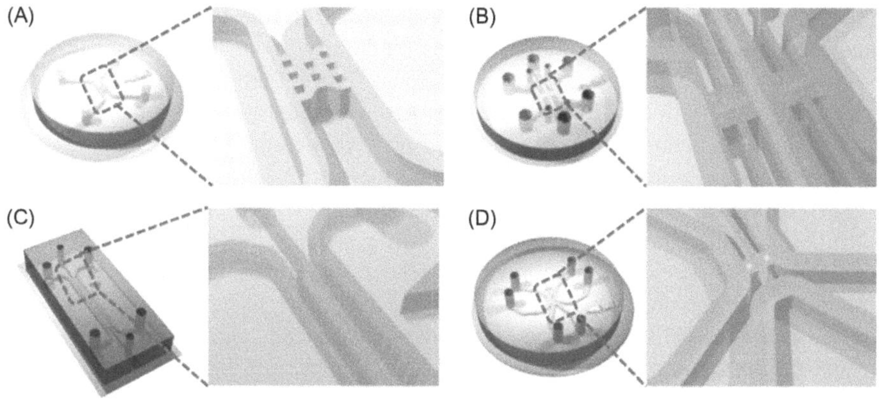

FIGURE 16.2

Examples of a microfluidic platform [13,15,16]. (A) Requires the hydrogel to be injected into the gel region prior to affixing the glass coverslip where as (B)–(D) have gel-filling ports to inject hydrogel. Design (B) provides for direct comparison of the effects of two different media compositions by having three channels, a central one where the endothelial cells are seeded and two side channels for medium, and design (C) incorporates a longer gel region to enable the acquisition of more experimental area. The T-shaped gel region in (D) allows for the study of directional guidance effects due to biochemical gradients.

hydrogel in a microfluidic system allows capillary growth in 3D, thereby producing the most realistic model for angiogenesis *in vivo* and at the same time, allowing spatiotemporal imaging.

Either of these approaches can also be used to recreate *in vitro* the process of vasculogenesis. The primary difference is that endothelial precursor cells would be seeded within the gel, and no endothelial cells would be seeded inside the channels. Channels might not even be necessary, other than to provide fresh media to the system, or to generate a flow through the formed vascular network if desired.

In the following discussion, we focus on the experiments performed in the devices depicted in Figure 16.2 or one of several variations on these basic designs. The first system (Figure 16.2A) is one in which the hydrogel needs to be injected into the gel region prior to affixing the glass coverslip. One advantage of this system is that a linear gradient can be maintained (without the artifacts associated with the gel-filling ports of the other systems) especially when the two channels are merged downstream of the gel region and a slow flow is maintained in both channels. Another advantage is that the gel can be "overfilled" so that it spreads out into the channel slightly, offering a smooth, continuous face to which the cells can adhere. Other systems have gel-filling ports (e.g., Figure 16.2B) that allow for the system to be fully assembled in advance with the gel added at a later time, making them somewhat easier to use. An additional advantage of the system shown in Figure 16.2B is that it provides for an internal control in the sense that the test agent (chemical or cell type) can be placed in one of the side channels, and the opposite side can contain regular medium without the agent. The third system shown (Figure 16.2C) provides for a longer gel region, allowing for greater data collection in a single device. Also, with the greater length, any perturbation in gradients caused by the filling ports at the ends are confined, leaving gradients relatively unperturbed over most of the system length.

For experiments in which another cell type is introduced either into the gel or into the opposite channel with the aim of using secreted factors from these cells to induce angiogenesis, migration of the second cell type into the gel and intermingling with the endothelial sprouts can be an undesired consequence. In such cases, where the cells need to be kept separated, the cells can be encapsulated in a more rigid, more slowly degradable gel, and introduced in encapsulated form either suspended in the gel or flowed into the channel.

For studying chemotaxis effects such as those related to a VEGF gradient in guiding angiogenic tip cells of sprouting networks, a T-shaped gel region can be advantageous (Figure 16.2D) [15] since the turning of the tip cell is then a direct consequence of the biochemical gradient. Chemical gradients can also be created in any of these designs but tend to be time dependent, both during start and as a result of diffusive or convective exchange of the chemoattractant across the gel region during normal operation, especially during media changes when small pressure differences might arise. These fluctuations can be avoided by using a design in which the channels are joined at the downstream end and a constant flow of medium is drawn through both channels simultaneously. In this case, there is a single start-up phase lasting a time that scales with the time for diffusion across the gel region (L^2/D, where L is the distance across the gel and D is the diffusivity of the chemoattractant), and the flows need to be maintained at a level that provides sufficiently high convective effects relative to diffusion, as characterized by the Peclet number, $Pe = VW/D$, where V is the mean flow velocity in the channel, W is the channel width, and D is the diffusivity of chemoattractant in the medium. When the goal is to observe as many sprouting events as possible in a single device, a longer gel area can be employed and as many as 30 experimental regions attained [22].

16.2.1.2 Endothelial cell culture

Methods for cell culture in microfluidic systems are similar across different cell types. Cells can either be seeded in the channels once the gel has been introduced and solidified, or mixed with gel solution prior to injection and gelation. Once seeded in the channel, cells attach to the channel walls including the gel surface and begin to form a confluent monolayer. This can take on the order of 24 h, depending on the nature of surface treatment and the cell seeding density. As a gel region separates the channels, different cell types cultured in independent channels can be observed and studied.

The most commonly used types of endothelial cells are human umbilical vascular endothelial cells (hUVECs) and human microvascular endothelial cells (hMVECs). On average, after 2 days of culture with 2×10^6 cells/ml seeding density, a continuous endothelial monolayer forms and covers the entire channel [13,23]. Initially a monolayer covering the gel scaffold, endothelial cells then later form sprouts into the hydrogel leading to angiogenesis.

The culture is maintained with daily replenishment of endothelial supplemental medium, and with application of chemical or physical stimuli, *in vivo* like angiogenesis is induced. Typically single or multiple growth factors can be introduced into the scaffold as chemical stimuli, causing the endothelial cells to sprout and eventually form a lumen structure. Factors controlling this process are discussed next.

16.2.2 Inducing angiogenesis

The first isolation of angiogenic factors from tumor cells occurred in 1970s. Since then, much effort in vascular biology has been directed toward identifying other possible biochemical as well as biophysical factors that are thought to promote angiogenesis and to understand their significance in forming vascular networks [24,25]. The following sections discuss different ways that angiogenesis can be induced and regulated.

16.2.2.1 Chemical factors

Creation and maintenance of chemical gradients in a microfluidic platform of the type described above has been confirmed [14]. Throughout the culture period, daily replenishment of medium containing the chemical factor of interest produces a stable linear gradient that allows for the systematic investigation of each factor. As mentioned earlier, active research to identify biochemical factors inducing angiogenesis resulted in great advancement of determining and analyzing both pro- and antiangiogenic factors. While the vascular endothelial growth factor (VEGF) family has been recognized as the most potent inducer of angiogenesis, there are many others including angiopoietins, fibroblast growth factors (FGF), transforming growth factors (TGF), platelet-derived growth factors (PDGF), tumor necrosis factor (TNF), and interleukins [26–32]. Table 16.1 is a partial list of factors that are known to be involved in angiogenesis whether their functions are activation or inhibition.

Among the angiogenic factors identified in Table 16.1, members of VEGF family are the best characterized [4,15,33,34]. VEGF stimulates VEGF receptor (VEGFR) which is a tyrosine kinase receptor causing endothelial cells to be activated. The signaling cascade then stimulates the secretion of factors leading to increased proliferation, maturation, migration, and vessel permeability [26,35]. These latter two are a reflection of the delocalization of the VE-cadherins that occurs, thereby loosening the cell–cell adhesions [36,37]. In addition, other factors such as angiopoietins are also known to be involved in the process of attracting supporting cells as well as in stabilizing newly formed blood vessels.

In microfluidic systems, it is possible to explore not only the effects of individual factors but also the orchestrated effects of multiple angiogenic factors. In particular, the effects have been studied for cases in which both VEGF and ANG-1 gradients are applied [15]. VEGF gradients alone induce the proliferation of tip cells that initiate angiogenic sprouting, but the addition of an ANG-1 gradient greatly enhances the tendency to form a stable, continuous new vessel in which the tip cell remains attached to the stalk cells.

16.2.2.2 Mechanical factors

Application of shear stress to an endothelial monolayer mimicking flow conditions *in vivo* is achieved in a microfluidic system by integrating some form of pumping, such as a syringe pump, peristaltic pump, or some form of on-chip pumping [38]. The fluidic access ports to the microchannels inherently support the addition of external flow sources to the existing platform.

Shear stress has been shown to be a major biophysical factor influencing angiogenesis, and that it initiates sprouting from a monolayer by stimulating endothelial migration as well as proliferation through an upregulation of VEGF expression [39,40]. In addition to increased production of growth factors, shear stress also promotes development of cell–cell and cell–matrix junctions and maturation of vascular networks [41–43]. However, it has also been shown that shear stress stabilizes the endothelial layer in the microfluidic devices [38,44,45]. The actual effect of shear stress is likely to be complex and is a topic of continuing study [46].

16.2.2.3 Microenvironmental factors

It is becoming increasingly recognized that the local microenvironment surrounding the endothelium affects angiogenesis in various ways and its importance in cell function has been of considerable recent interest. Microfluidic platforms that incorporate hydrogels offer a unique opportunity to vary the properties (e.g., composition, stiffness) or local conditions (e.g., hypoxia) of the ECM, and at the same time, make detailed observations of the resulting response.

Hypoxia arises from a number of pathological conditions, but especially in the case of rapid tissue growth as in a tumor, or impairment of the local circulation as a result of an infarct or stroke [47]. In hypoxia, low oxygen tensions lead to a delay in the inactivation of hypoxia inducible factor 1α (HIF-1α), a transcription factor for numerous angiogenic factors such as VEGF-A and stromal-derived factor, SDF-1 [34,48]. Methods to control the oxygen tension in a microfluidic system have not yet been reported, but their small size and the gas permeability of PDMS offer new opportunities.

Another feature that has been shown to influence angiogenesis and the formation of a vascular network is ECM stiffness. This can easily be varied in a microfluidic system simply by altering the gel concentration, the type of gel, or the degree of polymerization as in the case of photopolymerizable gels or gels such as collagen for which the stiffness of the formed gel can be controlled by varying the pH at which the gel is formed [49].

16.2.3 Coculture methods

16.2.3.1 Coculturing in microfluidic platform

An important advantage of microfluidic cell culture platforms is their ability to better mimic the conditions that exist *in vivo*. In the previous section, we discussed the feasibility of introducing chemical or physical factors and how these factors affect endothelial cell function in the context of angiogenesis. However, in order to truly replicate physiological conditions, we must also consider the effects of interactions between multiple cell types. Microfluidic systems with channels or gel regions that can be individually seeded allow a simple integration of heterotypic cell types in a single device and recent studies have shown this to be uniquely suited for the study of cell–cell interactions [16,22]. This can be also be accomplished through the addition of a porous membrane with different cells seeded on either side [50–52] or by seeding one cell type on top of the other as is readily done in a Trans-well® assay, but these methods can be constraining in terms of their ability to image the cell–cell interactions or the growth of the vascular networks. Microfluidics allow for both spatial and temporal organization. For example, multiple cell types can be seeded into the different channels or gel regions, separated by distances of less than one mm, similar to physiological length scales. Similarly, one cell type can be seeded initially, left to adhere and stabilize, with another added at a later time point. Using these capabilities, numerous biological conditions can be simulated including tumor angiogenesis and metastasis [13]. In all cases, chemical factors secreted by the cells are readily transported through the hydrogel region that is separating the channels.

16.2.3.2 Tumor angiogenesis model

Expanding tumors, in particular, actively promote the growth of new blood vessel. Their demands for gas exchange and nutrients lead to the conditions described above that promote angiogenesis. Not only do tumor cells secrete various growth factors such as VEGF and bFGF in response to the local hypoxia, but they also disrupt the production of antiangiogenic enzymes [53]. It has also been found that tumor cells can create their own blood vessels through a process in which tumor associated stem cells differentiate into endothelium [54,55].

The use of microfluidics in studying tumor angiogenesis enables a parametric *in vitro* study with tumor cells in close proximity to and actively signaling with endothelial cells. Studies have been conducted in microfluidic systems under coculture of human endothelium and various tumor cell lines (Figure 16.3) [12,13,20,50]. By designating one channel in a three-channel system as the "condition" channel and the opposite one as the "control," the effect of different types of cancers on endothelial migration and vascular sprouting can easily be monitored and quantitatively compared. Such studies could be used to investigate, for example, phenotypic variations sometimes found in endothelial cells in a tumor environment [56,57]. Tumor blood vessels have increased permeability due to vessel dilation and detachment of pericytes and their morphology is also distorted [58–60].

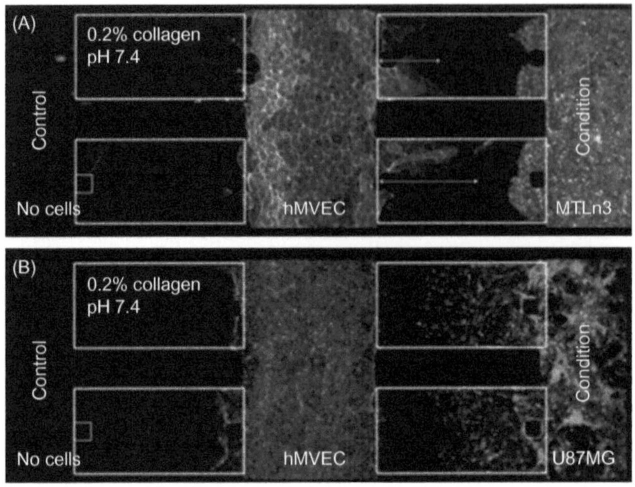

FIGURE 16.3

Coculture of endothelial cells with tumor cells in a microfluidic system. Two types of tumor cells, adenocarcinoma, MTLn3 (A), and glioblastoma, U87MG (B), are migrating toward hMVECs through a collagen gel region contained within the white rectangular regions [13].

Source: *Reproduced by permission of the Royal Society of Chemistry.*

16.2.4 Maturation/stabilization of blood vessels

The fundamental goal of tissue engineering in regenerative medicine does not end in the creation of neovasculature, but rather, in the construction of a stable vascular network. Once endothelial cells sprout, proliferate, and recruit new cells to the developing vascular network, naked tubular structures of endothelial cells initially form, but vessel maturation is required for stable, long-term perfusion of blood [2]. Fully functional blood vessels are characterized by a central conduit formed by endothelial cells that are adherent to basement membrane and encompassed by perivascular cells of various types, and the overall structure resides in ECM. Depending on the location or function of the vessel, ranging from arterioles to distal capillaries for example, their functional requirements in terms of size, permeability, or cell–cell adhesion strength or tightness all vary. Their stability as a fully functional microvascular network is governed by proper organization among the different cells and their environment [61]. Therefore, the importance of perivascular cells should not be underestimated. Perivascular cells include smooth muscle cells and pericytes, and they are recruited by endothelial cells once the tubular structures are formed [29,62]. Initially the naked vessel is highly permeable [37,63], and perivascular cells work to achieve stabilization and maturation by means of providing structural integrity and supplying necessary survival factors. They also secrete proteins that are used to create ECM [64] (Figure 16.4).

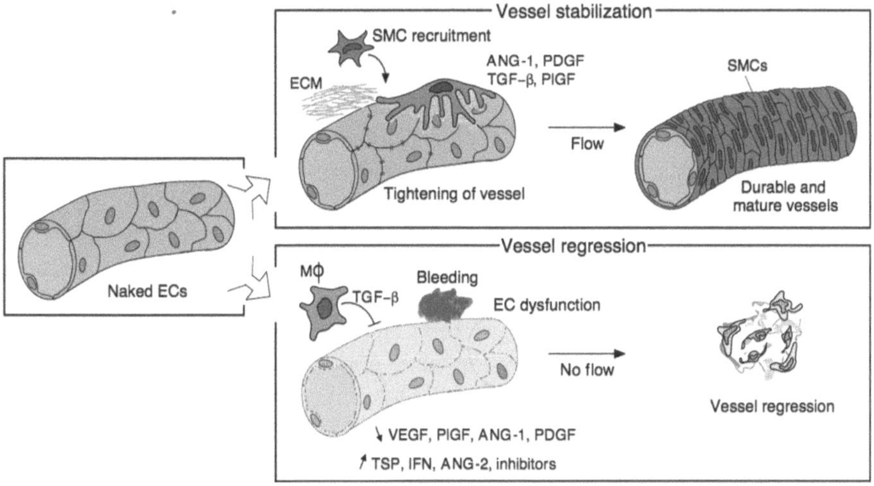

FIGURE 16.4

Maturation of naked endothelial cells (ECs) into a stable blood vessel is achieved with addition of cytokines, recruitment of smooth muscle cells (SMCs) and flow. Naked ECs will otherwise regress and lose functionality.

Source: *Adapted by permission from Macmillan Publishers Ltd: Ref. [65], copyright 2003.*

The importance of these accessory cells is underscored by the fact that the engineered blood vessels have often been found to be immature and unstable [36,66]. However, it has been reported that when endothelial cells were cocultured with appropriate perivascular cells, a stable and long-lasting vasculature could be formed *in vivo* [61,67,68]. The necessity of the presence of pericytes and smooth muscle cells for providing structural strength and regulation of perfusion on nascent endothelial vessels is once again confirmed.

Fluid shear stress has also been found to be a crucial factor in maturation of blood vessels in addition to its aforementioned effects in inducing angiogenesis. The growth factors induced by shear that are particularly involved with collateral growth through stimulation of endothelium and smooth muscle cells include TGF-β1, TNF-α, PIGF, and MCP-1 [28,69,70]. Under low- or no shear conditions, vessel regression has been observed, which indicates that fluid shear is an important factor in vessel maturation and maintenance [8,38,65,71].

16.2.5 Quantification

Interpretation of the results from experiments in angiogenesis using microfluidic cell culture relies heavily on imaging, thus far both qualitatively and quantitatively. Fortunately, the imaging capabilities of microfluidic platforms are among the major advantages of working with these systems. The thin glass coverslip provides optimal imaging access for observations using phase contrast or fluorescent microscopy up to several hundreds of microns in distance. When placed in an environmental chamber, real-time imaging is also made available, and confocal microscopy combined with microfluidics allows visualization of the 3D cellular response, which is a true highlight of this technology.

The majority of quantification methods from imaging can be considered as extensions from established 2D quantification methods. Particularly, as angiogenesis focuses on capillary sprouting events of endothelial cells, most of the established metrics are directed toward quantifying the change in sprout morphology. To capture the 3D nature of the response, an image that combines multiple images at different heights for the projected view should be created. Two different quantification methods that have been employed in a single phase contrast image are indicated in Figure 16.5; the white dotted lines delineate the perimeter of the sprouting region projected onto a 2D image, and the arrowheads both black and white indicate either attached or separated tip cells, respectively, and the total number of tip cells are summed for different sprouting conditions. Similarly, other quantification methods include but are not limited to counting of single cell migration events or counting the number of branch points, measuring the total migration area increase as well as measuring the individual tip cell length or the average tubule length [72]. Directional growth, as promoted by the presence of gradients in growth factors, can be quantified by measuring the turning angle of the sprouts, whereas a "directionality index matrix" has been used to quantify directional capillary sprouting [73]. Other measures not yet employed include the permeability of the vascular wall as a measure of how well the *in vitro* sprouts replicate *in vivo* behavior, or the flow resistance of vascular networks grown across gel regions.

16.2.6 Pros and cons of a microfluidic approach

Many of the advantages to be gained by using microfluidics have been discussed above. Included among these are the capability of simultaneous 2D and 3D culture, heterotypic culture methods, ease of imaging, ease of attaining physiological length scales to allow for realistic cell–cell communication, the capability to establish gradients in growth factors or fluid pressure, and for the application of shear stress. Although there is much to be gained through a microfluidic approach to simulating angiogenesis or vasculogenesis, there are also some drawbacks that need to be recognized. First, while the small volume is advantageous from the perspectives of facilitating high-throughput studies, keeping costs low, and ease of imaging, it can also pose problems, chief among them being that many of the more conventional assays of biological function (e.g., Western blots, Northern blots) are difficult or even impossible to perform with such small numbers of cells. Small volumes can also pose challenges in terms of the initial filling of the device and subsequent media changes. Surface properties of the devices are paramount since the wetting characteristics determine to a large extent the ease with which the gel solutions or media can be introduced into an initially dry system.

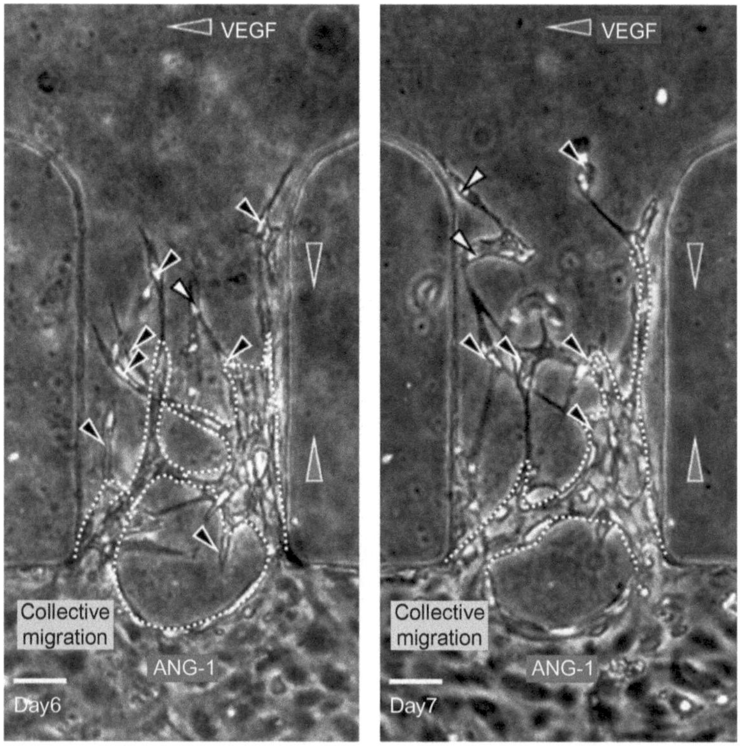

FIGURE 16.5

Ease of visualization and quantification of cellular behavior is one of advantages of using a microfluidic platform. Angiogenic sprouting is quantified by measuring sprouting area and counting the number of tip cells connected, leading collective migration of endothelial cells. The white dotted lines delineate the perimeter of the sprouting region projected onto a 2D image, and both black and white arrowheads indicate either attached or separated tip cells, respectively, and the total number of tip cells is summed for different sprouting conditions. The blue and red arrowheads indicate chemical gradient of ANG-1 and VEGF, respectively [15].

Source: *Reproduced by permission of the Royal Society of Chemistry.*

Small volumes also lead to large surface-to-volume ratios, meaning that conventional rules of thumb for frequency of medium changes, for example, need to be reevaluated. This topic has been addressed by Beebe and Young [74] through the introduction of the concepts of "effective culture time" and "critical perfusion rate," the latter being relevant to systems in which the cells are nourished by means of a slow, constant rate of flow of medium through the system channels. Another complication arises when cells or vascular structures experience a strong interaction with the device boundaries. As was seen in one set of experiments on angiogenesis, if cell adhesion to the walls of the system dominates over matrix adhesion, endothelial cells can migrate into the gel along the surface (sometimes referred to as migration in 2.5D) [14]. And even though surface migration can give way to the formation of tubular structures, the process by which these form and their morphology can differ considerably from the process that occurs *in vivo*. These concepts pertain not just to angiogenesis but to all cell culture methods in microfluidic systems.

16.3 Conclusion and future directions

Microfluidic systems enable various means of visualization and thereby quantification of cell–cell and cell–matrix interactions in 3D. Various designs of the platform are utilized to address the specific questions, and device designs for mimicking and evaluating angiogenesis have been reviewed in this chapter. Moreover, factors inducing angiogenesis and how they could be applied in microfluidics as well as the factors leading up to maturation of vessels are discussed.

There are yet some drawbacks in using microfluidics for cell culture platform as mentioned in the section above, and in addition, there are several fundamental discrepancies between *in vitro* and *in vivo* systems leading to heterogeneity in vascular permeability and microvessel density [75]. In spite of these limitations, microfluidic systems possess capabilities that extend beyond 2D modeling into 3D parametric cell culture studies.

Although the majority of microfluidic systems are currently fabricated with PDMS, hard plastics such as PMMA, COC, or PS may be used in the future through mass production of microfluidic chips for cell culture [76–79]. Through high-throughput and low-cost fabrication, microfluidic system may be used in commercial setting for applications such as pharmaceutical screening or patient-specific drug screening in clinical settings.

Finally, as methods become refined, systems can increase in complexity and achieve greater realism in the representation of organ morphology and function. As they do, applications of microfluidics to basic research and drug screening, for example, will certainly come into common practice. Due to their small size and since they can be fabricated from a variety of materials, microfluidics may also find increased use as *in vivo* implants. While these applications are limited at present, the field is poised for rapid growth into new areas.

References

[1] J. Folkman, Angiogenesis in cancer, vascular, rheumatoid and other disease, Nat. Med. 1 (1) (1995) 27–31.

[2] P. Carmeliet, Angiogenesis in life, disease and medicine, Nature 438 (7070) (2005) 932–936.

[3] R.H. Adams, K. Alitalo, Molecular regulation of angiogenesis and lymphangiogenesis, Nat. Rev. Mol. Cell Biol. 8 (6) (2007) 464–478.
[4] P. Carmeliet, Mechanisms of angiogenesis and arteriogenesis, Nat. Med. 6 (4) (2000) 389–395.
[5] P. Carmeliet, R.K. Jain, Angiogenesis in cancer and other diseases, Nature 407 (6801) (2000) 249–257.
[6] J. Folkman, Y. Shing, Angiogenesis, J. Biol. Chem. 267 (16) (1992) 10931–10934.
[7] P. Carmeliet, Y. Dor, J.M. Herbert, D. Fukumura, K. Brusselmans, M. Dewerchin, et al., Role of HIF-1 alpha in hypoxia-mediated apoptosis, cell proliferation and tumour angiogenesis (vol 394, pg 485, 1998), Nature 395 (6701) (1998) 525.
[8] R.K. Jain, Molecular regulation of vessel maturation, Nat. Med. 9 (6) (2003) 685–693.
[9] S.R. Quake, T. Thorsen, S.J. Maerkl, Microfluidic large-scale integration, Science 298 (5593) (2002) 580–584.
[10] S. Rajaraman, S-O. Choi, R.H. Shafer, J.D. Ross, J. Vukasinovic, Y. Choi, et al., Microfabrication technologies for a coupled three-dimensional microelectrode, microfluidic array, J. Micromech. Microeng. 17 (1) (2007) 163–171.
[11] C.R. Kothapalli, E. van Veen, S. de Valence, S. Chung, I.K. Zervantonakis, F.B. Gertler, et al., A high-throughput microfluidic assay to study neurite response to growth factor gradients, Lab Chip 11 (3) (2011) 497–507.
[12] A.D. Stroock, C. Fischbach, Microfluidic culture models of tumor angiogenesis, Tissue Eng. A 16 (7) (2010) 2143–2146.
[13] S. Chung, R. Sudo, P.J. Mack, C.R. Wan, V. Vickerman, R.D. Kamm, Cell migration into scaffolds under co-culture conditions in a microfluidic platform, Lab Chip 9 (2) (2009) 269–275.
[14] V. Vickerman, J. Blundo, S. Chung, R. Kamm, Design, fabrication and implementation of a novel multi-parameter control microfluidic platform for three-dimensional cell culture and real-time imaging, Lab Chip 8 (9) (2008) 1468–1477.
[15] Y. Shin, J.S. Jeon, S. Han, G.S. Jung, S. Shin, S.H. Lee, et al., In vitro 3D collective sprouting angiogenesis under orchestrated ANG-1 and VEGF gradients, Lab Chip 11 (13) (2011) 2175–2181.
[16] S. Chung, R. Sudo, V. Vickerman, I.K. Zervantonakis, R.D. Kamm, Microfluidic platforms for studies of angiogenesis, cell migration, and cell–cell interactions, Ann. Biomed. Eng. 38 (3) (2010) 1164–1177.
[17] C.P. Huang, J. Lu, H. Seon, A.P. Lee, L.A. Flanagan, H.Y. Kim, et al., Engineering microscale cellular niches for three-dimensional multicellular co-cultures, Lab Chip 9 (12) (2009) 1740–1748.
[18] K.C. Chaw, M. Manimaran, E.H. Tay, S. Swaminathan, Multi-step microfluidic device for studying cancer metastasis, Lab Chip 7 (8) (2007) 1041–1047.
[19] K.C. Chaw, M. Manimaran, F.E. Tay, S. Swaminathan, Matrigel coated polydimethylsiloxane based microfluidic devices for studying metastatic and non-metastatic cancer cell invasion and migration, Biomed. Microdevices 9 (4) (2007) 597–602.
[20] V.L. Cross, Y. Zheng, N. Won Choi, S.S. Verbridge, B.A. Sutermaster, L.J. Bonassar, et al., Dense type I collagen matrices that support cellular remodeling and microfabrication for studies of tumor angiogenesis and vasculogenesis in vitro, Biomaterials 31 (33) (2010) 8596–8607.

[21] O.C. Amadi, M.L. Steinhauser, Y. Nishi, S. Chung, R.D. Kamm, A.P. McMahon, et al., A low resistance microfluidic system for the creation of stable concentration gradients in a defined 3D microenvironment, Biomed. Microdevices 12 (6) (2010) 1027–1041.
[22] I.K. Zervantonakis, C.R. Kothapalli, S. Chung, R. Sudo, R.D. Kamm, Microfluidic devices for studying heterotypic cell-cell interactions and tissue specimen cultures under controlled microenvironments, Biomicrofluidics 5 (1) (2011). p.013406-1-13.
[23] R. Sudo, S. Chung, I.K. Zervantonakis, V. Vickerman, Y. Toshimitsu, L.G. Griffith, et al., Transport-mediated angiogenesis in 3D epithelial coculture, FASEB J. 23 (7) (2009) 2155–2164.
[24] J. Folkman, M. Klagsbrun, Angiogenic Fac., Science 235 (4787) (1987) 442–447.
[25] J. Folkman, M. Klagsbrun, A family of angiogenic peptides, Nature 329 (6141) (1987) 671–672.
[26] E. Giraudo, L. Primo, E. Audero, H.P. Gerber, P. Koolwijk, S. Soker, et al., Tumor necrosis factor-alpha regulates expression of vascular endothelial growth factor receptor-2 and of its co-receptor neuropilin-1 in human vascular endothelial cells, J. Biol. Chem. 273 (34) (1998) 22128–22135.
[27] P. Au, J. Tam, D.G. Duda, P.C. Lin, L.L. Munn, D. Fukumura, et al., Paradoxical effects of PDGF-BB overexpression in endothelial cells on engineered blood vessels in vivo, Am. J. Pathol. 175 (1) (2009) 294–302.
[28] R.H. Cao, E. Bråkenhielm, R. Pawliuk, D. Wariaro, M.J. Post, E. Wahlberg, et al., Angiogenic synergism, vascular stability and improvement of hind-limb ischemia by a combination of PDGF-BB and FGF-2, Nat. Med. 9 (5) (2003) 604–613.
[29] H. Gerhardt, C. Betsholtz, Endothelial–pericyte interactions in angiogenesis, Cell Tissue Res. 314 (1) (2003) 15–23.
[30] M. Klagsbrun, M.A. Moses, Molecular angiogenesis, Chem. Biol. 6 (8) (1999) R217–R224.
[31] H. Kamihata, H. Matsubara, T. Nishiue, S. Fujiyama, K. Amano, O. Iba, et al., Improvement of collateral perfusion and regional function by implantation of peripheral blood mononuclear cells into ischemic hibernating myocardium, Arterioscler. Thromb. Vasc. Biol. 22 (11) (2002) 1804–1810.
[32] S.S. Verbridge, N.W. Choi, Y. Zheng, D.J. Brooks, A.D. Stroock, C. Fischbach, Oxygen-controlled three-dimensional cultures to analyze tumor angiogenesis, Tissue Eng. A 16 (7) (2010) 2133–2141.
[33] H. Gerhardt, M. Golding, M. Fruttiger, C. Ruhrberg, A. Lundkvist, A. Abramsson, et al., VEGF guides angiogenic sprouting utilizing endothelial tip cell filopodia, J. Cell Biol. 161 (6) (2003) 1163–1177.
[34] P. Carmeliet, R.K. Jain, Molecular mechanisms and clinical applications of angiogenesis, Nature 473 (7347) (2011) 298–307.
[35] V. Joukov, K. Pajusola, A. Kaipainen, D. Chilov, I. Lahtinen, E. Kukk, et al., A novel vascular endothelial growth factor, VEGF-C, is a ligand for the Flt4 (VEGFR-3) and KDR (VEGFR-2) receptor tyrosine kinases, EMBO J. 15 (2) (1996) 290–298.
[36] S.M. Albelda, P.M. Sampson, F.R. Haselton, J.M. McNiff, S.N. Mueller, S.K. Williams, et al., Permeability characteristics of cultured endothelial-cell monolayers, J. Appl. Physiol. 64 (1) (1988) 308–322.
[37] A. Pettersson, J.A. Nagy, L.F Brown, C. Sundberg, E. Morgan, S. Jungles, et al., Heterogeneity of the angiogenic response induced in different normal adult tissues by

vascular permeability factor/vascular endothelial growth factor, Lab. Invest. 80 (1) (2000) 99–115.
[38] G.M. Price, K.H. Wong, J.G. Truslow, A.D. Leung, C. Acharya, J. Tien, Effect of mechanical factors on the function of engineered human blood microvessels in microfluidic collagen gels, Biomaterials 31 (24) (2010) 6182–6189.
[39] M. Milkiewicz, M.D. Brown, S. Egginton, O. Hudlicka, Association between shear stress, angiogenesis, and VEGF in skeletal muscles in vivo, Microcirculation 8 (4) (2001) 229–241.
[40] N. Resnick, H. Yahav, A. Shay-Salit, M. Shushy, S. Schubert, L.C. Zilberman, et al., Fluid shear stress and the vascular endothelium: for better and for worse, Prog. Biophys. Mol. Biol. 81 (3) (2003) 177–199.
[41] W. Risau, Mechanisms of angiogenesis, Nature 386 (6626) (1997) 671–674.
[42] H. Franke, Nature and significance of polypathy and multimorbidity in geriatrics, Internist 25 (8) (1984) 451–455.
[43] N. Resnick, M.A. Gimbrone, Hemodynamic forces are complex regulators of endothelial gene-expression, FASEB J. 9 (10) (1995) 874–882.
[44] O.C. Colgan, G. Ferguson, N.T. Collins, R.P. Murphy, G. Meade, P.A. Cahill, et al., Regulation of bovine brain microvascular endothelial tight junction assembly and barrier function by laminar shear stress, Am. J. Physiol. Heart Circ. Phys. 292 (6) (2007) H3190–H3197.
[45] J.W. Song, L.L. Munn, Fluid forces control endothelial sprouting, Proc. Nat. Acad. Sci. U.S.A. (2011).
[46] K. Yamamoto, T. Takahashi, T. Asahara, N. Ohura, T. Sokabe, A. Kamiya, et al., Proliferation, differentiation, and tube formation by endothelial progenitor cells in response to shear stress, J. Appl. Physiol. 95 (5) (2003) 2081–2088.
[47] W.R. Wilson, M.P. Hay, Targeting hypoxia in cancer therapy, Nat. Rev. Cancer 11 (6) (2011) 393–410.
[48] D.R. Mole, P.J. Ratcliffe, Cellular oxygen sensing in health and disease, Pediat. Nephrol. 23 (5) (2008) 681–694.
[49] N. Wang, D.E. Ingber, Control of cytoskeletal mechanics by extracellular-matrix, cell-shape, and mechanical tension, Biophys. J. 66 (6) (1994) 2181–2189.
[50] J.W. Song, S.P. Cavnar, A.C. Walker, K.E. Luker, M. Gupta, Y.C. Tung, et al., Microfluidic endothelium for studying the intravascular adhesion of metastatic breast cancer cells, Plos One 4 (6) (2009). p. e5756.
[51] B.H. Chueh, D. Huh, C.R. Kyrtsos, T. Houssin, N. Futai, S. Takayama, Leakage-free bonding of porous membranes into layered microfluidic array systems, Anal. Chem. 79 (9) (2007) 3504–3508.
[52] R.G.H. Lammertink, J. de Jong, M. Wessling, Membranes and microfluidics: a review, Lab Chip 6 (9) (2006) 1125–1139.
[53] L. Ying, L.J. Hofseth, An emerging role for endothelial nitric oxide synthase in chronic inflammation and cancer, Cancer Res. 67 (4) (2007) 1407–1410.
[54] R. Wang, K. Chadalavada, J. Wilshire, U. Kowalik, K.E. Hovinga, A. Geber, et al., Glioblastoma stem-like cells give rise to tumour endothelium, Nature 468 (7325) (2010) 829U128.
[55] L. Ricci-Vitiani, R. Pallini, M. Biffoni, M. Todaro, G. Invernici, T. Cenci, et al., Tumour vascularization via endothelial differentiation of glioblastoma stem-like cells, Nature 468 (7325) (2010) 824U121.

[56] K. Hida, D.N. Amin, A.F. Flint, D. Panigrahy, C.C. Morton, M. Klagsbrun, Tumor-associated endothelial cells with cytogenetic abnormalities, Cancer Res. 64 (22) (2004) 8249–8255.
[57] K. Hida, N. Klagsbrun, A new perspective on tumor endothelial cells: unexpected chromosome and centrosome abnormalities, Cancer Res. 65 (7) (2005) 2507–2510.
[58] E. Ryschich, J. Schmidt, G.J. Hämmerling, E. Klar, R. Ganss, Transformation of the microvascular system during multistage tumorigenesis, Int. J. Cancer 97 (6) (2002) 719–725.
[59] L.C.L. van Kempen, J-S. Rhee, K. Dehne, J. Lee, D.R. Edwards, L.M. Coussens, Epithelial carcinogenesis: dynamic interplay between neoplastic cells and their microenvironment, Differentiation 70 (9–10) (2002) 610–623.
[60] J. Chen, D. Brantley-Siders, J.S. Penn, Eph receptor tyrosine kinases: modulators of angiogenesis, Retinal Choroidal Angiogenesis (2008) 203–219.
[61] N. Koike, D. Fukumura, O. Gralla, P. Au, J.S. Schechner, R.K. Jain, Creation of long-lasting blood vessels, Nature 428 (6979) (2004) 138–139.
[62] K.K. Hirschi, P.A. DAmore, Pericytes in the microvasculature, Cardiovasc. Res. 32 (4) (1996) 687–698.
[63] P. Au, J. Tam, D. Fukumura, R.K. Jain, Bone marrow-derived mesenchymal stem cells facilitate engineering of long-lasting functional vasculature, Blood 111 (9) (2008) 4551–4558.
[64] K.K. Hirschi, T.C. Skalak, S.M. Peirce, C.D. Little, Vascular assembly in natural and engineered tissues, Ann. NY Acad. Sci. 961 (2002) 223–242.
[65] P. Carmeliet, Angiogenesis in health and disease, Nat. Med. 9 (6) (2003) 653–660.
[66] J.S. Schechner, A.K. Nath, L. Zheng, M.S. Kluger, C.C. Hughes, M.R. Sierra-Honigmann, et al., In vivo formation of complex microvessels lined by human endothelial cells in an immunodeficient mouse, Proc. Nat. Acad. Sci. U.S.A. 97 (16) (2000) 9191–9196.
[67] P. Au, L.M. Daheron, D.G. Duda, K.S. Cohen, J.A. Tyrrell, R.M. Lanning, et al., Differential in vivo potential of endothelial progenitor cells from human umbilical cord blood and adult peripheral blood to form functional long-lasting vessels, Blood 111 (3) (2008) 1302–1305.
[68] R.K. Jain, P. Au, J. Tam, D.G. Duda, D. Fukumura, Engineering vascularized tissue, Nat. Biotechnol. 23 (7) (2005) 821–823.
[69] F. Pipp, M. Heil, K. Issbrücker, T. Ziegelhoeffer, S. Martin, J. van den Heuvel, et al., VEGFR-1-selective VEGF homologue PlGF is arteriogenic—evidence for a monocyte-mediated mechanism, Circ. Res. 92 (4) (2003) 378–385.
[70] A. Luttun, M. Tjwa, L. Moons, Y. Wu, A. Angelillo-Scherrer, F. Liao, et al., Revascularization of ischemic tissues by PlGF treatment, and inhibition of tumor angiogenesis, arthritis and atherosclerosis by anti-Flt1, Nat. Med. 8 (8) (2002) 831–840.
[71] E.M. Conway, D. Collen, P. Carmeliet, Molecular mechanisms of blood vessel growth, Cardiovasc. Res. 49 (3) (2001) 507–521.
[72] D. Donovan, N.J. Brown, E.T. Bishop, C.E. Lewis, Comparison of three in vitro human angiogenesis assays with capillaries formed in vivo, Angiogenesis 4 (2) (2001) 113–121.
[73] T. Korff, H.G. Augustin, Tensional forces in fibrillar extracellular matrices control directional capillary sprouting, J. Cell Sci. 112 (19) (1999) 3249–3258.
[74] D.J. Beebe, E.W.K. Young, Fundamentals of microfluidic cell culture in controlled microenvironments, Chem. Soc. Rev. 39 (3) (2010) 1036–1048.

[75] A. Eberhard, Heterogeneity of angiogenesis and blood vessel maturation in human tumors: implications for antiangiogenic turner therapies (vol 59, pg 1388, 2000, Cancer Res. 60 (13) (2000) 3668.

[76] J.S. Jeon, S. Chung, R.D. Kamm, J.L. Charest, Hot embossing for fabrication of a microfluidic 3D cell culture platform, Biomed. Microdevices 13 (2) (2011) 325–333.

[77] L. Brown, T. Koerner, R. Oleschuk, Fabrication of epoxy stamps for hot embossing microfluidic devices and sub-micron structures, Micro. Total Anal. Syst. 2 (297) (2004) 52–54 (2005)

[78] G. Chen, Y. Chen, L.Y. Zhang, Fabrication, modification, and application of department of analytical poly(methyl methacrylate) microfluidic chips, Electrophoresis 29 (9) (2008) 1801–1814.

[79] J.T. Borenstein, M.M. Tupper, P.J. Mack, E.J. Weinberg, A.S. Khalil, J. Hsiao, et al., Functional endothelialized microvascular networks with circular cross-sections in a tissue culture substrate, Biomed. Microdevices 12 (1) (2010) 71–79.

CHAPTER

Cardiovascular Disease/ Discovery Models

17

Vivienne Williams, Dmitry Kashanin, Toby Paul and Sara Pizzato
Cellix Ltd., Ireland

17.1 Introduction

According to the World Health Organization, cardiovascular diseases kill more people each year than any others: *"In 2008, 7.3 million people died of ischemic heart disease, 6.2 million from stroke or another form of cerebrovascular disease."*

Atherosclerosis is a chronic inflammatory disease that constitutes the primary cause of heart disease and stroke. It is a progressive condition characterized by the gradual accumulation of lipid material in the artery wall [1]. Traditionally atherosclerosis was considered a simple lipid disorder, but now inflammation is recognized as an important factor in all stages of the disease, from its genesis to plaque rupture and associated thrombotic complications. Fatty streaks, the hallmark of early atherosclerotic lesions, are composed of lipid-laden macrophages called foam cells, which originate from circulating blood monocytes (Figure 17.1).

Therefore, the recruitment of leukocytes, their adhesion to the arterial endothelium and subsequent migration into the intima are central events in the pathogenesis of atherosclerosis. This process is triggered by local production of chemokines: endothelial cells that are activated by inflammatory cytokines express adhesion molecules *ex novo* and synthesize chemokines and lipid chemoattractants. The responsiveness of a leukocyte to a particular chemokine is determined by its set of chemokine receptors. Chemokine binding activates a signal transduction cascade that leads to various effects, such as integrin activation and subsequent cell adhesion.

Monocyte chemoattractant protein-1 (MCP-1/CCL2) has been indicated as a key player in the recruitment of monocytes from the bloodstream into early atherosclerotic lesions and may be involved in other processes leading to thrombotic events. MCP-1 has been shown to be present in macrophage-rich atherosclerotic plaques in humans [2]. Furthermore, studies reported that MCP-1 receptor CCR2 expression is greatly enhanced in monocytes from hypercholesterolemic patients compared with normal controls [3]. Inducers of MCP-1 expression include *oxidized low-density lipoprotein (OxLDL)* and *shear stress*, two well-documented atherogenic stimuli [4].

Increasing relevance in different steps of atherogenesis is being directed to Fractalkine (Fkn/CX3CL1), the only known member of the CX3C chemokine

Microfluidic Cell Culture Systems. ISBN: 978-1-4377-3459-1

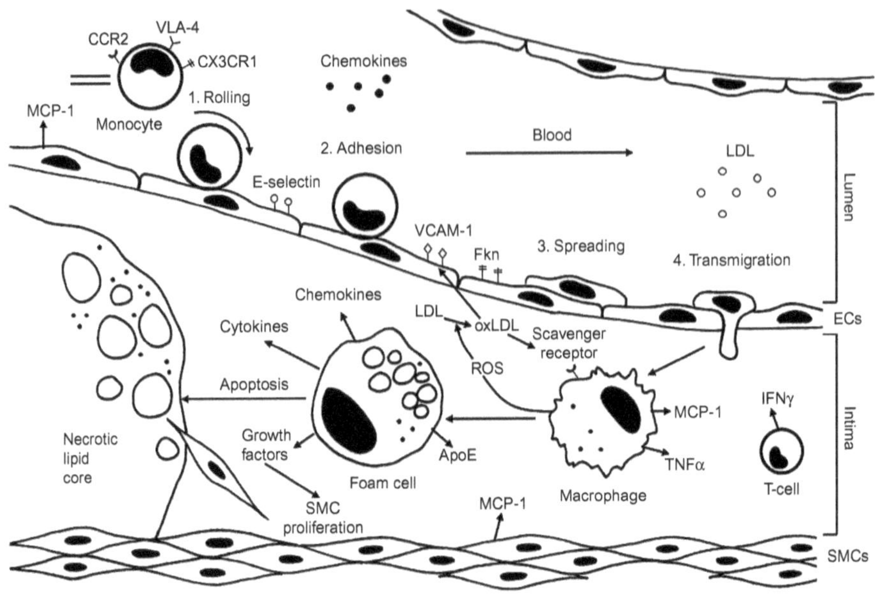

FIGURE 17.1

Interaction of monocytes with endothelial cells and lesion initiation. Monocytes leave the bloodstream to enter the subendothelial tissue (transmigration), after rolling, adhering, and spreading on the endothelium. This multistep process involves the joint action of numerous chemokines and adhesion molecules. Of particular interest in atherosclerosis are MCP-1, a potent chemoattractant for monocytes, and Fractalkine, which may act both as an adhesion molecule and a soluble chemokine. After transmigration, monocytes differentiate into macrophages, and the continuous uptake of OxLDL via scavenger receptors leads to the formation of foam cells, which will gradually build up the necrotic lipid core typical of atherosclerotic lesions.

family. It consists of a transmembrane domain and an extended mucin-like stalk with a chemokine domain on top. Fkn is bounded directly to the cell membrane, but a soluble form also exists through cleavage with an endogenous protease, and this form exhibits chemoattractant activity for T cells, monocytes, and NK cells [5,6]. The full-length cell-bound chemokine promotes strong adhesion of leukocytes to activated endothelial cells under physiological flow conditions [7].

High levels of Fkn expression have been observed in human arteries with atherosclerotic lesions, adding evidence to its potential role in atherogenesis and cardiovascular pathophysiology [8].

Another important factor to be considered is oxidative stress, which is caused by an excess of reactive oxygen species (ROS) or diminished antioxidant ability against them. Oxidative stress increases with atherosclerosis risk factors such as obesity, diabetes, hypertension, hyperlipidemia, smoking, and is considered to play a key role in the pathogenesis of atherosclerosis [9].

An initial event in atherosclerosis is LDL conversion into modified LDL or OxLDL (oxidized LDL) by factors including radicals, transition metals, and lipoxygenases, in the subendothelial microenvironment. OxLDL has been shown to upregulate adhesion molecule expression on endothelial cells, increasing the recruitment of leukocytes at the site of the atherosclerotic lesion [10–12]. It has also been reported that oxLDL, but not native LDL, induces MCP-1 production in vascular endothelial cells and smooth muscle cells.

Since *shear stress* has been cited as one of atherogenic stimuli, it is necessary to consider this factor, when studying the disease, designing and developing physiologically relevant *in vitro* models. To investigate the role of different stimuli recognized to play an important part in atherosclerosis development (chemokines, cytokines, oxidized lipoproteins) on leukocyte adhesion to purified adhesion molecules or endothelial cells under physiological flow.

Despite continued developments in microfluidic technologies with numerous researchers worldwide publishing papers on a vast array of different microfluidic biochip designs and varying culture techniques, the pharmaceutical and biotechnology industry has been slow to replace standard static cell culture techniques in favor of microfluidic devices. Industry is very conservative and will remain slow to adopt such new technologies unless a significant body of validated data illustrates robustness and reliability of these commercially available microfluidic devices.

The advantages of microfluidic cell culture are well known and those often cited include more physiologically relevant models, reduced sample/reagent volumes, greater control of cell patterning/seeding, and the ability to multiplex [13]. Challenges reported include the material of choice for such microdevices, polydimethylsiloxane (PDMS), which has been known to have side effects on cell behavior, difficulties in cell seeding resulting in cell damage [14], the frequent culture media change, or evaporation issues associated with small media sample volumes. Solutions to the latter challenges include integrating pumps on-chip or interfacing with external pumping devices including syringe-driven pumps, peristaltic pumps which can provide a continuous flow or pulses of fresh media to provide nutrients to cells while simultaneously removing waste. Such solutions also pose challenges as the effect of shear stress on the cells being cultured must be taken into account. Depending on the type, cells may be shear sensitive and so precise and accurate delivery of the culture media for feeding purposes must be taken into consideration [15]. Shear stress levels need to be optimized [16] to enable cells to proliferate without washing them out of the microfluidic channel or chamber during cell division. Additionally, perfusion may hamper cell-to-cell communication and cell proliferation by washing away signaling and conditioning metabolites excreted by the cells themselves required to maintain the cell culture [17].

Perfusion of culture medium can be achieved in a number of ways including pressure-driven flow via syringe pumps [18], peristaltic pumps [19], gravity-driven [20], or electro-osmotic flow [21].

There has been some success with gravity-driven flow but it is less precise than pressure-driven systems. One study described how alveolar epithelial cells (A549) were seeded on a membrane and media was constantly replenished via gravity-driven flow. However, once a confluent layer was achieved, fresh media was maintained via a syringe pump where cultures were successfully maintained for up to 3 weeks [22].

In the case of electro-osmotic flow, a key advantage of this compared to pressure-driven flow is the lack of moving parts. However, there are also a number of disadvantages, namely, the potentially high electric fields produced which can damage the cell membrane making it a less likely choice for microfluidic cell-based assays [23]. It also lacks stability during continuous operation due to electrolysis at the electrodes rendering it an unlikely choice for long-term cell culture [24].

Syringe-driven flow is becoming more popular especially given the wide variety of commercial pumping systems available, some with user-friendly interfaces and easily programmable LCDs. However, many of these pumps are unsuitable for precise delivery of media or other reagents into microfluidic channels where sudden pulses, e.g., due to the syringe refilling, can result in the detachment of cells from the microchannel. In addition, their size is also a limiting factor as some studies have reported putting the whole syringe pump into the incubator with the microfluidic device to achieve long-term cultures. Despite this, there are a number of successful studies incorporating microfluidic cell bioreactors, chamber, or microchannels facilitating the culture of endothelial cells [16,25–37] and in some cases as cocultures [38–42].

While there is currently no commercial gold standard, Cellix has developed an elegant solution which enables researchers to culture endothelial cells in their Vena8 Endothelial + ™ microfluidic biochips under static or perfusion conditions using the Kima pump, controlled by the iKima Application for use with the iPod Touch and iPhone.

In this chapter, Cellix presents a validated, reliable, and robust method of 2D culturing of primary human umbilical vein endothelial cells (HUVECs) within microfluidic biochips (Vena8 Endothelial + biochips) using static and perfusion-based methods with Cellix's Kima™ pump. Adhesion molecules coating Vena8 Fluoro + ™ and HUVECs cultured inside Cellix's Vena8 Endothelial + biochips serve as robust and reliable models for studying atherosclerosis as we present validated results for cell–ligand and cell–cell analysis, respectively, under shear stress conditions as Cellix's Mirus Nanopump™ is used to perfuse THP-1 monocytic leukemia cells over rhVCAM-1 and primary HUVECs.

17.2 Cell culture in cellix's Vena8 Endothelial + microfluidic biochips

17.2.1 Validation of Vena8 Endothelial + biochip model)

17.2.1.1 Materials

The adhesion protein used to coat the inside of the microcapillaries of the biochips and promote adhesion of HUVECs to the surface was either Laminin

(Sigma L-6274) or Fibronectin (Sigma). rhTNFα (Cat. No. 210-TA, R&D Systems) was used to stimulate the HUVECs. HUVECs were primary cells from human umbilical veins grown adherent in flasks with an Endothelial Medium Kit (Promocell C-22110), Detach Kit-30 (Promocell C-41200), and Accutase Solution (Promocell C-41310). THP-1 cells, which were subsequently perfused over the HUVECs inside the biochips, were acute monocytic leukemia cells from peripheral blood (human) grown in suspension. Cells were maintained using standard cell culture reagents RPMI 1640 (Gibco 31870), FBS at 10%, 2 mM L-glutamine and 100 μg/ml penicillin/streptomycin (Sigma G6784) (Figure 17.2, Table 17.1).

17.2.1.2 Methods

Static culture

Static culture involved culturing HUVECs in a standard T75 cm^2 flask before seeding at high density inside the Vena8 Endothelial + biochips and enabling them to attach and spread. To achieve this, primary HUVECs were allowed to reach 90% confluency before being stimulated with 10 ng/ml rhTNFα overnight (16–18 h) in T75 cm^2 flask. The Vena8 Endothelial + biochip was kept under UV for 20–30 min and all the channels were coated with 12 μl of 100 μg/ml Laminin (Sigma, Cat. No. L-6274). The biochip was then kept in an opened humidified chamber in CO_2 incubator for 1–1.5 h. The primary HUVECs were

FIGURE 17.2

Vena8 Endothelial + biochip.

Table 17.1 Technical Specifications of Vena8 Endothelial + Biochip	
Technical Specifications	
Number of channels per biochip	8
Volume of each channel	1.96 μl
Depth of channel	120 μm/0.12 mm
Width of channel	800 μm/0.8 mm
Length of channel	28 mm
Dead volume at input port	0.1 μl
Vena8 biochip dimensions	35 mm (W) × 2.5 mm (D) × 40 mm (L)
Thickness of bottom substrate	0.5 mm

prepared by maintaining the cells in the recommended growth medium. A cell suspension of 1.5×10^6 per 100 μl was prepared for seeding in the channels. Following the incubation period, 5 μl of 1.5×10^6 per 100 μl of primary endothelial cells were gently added, via pipette, into each channel of Vena8 Endothelial + biochip. The biochip was kept in the CO_2 incubator for 15–20 min. The biochip was then observed under the microscope and all the reservoirs were topped up with 40 μl of media. The biochip was kept for a further 1.5–2 h in the CO_2 incubator to obtain a primary cell monolayer inside the channels of the Vena8 Endothelial + biochip.

Culture under perfusion with Kima pump and iKima App

Culturing HUVECs under perfusion involved seeding the cells within the microcapillaries of the biochip and perfusing fresh media via the Kima pump, thereby enabling the cells to firmly attach and culture within the biochip. In this case, the Vena8 Endothelial + biochip was kept under UV for 20–30 min and all the channels were coated with 12 μl of 100 μg/ml Fibronectin (Sigma). The biochip was then kept in a humidified sterile petri dish at 4°C overnight. The primary HUVECs were maintained in the recommended growth medium and a cell suspension of 2×10^6 per 100 μl (or 20×10^6 cells/ml) was prepared for seeding in the channels. Following the incubation period, 5 μl of 2×10^6 per 100 μl of primary endothelial cells were gently added, via pipette, into each channel of the Vena8 Endothelial + biochip. Biochip was kept in the CO_2 incubator for 15–20 min. The biochip was then observed under the microscope and all the reservoirs were topped up with 40 μl of media. The biochip was kept for a further 1.5–2 h in the CO_2 incubator and subsequently connected to the Kima pump (Figure 17.3A), which had been prewashed with 70% ethanol and sterile distilled water. Once connected to the Kima pump, perfusion is started for 2 min via iKima, the Kima pump App (Figure 17.3B), to check the flow and tubing connected to the Vena8 Endothelial + biochip. Following this, the Kima pump together with the Vena8 Endothelial + biochip is transferred to the CO_2 incubator.

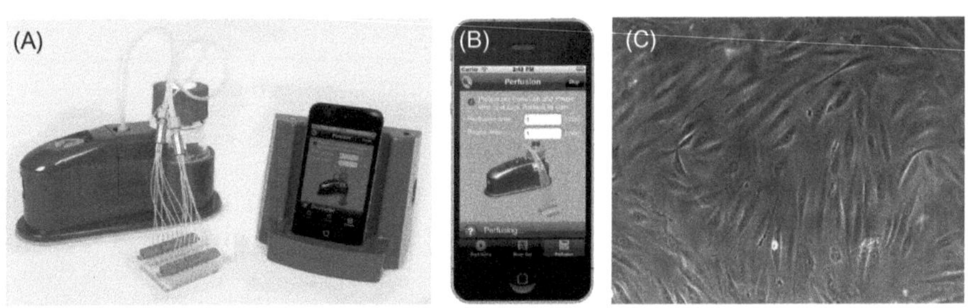

FIGURE 17.3

(A) Kima pump connected to the Vena8 Endothelial + biochip. (B) iKima App. (C) Primary HUVEC cells in Vena8 Endothelial + biochip, 72 hours post-perfusion with Kima pump.

THP-1 cells were maintained in the recommended growth media. A cell suspension of 5×10^6/ml cells was prepared for the experiment. After the incubation period, THP-1 cells were perfused over the HUVEC monolayer inside the channels at a shear stress of 0.5 dynes/cm^2 using VenaFlux™ Assay software controlling the Mirus Nanopump. The THP-1 cells were perfused for 3 min per channel. Images were acquired at positions 3, 4, and 5 after 2 min 40 s of perfusion in every channel. Ten images per position were captured using a high-definition *QImaging* camera with the help of the *Marzhauser IM* series motorized stage. Five Vena8 Endothelial + biochips were used for the assay and all the experiments were performed at 37°C using *Oko-lab's Microscope Cage incubator*.

The THP-1 cells which adhered to the primary HUVEC were counted using the *DucoCell* application software and *Image Pro Analyser 7.0*. The data was exported to Excel for calculations and interpretations.

17.2.1.3 Results

Described experiments were carried out to evaluate reproducibility of the cell culture model in respect to adhesion molecules and selectin molecules expression. This is further important for investigation of monocyte adhesion to endothelial layers under flow conditions, mimicking onset of atherosclerosis. In order to evaluate functionality of endothelial layer and to access variability of the model, adhesion of monocytic cell line (THP-1) was measured in adhesion flow assay.

An average of *100* cells (THP-1) adhered on primary HUVEC monolayer per microscopic frame. The channel to channel variation was less than *10%* and chip to chip variation was *4%*. Such level of variability proves that current model is reproducible, sensitive to proinflammatory stimulation of endothelial cells, and can be used in investigations of atherosclerosis (Figure 17.4, Table 17.2).

17.3 Microfluidic cell culture biochip model for atherosclerosis

17.3.1 Monocyte (THP-1) adhesion to HUVECs in Cellix's Vena8 Endothelial + microfluidic biochips

17.3.1.1 Materials

Recombinant proteins rhVCAM-1 and rhCX3CL1/Fractalkine were used to coat the microcapillaries of the biochips prior to perfusion of THP-1 and peripheral blood mononuclear cells (PBMCs). Monoclonal antihuman integrin α4 was subsequently used to treat THP-1 cells in order to investigate its effect on adhesion to prestimulated HUVECs. In this case, rhTNFα and rhIFNγ from Sigma were used to stimulate HUVECs. Chemotactic cytokine rhCCL2/MCP-1 was used to treat PBMCs prior to perfusing cells over purified human VCAM-1 and rhCX3CL1/Fractalkine.

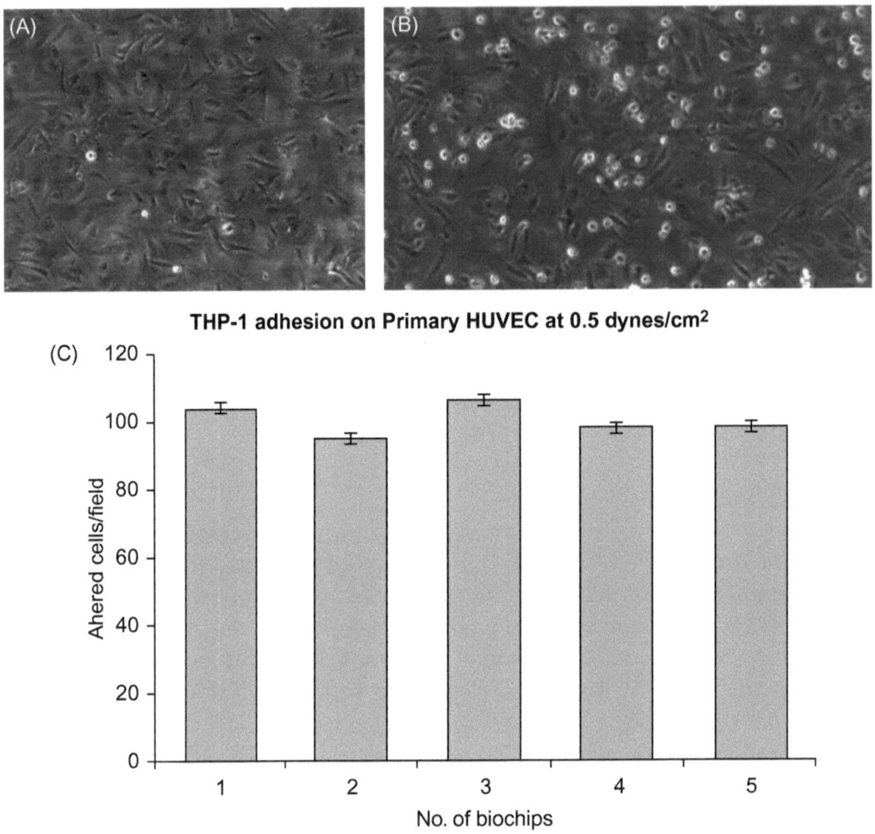

FIGURE 17.4

(A) Primary HUVEC cultured inside Vena8 Endothelial + biochip channels. (B) Adhesion of THP-1 monocytes at a shear stress of 0.5 dyne/cm^2 to 10 ng/ml rhTNFα stimulated primary HUVEC cultured in channel. (C) THP-1 cell adhesion to 10 ng/ml rhTNFα stimulated primary HUVEC at a constant shear stress of 0.5 dynes/cm^2 on Vena8 Endothelial + biochip.

oxLDL, oxidized with copper II sulfate (LDL from human plasma) from Kalen Biomedicals, was incubated with HUVECs. A subsequent incubation with rhTNFα was also executed. Fluorescein-conjugated mouse antihuman VCAM-1, ICAM-1, E-selectin mAbs, fluorescein-conjugated mouse IgG matched isotype control (all from R&D) were used to treat stimulated HUVECs in order to investigate the knockdown of particular expression levels of the different proteins.

THP-1 cells were acute monocytic leukemia cell from peripheral blood (human) grown in suspension from European Collection of Cell Cultures. PBMCs were isolated from healthy donors and HUVECs were isolated from patients giving written consent.

Table 17.2 Results of Standard Deviation When Comparing Results from Channel to Channel and Chip to Chip for Vena8 Endothelial + Biochips

Channel to Channel Analysis					Chip to Chip Analysis			
Chip No.	Average Number of Cells (THP-1)/8 Channels	STDEV	% CV	SEM	Average Number of Cells	STDEV	% CV	SEM
1	104	7.76	7.4	1.6	100	4.41	4	0.4
2	95	7.98	8.4	1.6				
3	106	7.59	7.2	1.5				
4	98	8.81	9.0	1.9				
5	98	5.57	5.7	1.2				

17.3.1.2 Methods

Vena8 Fluoro + Biochips were coated with 10 μg/ml of recombinant protein overnight, at 4°C. THP-1 cells were incubated with anti-integrin α4 mAb for 15 min at 37°C. PBMCs were incubated with Fractalkine (chemokine domain) or MCP-1 for 30 min at 37°C. The perfusion of cells in the microcapillaries of the biochips was performed using Cellix's Mirus Nanopump at a shear stress of 0.5 dyn/cm^2 for 3 min.

Vena8 Endothelial + biochips were kept under UV for 20–30 min and all the channels were coated with 12 μl of 100 μg/ml Laminin (Sigma, Cat. No. L-6274). The biochips were then kept in an opened humidified chamber in CO_2 incubator for 1–1.5 h. Primary HUVECs were maintained in the recommended growth medium and a cell suspension of 1.5×10^6 per 100 μl was prepared for seeding in the channels. Seeding of the cells in the channels is as previously described. After 48 h, HUVECS were incubated with oxLDL (150 μg/ml) for 16 h at 37°C and stimulated with TNFα (50 ng/ml) or IFNγ (500 units/ml) for 4 h at 37°C. The perfusion of cells in the microcapillaries of the biochips was performed using Cellix's Mirus Nanopump at a shear stress of 0.5 dyne/cm^2 for 5 min. Cell suspensions were always at a density of 2×10^6 cells/ml.

All images acquired during flow assays were analyzed with Cellix's DucoCell™ analysis software. All data are represented as mean + SEM ($n = 3$ or 4). Statistical analysis was performed with GraphPad Prism®5. ns: $P > 0.05$; *: $0.01 < P < 0.05$; **: $0.001 < P < 0.01$; ***: $P < 0.001$.

17.3.1.3 Results and discussion

Atherosclerosis is a complex multifactorial inflammatory disease, with signaling cascades involving different adhesion molecules, chemokines, lipids, etc.

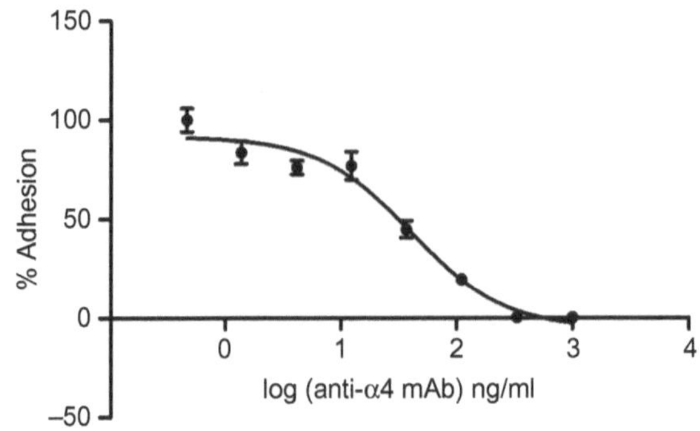

FIGURE 17.5

Adhesion blockade to VCAM-1. The inhibition profile of THP-1 adhesion to VCAM-1 was measured using an anti-α4 mAb to block the α4β1 (VLA-4) receptor for VCAM-1. A nonlinear regression was used to interpolate the data, with a hill slope of −1.076. Best fit IC50 value: 41.09 ng/ml.

However, their interactions, their independent or complimentary and synergistic roles are still unclear. Cellix's technology offers the possibility to perform a simple and physiologically relevant functional assay to study individual mechanisms or the combined effects of more stimuli, in relation to the initial steps in atherogenesis of cell recruitment and adhesion to the vascular wall.

Adhesion to VCAM-1, a major adhesion molecule expressed on inflamed endothelium, was completely inhibited when blocking the integrin α4 on THP-1 cells (Figure 17.5). Interestingly, adhesion was not significantly reduced when looking at adhesion to endothelial cells, even with a concentration of anti-α4 mAb 100 times higher (Figure 17.6). This suggests the idea that blocking adhesion to VCAM-1 only might not be sufficient to achieve a satisfactory decrease in leukocyte infiltration. Adhesion blockade to full-length Fractalkine was also studied using the soluble form of the chemokine to block CX3CR1 receptors on PBMCs (Figure 17.7).

To verify if a synergistic effect existed between VCAM-1 and the chemotactic cytokine MCP-1 (CCL2), we studied the adhesion of PBMCs to purified human VCAM-1 using different concentrations of the chemokine. An increased affinity for the binding was manifested from a more consistent adhesion at the higher concentrations of MCP-1 (Figure 17.8).

Different considerations should be made for Fractalkine since MCP-1 was shown to cause no effect in relation to PBMC adhesion to the immobilized chemokine (Figure 17.9). Consistently with studies on ApoE$^{-/-}$ mice, MCP-1 and Fkn, two important players in atherosclerosis initiation and development, were proved to have independent roles in promoting monocyte recruitment [43]. It was shown that genetic deletions of MCP-1, Fkn, or their associated receptors, CCR2 and CX3CR1, respectively, considerably reduced lesion size in murine models,

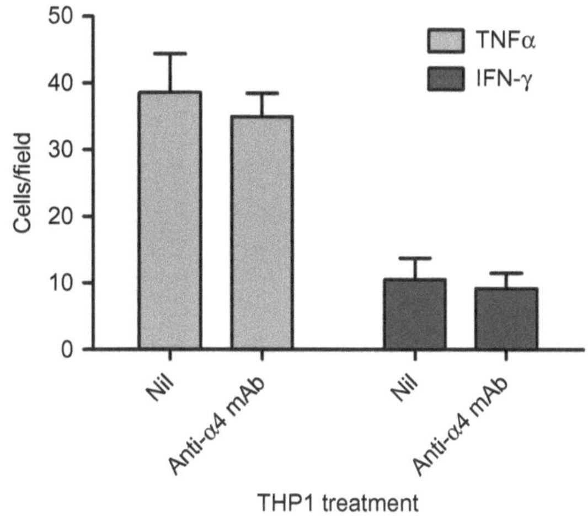

FIGURE 17.6

THP-1 adhesion to HUVECs. Endothelial cells were stimulated with inflammatory cytokines, either TNFα or IFNγ, and adhesion of THP-1 was measured. THP-1 were also treated with an anti-α4 mAb to block the integrin, but a nonsignificant inhibition of adhesion was shown. To assess the statistical difference, unpaired *t*-test (two-tailed *P*-value) was performed, showing a nonsignificant difference between the samples treated or not with the anti-α4 mAb, for both TNFα and IFNγ stimulated endothelial cells.

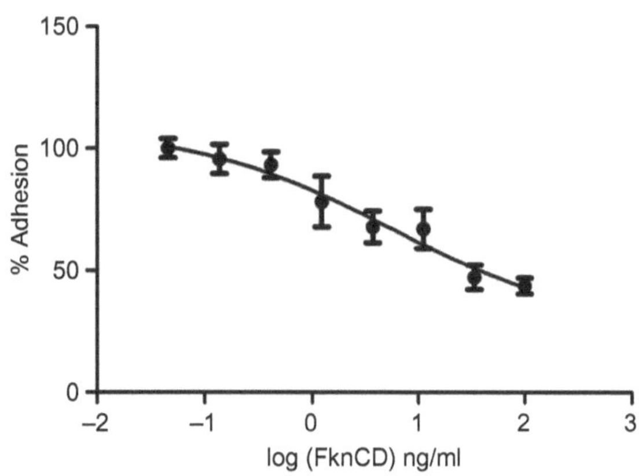

FIGURE 17.7

Adhesion blockade to Fractalkine (full length). The inhibition profile of PBMCs adhesion to Fkn was measured using the soluble form of Fkn (chemokine domain) to block the CX3CR1 receptor for Fkn. A nonlinear regression was used to interpolate the data, with a hill slope of −0.4595. Best fit IC50 value: 5.543 ng/ml.

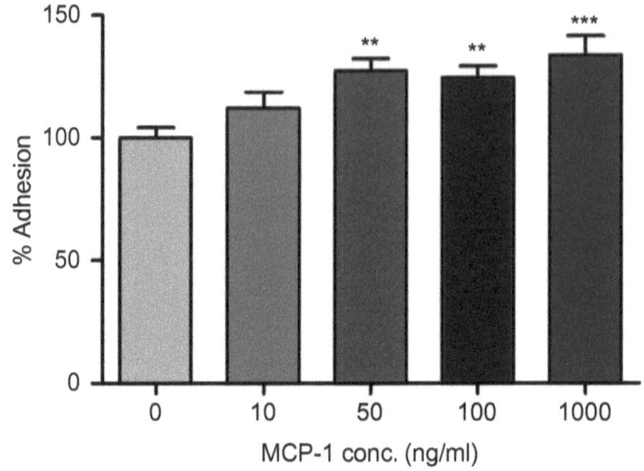

FIGURE 17.8

Effect of MCP-1 adhesion to VCAM-1. A significant increase in PBMC adhesion to VCAM-1 was shown upon stimulation with MCP-1, which therefore may be involved in integrin activation and enhanced affinity in the VCAM-1/VLA-4 axis. To assess the statistical difference, one-way ANOVA was performed, followed by Dunnett's posttest to compare all columns to control (MCP-1 concentration = 0).

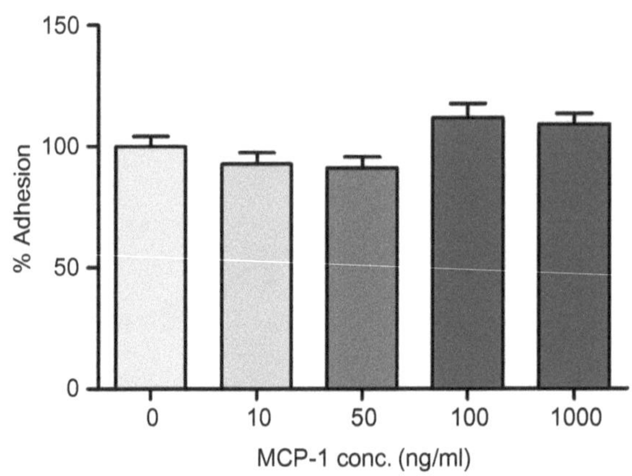

FIGURE 17.9

Effect of MCP-1 on adhesion to Fractalkine. No significant effect of MCP-1 stimulation was detected in relation to PBMC adhesion to Fkn, suggesting that MCP-1 and Fkn do not act in concert, but independently promote monocyte recruitment and atherogenesis. To assess the statistical difference, one-way ANOVA was performed, followed by Dunnett's posttest to compare all columns to control (MCP-1 concentration = 0).

but the combined genetic deletion of both MCP-1 and Fkn, on an $ApoE^{-/-}$ background, resulted in a much greater reduction of atherosclerotic lesions. All these data suggest that, from a clinical perspective, a double blocking could lead to a significant improvement in atheroprotection.

We also found that Fkn increased PBMC adhesion to VCAM-1 (data not shown), adding evidence to the importance of this chemokine in leukocyte recruitment and lesion initiation.

In relation to oxidative stress, a central factor in atherosclerosis development, we studied the effects of OxLDL on adhesion molecule expression on endothelial cells and subsequent leukocyte adhesion. Three different levels of oxidation were considered, and the result of oxLDL incubation with regard to adhesion molecule upregulation on HUVECs was assessed via flow cytometric analysis (Figure 17.10). Only the highly OxLDL (Hox) was shown to cause a significant increase in VCAM-1, ICAM-1, and E-selectin expression. When endothelial cells were incubated with TNFα on top of oxLDL, this result was not seen.

A small but significant increase in THP-1 adhesion was seen on Hox stimulated HUVECs compared to the nontreated control (Figure 17.11) once the monocytic cells were perfused over the endothelial layer at physiological flow conditions (Figure 17.12).

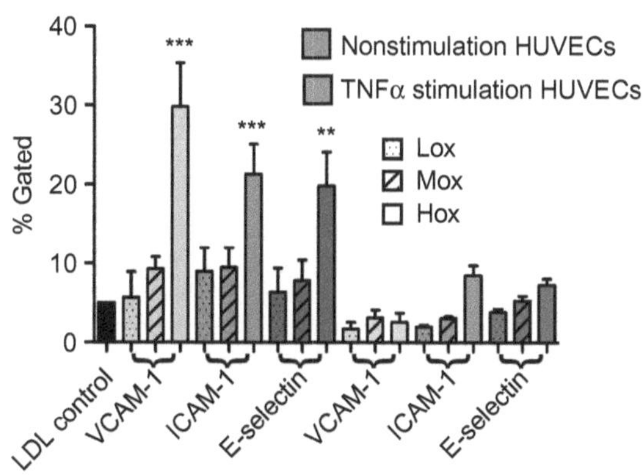

FIGURE 17.10

Adhesion molecule expression on endothelial cells upon stimulation with OxLDL. Flow cytometry analysis was performed to assess the level of VCAM-1, ICAM-1, and E-selectin expression on HUVECs, after a 16 h incubation with low-OxLDL (Lox), standard (medium) OxLDL (Mox), and high-OxLDL (Hox), in the presence or not of TNFα. Only Hox stimulation without TNFα induced a significant upregulation of adhesion molecule expression. The reference is HUVECs treated with native LDL, or native LDL plus TNFα. To assess the statistical difference, two-way ANOVA was performed, followed by Dunnett's posttest to compare all columns to LDL control.

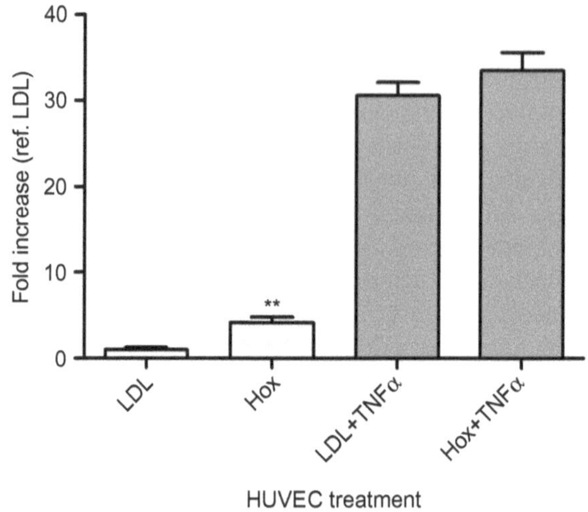

FIGURE 17.11

THP-1 adhesion to endothelial cells: effect of oxLDL. HUVECs were treated with native LDL or high-OxLDL (Hox), with or without costimulation with TNFα. The upregulation of adhesion molecule expression in endothelial cells due to oxLDL is much lower in comparison to TNFα effect, but anyway it corresponded to a small but significant increase in THP-1 adhesion. To assess the statistical difference, unpaired *t*-test (two-tailed *P*-value) was performed.

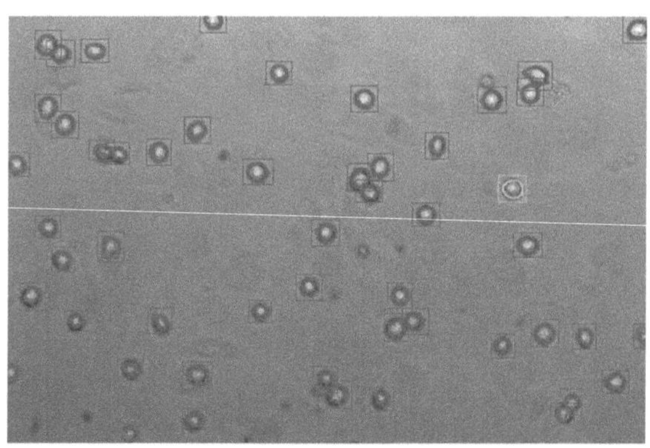

FIGURE 17.12

Detection of THP-1 adhesion to TNFα stimulated endothelial cells. HUVECs were seeded on a VenaEC biochip, cultured for 48 h in static conditions and then treated with TNFα 4 h prior to flow assay. THP-1 (2×10^6 cells/ml) were perfused over the endothelial layer for 5 min at 0.5 dyn/cm^2. Detection was performed with DucoCell analysis software, which returned the cell count and other relevant shape parameters.

17.4 Conclusion

This chapter describes *in vitro* studies, closely mimicking the human *in vivo* vascular microenvironment, that represent a useful model to dissect different mechanisms involved in atherogenesis. We have addressed two key factors necessary for physiological model to study cardiovascular disease: endothelial cell culture in microfluidic biochips both in static and flow conditions; validation of microfluidic biochips with cultured endothelial cells for studying involvement of adhesion molecules, selectins, and chemokines in development of atherosclerosis. In particular, roles of Fractalkine and MCP-1 chemokines have been investigated and results found are consistent with animal models reported; cultured endothelial cells were subjected to increasing levels of oxidative stress in order to elucidate its effects on adhesion molecule expression and therefore atherogenic potential.

The developed microfluidic biochips, instrumentation, and cardiovascular discovery model allow for a design of more focused experiments involving animal models, and for a more realistic prediction of the responses in humans, considering the difficulty in engaging into clinical trials, due to the dramatic end points of the disease (such as myocardial infarction, stroke, or death), and the lack of specific markers.

References

[1] A.J. Lusis, Atherosclerosis, Nature 407 (2000) 233–241.

[2] I.F. Charo, M.B. Taubman, Chemokines in the pathogenesis of vascular disease, Circ. Res. 95 (2004) 858–866.

[3] K.H. Han, K.O. Han, S.R. Green, O. Quehenberger, Expression of the monocyte chemoattractant protein-1 receptor CCR2 is increased in hypercholesterolemia: differential effects of plasma lipoproteins on monocyte function, J. Lipid. Res. 40 (1999) 1053–1063.

[4] Y.J. Shyy, H.J. Hsieh, S. Usami, S. Chien, Fluid shear stress induces a biphasic response of human monocyte chemotactic protein 1 gene expression in vascular endothelium, Proc. Natl. Acad. Sci. U.S.A. 91 (11) (1994) 4678–4682.

[5] S. Mahalingam, G. Karupiah, Chemokines and chemokine receptors in infectious diseases, Immunol. Cell Biol. 77 (1999) 469–475.

[6] O. Yoneda, T. Imai, S. Goda, H. Inoue, A. Yamauchi, T. Okazaki, et al., Fractalkine-mediated endothelial cell injury by NK cells, J. Immunol. 164 (2000) 4055–4062.

[7] C. Schulz, A. Schafer, M. Stolla, S. Kerstan, M. Lorenz, M.-L. von Bruhl, et al., Chemokine fractalkine mediates leukocyte recruitment to inflammatory endothelial cells in flowing whole blood, Circulation 116 (2007) 764–773.

[8] X.P. Yang, S. Mattagajasingh, S. Su, G. Chen, Z. Cai, K. Fox-Talbot, et al., Fractalkine upgregulates intercellular adhesion molecule-1 in endothelial cells through CX3CR1 and the Jak-Stat5 pathway, Circ. Res. 101 (2007) 1001–1008.

[9] M. Bakic, Pathogenetic aspects of atherosclerosis, Acta Medica Medianae 46 (2007) 25–29.

[10] D. Zapolska-Downar, Mononuclear cell recruitment and inflammation in atherosclerosis, Vasc. Dis. Prevent. 1 (2004) 41–51.

[11] C.A. Gleissner, N. Leitiniger, K. Ley, Effects of native and modified low-density lipoproteins on monocyte recruitment in atherosclerosis, Hypertension 50 (2007) 1–8.

[12] Z. Chi, A.J. Melendez, Role of cell adhesion molecules and immune-cell migration in the initiation, onset and development of atherosclerosis, Cell Adh. Migr. 1 (4) (2007) 171–175.

[13] I.K. Zervantonakis, C.R. Kothapalli, S. Chung, R. Sudo, R.D. Kamm, Microfluidic devices for studying heterotypic cell–cell interactions and tissue specimen cultures under controlled microenvironments, Biomicrofluidics 5 (1) (2011).

[14] M.S. Kim, H. Hwang, Y.-S. Choi, J.-K. Park, Microfluidic micropillar arrays for 3D cell culture, Open Biotechnol. J. 2 (2008) 224–228.

[15] L.L. Sohn, A Microfluidics Approach to Investigate Axon Guidance, Report for the U.S. Army Research Office, 2007.

[16] M. S. Ke Liu, Investigation of Low-Shear Design in Cell Culture Applications. A Dissertation in Analytical Chemistry submitted to the Graduate Faculty of Texas Tech University for the Degree of Doctor of Philosophy, 2009.

[17] A.W. Blau, C.M. Ziegler, Prototype of a novel autonomous perfusion chamber for long-term culturing and *in situ* investigation of various cell types, J. Biochem. Biophys. Methods 50 (2001) 15–27.

[18] L. Kim, M.D. Vahey, H.Y. Lee, J. Voldman, Microfluidic arrays for logarithmically perfused embryonic stem cell culture, Lab Chip 6 (2006) 394–406.

[19] E. Leclerc, Y. Sakai, T. Fujii, Microfluidic PDMS (polydimethylsiloxane) bioreactor for large-scale culture of hepatocytes, Biotechnol. Prog. 20 (2004) 750–755.

[20] X.Y. Zhu, L.Y. Chu, B.H. Chuech, M.W. Shen, B. Hazarika, N. Phadke, et al., Arrays of horizontally-oriented mini-reservoirs generate steady microfluidic flow for continuous perfusion cell culture and gradient generation, Analyst 129 (2004) 1026–1031.

[21] T. Glawdel, C.L. Ren, Electro-osmotic flow control for living cell analysis in microfluidic PDMS chips, Mech. Res. Commun. 36 (2009) 75–81.

[22] D.D. Nalayanda, C. Puleo, W.B. Bulton, L.M. Sharpe, T.-H. Wang, F. Abdullah, An open-access microfluidic model for lung-specific functional studies at an air–liquid interface, Biomed. Microdevices 11 (2009) 1081–1089.

[23] D.W. Lee, Y.H. Cho, A continuous electrical cell lysis device using a low dc voltage for a cell transport and rupture, *Sensor Actuator* B 124 (2007) 84–89.

[24] A. Brask, J.P. Kutter, H. Bruus, Long-term stable electroosmotic pump with ion exchange membranes, Lab Chip 5 (2005) 730–738.

[25] E.W.K., Young, Accessible Microfluidic Devices for Studying Endothelial Cell Biology, A thesis submitted to the Graduate Department of Mechanical and Industrial Engineering University of Toronto for the Degree of Doctor of Philosophy, 2009.

[26] P.J. Lee, P.J. Hung, V.M. Rao, L.P. Lee, Nanoliter scale microbioreactor array for quantitative cell biology, Biotechnol. Bioeng. 94 (1) (2006).

[27] J.W. Song, W. Gu, N. Futai, K.A. Warner, J.E. Nor, S. Takayama, Computer-controlled microcirculatory support system for endothelial cell culture and shearing, Anal. Chem. 77 (2005) 3993–3999.

[28] B. Zhang, M. Kim, T. Thorsen, Z. Wang, A self-contained microfluidic cell culture system, Biomed. Microdevices (2009).
[29] J.T. Borenstein, H. Terai, K.R. King, E.J. Weinberg, M.R. Kaazempur-Mofrad, J.P. Vacanti, Microfabrication technology for vascularized tissue engineering, Biomed. Microdevices 4 (3) (2002) 167–175.
[30] N.J. Douville, Y.-C. Tung, R. Li, J.D. Wang, M.E.H. El-Sayed, S. Takayama, Fabrication of two-layered channel system with embedded electrodes to measure resistance across epithelial and endothelial barriers, Anal. Chem. 82 (2010) 2505–2511.
[31] U.Y. Schaff, M.M. Xing, K.K. Lin, N. Pan, N.L. Jeon, S.I. Simon, Vascular mimetics based on microfluidics for imaging the leukocyte-endothelial inflammatory response, Lab Chip 7 (2007) 448–456.
[32] E.W.K. Young, A.R. Wheeler, C.A. Simmons, Matrix-dependent adhesion of vascular and valvular endothelial cells in microfluidic channels, Lab Chip 7 (2007) 1759–1766.
[33] J. Shao, L. Wu, J. Wu, Y. Zheng, H. Zhao, Q. Jin, et al., Integrated microfluidic chip for endothelial cells culture and analysis exposed to a pulsatile and oscillatory shear stress, Lab Chip 9 (2009) 3118–3125.
[34] J.Y. Park, J.B. White, N. Walker, C.-H. Kuo, W. cha, M.E. Meyerhoff, S. Takayama, Responses of endothelial cells to extremely slow flows, Biomicrofluidics 5 (2) (2011).
[35] Y.S. Heo, L.M. Cabrera, J.W. Song, N. Futai, Y.-C. Tung, G.D. Smith, et al., Characterization and resolution of evaporation-mediated osmolality shifts that constrain microfluidic cell culture in poly(dimethylsiloxane) devices, Anal. Chem. 79 (3) (2007) 1126–11341
[36] K.M. Chrobak, D.R. Potter, J. Tien, Formation of perfused, functional microvascular tubes *in vitro*, Microvasc. Res. 71 (2006) 185–196.
[37] M. Shin, K. Matsuda, O. Ishii, H. Terai, M. Kaazempur-Mofrad, J. Borenstein, et al., Endothelialized networks with a vascular geometry in microfabricated poly(dimethyl siloxane), Biomed. Microdevices 6 (4) (2004) 269–278.
[38] D. Mazzei, F. Vozzi, A. Cisternino, G. Vozzi, A. Ahluwalia, A high-throughput bioreactor system for stimulating physiological environments, IEEE Trans. Ind. Electron. 55 (9) (2008).
[39] Y. Du, M. Ghodousi, H. Qi, N. Haas, W. Xiao, A. Khademhosseini, Sequential assembly of cell-laden hydrogel constructs to engineer vascular-like microchannels, Biotechnol. Bioeng. (2011).
[40] P.A. Jones, Construction of an artificial blood vessel wall from cultured endothelial and smooth muscle cells, Proc. Natl. Acad. Sci. U.S.A. 76 (4) (1979) 1882–1886.
[41] W. Tan, T.A. Desai, Microscale multiplayer cocultures for biomimetic blood vessels, J. Biomed. Mater. Res. A 72 (2005) 146–160.
[42] S.H. Ma, L.A. Lepak, R.J. Hussain, W. Shain, M.L. Shuler, An endothelial and astrocyte co-culture model of the blood–brain barrier utilizing an ultra-thin, nanofabricated silicon nitride membrane, Lab Chip 5 (2005) 74–85.
[43] N. Saederup, L. Chan, S.A. Lira, I.F. Charo, Fractalkine deficiency markedly reduces macrophage accumulation and atherosclerotic lesion formation in CCR2 −/− mice: evidence for independent chemokine functions in atherogenesis, Circulation 117 (13) (2008) 1642–1648.

Index

Note: Page numbers followed by "*f*" and "*t*" refer to figures and tables respectively.

D

N

O

P

W

X

FSC
www.fsc.org
MIX
Papier | Fördert
gute Waldnutzung
FSC® C083411

Kolibri 360 GmbH
Ferdinand-Jühlke-Straße 7
99095 Erfurt, Deutschland
produktsicherheit@kolibri360.de